Sea Otter Conservation

Sea Otter Conservation

Edited by

Shawn E. Larson
Seattle Aquarium, Seattle, WA, USA

James L. Bodkin
Scientist Emeritus, Alaska Science Center, USGS, Anchorage, AK, USA

Glenn R. VanBlaricom
Washington Cooperative Fish and Wildlife Research Unit, USGS, and School of Aquatic and Fishery Sciences, University of Washington, Seattle, WA, USA

AMSTERDAM • BOSTON • HEIDELBERG • LONDON
NEW YORK • OXFORD • PARIS • SAN DIEGO
SAN FRANCISCO • SINGAPORE • SYDNEY • TOKYO
Academic Press is an imprint of Elsevier

Academic Press is an imprint of Elsevier
32 Jamestown Road, London NW1 7BY, UK
525 B Street, Suite 1800, San Diego, CA 92101-4495, USA
225 Wyman Street, Waltham, MA 02451, USA
The Boulevard, Langford Lane, Kidlington, Oxford OX5 1GB, UK

ISBN: 978-0-12-801402-8

British Library Cataloguing-in-Publication Data
A catalogue record for this book is available from the British Library

Library of Congress Cataloging-in-Publication Data
A catalog record for this book is available from the Library of Congress

For Information on all Academic Press publications
visit our website at http://store.elsevier.com/

Typeset by MPS Limited, Chennai, India
www.adi-mps.com

Printed and bound in United States of America

Dedication

To "Annie," an Exxon Valdez oil spill orphan,
who captured my heart and
inspired my interest in sea otters.

Shawn E. Larson

Contents

5. Sea Otter Conservation Genetics

Shawn E. Larson, Katherine Ralls and Holly Ernest

6. Evaluating the Status of Individuals and Populations: Advantages of Multiple Approaches and Time Scales

Daniel H. Monson and Lizabeth Bowen

7. Veterinary Medicine and Sea Otter Conservation

Michael J. Murray

List of Contributors

Brenda E. Ballachey, US Geological Survey, Alaska Science Center, Anchorage, AK, USA

Traci F. Belting, Department of Life Sciences, The Seattle Aquarium, Seattle, WA, USA

James L. Bodkin, US Geological Survey, Alaska Science Center, Anchorage, AK, USA

Lizabeth Bowen, US Geological Survey, Western Ecological Research Center, Davis Field Station, UC Davis, Davis, CA, USA

Lilian P. Carswell, US Fish and Wildlife Service, Ventura Fish and Wildlife Office, Ventura, CA, USA

Holly Ernest, Department of Veterinary Sciences, University of Wyoming, Laramie, WY, USA

James A. Estes, Department of Ecology and Evolutionary Biology, University of California, Santa Cruz, CA, USA

Verena A. Gill, US Fish and Wildlife Service, Marine Mammals Management, Anchorage, AK, USA

Tom Mexsis Happynook, Uu-a-thluk Council of Ha'wiih, Huu-ay-aht, BC, Canada

Andrew Johnson, Monterey Bay Aquarium, Monterey, CA, USA

Shawn E. Larson, Department of Life Sciences, The Seattle Aquarium, Seattle, WA, USA

Karl Mayer, Monterey Bay Aquarium, Monterey, CA, USA

Daniel H. Monson, US Geological Survey, Alaska Science Center, Anchorage, AK, USA

Michael J. Murray, Monterey Bay Aquarium, Monterey, CA, USA

Linda M. Nichol, Fisheries and Oceans Canada, Pacific Biological Station, Nanaimo, BC, Canada

Katherine Ralls, Smithsonian Institution, Washington, DC, USA

Anne K. Salomon, School of Resource and Environmental Management, Simon Fraser University, Burnaby, BC, Canada

Suzann G. Speckman, PO Box 244145, Anchorage, AK, USA

Nick Tanape Sr., Nanwalek, AK, USA

M. Tim Tinker, US Geological Survey, Western Ecological Research Center, Long Marine Laboratory, Santa Cruz, CA, USA

Lisa H. Triggs, Aquatic Animal Department, Point Defiance Zoo & Aquarium, Tacoma, WA, USA

Glenn R. VanBlaricom, Washington Cooperative Fish and Wildlife Research Unit, USGS, and School of Aquatic and Fishery Sciences, College of the Environment, University of Washington, Seattle, WA, USA

Xanius Elroy White, Bella Bella, BC, Canada

K̲ii'iljuus Barb J. Wilson, Skidegate, Haida Gwaii, BC, Canada

Chapter Reviewers

Chapter 1: Editor: Glenn R. VanBlaricom

1. Mike Kenner, Research Scientist, US Geological Survey
2. Dr. Jim Estes, Professor, University of California, Santa Cruz

Chapter 2: Editor: James L. Bodkin

1. Dr. Keith Miles, Research Scientist, US Geological Survey

Chapter 3: Editor: Glenn R. VanBlaricom

1. Anthony Degange, Scientist, US Geological Survey, retired
2. Dr. Shawn E. Larson, Seattle Aquarium

Chapter 4: Editor: Shawn E. Larson

1. Dr. Tom Gelatt, Program Leader, Alaska Ecosystem Program, NOAA
2. Dr. Dan Esler, Research Scientist, US Geological Survey

Chapter 5: Editor: James L. Bodkin

1. Dr. Kim Scribner, Professor, University of Michigan
2. Dr. Brenda Ballachey, Research Scientist, US Geological Survey

Chapter 6: Editor: James L. Bodkin

1. Dr. Tim Tinker, Research Scientist, US Geological Survey

Chapter 7: Editor: Shawn E. Larson

1. Dr. Pam Tuomi, Veterinarian, Seward Sealife Center
2. James L. Bodkin, Scientist Emeritus, US Geological Survey

Chapter 8: Editor: Shawn E. Larson

1. Dr. Lesanna Lahner, Staff Veterinarian, Seattle Aquarium
2. Caroline Hempstead, Biologist, Seattle Aquarium
3. James L. Bodkin, Scientist Emeritus, US Geological Survey

Chapter 9: Editor: Shawn E. Larson

1. Dr. Mike Murray, Veterinarian, Monterey Bay Aquarium
2. James L. Bodkin, Scientist Emeritus, US Geological Survey

Chapter 10: Editor: James L. Bodkin

1. Dr. Dan Doak, University of Colorado, Boulder

Chapter 11: Editor: Shawn E. Larson

1. Dr. Norm Sloan, Marine Ecologist, Gwaii Haanas, National Park Reserve and Haida Heritage Site
2. James L. Bodkin, Scientist Emeritus, US Geological Survey

Chapter 12: Editor: Shawn E. Larson

1. Angela Doroff, Research Coordinator, Kachemak Bay Research Reserve
2. James L. Bodkin, Scientist Emeritus, US Geological Survey

Chapter 13: Editor: James L. Bodkin

1. Dr. John Ford, Research Scientist, Department of Fisheries and Oceans, Canada
2. Dr. Shawn E. Larson, Seattle Aquarium

Chapter 14: Editor: James L. Bodkin

1. Ancel Johnson, Research Scientist, US Fish and Wildlife Service, Retired
2. Doug Burn, Wildlife Biologist, US Fish and Wildlife Service
3. Dr. Shawn E. Larson, Seattle Aquarium

Preface

For more than a million years, sea otters (*Enhydra lutris*) have existed along the shores of the north Pacific. They are closely related to the large family of otters that occupy primarily freshwater habitats around much of the world. Several ancestors of the sea otter, known from fossils worldwide, utilized both freshwater and coastal marine habitats. Confined exclusively to marine waters, the modern sea otter is the only species in the lineage that survived the dramatic climatic and oceanographic shifts of the Pleistocene period.

With their excursion into marine waters came adaptations that allowed the sea otter to exploit habitats and compete successfully for prey uncommon for other marine mammals such as cetaceans and pinnipeds. Among these adaptations are a luxurious pelt for insulation and warmth and the utilization of a diverse assemblage of benthic marine invertebrates as prey to satisfy an extraordinary demand for calories, also necessary to support an existence in cold marine environments. Because of their near-exclusive prey base of large marine invertebrates such as crabs, snails, mussels, and urchins that reside on the sea floor, and a diving capacity limited to about 100 m, the sea otter is entirely dependent on the relatively shallow coastal marine habitats adjacent to the shorelines of the North Pacific. Over the course of history, sea otters developed profound ecological relationships with the nearshore ecosystems they occupied.

Approximately 15,000 years ago humans began occupying North America. By some accounts, it was the kelp forests and associated species of invertebrates, fishes, birds, and mammals that facilitated the exploration and eventual settlement of coastal environments as humans expanded their range in the New World. Although uncertainty remains regarding the role of sea otters in facilitation of human population expansion by virtue of ecosystem regulation, it seems clear that among those species that aided human expansion was the sea otter, which provided furs for warmth and food for nourishment. Thus began the development of a complex history between sea otters and humans. We can view the sea otter–human relationship in a variety of terms that include (1) exploitation by humans for fur and food, (2) competition over the marine invertebrates used by both species, (3) ecological consequences of sea otter presence and absence on the structure and function of nearshore ecosystems, and (4) the real and potential adverse effects of escalating human activities in the nearshore and adjacent watersheds on sea otter populations.

The goal of this book is to tell the story of sea otter conservation in a context that will inform and guide those interested in sea otters specifically, as well as those engaged in the broader field of conservation science. In this book we explore the science behind sea otter conservation and management and highlight lessons learned that may benefit other species and ecosystems. Chapters assess not only the biology and ecology of this charismatic marine mammal but many of the societal issues that are influencing prospects for continued conservation and recovery of sea otter populations, some of which remain threatened.

Sea otters have drawn the attention and fascination of the general public since the earliest emergence of conservation and environmental awareness in our culture. The result has been intensive social and political interest in the status and preservation of sea otter populations. In the design of this book we cover key aspects of sea otter conservation based on the last 80 years of scientific effort and discovery focused on this nearshore marine mammal. Literally millions of dollars have been spent and thousands of scientific papers have been published on sea otter biology, ecology, and conservation. Each chapter is written by recognized leaders in their field(s) and in the sea otter community, and each strives to focus scientific attention on the key issues embedded in the practice and concept of conservation. We believe that ongoing public interest in sea otters and their conservation will be well served by this book, which presents the reality of science-based sea otter conservation in a way that engages the informed lay public as well as the scientific community.

Shawn E. Larson
James L. Bodkin
Glenn R. VanBlaricom

Chapter 1

The Conservation of Sea Otters: A Prelude

Shawn E. Larson[1] **and James L. Bodkin**[2]
[1]Department of Life Sciences, The Seattle Aquarium, Seattle, WA, USA, [2]US Geological Survey, Alaska Science Center, Anchorage, AK, USA

This photo is from Lisa Triggs.

> *Sea otters have come to symbolize the wonderful exuberance of nature as well as the tension among species competing for survival. Their remarkable hunting skills and huge appetites give them the capacity to alter the ecological balance of a small bay or inlet, while they are themselves vulnerable to the larger impacts of other predators, most notably humans. Our fascination with this remarkable species draws us to better understand them and the oceans on whose health we all depend.*
>
> Bob Davidson, CEO, Seattle Aquarium

Sea Otter Conservation. DOI: http://dx.doi.org/10.1016/B978-0-12-801402-8.00001-9

INTRODUCTION

Over the past several centuries, expanding human populations have played an ever-increasing role in the diminishment and extinction of species, primarily through direct exploitation and alteration of habitats and ecosystem structure. Virtually nowhere on earth remains without the footprint of humans and, in most instances, that footprint comes with adverse consequences to resident species and ecosystems. The quest to regain functioning ecosystems and restore species presents one of the great challenges to humanity, and provides an opportunity for leadership by the science and conservation communities.

The sea otter, *Enhydra lutris* [L., 1758], and the coastal waters of the North Pacific provide an excellent example of both the adverse effects of human intervention and the positive effects of conservation and management directed toward restoration of species. Sea otters face many challenges because their unique history, biology, and nearshore habitat place them in close proximity to, and often in adverse interactions with, humans. The story of the sea otter ranges from extreme population decimation due to overharvest during the maritime fur trade to near complete protection and active conservation efforts, resulting in recovery of many sea otter populations over the past century. The sea otter may be one of the most widely studied and intensively managed marine mammals. It has been said that "if science can't save the California sea otter population, science probably can't save anything" (VanBlaricom, 1996). In this book we explore the science behind sea otter conservation and management and highlight lessons learned that may benefit other species and ecosystems.

Following more than two centuries of a largely unregulated harvest for their fur, sea otters were on the precipice of extinction in the early twentieth century. By this time their population numbers were so low (estimated at <1% of pre-harvest abundance) and so widely dispersed that they could no longer support commercial harvest at any level (Kenyon, 1969). A population that once numbered perhaps several hundred thousand and extended from Japan along the North Pacific Rim to Mexico was reduced to perhaps a few hundred individuals in isolated groups, mostly in the far north of their range. Although occasional illegal and legal harvests are noted in the late nineteenth and the early twentieth century (Hooper, 1897; Anonymous, 1939; Lensink, 1960), the international Pacific maritime fur trade is widely recognized as ending in 1910 (Kenyon, 1969; Chapter 3). Subsequently, sea otter populations began the slow process of recovery in the early twentieth century, (Lensink, 1962; Kenyon, 1969; Chapter 14). In 1965, Kenyon (1969) estimated the global sea otter population at about 35,000 animals, mostly in Alaska. At that time, more than 3000 km of habitat remained unoccupied between California and the Gulf of Alaska (Figure 1.1). Early attempts to translocate otters into unoccupied habitat in Russia and the United

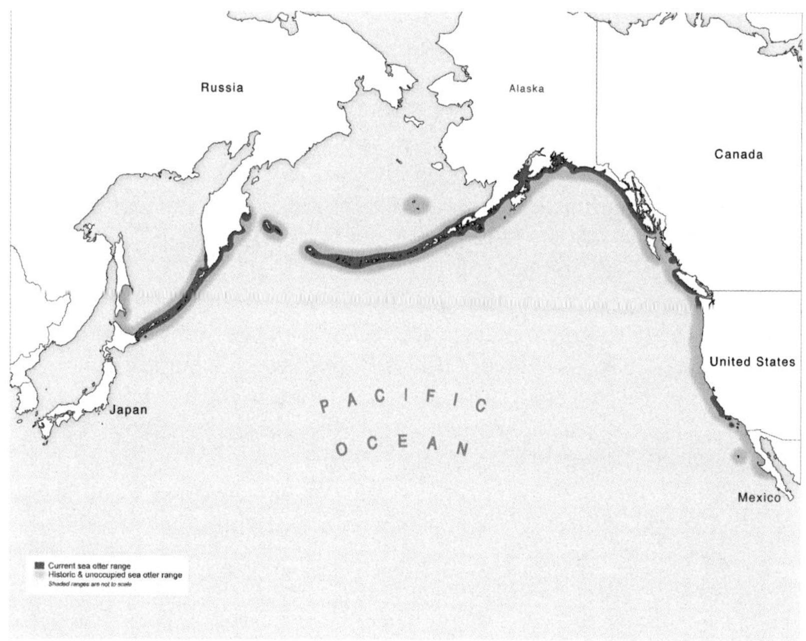

FIGURE 1.1 Historic (yellow) and current (purple) range of the sea otter in the north Pacific. *Illustration by Cecelia Azhderian.*

States beginning in 1937 were largely unsuccessful (Barabash-Nikiforov, 1947; Kenyon, 1969) but provided valuable experience that would aid in the success of future reintroductions (Chapter 8). From the 1960s through the 1980s, independent and international efforts were conducted to aid the recovery of sea otters through a series of translocations from Alaska to Oregon, Washington, and British Columbia, and within Alaska and California (Chapter 3). The translocations re-established several populations and contributed to restoring some of the genetic diversity lost from harvest-induced population bottlenecks (Chapter 5).

The eventual conservation and restoration of sea otters was facilitated by three events that took place in the twentieth century: (1) the cessation of nearly all commercial-scale fur harvest early in the 1900s, (2) increased legal protections at state and national levels, and (3) the establishment of several translocated colonies along the west coast of North America late in the twentieth century that in 2013 represented more than one-third of the global sea otter population. In chapter 3 Bodkin provides a more complete description of the maritime fur trade and the process of recovery early into the twenty-first century. While Nichol (Chapter 13) and VanBlaricom (Chapter 14) describe sea otter conservation in practice in North American and the various legal protections enacted to restore sea otters in US waters.

In the twentieth century the spatial and temporal pattern of sea otter recovery provided significant research opportunities, as prey assemblages and coastal food webs were transformed during recovery of sea otter populations. Observations acquired over decades of research led to improved understanding of the role of sea otters, and by extension other apex or top predators, in supporting the form and function of ecosystems (Chapter 2). The pattern of presence and absence, resulting from spatial variation in population recovery, also afforded the opportunity to closely explore those biological processes that govern the birth and death rates of sea otters that, in turn, ultimately dictate population abundance (Chapter 6).

Entering the twenty-first century, the recovery of sea otters and the restoration of nearshore ecosystems seemed to be proceeding unimpeded throughout most of the North Pacific. However, late in the twentieth century it was discovered that across a vast portion of their northern range, most sea otter populations had collapsed (Estes et al., 1998). This turn of events provided new challenges to sea otter conservation and new opportunities to further an understanding of the functional complexity of ecosystems of which sea otters are an integral part (Chapter 4).

As a result of broad human interest in sea otters, both in terms of conservation and management, increasingly intensive research has been conducted over the last eight decades. Research questions have been diverse, embracing basic biology and life history, population biology and demography, behavior, physiology, genetics, community ecology, and interactions with marine resources such as fisheries and offshore petroleum deposits. Other lines of inquiry have included husbandry, veterinary medicine, pathology, and human-related sources of mortality. Much of the research has been focused on the recovery, conservation, and management of sea otters, while some has been directed at an improved understanding of the role of sea otters as a keystone species in nearshore ecosystems. While much of the research has been directed specifically at sea otters, many of the efforts and results are applicable to the conservation and restoration of other species and ecosystems. Perhaps one of the best examples of the concept of "keystone" predators in structuring communities and ecosystems came from research on the effects of sea otter foraging on sea urchins and the subsequent development of kelp forest communities (McLean, 1962; Estes and Palmisano, 1974).

For a variety of reasons, sea otters and their conservation have been the beneficiaries of long-term and sustained human investment rarely available to the conservation and recovery of a single species. As a consequence, an unusual level of knowledge of both basic biology and ecology exists for this species. This volume brings together many of the scientists who have been responsible for the design, implementation, and interpretation of decades of sea otter research and conservation activity. Here they share the lessons learned about sea otters that may benefit conservation and restoration of other species and systems. Contributing authors are internationally recognized experts in their fields and collectively account for many hundreds of papers in the primary scientific literature. Following a brief introduction to the life history and ecology of the sea otter, we will discuss the lessons learned from sea otter research and conservation and suggest how those lessons may be transferred to improve the conservation and restoration of degraded species and ecosystems. We also discuss persistent impediments to future conservation of sea otters and use examples from specific populations to illustrate where additional research will be of benefit. Embedded within this introduction are references to those chapters that will provide the reader with greater detail on specific topics.

NATURAL HISTORY

The sea otter is a member of the subfamily Lutrinae (otters) of the family Mustelidae. They are one of 13 species of otter that occur worldwide in tropical to subpolar aquatic habitats. All the otter species are recipients of protective classifications by the International Union for the Conservation of Nature (IUCN) or the Committee on International Trade of Endangered Species (CITES) (IUCN, 2013). In addition, most populations of otters receive protective classifications under local, regional, or national laws.

Contrasts between the sea otter and freshwater species of otter provide a good example of the value in understanding the underlying causes of population decline in developing conservation strategies of species in general (Estes et al., 2008). All otter species share a common relationship with humans via the intersection and shared use of preferred habitats that consist of fresh and marine waters and adjacent watersheds. Most otters occur along freshwater rivers, streams, estuaries, and lakes, where there is potential for human alteration and pollution of waterways as well as some instances of over harvesting. In the case of the sea otter, that intersection is along the continental margin and the coastal oceans of the North Pacific, another region of preferred human habitation. Perhaps a lesson to be learned from relations between all otters and humans is that habitat modification and degradation can have pervasive and long-lasting conservation consequences for many species that can be difficult to remedy. Alternatively, when habitat is relatively unchanged and ecosystems are fairly intact, as is the case for much of the sea otter's habitat, conservation can be achieved through directed species-specific management and conservation practices.

Sea otters are a recently evolved marine mammal. Their adaptations for marine existence include relatively shallow diving capabilities and short breath hold capacities that in turn limit their foraging habitat to water depths of less than about 100 m (Berta and Morgan, 1986; Bodkin et al., 2004). This characteristic renders the sea otter exclusively a nearshore, or shallow water, species and puts it in direct proximity to humans and coastal influences. Estes in Chapter 2 develops a more complete description of the underlying biology and ecology of sea otters, providing context for considering their conservation and management, and lays the foundation for applying the lessons learned to other species and ecosystems.

Availability of adequate food and energy resources in the sea otter's habitat may be one of the most important factors in regulating population growth rates and abundance (Kenyon, 1969; Monson et al., 2000; Tinker et al., 2008). Prey availability has often been considered unimportant, as sea otters theoretically had room to re-expand into vacant habitat where preferred prey had remained abundant. Such options were available because the sea otter, an apex predator with a high metabolic rate and substantial need for high rates of food intake, had been absent in these systems for decades, and in some cases for centuries, allowing for largely unregulated population growth of benthic invertebrates. However, as sea otters continue to expand their range and as their ability to move or disperse into unoccupied habitat diminishes, limitation of food resources within occupied habitat becomes more important and thus can modify population growth patterns. As a result of diminished prey resources, competition among individuals for food may increase. In areas where sea otters have remained present for the longest time periods, some individuals have been known to specialize in certain prey items that other otters do not exploit, allowing for fuller utilization of all prey resources (Estes et al., 2003; Tinker et al., 2008; Chapter 10). This may be particularly evident as recovering populations reach equilibrium density and otters must compete with one another for calories where food or energy is the limiting factor and density-dependent processes become increasingly important in regulating population growth (Bodkin et al., 2000). Thus a high caloric need that requires specialization on a few prey types may benefit those animals that have "local knowledge" of prey and their associated distribution within a small, well-defined area.

Sea otters are social animals that often gather in large groups called rafts when not actively foraging or traveling (Kenyon, 1969). Adult sea otters generally have relatively small annual home ranges, from a few to a few tens of square kilometers (Loughlin, 1980; Garshelis and Garshelis, 1984; Jameson, 1989). Adult females can frequently be found resting and foraging in the same locations over days to years, often in association with the same females and within the same male territories on annual time scales. Pups are weaned at about 6 months (Jameson and Johnson, 1993; Monson et al., 1995) and typically do not establish residence within their mother's home range. Juvenile males appear to disperse greater distances after weaning, and adult males generally travel greater distances over time than females.

Sea otters tend to exhibit sexual segregation, with rafts of males and females often resting in different areas (Kenyon, 1969; Loughlin, 1980). Even though most rafts are made up of primarily one sex (except for immature males in female rafts), a single territorial male may often be found in close proximity to female rafts to exploit potential mating opportunities (Loughlin, 1980).

Adult and subadult males not on territories often aggregate in large rafts in specific well-defined locations called "male areas" (Kenyon, 1969) that have been known to persist for more than 50 years (Bodkin, personal observation). Thus, sea otters tend to have a high degree of population structuring based on their social and reproductive systems (Bodkin and Ballachey, 2010) that may be stable for decades, as long as they have access to adequate food resources.

CONSERVATION SUCCESSES AND CHALLENGES

The Fur Trade

The Pacific maritime fur trade for sea otter pelts, while lucrative for humans over most of two centuries, was clearly a failure in terms of sustainable management (or conservation) of a resource. At least three explanations are evident. The first was the common and repeated phenomenon of overexploitation of a resource by competing interests, often referred to as the "tragedy of the commons" (Hardin, 1968). In the early period of the harvest, Russians managed the harvest exclusively. In the middle nineteenth century it was determined that harvest rates were diminishing, and corrective restrictions were imposed (Lensink, 1962). However, ships flagged from Japan, Europe, and America soon entered the harvest and by the late nineteenth century the demise of the sea otter was evident and nearly complete, despite the protections afforded in 1868 in Alaska (Chapters 13 and 14; Kenyon, 1969). The second reason behind the overharvest was ignorance of the spatial scale at which sea otter populations are structured. For example, if the rather modest harvest of less than 1.5% per year had been applied proportionately throughout the global population, it is likely that there would have been little, if any, decline of the sea otter, at least from the human harvest (Gorbics and Bodkin, 2001). But rather, most likely due to human nature and economic realities, the harvest was spatially allocated in a way that optimized efficiency, decimating sea otter populations through a process of "serial depletion" (Chapter 3), as opposed to a sustainable harvest. And third, it is clear that the lack of effective harvest management during the fur trade, due to the lack of understanding of sea otter demographics and spatial structuring and inadequate enforcement of existing conservation legislation (Chapter 14), resulted in the catastrophic population declines that essentially ended the fur trade due to scarcity before any effective conservation management strategies could be employed.

Given the emerging recognition by many humans of the intrinsic value in sustaining resources and the advances in methods of conservation, management, and protection of populations, we suspect it is highly unlikely that sea otters globally will once again be threatened by human harvest (Chapters 12 and 13). While sea otter populations in Alaska continue to be legally harvested under the exemption to the Marine Mammal Protection Act (MMPA) afforded to Alaskan

Natives, the potential negative impact on those populations are unknown at this time (Chapter 4, Chapter 12). However, regulatory mechanisms to prevent depletion exist under both MMPA and the U.S. Endangered Species Act.

Recovery

The road to sea otter recovery began with the end of the widespread commercial fur harvest when sea otter population abundance fell below that necessary for profitable hunting. This event and the beginning of partial legal protection internationally and nationally (Chapters 13 and 14), combined with the scarcity of sea otters at the time and a changing public attitude toward wildlife (Chapter 12), contributed to early recovery. Recovery rates of remnant populations in the early twentieth century varied but never attained rates approaching the theoretical maximum of near 24% annually (Estes, 1990). Despite limited chronic harvest-related mortality and the narrow scope of protections afforded by various laws that extended well into the middle twentieth century (Chapter 14), the cessation of wide-scale unregulated hunting was adequate to allow recovery of sea otters where remnant populations survived.

The recovery of the scattered remnant groups was largely unmanaged and unmonitored during the first half of the twentieth century. In addition, sea otters remained absent from much of the west coast of the Unites States and Canada. The fragmentation of sea otter populations stimulated the first major conservation management action of sea otters in the second half of the twentieth century: translocations of sea otters from recovered populations to unoccupied areas within their historical range in the Northeast Pacific (Figure 1.1). Collectively, translocated sea otter populations accounted for 35% of all sea otters in the world in 2013; arguably one of the most profound conservation success stories of the twenty-first century (Chapter 3). Additionally, in populations that were founded from two distinct stocks, we see the highest levels of genetic diversity since the end of the fur trade (Larson et al., 2002a,b).

Given the contribution of past translocations to the conservation of sea otters, we consider it worth exploring future opportunities for additional translocations into unoccupied habitat. We argue that due to the demonstrated success of translocations in the recovery of sea otters and the subsequent restoration of affected coastal ecosystems, translocations can be a valuable tool for wildlife conservation in general and should be considered specifically to further both sea otter conservation and coastal ecosystem restoration.

Oil Spills

Although multiple protections and translocations contributed to sea otter recovery well into the twentieth century, the vulnerability of sea otters was driven home on March 24, 1989, when the T/V *Exxon Valdez* ran aground and spilled 11 million gallons of crude oil into Prince William Sound, AL (Chapter 4). The spill resulted in acute catastrophic mortality across a wide range of taxa ranging from marine invertebrates to fishes, birds, and mammals. Because sea otters rely on their dense fur for thermoregulation, they are extremely vulnerable to oil spills. Immediate mortality to sea otters numbered into the thousands and chronic effects lasted for at least two decades where spilled oil was greatest and persisted in intertidal sediments and contaminated prey resources. Several important conservation lessons were learned from this catastrophe (Monson et al., 2011; Peterson et al., 2003, Bodkin et al., 2014). First, the importance of pre-spill data on population size, distribution, and status (increasing, decreasing, or stable) as well as demography, behavior, and diet were essential to documenting the magnitude of effects. Second, long-term effects from acute and chronic exposure to contaminants can equal or exceed acute effects, delaying recovery from complicated and unanticipated sources that are very difficult to mitigate and manage. Third, once a large spill occurs there is relatively little than can be done to mitigate effects. Finally, the need for proactive (i.e., before oil spills occur) development of adequate and trained spill response staff and resources for effective rescue and treatment of affected wildlife was evident (Chapter 14).

Predation

In addition to vulnerability to oil spills, sea otters are also subjected to various sources of predation, including marine and terrestrial carnivores (Chapter 4). Except at small local scales, predation was not widely considered a major factor limiting population size or growth (Chapter 4). However, in the Aleutian archipelago, where local sea otter populations were deemed recovered to pre-exploitation numbers by the middle 1960s, sea otter abundance unexpectedly declined by more than 90% in the 1990s (Kenyon, 1969; Estes et al., 1998; Doroff et al., 2003). This dramatic decline was attributed to predation by killer whales (*Orcinus orca*) (Estes et al., 1998; Chapter 4). The lesson learned here was that unexpected sources of mortality can affect conservation at extremely large numerical and spatial scales. Also, here was a case where pre-event and real-time data were essential to documenting the decline as well as in determining the probable cause. Following work on the sea otter decline, Springer et al. presented a theory that the decline was ultimately fueled by industrial whaling, which precipitated shifts in killer whale diets from large whales to pinnipeds and ultimately to sea otters, thus linking oceanic food webs with the nearshore food web, where sea otters had once been considered apex predators (Springer et al., 2003). Although controversial, this theory serves to illustrate the potential complexities inherent in food webs and the potential for unexpected cascading ecological effects resulting from large-scale population reductions such as the collapse of the large whales following industrial whaling (DeMaster et al., 2006; Wade et al., 2007).

Genetic Diversity

All sea otter populations have experienced at least one bottleneck due to exploitation during the fur trade. Some also experience secondary bottlenecks from translocations and natural emigration from newly established populations such that few animals actually survive to become successful founders (Kenyon, 1969; Bodkin et al., 1999; Larson et al., 2002b; Aguilar et al., 2008). The population bottleneck from the fur trade extirpations resulted in a loss of >99% of original sea otter numbers and resulted in a loss of over half their original genetic diversity (Larson et al., 2002b). As a result, sea otters now have relatively low genetic diversity throughout their genome, including the genes that control immune function and disease response (Bowen et al., 2006).

The effect of reduced genetic variation on contemporary sea otter population growth and viability remains uncertain. The pathway to resolution of the matter is complex and not entirely clear (Chapter 5). Many sea otter populations appear to be in a precarious balance. Because they inhabit coastal areas, they are often in contact with humans and can suffer negative interactions associated with fishing activities and exposure to shoreline and nearshore pollution sources. Add to these difficulties the potential for negative effects associated with the loss of genetic diversity and it becomes more apparent why some sea otter populations fail to thrive and do not approach expected growth rates. A consequence of low population size and slow growth includes higher probabilities of further population reductions due to stochastic events, which can lead to further declines in genetic diversity due to drift and potentially to further population declines.

Subspecific Taxonomy, Stocks, and Management

Currently three subspecies of sea otters are recognized based on skull morphology (Wilson et al., 1991): Russian (*E.l. lutris*), Northern or Alaska (*E.l. kenyoni*), and Southern or California (*E.l. nereis*). Within the Northern subspecies (*E.l. kenyoni*), three genetic stocks are recognized: a Southwest stock (SW) including the Aleutian Islands and Kodiak Island; a Southcentral (SC) stock including Prince William Sound, the Kenai Peninsula, and Cordova; and a Southeastern (SE) stock including the Alexander Archipelago (Cronin et al., 2002). Within the SW Alaska stock, listed as "threatened" under the US Endangered Species Act (ESA), at least five distinguished population segments (DPS) are identified (Burn and Doroff, 2005). The recent designation of multiple DPS in SW Alaska clearly recognizes population structuring at spatial scales smaller than current subspecies classification. In our view it is appropriate to consider current and future conservation and management of sea otters at spatial scales that are consistent with the underlying biology that contributes to those fundamental demographic processes (births and deaths) that, in concert with individual movements, result in the relatively small spatial scales that define populations (perhaps <100 s of km^2) and not the much larger scale of subspecies (Gorbics and Bodkin, 2001; Bodkin and Ballachey, 2010; Chapter 3).

Sea otter populations throughout their range remain fragmented and in various stages of recovery. Some populations are increasing and others are stable or in decline; in some cases populations separated by as little as a few tens of kilometers exhibit different population trajectories (Bodkin and Ballachey, 2010). Telemetry data suggest that differential movements cannot explain geographic differences in growth rates that have been observed in California and Alaska (Bodkin et al., 2011). Future management of sea otters must include explicit consideration of the spatial scales at which management or conservation actions should be taken. For example, should we expect the entire population to achieve a particular growth rate when not all segments of the population have equal access to food or habitat that may be required for population expansion and growth? If so, we may be looking for causes of low rates of increase when higher rates actually should not be expected. Similarly, if any subsistence harvests occur without consideration of population growth rate or spatial structuring then we might expect the process of localized serial depletion to potentially repeat itself. Implications of spatial structuring to management, conservation and recovery are far reaching and speak to the critical need to better delineate spatial scales of population structuring (Chapters 2, 3, 4, and 10). This question is one that is certainly not unique to sea otters, but is broadly relevant to conservation and management.

Apex Predator, Keystone Species, and Food Limitation

The sea otter is an effective predator, consuming large quantities of food and structuring the nearshore environment. It is widely recognized as a keystone species primarily because of an ability to limit prey populations. Sea otters are often thought to be beneficial because they structure and encourage the diverse and complex kelp forest community that provides habitat for juvenile fish as well as many other aquatic animals (McLean, 1962; Estes and Palmisano, 1974; Chapter 2). However, because of their diet and caloric needs, sea otters compete directly with people for commercially valuable shellfish resources, and thus may not be seen as a beneficial addition in some areas (Chapter 12). Indeed in some areas sea otter population expansions may be limited and restricted because of direct conflicts with fishers and fisheries (Chapter 4, Chapter 12).

Because of the sea otter's need for abundant food, significant effort has gone into developing tools to assess the status of sea otter populations in terms related to their food resources (Chapter 6). Several tools have been developed for the monitoring of sea otter activity and foraging success. Time depth recorders (TDRs) measure the amount of time an otter spends diving and can be used to calculate the amount of time spent foraging over months or years, which turns out to be an effective measure of the status of an individual (or population) relative to food availability (Bodkin et al., 2007). Activity budgets can also be estimated by biologists that record the prey type and number of prey an otter brings to the surface to consume. These data then go into a model developed to estimate the number of calories an animal consumes in a typical foraging session or "bout," which can then be converted into an energy recovery rate (Chapter 10). Such tools were developed specifically to effectively monitor sea otter activity and food availability and are an effective management tool that could be applied to monitoring activity of other wildlife as well.

Sources of Mortality

Present challenges to the recovery of sea otter populations are multifaceted. The slow growth rates of some populations, such as the one in California, are thought to result from a combination of multiple factors. These include food limitation, predation, and relatively high mortality due to exposure to potential contaminants and pathogens resulting in a high incidence of death by disease (Thomas and Cole, 1996; Estes et al., 2003). In recent years the wildlife research community has devoted significant effort and resources to documenting the causes of mortality in various sea otter populations. As a result we have an improved understanding of the sources of mortality in a subset of animals in several populations of sea otters. However, the sampled subset almost certainly does not represent the living population and probably also does not represent the entire dying population. From these necropsy programs we now can partition death among a variety of proximate causes including various diseases, starvation, predation, and acute injury. What we have not been able to demonstrate effectively in many cases is the ultimate underlying cause of death and, more importantly, how those deaths contribute to the rate of change in the population.

In Chapter 7, Murray discusses the disease risk in sea otters and the fact that mortality is an essential component of healthy populations. He argues that health should be measured at the level of the population rather than the individual. Furthermore, we have little direct evidence to conclude that specific (or even cumulative) disease rates result in declining sea otter abundance or growth rates that are lower than expected. The lesson here is that data-based assessment of the status of populations is essential and that perhaps extensive survival studies including disease screening should be a priority for all populations, particularly those in decline.

Rehabilitation

The need for rehabilitation capabilities is a challenge associated with stranding of sea otters due to food limitation, disease, trauma, or environmental disasters such as oil spills. In Chapter 9, Johnson and Mayer describe issues relating to rehabilitation, from the care of stranded newborn pups to emergency care of adults including triage and euthanasia. The chapter also discusses the philosophical question of the rescue, rehabilitation, and reintroduction of injured or less fit individuals back into the wild population, potentially taking valuable resources from animals possibly without contributing reproductively to the population as a whole; this topic is also discussed by Estes in Chapter 2. Furthermore VanBlaricom et al. (Chapter 8) describe some of the difficulties associated with decisions regarding rescue and rehabilitation of impaired sea otters, particularly in high-profile emergency circumstances.

The California or southern sea otter population (*E.l. nereis*) has received by far the lion's share of attention when it comes to the care and release of stranded sea otters. Following on efforts by the Society for the Prevention of Cruelty to Animals to rehabilitate sea otters in the late 1970s and early 1980s (Chapter 14), the Monterey Bay Aquarium assumed and expanded a program devoted to the rescue and rehabilitation of southern sea otters, the Sea Otter Research and Conservation program (SORAC). In the last decade SORAC have developed an improved method of reintroducing stranded pups into the wild, using surrogate sea otter mothers rather than humans to raise the orphaned pups (Chapter 9). These efforts on behalf of sea otter rescue and rehabilitation both in California and during oil spills have been extensive and expensive, and generate significant questions such as: Is there a realized benefit to the population, or are these efforts mainly to satisfy our cultural and social values? Are there potential costs to the population, in terms of reduced growth through our intervention on behalf of stranded individuals? Would resources allocated to rehabilitation be better spent on other conservation actions, such as translocation or mitigation of mortality associated with other risk factors?

Interactions with People

Sea otters occur in the nearshore habitat along the northwest Pacific coastline in areas that people have inhabited for millennia. In Chapter 11, Salomon et al. address the First Nations/Native American perspective on sea otter conservation. The authors delve into the historical views and uses of sea otters by First Nations people to understand how sea otters were managed by native people in an effort to inform contemporary ecosystem approaches to sea otter management in the modern world. They also argue that people co-existed with sea otters and the nearshore environment prior to contact with Europeans, and that the indigenous people effectively managed sea otter populations to maintain an optimal balance of nearshore apex predators and shellfish resources (Chapter 11).

Sea otters today often live adjacent to areas of high human occupancy. In some areas their charismatic nature stimulates admiration and public concern for their protection. In other areas there is concern about sea otter expansion and reintroductions since their high caloric requirements and the resulting impact on invertebrate prey populations could potentially collapse and close significant commercial fisheries for crabs, clams, abalone, and urchins. Chapters 12 (Carswell et al.) and 13 (Nichol) explore the complex relationships between people and sea otters, both positive and negative, as sea otters have experienced recovery and established interactions with humans.

CONCLUSION

The dedicated support and effort at local to international scales from governmental and non-governmental organizations including zoos and aquaria over many decades has provided a deep understanding of the biology of sea otters and the function of their coastal ecosystems. It is our goal in this book to translate some of that current understanding of sea otter biology, ecology, and human perceptions and share the lessons learned in an attempt to enlighten those interested not only in sea otters and their environment but in other species and ecosystems as well. We bring together in this volume the collective knowledge and experience of scientists, many of whom have dedicated entire careers to this species and their ecosystem, and share their thoughts on sea otter conservation with the larger community engaged in the conservation, management, and restoration of species and ecosystems. The following chapters discuss in detail the major conservation successes, challenges, and lessons learned in the ongoing quest to conserve and manage sea otter populations. Conservation successes include, but are not limited to, protection from widespread commercial harvest, re-establishment of populations into native habitats through translocation, and a deepening understanding of coastal ecosystems. Conservation challenges include the inability to increase the growth rate of the southern sea otter population, the failure to recognize the process and spatial scale of population structuring, and the current uncertainty about population consequences of increasing levels of subsistence harvest in some parts of Alaska (Chapters 4 and 12). Some of the lessons learned include the realization that humans can restore populations if habitats are not irrevocably modified and that conservation and management for restoration of populations and ecosystems requires dedication and patience. The sea otter research community has learned and written volumes about sea otter conservation but we still have much to learn about science and effective conservation. The information reflected in this volume can serve to synthesize and support continuing efforts to understand, conserve, and restore sea otter populations and coastal marine ecosystems.

REFERENCES

Aguilar, A., Jessup, D.A., Estes, J., Garza, J.C., 2008. The distribution of nuclear generation variation and historical demography of sea otters. Anim. Conserv. 11, 35–45. Available from: http://dx.doi.org/doi:10.1111/j.1469-1795.2007.00144.x.

Anonymous, 1939. Comment and news. J. Mammal. 20, 407.

Barabash-Nikiforov, I.I., 1947. Kalan (The sea otter). Soviet Ministry RSFSR. (Translated from Russian by Israel Program for Scientific Translation, Jerusalem, Israel, 1962.) 227 pages.

Berta, A., Morgan, G.S., 1986. A new sea otter (Carnivora: Mustelidae) from the late Miocene and early Pliocene (Hemphillian) of North America. J. Paleoentol. 59, 809–819.

Bodkin, J.L., Ballachey, B.E., 2010. Modeling the effects of mortality on sea otter populations. USGS Scientific Investigation Report 2010-5096. 12 pages.

Bodkin, J.L., Esler, D., Rice, S.D., Matkin, C.O., Ballachey, B.E., 2014. The effects of spilled oil on coastal ecosystems: lessons from the Exxon Valdez spill. In: Maslow, B., Lockwood, J.L. (Eds.), Coastal Conservation. Cambridge University Press, New York, NY, pp. 311–346.

Bodkin, J.L., Ballachey, B.E., Cronin, M.A., Scribner, K.T., 1999. Population demographics and genetic diversity in remnant and re-established populations of sea otters. Conserv. Biol. 13 (6), 1278–1385.

Bodkin, J.L., Burdin, A.M., Ryzanov, D.A., 2000. Age and sex specific mortality and population structure in sea otters. Mar. Mammal Sci. 16 (1), 201–219.

Bodkin, J.L., Esslinger, G.G., Monson, D.H., 2004. Foraging depths of sea otters and implications to coastal marine communities. Mar. Mammal Sci. 20 (2), 305–321.

Bodkin, J.L., Monson, D.H., Esslinger, G.G., 2007. Activity budgets derived from time–depth recorders in a diving mammal. J. Wildl. Manage. 71 (6), 2034–2044. Available from: http://dx.doi.org/doi:10.2193/2006-258.

Bodkin, J.L., Ballachey, B.E., Esslinger, G.G., 2011. Trends in sea otter population abundance in western Prince William Sound, Alaska: progress toward recovery following the 1989 Exxon Valdez oil spill. US Geological Survey Scientific Investigations Report 2011-5213. 14 pages.

Bowen, L., Aldridge, B., Miles, A.K., Stott, J.L., 2006. Partial characterization of MHC genes in geographically disparate populations of sea otters (*Enhydra lutris*). Tissue Antigens 67 (5), 402–408.

Burn, D., Doroff, A., 2005. Endangered and threatened wildlife and plants; determination of threatened status for the southwest Alaska distinct population segment of the northern sea otter (*Enhydra lutris kenyoni*). Federal Register 70:46366.

Cronin, M.A., Jack, L., Buchholz, W.G., 2002. Microsatellite DNA and mitochondrial DNA variation in Alaskan sea otters. Final report prepared for US Fish and Wildlife Service by LGL Alaska Research Associates. 17 pages.

DeMaster, D.P., Trites, A.W., Clapham, P., Mizroch, S., Wade, P., Small, R.J., et al., 2006. The sequential megafaunal collapse hypothesis: testing with existing data. Prog. Oceanogr. 68, 329–342.

Doroff, A.M., Estes, J.A., Tinker, M.T., Burn, D.M., Evans, T.J., 2003. Sea otter population declines in the Aleutian archipelago. J. Mammal. 84 (1), 55–64.

Estes, J.A., 1990. Growth and equilibrium in sea otter populations. J. Anim. Ecol. 59, 385–401.

Estes, J.A., Palmisano, J.F., 1974. Sea otters: their role in structuring nearshore communities. Science 185, 1058–1060.

Estes, J.A., Tinker, M.T., Williams, T.M., Doak, D.F., 1998. Killer whale predation on sea otters linking oceanic and nearshore ecosystems. Science 282, 473–476.

Estes, J.A., Riedman, M.L., Staedler, M.M., Tinker, M.T., Lyon, B.E., 2003. Individual variation in prey selection by sea otters: patterns, causes and implications. J. Anim. Ecol. 72, 144–155.

Estes, J.A., Bodkin, J.L., Ben-David, M., 2008. Marine otters. In: Perrin, W.F., Wursig, B., Thewissen, J.G.M., Crumly, C.R. (Eds.), Encyclopedia of Marine Mammals, second ed. Academic Press, Waltham, MA, pp. 797–806.

Garshelis, D.L., Garshelis, J.A., 1984. Movements and management of sea otters in Alaska. J. Wildl. Manage. 48, 665–677.

Gorbics, C., Bodkin, J.L., 2001. Stock identity of sea otters in Alaska. Mar. Mammal Sci. 17 (3), 632–647.

Hardin, G., 1968. The tragedy of the commons. Science 162 (3859), 1243–1248.

Hooper, C.L., 1897. Report on the Sea Otter banks of Alaska. No 1977. Government printing office, Treasury Department, Washington.

IUCN, 2013. IUCN Red List of Threatened Species. Version 2013.2. <www.iucnredlist.org>. Downloaded on 22 April 2014.

Jameson, R.J., 1989. Movements, home ranges, and territories of male sea otters off central California. Mar. Mammal Sci. 5, 159–172.

Jameson, R.J., Johnson, A.M., 1993. Reproductive characteristics of female sea otters. Mar. Mammal Sci. 9 (2), 156–167.

Kenyon, K.W., 1969. The Sea Otter in the Eastern Pacific Ocean. North American Fauna 68. U.S. Department of the Interior, Washington, DC.

Larson, S.E., Jameson, R., Bodkin, J., Staedler, M., Bentzen, P., 2002a. Microsatellite and mitochondrial DNA variation in remnant and translocated sea otter (*Enhydra lutris*) populations. J. Mammal. 83, 893–906.

Larson, S., Jameson, R., Etnier, M., Fleming, M., Bentzen, P., 2002b. Loss of genetic diversity in sea otters (*Enhydra lutris*) associated with the fur trade of the 18th and 19th centuries. Mol. Ecol. 11, 1899–1903.

Lensink, C.J., 1960. Status and distribution of sea otters in Alaska. J. Mammal. 41, 172–182.

Lensink, C.J., 1962. The History and Status of Sea Otters in Alaska. PhD dissertation. Purdue University, New York.

Loughlin, T.R., 1980. Home range and territoriality of sea otters near Monterey, California. J. Wildl. Manage. 44 (3), 576–582.

McLean, J.H., 1962. Sublittoral ecology of kelp beds of the open coast area near Carmel, California. Biol. Bull. (Woods Hole) 122, 95–114.

Monson, D.H., DeGange, A.R., 1995. Reproduction, preweaning survival, and survival of adult sea otters at Kodiak Island, Alaska. Can. J. Zool 73, 1161–1169.

Monson, D.H., Estes, J.A., Bodkin, J.L., Siniff, D.B., 2000. Life history plasticity and population regulation in sea otters. Oikos 90, 457–468.

Monson, D.H., Doak, D.F., Ballachey, B.E., Bodkin, J.L., 2011. Effect of the Exxon Valdez oil spill on the sea otter population of Prince William Sound, Alaska: do lingering oil and source-sink dynamics explain the long-term population trajectory? Ecol. Appl. 21 (8), 2917–2932.

Peterson, C.H., Rice, S.D., Short, J.W., Esler, D., Bodkin, J.L., Ballachey, B.E., et al., 2003. Long-term ecosystem response to the Exxon Valdez oil spill. Science 302 (5653), 2082–2086.

Springer, A.M., Estes, J.A., Van Vliet, G.B., Williams, T.M., Doak, D.F., Danner, E.M., et al., 2003. Sequential megafaunal collapse in the North Pacific Ocean: an ongoing legacy of industrial whaling? PNAS 100 (21), 12223–12228. Available from: http://dx.doi.org/doi:10.1073/pnas.1635156100.

Thomas, N.J., Cole, R.A., 1996. The risk of disease and threats to the wild population. Endangered Species Update 13 (2), 23–27.

Tinker, M.T., Bentall, G., Estes, J.A., 2008. Food limitation leads to behavioral diversification and dietary specialization in sea otters. Proc. Natl. Acad. Sci. USA 105, 560–565.

VanBlaricom, G.R., 1996. Saving the sea otter population in California: contemporary problems and future pitfalls. Endangered Species Update 13 (2), 85–90.

Wade, P.R., Barrett-Lennard, L.G., Black, N.A., Burkanov, V.N., Burdin, A.M., Calambokidis, J., et al., 2007. Killer whales and marine mammal trends in the North Pacific—a re-examination of evidence for sequential megafauna collapse and the prey-switching hypothesis. Mar. Mammal Sci. 23, 766–802.

Wilson, D.E., Bogan, M.A., Brownell Jr., R.L., Burdin, A.M., Maminov, M.K., 1991. Geographic variation in sea otters, *Enhydra lutris*. J. Mammal. 72 (1), 22–36.

Chapter 2

Natural History, Ecology, and the Conservation and Management of Sea Otters

James A. Estes
Department of Ecology and Evolutionary Biology, University of California, Santa Cruz, CA, USA

Most if not all of our planet's 1.2 million described extant species (and the 10 million or more others that remain unknown or undescribed—Mora et al., 2011) have been affected by humankind's sundry and pervasive influences. These influences have resulted in a 100- to 1000-fold increase in extinction rates (Lawton and May, 1995) with the anticipated loss of roughly half of the extant species over the next 100 years (Wilson, 2002). Even our

Sea Otter Conservation. DOI: http://dx.doi.org/10.1016/B978-0-12-801402-8.00002-0

own species' survival to the end of the twenty-first century has been questioned (Rees, 2003).

Given the enormity of this situation and the fact that many species would seem to be in graver danger of impending extinction than are sea otters, why would an entire volume be devoted to the conservation of this one species? There are at least three reasons. First, the sea otters' high trophic status, requirement for relatively large areas of habitat, and keystone role (sensu Paine, 1969) in ecosystem function (Estes, 2008) make them umbrellas for the conservation of other associated species (Soulé and Terborgh, 1999). Second, sea otters conflict with several human interests and activities. In particular, they are vulnerable to environmental pollution (especially spilled oil; see Chapter 4), and they compete with people for shellfish resources. Last, sea otters command inordinate public attention simply because many people find them appealing (see Chapter 12).

Although this volume is explicitly about conservation (which I think of as the endeavor to preserve biodiversity and maintain the functional integrity of ecosystems), I am including management (which I think of as the manipulation of species or their habitats so as to enhance their benefits or reduce their costs to human societies) in the scope of my discussion because these endeavors are so intimately intertwined. I will begin by identifying the main issues surrounding the conservation and management of sea otters (Table 2.1). This list can be divided into those issues that have actually resulted in conservation or management actions and those for which actions have only been proposed or imagined. Both the understanding and resolution of these issues are defined to an important extent by the features of sea otters, the species with which sea otters interact, and the environments in

TABLE 2.1 Issues and Actions Concerning the Conservation and Management of Sea Otters

For Which Actions Have Been Taken

- Protection from overharvest
- Translocations to reestablish populations
- Protection from incidental mortality in fisheries
- Protection from terrestrial pathogens
- Rehabilitation

Issues Not Yet Actionable

- Shellfish conflicts
- Indirect effects of predation on other species and ecosystem processes
- Pollution (chemical and biological)
- Genetic diversity
- Human disturbance
- Resource economics and tourism

which sea otters live. My purpose in this chapter is to identify these features and discuss why they are important to conservation and management.

HISTORY

The present is inextricably linked to and thus an inevitable product of the past. A consideration of history is therefore integral to understanding the conservation and management of any species. In this section I will briefly recount the history of sea otters and their environment on two time scales—pre- and post-contact with nonindigenous peoples.

Where should a pre-contact history of sea otters begin? One might look back a half billion years to the Cambrian explosion, which marks the beginnings of metazoan evolution. That's far too long for my purposes and those of most people who will read this volume. Instead, I will begin in the late Miocene or early Pliocene, some 3 to 5 million years ago, at the onset of the current glacial age (Pielou, 1992). Low productivity tropical/subtropical conditions characterized higher latitude oceans of the northern hemisphere before that time. Increased production associated with the onset of polar cooling likely drew the sea otters' first marine ancestors from their primitive freshwater ecosystems into the coastal oceans of the North Pacific. A species not unlike the modern sea otter dispersed into the North Atlantic Ocean via the transarctic interchange (Willemsen, 1992), but subsequently became extinct. Otherwise, the *Enhydra* lineage (see Box 2.1 for a synopsis of nomenclature and phylogeny) has been restricted to the North Pacific Ocean

Box 2.1 Nomenclature

Common name. Sea Otter
Order: Carnivora
Family: Mustelidae
Scientific name: *Enhydra lutris*
Subspecies. *E.l. nereis* (California), *E.l. kenyoni* (Washington to Alaska), and *E.l. lutris* (Asia) (Wilson, 1991)

Phylogeny, Classification, and Population Structure

The modern sea otter is the only completely marine species in the family Mustelidae (skunks, weasels, minks, badgers, and honey badgers) (Wozencraft, 1993). The sub-specific taxonomy of the sea otter was based largely on morphological analyses (Wilson, 1991) that is at least partially supported by subsequent molecular genetic data. Scribner et al. (1997) and Larson (see Chapter 5) conclude that although populations today are differentiated genetically, limited sequence divergence and lack of phylogeographic concordance suggest an evolutionarily recent common ancestor and some degree of gene flow throughout their range.

and southern Bering Sea by frozen oceans to the north, New and Old World land masses to the east and west, and tropical conditions to the south.

As the earliest sea otters radiated into marine environments, they were confronted with a choice—whether to feed on shellfishes or finfishes. Seemingly they could not do both because of the different morphological and behavioral traits required for efficient exploitation of these two prey types. As it was, they became shellfish predators, possibly because competition with pinnipeds and other piscivores precluded them from exploiting finfishes or possibly simply because this is what their freshwater ancestors did. Whatever the reason, sea otters evolved to consume a wide array of benthic macroinvertebrates, including such groups as echinoderms, gastropods, bivalves, and crustaceans (Riedman and Estes, 1990; see also Box 2.2 for further details on diet). Sea otters developed effective means for the capture and consumption of these prey, such as forelimb strength and sensitivity, bunodont dentition for crushing the exoskeletons of their invertebrate prey, and tool use as an aid to breaking through the exoskeletal defenses of especially well-armored species such as bivalves and gastropods.

Sea otters evolved into a species with strong limiting effects on their benthic invertebrate prey, in turn affecting many other species through indirect interactions. The kelps (Laminariales) for example, which also had a North Pacific center of origin (Estes and Steinberg, 1988), evolved in an environment in which their principal enemies—herbivorous sea urchins—were limited to cryptic habitats and held at small population sizes by prototypical sea otter predation (see Box 2.3 and food web section [below] for additional detail). This situation decoupled the potential for an evolutionary arms race (i.e., the co-evolution of defense and resistance) between kelps and their herbivores, in turn facilitating the development of a modern kelp flora that is poorly defended (chemically and morphologically) against herbivory (Steinberg et al., 1995). That co-evolutionary scenario might explain the extreme vulnerability of Pacific kelps to sea urchin grazing and the dramatic phase shifts between kelp forests and urchin barrens that have followed the loss or recovery of sea otters across much of the North Pacific Ocean (Estes and Duggins, 1995; Watson and Estes, 2011; Estes et al., 2013). It might also explain why the kelps, freed from the burden of investing resources into defenses against their herbivores, have become so highly productive (Reed and Brzezinski, 2009).

Several other evolutionary influences of the "trophic cascade" (sensu Paine, 1980) from sea otters to urchins to kelps are known or suspected. For example, the late Cenozoic radiation of hydrodamaline sirenians (the lineage that ultimately led to Steller's sea cow [*Hydrodamalis gigas*]—a kelp consumer) from their tropical, sea grass-consuming ancestors into the North Pacific Ocean (Domning, 1978) must have been facilitated, if not primarily driven, by the great abundance and high nutritional quality of North Pacific kelps (Estes and Steinberg, 1988). Similarly, the diversity and large body size of North Pacific abalones may have been facilitated by the sea otter's

Box 2.2 Diet and Foraging Behavior

The sea otter is a generalist predator, known to consume more than 150 different prey species (Kenyon, 1969; Riedman and Estes, 1990; Estes and Bodkin, 2002). With few exceptions, their prey consist of sessile or slow moving benthic invertebrates such as mollusks, crustaceans, and echinoderms. Foraging occurs in habitats with rocky and mixed-sediment substrates between the high intertidal to depths slightly in excess of 100 m. Preferred foraging habitat is generally in depths less than 25 or 40 m (Riedman and Estes, 1990), although studies in southeast Alaska have found that some animals forage mostly at depths from 40 to 80 m (Bodkin et al., 2004; see Chapter 3).

The diet of sea otters is usually studied by observing prey items brought to the surface for consumption, and therefore diet composition is usually expressed as a percentage of all identified prey that belong to a particular prey species or type. Although the sea otter is known to prey on a large number of species, only a few tend to predominate in the diet in any particular area. Prey type and size depends on location, habitat type, season, and length of occupation (see Chapters 6 and 10 for further analyses of sea otter diets). In California, otters foraging over rocky substrates and in kelp forests mainly consume decapod crustaceans, gastropod and bivalve mollusks, and echinoderms (Ebert, 1968; Estes et al., 1981). In protected bays with mixed sediments, otters mainly consume infaunal clams (*Saxidomus nuttallii* and *Tresus nuttallii*; Kvitek and Oliver, 1988). Along exposed coasts of mixed sediments, the Pismo clam (*Tivela stultorum*) is a common prey (Stevenson, 1977). Important prey in Washington State include crabs (*Cancer* spp., *Pugettia* spp.), octopus (*Octopus* spp.), intertidal clams (*Protothaca* spp.), sea cucumbers (*Cucumaria miniata*), and the red sea urchin (*Strongylocentrotus franciscanus*) (Kvitek et al., 1989). The predominately mixed-sediment habitats of southeast Alaska, Prince William Sound, and Kodiak Island support populations of clams that are the primary prey of sea otters. Throughout most of southeast Alaska, burrowing clams (species of *Saxidomus, Protothaca, Macoma,* and *Mya*) predominate in the sea otter's diet (Kvitek et al., 1993). They account for more than 50% of the identified prey, although urchins (*S. droebachiensis*) and mussels (*Modiolis modiolis, Mytilus* spp., and *Musculus* spp.) can also be important. In Prince William Sound and Kodiak Island, clams account for 34–100% of the otter's prey (Calkins, 1978; Doroff and Bodkin, 1994; Doroff and DeGange, 1994). Mussels (*Mytilus trossulus*) apparently become more important as the length of occupation by sea otters increases, ranging from 0% at newly occupied sites at Kodiak to 22% in long-occupied areas (Doroff and DeGange, 1994). Crabs (*C. magister*) were once important sea otter prey in eastern Prince William Sound, but apparently have been depleted by otter predation and are no longer eaten in large numbers (Garshelis et al., 1986). Sea urchins are minor components of the sea otter's diet in Prince William Sound and the Kodiak archipelago. In contrast, diet in the Aleutian, Commander, and Kuril islands is dominated by sea urchins and a variety of finfish (including hexagrammids, gadids, cottids, perciformes, cyclopterids, and scorpaenids; Kenyon, 1969; Estes et al., 1982). Sea urchins tend to dominate the

(*Continued*)

Box 2.2 (Continued)

diet of low-density sea otter populations, whereas more fishes are consumed in populations near equilibrium density (Estes et al., 1982). For unknown reasons, fish are rarely consumed by sea otters in regions east of the Aleutian Islands.

Sea otters also exploit episodically abundant prey such as squid (*Loligo* spp.) and pelagic red crabs (*Pleuroncodes planipes*) in California and smooth lump-suckers (*Aptocyclus ventricosus*) in the Aleutian Islands and the Gulf of Alaska. On occasion, sea otters attack and consume sea birds, including teal (*Anas crecca*), scoters (*Melanita perspicillata*), loons (*Gavia immer*), gulls (*Larus* spp.), grebes (*Aechmophorus occidentalis*), and cormorants (*Phalacrocorax* spp.) (Kenyon, 1969; Riedman and Estes, 1990).

Box 2.3 Sea Otters, Sea Urchins, and Kelp—A Trophic Cascade

Sea otters eat sea urchins and sea urchins eat kelp. Joined together, these direct consumer–prey interactions create a positive indirect effect of sea otters on kelp (see illustrations and images below). Indirect interaction chains that begin with high trophic level species and radiate downward through the food web via top-down forcing processes are known as "trophic cascades" (Paine, 1980). This effect was discovered in kelp forest ecosystems by comparing islands with and without sea otters (created by the fortuitous survival of otters at some islands and their extinction at others following the maritime fur trade) in the western Aleutian archipelago (Estes and Palmisano, 1974). Subsequent research has shown this otter–urchin–kelp cascade to occur broadly and predictably across much of the North Pacific Ocean (Estes and Duggins, 1995; Kvitek et al., 1998; Watson and Estes, 2011). The biotic communities living on and around the shallow rocky reefs exist in two distinct phase states, one of which is characterized by an abundance of kelp and few urchins and the other of which is characterized by abundant urchins and few kelps. The transition between these phase states as otter densities change occurs as an abrupt phase shift (Steneck et al., 2002; Estes et al., 2010), which varies as a function of sea otter density with geographic region and depending upon whether otter populations are increasing or declining (i.e., the system displays "hysteresis," Scheffer, 2009). The mechanisms responsible for this behavior are now fairly well understood. When a system is in the kelp-dominated state, it tends to remain in that state because detrital fall-out from the overlying kelp canopy provides adequate nutritional resources for sea urchins without the urchins having to move in search of food. Under this circumstance the urchins have little or no destructive effect on the living kelps. Moreover, the kelp forest, once established, interacts with wave surge to physically repel invading sea urchins through a whiplash effect (Konar, 2000).

(*Continued*)

Box 2.3 (Continued)

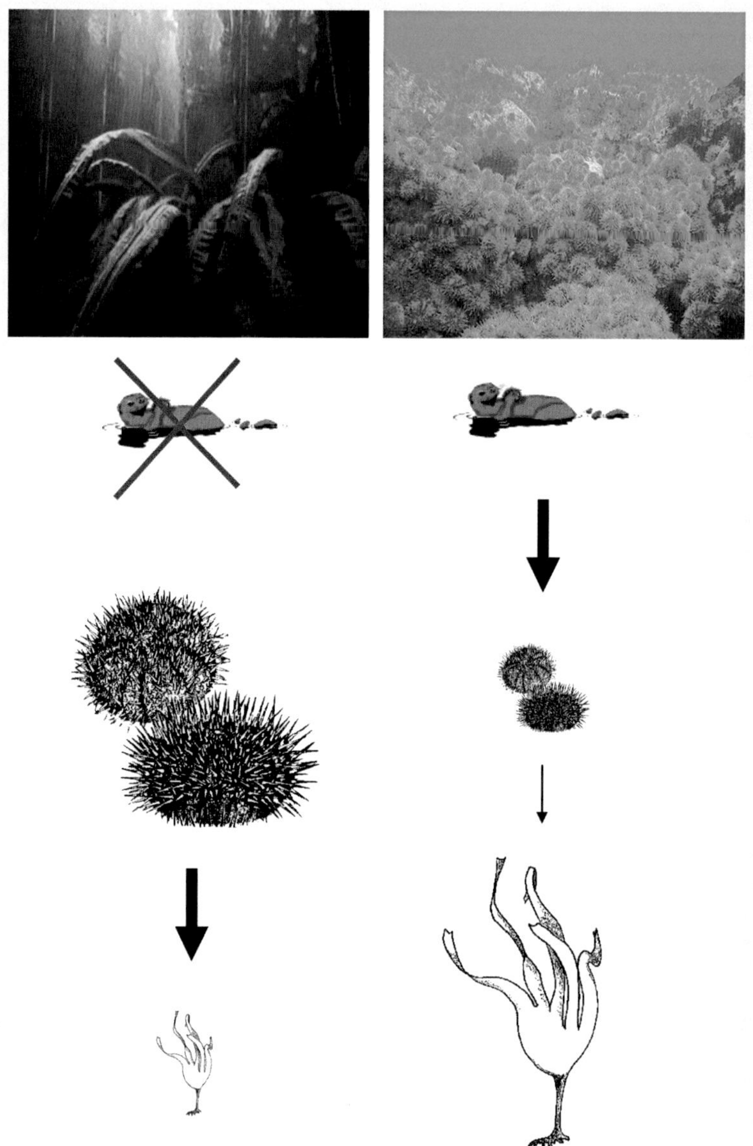

The drawing illustrates the mechanisms by which sea otter predation on sea urchins cascades downward through the kelp forest food web to create either kelp-dominated (left) or urchin-dominated (right) communities. *Photos at the top are from islands in the western Aleutian archipelago with (left) and without (right) sea otters.*

(Continued)

Box 2.3 (Continued)

When a system is in the urchin-dominated state, it also tends to remain there because the hungry urchins are continually on the move in search of food, thereby destructively grazing any newly settled kelp plant and preventing kelp from becoming established (Konar and Estes, 2003).

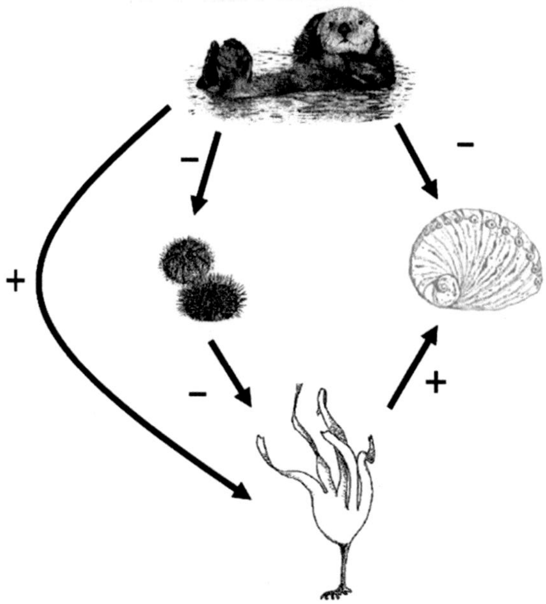

FIGURE 2.1 A food web diagram showing the potential conflicting influences of the direct (negative) and indirect (positive) influences of sea otters on abalones. See text for further explanations.

evolutionary role in creating such a productive environment with high algal food quality (Estes et al., 2005). Although the direct effect of sea otters on abalone limits their size and abundance, the indirect effect through the otter–urchin–kelp trophic cascade promoted increased growth and body size (Figure 2.1). I suspect this evolutionary process played out not on the open sea floor but instead within the many substrate cracks and crevices that have provided safe haven to abalone from sea otter predation (Lowry and Pearse, 1973) since the Miocene.

The post-contact history of sea otters began in 1741 with the Bering Expedition's discovery of an abundant otter population in the Commander Islands. This discovery initiated the Pacific maritime fur trade, which reduced once-abundant otter populations to near extinction over the ensuing 160 years (see Chapter 4). The quest for sea otter pelts strongly influenced patterns of human colonization and the political history of the Northeast Pacific region

(Meinig, 1986). For instance, the depletion of sea otters was an integral factor underlying Russia's willingness to sell Alaska to the United States (Gibson, 1976). The near extinction of sea otters also caused the standing biomass of their benthic invertebrate prey to increase, which in turn fueled the development of coastal North Pacific commercial shellfisheries (Estes and VanBlaricom, 1985). Following protection of sea otters in the early 1900s and the regrowth of populations in more recent decades, conflicts with these fisheries have become a point of extreme contention and debate in the conservation and management of sea otters (see Chapters 11 and 12).

FORM AND FUNCTION

In this section I will briefly describe some of the sea otter's basic biology and discuss how several aspects of the species' form and function are integral to what I see as the three main dimensions of their conservation and management—exploitation and recovery, fishery conflicts, and marine pollution.

Insulation

Homeothermy at or around 37°C is a feature of all birds and mammals, demanding in turn a careful balance between heat production and heat loss. In contrast with pinnipeds and cetaceans, which use blubber for insulation against heat loss to the cold aquatic medium in which they live, sea otters depend entirely on their fur. The sea otter's exclusive dependency on fur promoted the evolution of their luxuriant pelt, thus leading to the species' demise in the Pacific maritime fur trade. Even today, the value of sea otter fur is the principal motivation for their exploitation by Alaska natives, again raising questions about the sustainability of this endeavor in some areas (Esslinger and Bodkin, 2009; Bodkin and Ballachey, 2010). The sea otter's reliance on fur for insulation also makes it especially vulnerable to the fouling effects of oil spills (Bodkin et al., 2002), which have increased in frequency in recent decades and will likely continue to pose a threat to sea otters so long as human societies extract fossil fuels and transport them across the oceans.

Metabolic Rate

High metabolic rate (Morrison et al., 1975; Davis et al., 1988; Costa and Kooyman, 1984; Yeates et al., 2007) is another feature of sea otter form and function that is germane to their conservation and management. This high metabolic rate is due in part to their small body size and the scaling of metabolic rate with body size (Kleiber, 1961; Brown et al., 2004), in part to their mustelid ancestry (which as a group has an elevated metabolism—Iverson, 1972; McNab, 1989), and in part to their need to generate heat for the maintenance of thermal neutrality in a cold aquatic environment. A high metabolic rate coupled with high population densities (compared with other mammalian carnivores of similar body size) underlies the conflict between sea otters and shellfisheries (see Chapters 4, 11, and 12).

Life History

Yet another important feature of sea otter form and function that is relevant to the species' conservation and management is life history strategy, or the manner in which they acquire and allocate resources to reproduction and survival in order to maximize fitness (Stearns, 1989). This strategy in sea otters is a product of two forces—the constraints of their evolutionary history and selection imposed by life in the sea. An understanding of the consequences of these forces can be gleaned by comparing sea otters with other mustelids (with whom they share a recent common ancestor), and with other marine mammals (with whom they share a common environment). Such a comparison reveals two important features of the sea otter's life history—litter size and reproductive tactics (Box 2.4).

In contrast with all other mustelids (which have multiple young litters), sea otters are like marine mammals in having single-offspring pregnancies. This simple observation suggests that litter size in sea otters has been under strong selection by life in the sea. The net effect on their population biology is an r_{max} value of about 20% per year (Estes, 1990; see also Chapter 10 and Box 2.4). This one feature of the sea otter's life history limits the rate at which populations can be sustainably harvested and the rate at which depleted populations can recover (Bodkin and Ballachey, 2010).

Most marine mammals are extreme "investment strategists" (Stearns, 1989) in that they store energy by accumulating blubber during good times and expend this stored energy to fuel metabolism during bad times. These animals often feed in one part of their range and give birth to and nourish their young while living at an energy deficit elsewhere. Because individual females of such investment strategists can assess resource availability well in advance of giving birth (based on blubber storage), they have the innate capacity to make well-informed pre-partum (i.e., at the time of copulation and during pregnancy) decisions on whether or not to reproduce. The sea otters' high metabolic rate and inability to store energy defines the species as an extreme "income strategist," which means that they must feed regularly to survive and reproduce, and that they have no means of assessing pre-partum resource availability at the time of fertilization. Perhaps for this reason, sea otters have adopted a reproductive strategy known as "bet-hedging" (Cohen, 1966; Monson et al., 2000). That is, adult females become pregnant

Box 2.4 More on Sea Otter Life History

In addition to the limits placed by a single offspring, other traits limit population growth. Females usually attain sexual maturity at age 3–4 (Bodkin et al., 1993). von Biela et al. (2009) reported age of first reproduction varies by a year or less among populations, depending on the degree of resource limitation. Adult female reproductive rates are consistently high and range from 0.80 to 0.94 (Siniff and Ralls, 1991; Bodkin et al., 1993; Jameson and Johnson, 1993; Riedman et al., 1994; Monson and DeGange, 1995; Monson et al., 2000). Whereas reproductive output remains relatively constant over a broad range of ecological conditions, pup survival appears to be more strongly influenced by resource availability. At Amchitka Island, a population that was at or near equilibrium density, dependent pup survival ranged from 0.22 to 0.40, compared to nearly 0.85 at Kodiak Island, where food was not limiting and the population was increasing (Monson et al., 2000). Various factors can contribute to variation in pup survival, including female experience (Riedman et al., 1994; Monson and DeGange, 1995) and body condition (weight/unit length) (Bodkin and Monson, 2002; see Chapter 6). Post-weaning juvenile survival appears to be a primary mechanism of regulation in many sea otter populations, and post-weaning survival rates, which can vary annually from 0.18 to 0.86 (Monson et al., 2000; Ballachey et al., 2003), also appear to be affected by the status of the population relative to food availability. Factors affecting survival of young otters (both pre- and post-weaning) appear to be more important than reproductive rates in regulating abundance in many populations. Once a sea otter survives its first year of life, there appears to be a relatively high probability of surviving to senescence where density dependent mechanisms are structuring population abundance. Such may not be the case when density independent factors such as predation are important. Survival of sea otters more than 2 years of age is generally high, approaching 0.90, gradually declining over time (Bodkin and Jameson, 1991; Monson et al., 2000). Maximum ages, based on tooth annuli, are about 22 years for females and 15 years for males. A female at the Seattle Aquarium lived to be 27 years old.

Causes of mortality in sea otter populations are inherently difficult to determine, with the probability of detecting and assigning cause of death dependent on the cause. For example, the carcass of a sea otter that dies of starvation is more likely to be recovered than one killed by a predator. Documented sources of mortality include predation, starvation, disease, contaminants, incidental take by humans, intentional human harvest, and intraspecific aggression. Recognized sea otter predators include the white shark (*Carcharadon carcharias*), brown bear (*Ursus arctos*), wolf (*Canis lupus*), red fox (*Vulpes vulpes*), wolverine (*Gulo gulo*), killer whale (*Orcinus orca*), and bald eagle (*Haliaeetus leucocephalus*) (Kenyon, 1969; Ames and Morejohn, 1980; Riedman and Estes, 1990; Monson and DeGange, 1995; Hatfield et al., 1998; Bodkin, 2001).

immediately after parturition and almost always carry their pregnancies to term. The reproductive decision is not whether to become pregnant but rather if or when to abandon their pups. Mothers apparently make that decision shortly following parturition. The combined effects of their income and bet-hedging strategies may thus account for the large number of small, abandoned pups, which in California are often discovered by the public and raised in captivity for later release to the wild.

Important dimensions to the conservation and management of sea otters have arisen from the combined effects of an extreme income strategy and their dependency on benthic invertebrates. These features hold the species close to shore where it has been easy to observe and study but where the animals also must endure a close association with human impacts. Their inability to store energy together with their nutritional dependency on benthic invertebrates has limited population recovery in some areas because of their inability to disperse across expanses of deep water (between islands, for example).

ASSOCIATED SPECIES

Sea otters associate and interact with numerous other species. In this section I will discuss how the characteristics of species in the coastal ecosystem help define the nature of species interactions, and how these interactions affect the conservation and management of sea otters. I will focus this discussion on three functional groups—autotrophs, free-living macroinvertebrates, and parasites—because of their importance to the way sea otters and their coastal ecosystems interact and operate.

Autotrophs

Coastal ecosystems contain two main groups of autotrophs—single-celled phytoplankton and multicellular macrophytes (algae and vascular plants). Together these organisms fuel production of coastal oceans across the sea otter's range.

Phytoplankton have short generation times, which means that phytoplankton populations can grow and decline rapidly. Such changes are typically responses to physical forcing (e.g., wind, upwelling, changes in the strength and direction of currents) and grazing by herbivorous zooplankton. Phytoplankton-dominated systems often shift quickly between high and low rates of production. The influences of these shifts on sea otters and the influences of sea otters on phytoplankton through indirect species interactions are largely unknown.

Macroalgae (i.e., kelps [Laminariales], rockweeds [Fucales], and other groups) comprise what are often referred to as kelp forests. In contrast with phytoplankton, the interactions between sea otters and macroalgae have been extensively studied and are reasonably well known. Although the life span (and thus the generation time) of a typical macroalga (typically one to a few years) is much longer than that of a phytoplankter (typically days to weeks), it is short compared with those of terrestrial plants (commonly decades to centuries, especially in forest systems). Macroalgae also have complex life histories that involve gametophyte and sporophyte life stages. Both the spores and gametes are free-living and mobile, thus creating the possibility for rapid, long-range dispersal (Reed et al., 1988). Gametophytes function much as terrestrial seed banks (Edwards, 2000) during periods that are unfavorable for the larger and more conspicuous sporophyte life states. Macroalgal sporophytes often die rapidly and en masse over large areas, either because of unfavorable ocean conditions (i.e., large ocean waves or warm, nutrient-impoverished water—Dayton et al., 1992; Edwards and Estes, 2006) or intense grazing (Steneck et al., 2002). Most macroalgal species are susceptible to grazing and can appear or disappear from an area quickly in response to changes in the intensity of herbivory (Steneck et al., 2002).

Kelp forests also are characterized by high rates of primary production, owing to the fact that their essential resources—water, nutrients, and light—are seldom limiting. This high rate of primary production fuels high levels of secondary production through a detritus-based food web (Duggins et al., 1989).

Macroinvertebrates

Like macroalgae, most of the sea otter's invertebrate prey have complex life histories with large, sessile, or weakly motile adult stages and small, highly dispersive larval stages. These characteristics create the capacity for long-range dispersal, rapid but sometimes highly episodic recruitment, and high rates of population growth. Larval dispersal and recruitment are influenced by the vagaries of coastal ocean currents, and high rates of population growth in the post-settlement adults are fueled in significant part by macroalgae and the detritus-based food web. Many of the sea otter's macroinvertebrate prey are valued by human consumers, thus resulting in conflicts between sea otters and shellfisheries in some areas (see Chapters 4, 11, and 12).

Parasites and Pathogens

Like most species, sea otters are parasitized by a variety of viruses, bacteria, protozoa, and multicellular organisms. And like macroalgae and invertebrates, most of these forms have complex life histories with dispersive life stages. However, unlike nonparasitic species, many of these life stages are not free-living but instead occur within the bodies of their hosts. Mostly for this reason, the life stages of parasites and pathogens are cryptic and their population dynamics are difficult to study and understand. These difficulties are exacerbated for sea otters in the coastal oceans by potential linkages between the land and the sea. For example, the eggs of *Toxoplasma gondii*, a protozoal parasite that is nonreproductive but pathogenic in sea otters, are believed to infect sea otters through runoff from land to the adjacent coastal ocean (Conrad et al., 2005).

FOOD WEB

A food web can be thought of as a roadmap of who is eaten by whom. Species are the elements of this map (analogous to towns and cities) and consumer–prey interactions are the linkages among the species (analogous to roads that connect the towns and cities). It is the nature of these linkages among species that determine the dynamic properties of food webs. Coastal food webs in general and the functional importance of sea otters to coastal food web structure and organization in particular have been extensively studied and are relatively well understood. My focus in this section will be on two properties of these coastal food webs: the linkages among species, and the resulting dynamical properties of populations and functional groups.

Consumer–prey interactions are by definition +/− in nature, which is to say that one species (the consumer) benefits while the other species (the prey) incurs a cost. The magnitude of this cost determines the *interaction strength* (Paine, 1992), defined as the difference in prey population size when the consumer is present or absent (Berlow et al., 1999). A second property of consumer–prey interactions is the extent to which they link up with one another to provide functional connectivity through the food web.

Sea otters are strong interactors in that they can have large negative effects on their prey populations. Sea urchins also are strong interactors because, when present in sufficient numbers, they have large negative effects on macroalgal populations. These two sets of strong interactions link together in kelp forest ecosystems to form a "trophic cascade" (Paine, 1980) in which sea otters exert an indirect positive effect on macroalgal populations (Estes and Palmisano, 1974; also see Box 2.3), in turn affecting other ecological processes and numerous associated species (Estes, 2005). Consumer–prey linkages also connect kelp forests with the open ocean (e.g., the episodic spawning migration of oceanic fishes to the coastal zone that in turn provide an important food resource to sea otters [Watt et al., 2000]; predation by the largely oceanic killer whales on sea otters [Estes et al., 1998]).

The aforementioned dynamical features of kelp forest food webs have important implications for sea otter conservation and management. Strong interactions between sea otters and their invertebrate prey have produced conflicts between sea otters and shellfisheries. However, follow-on linkages from otter–prey interactions to other species and ecosystem processes produce a broader array of influences by sea otters on coastal ecosystems, some of which are beneficial (e.g., kelp forest enhancement and related benefits to nearshore and oceanic fishes) to human welfare. Finally, the nonlinear dynamics of these consumer–prey interactions result in a system for which the critical transitions between phase states are abrupt and variable across regions in their functional forms (i.e., the way in which they vary with sea otter density).

ECOSYSTEMS

Ecosystems are communities of living organisms together with the nonliving (physical and chemical) elements of the surrounding environment. High productivity, a defining feature of higher latitude marine ecosystems, is accounted for by an abundance of the physical and chemical requisites for photosynthesis—water, nutrients, carbon, and light. These conditions result in the remarkably high rates of primary production by kelp forest ecosystems in general (Reed and Brzezinski, 2009) and by sea otter-dominated kelp forest ecosystems in particular (Duggins et al., 1989; Wilmers et al., 2012). Another possible contributing factor to high coastal production is the physical energy provided by tidally generated water movement. Although such kinetic energy is not in itself usable in biosynthesis, it does move large amounts of organic material across the sea floor where it can fuel secondary production in suspension-feeding organisms. This process may explain the high production and high associated sea otter densities in places like lower Cook Inlet and Glacier Bay in the northeast Gulf of Alaska, where tidal amplitudes and rates of tidally generated water flow are extreme.

A second important feature of coastal marine ecosystems is the diverse ways in which they are linked to land on one side, the open sea on the other, and the atmosphere above. Several of these linkages involve the consumer–prey interactions that were discussed earlier in the section on food web dynamics. However, land–sea–atmosphere linkages are more complex than this. The bulleted examples in Box 2.5 illustrate that complexity.

A third important characteristic of modern coastal ecosystems is their proximity to densely populated human settlements. The large numbers of people living near ocean margins use and modify these coastal environments in myriad ways.

The aforementioned features of coastal ecosystems influence their conservation and management in several obvious ways. For instance, their high productivity leads to the human exploitation of fish and other biological resources. The complex linkages among land, the coastal oceans, the open sea, and the atmosphere create an "openness" to the dynamics of all four systems that complicates the manner in which they work and thus the ease with which they can be conserved and managed. Some of the materials that are vectored by runoff from land to the coastal ocean constitute health risks to wildlife and humans.

Box 2.5 Inter-Ecosystem Linkages

- Runoff vectors' sediments, nutrients, toxins, pathogens, and perhaps other materials from land into the coastal ocean.
- Coastal marine production can in turn facilitate terrestrial productivity, especially in places where nutrients or water are otherwise limiting (Polis and Hurd, 1996; Maron et al., 2006).
- The larvae of coastal invertebrates and fishes may spend weeks or months in the water column, sometimes at great distances from shore, before returning to the coastal ocean as recruits to adult populations (Roughgarden et al., 1988). Ocean currents may transport these dispersive life stages long distances from their parent stocks (Palumbi, 2003).
- The open sea can also subsidize coastal ocean predators and thus increase their impacts on coastal ecosystems (Estes et al., 1998; McCaughley et al., 2012).
- Atmospheric CO_2 diffuses into coastal ocean water where it acidifies the aquatic environment and is consumed by photosynthesis. The high rate of production by kelps serves to both sequester atmospheric carbon and to reduce ocean acidification (Reed and Brzezinski, 2009; Wilmers et al., 2012).

The highly productive coastal ocean together with the sea otter's inability to feed beyond the shallow coastal zone is responsible for remarkably high sea otter population densities in some areas. The species' polygynous mating system, which is characterized in part by male territoriality, probably evolved because of these high population densities and the resulting ability of males to compete for females. In this regard sea otters are more like pinnipeds (Bartholomew, 1970) than other species of otters (Estes, 1989).

Because sea otters are carnivores, high population densities depend not only on high net primary productivity (NPP) but also on an effective trophic linkage between that NPP and animals that are suitable prey. Since most of the sea otter's prey species are invertebrates with larval life histories and open populations, their abundance and sustainability as a sea otter prey resource depends on larval delivery and settlement. Regional variation in larval delivery (probably caused by local oceanographic processes) might thus cause sea otter population densities to vary in ways that are not directly linked to NPP.

HUMAN EMOTIONS

Decisions concerning the conservation and management of natural resources are ultimately driven by two factors—logic and emotion. Sea otters engender strong emotions by both their advocates and detractors. The emotional responses to sea otters by their detractors (nearly all of whom are people who value the shellfish that otters eat) are fairly straightforward and easy to understand. The same cannot be said for those of their advocates.

Sea otters are unarguably among the world's most appealing and charismatic animals. Many people love them in ways they could never love a worm or an insect or even a fish. This emotional reaction, which is entirely visceral in nature, has led to what I see as a dichotomy of values that revolves around the love of animals on the one hand and the love of nature on the other. A love of animals places high value on the welfare of individuals, just as we relate to our family, friends, and pets. A love of nature places value instead on the preservation of species, biodiversity, and the functional integrity of ecosystems. These values aren't necessarily mutually exclusive but they often lead to conflicting views of and objectives for conservation and management. Such a scenario may be unfolding in California where there is an ongoing effort to save, rehabilitate, and release orphaned sea otter pups back to the wild population (see Chapter 9). This effort is seen by many as complementary to the goal of recovering the threatened southern sea otter population and maintaining or restoring the functional integrity of its ecosystem. However, others see the rehabilitation of orphaned sea otter pups and the conservation of sea otter populations as being at cross purposes. The argument by this latter group is that the orphaned pups that were destined to die (because of the sea otter's basic reproductive strategy and their mothers' choice to abandon them—as explained previously in the section on function and form), intraspecific competition is an important limiting influence on the California sea otter population, and thus rehabilitating and returning orphaned pups to nature may be detrimental to the wild animals that were otherwise destined to live. The other is that such efforts to rehabilitate orphaned sea otter pups divert time, attention, and resources from the conservation and management of sea otter populations and their associated communities and ecosystems.

DISCUSSION

It would be convenient if the approach to wildlife conservation and management was similar to that of most other science-based endeavors, such as building a bridge or designing a piece of software. But it isn't. Both latter activities can turn to well-established principles and working algorithms, which designers and engineers use with relative ease and a high degree of confidence to achieve their desired goals. General methods and models for conservation and management are in their infancy and most of those that do exist cannot be employed with a high degree of confidence that they will really work. The difference is that the infrastructural complexity of ecosystems is vastly greater than that of human constructs. It is for this reason that wildlife conservationists and managers have been forced to turn to the details of the ecology and natural history for guidance on the species or systems they are trying to conserve and manage.

The purpose of this chapter was to explore some of the linkages between ecology and natural history, conservation and management of sea otters, and coastal marine ecosystems. Toward that end, we have seen how a view of history was essential to understanding why the sea otter became what it is today and how the species functions as a keystone predator. The historical perspective helps us understand why fisheries conflicts have arisen as an inevitable consequence of the near extinction and recovery of sea otters. We have seen how the sea otter's life in a cold aquatic environment, its small body size, and its dependence on fur have together led to an extreme income strategy, which in turn affects the life history tactics employed by sea otters to maximize their fitness and lifetime reproductive success. And we have seen how this life history strategy has led to bet-hedging and pup abandonment as a means of population regulation and how the rehabilitation of these abandoned pups has taken sea otter conservation and management in a direction that appeals to human emotions but is not necessarily in the best interest of populations and their ecosystems.

The understanding of sea otter natural history and ecology adds dimension and depth to our thinking about two especially important aspects of sea otter management—the consequences of harvest and the criteria for delisting depleted populations. For example, the same extreme income strategy that prevents long-range dispersal and holds individual sea otters to small home ranges creates extreme spatial structure in populations, in turn making the species more vulnerable to exploitation than it would otherwise be if individuals moved and intermingled over large areas. The other side to this aspect of the sea otter's spatial ecology, which is of potential human benefit, is the increased ability to manage populations through selective harvest or removal over small spatial scales, if ever that is deemed necessary and desirable. Similarly, our understanding of the sea otter's keystone role in kelp forest ecosystem structure raises interesting questions about the targets and goals of conservation and management (such as the establishment of delisting criteria under the US Endangered Species Act). The maintenance of viable populations is presently the gold standard for conservation targets (Soulé, 1987; Morris and Doak, 2002). While that approach makes sense at one level, species do not live in isolation from their surrounding communities and ecosystems and we have seen that a minimum viable population is not sufficient to maintain the sea otter's keystone role in kelp forest ecosystems (Estes et al., 2010). The same argument holds in the establishment of conservation targets for any strongly interacting species (Soulé et al., 2003, 2004).

Much of what I have written reflects my own experiences and beliefs. Someone else, given the same charge, might have written something very different. Part of the reason for this is that much has been left unsaid. As reflected by the chapter's title, I have endeavored to establish the more important connections between ecology and natural history on the one hand and the conservation and management of sea otters and their coastal ecosystems on the other. The degree to which I have succeeded in this undertaking will be evident from the extent to which the various points and arguments surface again in the chapters that follow.

REFERENCES

Ames, J.A., Morejohn, G.V., 1980. Evidence of white shark, *Carcharodon carcharias*, attacks on sea otters, *Enhydra lutris*. Calif. Fish Game 66, 196–209.

Ballachey, B.E., Bodkin, J.L., Howlin, S., Howlin, A.M., Rebar, A.J., 2003. Correlates to survival of juvenile sea otters in Prince William Sound, Alaska, 1992–1993. Can. J. Zool. 81, 1494–1510.

Bartholomew, G., 1970. A model for the evolution of polygyny in pinnipeds. Evolution 24, 546–559.

Berlow, E.L., Navarrete, S.A., Briggs, C.J., Power, M.E., Menge, B.A., 1999. Quantifying variation in the strengths of species interactions. Ecology 80, 2206–2224.

Bodkin, J.L., 2001. Marine mammals: sea otters. In: Steele, J., Thorpe, S., Turekian, K. (Eds.), Encyclopedia of Ocean Sciences. Academic Press, London, pp. 2614–2621.

Bodkin, J.L., Ballachey, B.E., 2010. Modeling the effects of mortality on sea otter populations. USGS Scientific Investigation Report 2010–5096, 12p.

Bodkin, J.L., Jameson, R.J., 1991. Patterns of sea bird and marine mammal carcass deposition along the central California coast, 1980–1986. Can. J. Zool. 69, 1149–1155.

Bodkin, J.L., Monson, D.H., 2002. Sea otter population structure and ecology in Alaska. Arctic Res. 16, 31–35.

Bodkin, J.L., Mulcahy, D., Lensink, C.J., 1993. Age specific reproduction in the sea otter (*Enhydra lutris*): an analysis of reproductive tracts. Can. J. Zool. 71, 1811–1815.

Bodkin, J.L., Ballachey, B.E., Dean, T.A., Fukuyama, A.K., Jewett, S.C., McDonald, L.M., et al., 2002. Sea otter population status and the process of recovery from the Exxon Valdez oil spill. Mar. Ecol. Prog. Ser. 241, 237–253.

Bodkin, J.L., Esslinger, G.G., Monson, D.H., 2004. Foraging depths of sea otters and implications to coastal marine communities. Mar. Mammal Sci. 20, 305–321.

Brown, J.H., Gillooly, J.F., Allen, A.P., Savage, V.M., West, J.M., 2004. Toward a metabolic theory of ecology. Ecology 85, 1771–1789.

Calkins, D.G., 1978. Feeding behavior and major prey species of the sea otter, *Enhydra lutris*, in Montague Strait, Prince William Sound, Alaska. Fishery Bull. 76, 125–131.

Cohen, D., 1966. Optimizing reproduction in a randomly varying environment. J. Theor. Biol. 12, 119–129.

Conrad, P.A., Miller, M.A., Kreuder, C., James, E.R., Mazet, J., Dabritz, H., et al., 2005. Transmission of *Toxoplasma*: clues from the study of sea otters as sentinels of *Toxoplasma gondii* flow into the marine environment. Int. J. Parasitol. 35, 1155–1168.

Costa, D.P., Kooyman, G.L., 1984. Contribution of specific dynamic action to heat balance and thermoregulation in the sea otter *Enhydra lutris*. Physiol. Zool. 57, 199–203.

Davis, R.W., Williams, T.M., Thomas, J.A., Kastelein, R.A., Cornell, L.H., 1988. The effects of oil contamination and cleaning on sea otters (*Enhydra lutris*). II Metabolism, thermoregulation, and behavior. Can. J. Zool. 66, 2782–2790.

Dayton, P.K., Tegner, M.J., Parnell, P.E., Edwards, P.B., 1992. Temporal and spatial patterns of disturbance and recovery in a kelp forest community. Ecol. Monogr. 62, 421–445.

Domning, D.P., 1978. Sirenian evolution in the North Pacific Ocean. University of California Publications in Geological Science 118, 176 pp.

Doroff, A.M., Bodkin, J.L., 1994. Sea otter foraging behavior and hydrocarbon levels in prey following the *Exxon Valdez* oil spill in Prince William Sound, Alaska. In: Loughlin, T. (Ed.), Marine Mammals and the *Exxon Valdez*. Academic Press, San Diego, CA, pp. 193–208.

Doroff, A.M., DeGange, A.R., 1994. Sea otter, *Enhydra lutris*, prey composition and foraging success in the northern Kodiak archipelago. Fishery Bull. 92, 704–710.

Duggins, D.O., Simenstad, C.A., Estes, J.A., 1989. Magnification of secondary production by kelp detritus in coastal marine ecosystems. Science 245, 170–173.

Ebert, E.E., 1968. A food habits study of the southern sea otter, *Enhydra lutris nereis*. Calif. Fish Game 54, 33–42.

Edwards, M.S., 2000. The role of alternate life-history stages of a marine macroalga: a seed-bank analogue? Ecology 81, 2404–2415.

Edwards, M.S., Estes, J.A., 2006. Catastrophe, recovery and range limitation in Northeast Pacific kelp forests: a large-scale perspective. Mar. Ecol. Prog. Ser. 320, 79–87.

Esslinger, G.G., Bodkin, J.L., 2009. Trends in Southeast Alaska sea otter populations; 1969–2003. USGS Scientific Investigations 2009–5045, 18pp.

Estes, J.A., 1989. Adaptations for aquatic living in carnivores. In: Gittleman, J.L. (Ed.), Carnivore Behavior Ecology and Evolution. Cornell University Press, Ithaca, NY, pp. 242–282.

Estes, J.A., 1990. Growth and equilibrium in sea otter populations. J. Anim. Ecol. 59, 385–400.

Estes, J.A., 2005. Carnivory and trophic connectivity in kelp forests. In: Ray, J.C., Redford, K.H., Steneck, R.S., Berger, J. (Eds.), Large Carnivores and the Conservation of Biodiversity. Island Press, Washington, DC, pp. 61–81.

Estes, J.A., 2008. Kelp forest food webs in the Aleutian archipelago. In: McClahanan, T.R., Branch, G.M. (Eds.), Food Webs and the Dynamics of Marine Reefs. Oxford University Press, New York, NY, pp. 29–49.

Estes, J.A., Bodkin, J.L., 2002. Marine otters. In: Perrin, W.F., Würsig, B., Thewissen, J.G.M. (Eds.), Encyclopedia of Marine Mammals. Academic Press, San Diego, CA, pp. 842–858.

Estes, J.A., Duggins, D.O., 1995. Sea otters and kelp forests in Alaska: generality and variation in a community ecological paradigm. Ecol. Monogr. 65, 75–100.

Estes, J.A., Palmisano, J.F., 1974. Sea otters: their role in structuring nearshore communities. Science 185, 1058–1060.

Estes, J.A., Steinberg, P.D., 1988. Predation, herbivory and kelp evolution. Paleobiology 14, 19–36.

Estes, J.A., VanBlaricom, G.R., 1985. Sea otters and shellfisheries. In: Beverton, R.H., Lavigne, D., Beddington, J. (Eds.), Conflicts between Marine Mammals and Fisheries. Allen and Unwin, London, pp. 187–235.

Estes, J.A., Jameson, R.J., Johnson, A.M., 1981. Food selection and some foraging tactics of sea otters. In: Chapman, J.A., Pursley, D. (Eds.), Worldwide Furbearer Conference Proceedings. Worldwide Furbearer Conference, Inc., Frostburg, MD, pp. 606–641.

Estes, J.A., Jameson, R.J., Rhode, E.B., 1982. Activity and prey selection in the sea otter: influence of population status on community structure. Am. Nat. 120, 242–258.

Estes, J.A., Tinker, M.T., Williams, T.M., Doak, D.F., 1998. Killer whale predation on sea otters linking coastal with oceanic ecosystems. Science 282, 473–476.

Estes, J.A., Lindberg, D.R., Wray, C., 2005. Evolution of large body size in abalones (*Haliotis*): patterns and implications. Paleobiology 31, 591–606.

Estes, J.A., Tinker, M.T., Bodkin, J.L., 2010. Using ecological function to develop recovery criteria for depleted species: sea otters and kelp forests in the Aleutian archipelago. Conserv. Biol. 24, 852–860.

Estes, J.A., Brashares, J.S., Power, M.E., 2013. Predicting and detecting reciprocity between indirect ecological interactions and evolution. Am. Nat. 181, S76–S99.

Garshelis, D.L., Garshelis, J.A., Kimker, A.T., 1986. Sea otter time budgets and prey relationships in Alaska. J. Wildl. Manage. 50, 637–647.

Gibson, J.R., 1976. Imperial Russia in Frontier America: The Changing Geography of Supply of Russian America, 1784–1867. Oxford University Press, New York, NY.

Hatfield, B.B., Marks, D., Tinker, M.T., Nolar, K., Peirce, J., 1998. Attacks on sea otters by killer whales. Mar. Mammal Sci. 14, 888–894.

Iverson, J.A., 1972. Basal energy metabolism of mustelids. J. Comp. Physiol. 81, 341–344.

Jameson, R.J., Johnson, A.M., 1993. Reproductive characteristics of female sea otters. Mar. Mammal Sci. 9, 156–167.

Kenyon, K.W., 1969. The Sea Otter in the North Pacific Ocean. North American Fauna 68. US Department of the Interior, Washington, DC.

Kleiber, M., 1961. The Fire of Life. Wiley, New York, NY.

Konar, B., 2000. Seasonal inhibitory effects of marine plants on sea urchins: structuring communities the algal way. Oecologia 125, 208–217.

Konar, B., Estes, J.A., 2003. The stability of boundary regions between kelp beds and deforested areas. Ecology 84, 174–185.

Kvitek, R.G., Oliver, J.S., 1988. Sea otter foraging habits and effects on prey populations and communities in soft-bottom environments. In: VanBlaricom, G.R., Estes, J.A. (Eds.), The Community Ecology of Sea Otters. Springer Verlag, New York, NY, pp. 22–47.

Kvitek, R.G., Shull, D., Canestro, D., Bowlby, C.E., Troutman, B.L., 1989. Sea otters and benthic prey communities in Washington State. Mar. Mammal Sci. 5, 266–280.

Kvitek, R.G., Bowlby, C.E., Staedler, M., 1993. Diet and foraging behavior of sea otters in southeast Alaska. Mar. Mammal Sci. 9, 168–181.

Kvitek, R.G., Iampietro, P., Bowlby, C.E., 1998. Sea otters and benthic prey communities: a direct test of the sea otter as keystone predator in Washington State. Mar. Mammal Sci. 14, 895–902.

Lawton, J.H., May, R.M., 1995. Extinction Rates. Oxford University Press, Oxford.

Lowry, L.F., Pearse, J.S., 1973. Abalones and sea urchins in an area inhabited by sea otters. Mar. Biol. 23, 213–219.

Maron, J.L., Estes, J.A., Croll, D.A., Danner, E.M., Elmendorf, S.C., Buckelew, S.L., 2006. An introduced predator alters Aleutian Island plant communities by thwarting nutrient subsidies. Ecol. Monogr. 76, 3–24.

McCaughley, D.J., Young, H.S., Dunbar, R.B., Estes, J.A., Semmens, B.X., Michelli, F., 2012. Assessing the effects of large mobile predators on ecosystem connectivity. Ecol. Appl. 22, 1711–1717.

McNab, B.K., 1989. Basal rate of metabolism, body size and food habits in the order Carnivora. In: Gittleman, J.L. (Ed.), Carnivore Behavior, Ecology and Evolution. Cornell University Press, Ithaca, NY, pp. 335–354.

Meinig, D.W., 1986. The Shaping of America: A Geographical Perspective on 500 Years of History, vol. 1: Atlantic America, 1492–1800. Yale University Press, New Haven, CT.

Monson, D., Estes, J.A., Siniff, D.B., Bodkin, J.L., 2000. Life history plasticity and population regulation in sea otters. Oikos 90, 457–468.

Monson, D.H., DeGange, A.R., 1995. Reproduction, preweaning survival, and survival of adult sea otters at Kodiak Island, Alaska. Can. J. Zool. 73, 1161–1169.

Mora, C., Tittensor, D.P., Adl, S., Simpson, A.G.B., Worm, B., 2011. How many species are there on earth and in the ocean? Plos Biol. 9, 1–8.

Morris, W.F., Doak, D.F., 2002. Quantitative Population Biology: Theory and Practice of Population Viability Analysis. Sinauer, Sunderland, MA.

Morrison, P., Rosenman, M., Estes, J.A., 1975. Metabolism and thermoregulation in the sea otter. Physiol. Zool. 47, 218–229.

Paine, R.T., 1969. A note on trophic complexity and community stability. Am. Nat. 103, 91–93.

Paine, R.T., 1980. Food webs: linkage, interaction strength, and community infrastructure. J. Anim. Ecol. 49, 667–685.

Paine, R.T., 1992. Food-web analysis through field measurement of per capita interaction strength. Nature 355, 73–75.

Palumbi, S.R., 2003. Population genetics, demographic connectivity, and the design of marine reserves. Ecol. Appl. 13, S146–S158.

Pielou, E.C., 1992. After the Ice Age. University of Chicago Press, Chicago, IL.

Polis, G.A., Hurd, S.D., 1996. Linking marine and terrestrial food webs: allochotonous input from the ocean supports high secondary productivity on small islands and coastal land communities. Am. Nat. 197, 396–423.

Reed, D.C., Brzezinski, M.A., 2009. Kelp forests. In: Laffoley, D., Grimsditch, G. (Eds.), The Management of Natural Coastal Carbon Sinks. IUCN, Gland, Switzerland, pp. 31–37.

Reed, D.C., Laur, D.R., Ebeling, A.W., 1988. Variation in algal dispersal and recruitment: the importance of episodic events. Ecol. Monogr. 58, 321–335.

Rees, M., 2003. Our Final Hour. Basic Books, New York, NY.

Riedman, M.L., Estes, J.A., 1990. The sea otter (*Enhydra lutris*): behavior, ecology, and natural history. Biol. Rep. 90 (14), 126, US Fish and Wildlife Service.

Riedman, M.L., Estes, J.A., Staedler, M., Giles, A., Carlson, D., 1994. Breeding patterns and reproductive success of sea otters in California. J. Wildl. Manage. 58, 391–399.

Roughgarden, J., Gaines, S., Possingham, H., 1988. Recruitment dynamics in complex life cycles. Science 241, 1460–1466.

Scheffer, M., 2009. Critical Transitions in Nature and Society. Princeton University Press, Princeton, NJ.

Scribner, K.T., Bodkin, J.L., Bellachey, B.E., Fain, S.R., Chronin, M.A., Sanchez, M., 1997. Population and genetic studies of sea otter (*Enhydra lutris*): a review and interpretation of available data. In: Dizon, A.E., Chivers, S.J., Perrin, W.F. (Eds.), Molecular Genetics of Marine Mammals. Society for Marine Mammalogy, Lawrence, KS, pp. 197–208 [Special Publication 3].

Siniff, D.B., Ralls, K., 1991. Reproduction, survival and tag loss in California sea otters. Mar. Mammal Sci. 7, 211–229.

Soulé, M.E., 1987. Viable Populations for Conservation. Cambridge University Press, New York, NY.

Soulé, M.E., Terborgh, J. (Eds.), 1999. Continental Conservation. Island Press, Washington, DC.

Soulé, M.E., Estes, J.A., Berger, J., Martinez del Rio, C., 2003. Recovery goals for ecologically effective numbers of endangered keystone species. Conserv. Biol. 17, 1238–1250.

Soulé, M.E., Estes, J.A., Miller, B., Honnold, D.A., 2004. Strongly interacting species: conservation policy, management, and ethics. BioScience 55, 168–176.

Stearns, S.C., 1989. Trade-offs in life-history evolution. Funct. Ecol. 3, 259–268.

Steinberg, P.D., Estes, J.A., Winter, F.C., 1995. Evolutionary consequences of food chain length in kelp forest communities. Proc. Natl. Acad. Sci. USA 92, 8145–8148.

Steneck, R.S., Graham, M.H., Bourque, B.J., Corbett, D., Erlandson, J.M., Estes, J.A., et al., 2002. Kelp forest ecosystem: biodiversity, stability, resilience and future. Environ. Conserv. 29, 436–459.

Stevenson, M.D., 1977. Sea otter predation on Pismo clams in Monterey Bay. Calif. Fish Game 63, 117–120.

von Biela, V.R., Gill, V.A., Bodkin, J.L., Burns, J.M., 2009. Evidence of phenotypic plasticity in the average age at first reproduction of northern sea otters (*Enhydra lutris kenyoni*) in Alaska. J. Mammal 90, 1224–1231.

Watson, J., Estes, J.A., 2011. Stability, resilience, and phase shifts in kelp forest communities along the west coast of Vancouver Island, Canada. Ecol. Monogr. 81, 215–239.

Watt, J., Siniff, D.B., Estes, J.A., 2000. Interdecadal change in diet and population of sea otters at Amchitka Island, Alaska. Oecologia 124, 289–298.

Willemsen, G.F., 1992. A revision of the Pliocene and Quaternary Lutrinae from Europe. Scripta Geol. 101, Nationaal Natuurhistorisch Museum, Leiden.

Wilmers, C.C., Estes, J.A., Edwards, M., Laidre, C.L., Konar, B., 2012. Do trophic cascades affect the storage and flux of atmospheric carbon? An analysis for sea otters and kelp forests. Front. Ecol. Environ. 10, 409–415.

Wilson, D.E., 1991. Geographic variation in sea otters *Enhydra lutris*. J. Mammal 72, 22–36.

Wilson, E.O., 2002. The Future of Life. Knoph, Borzoi Books, New York, NY.

Wozencraft, W.C., 1993. Mammal Species of the World: A Taxonomic and Geographic Reference. Smithsonian Institution Press, Washington, DC.

Yeates, L.C., Williams, T.M., Fink, T.L., 2007. Diving and foraging energetics of the smallest marine mammal, the sea otter (*Enhydra lutris*). J. Exp. Biol. 210, 1960–1970.

Chapter 3

Historic and Contemporary Status of Sea Otters in the North Pacific

James L. Bodkin
US Geological Survey, Alaska Science Center, Anchorage, AK, USA

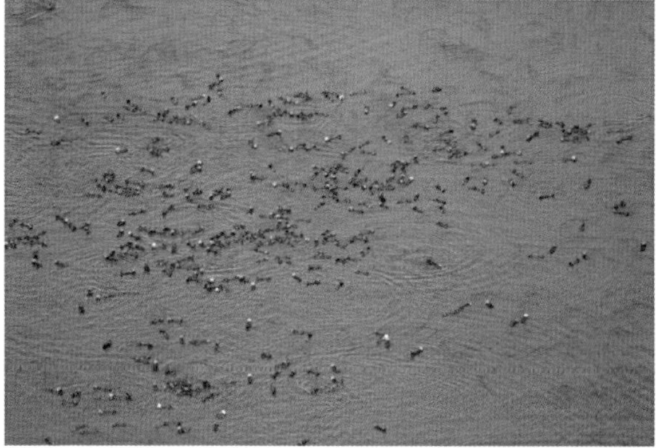

Sea Otter Conservation. DOI: http://dx.doi.org/10.1016/B978-0-12-801402-8.00003-2

INTRODUCTION

Sea otters and our collective attempts to conserve and quantify their numbers over the past century provide perhaps an unprecedented opportunity to explore how and why their abundance changes over time. Their distribution relative to the coast is defined by their need and ability to frequently dive to the sea floor to obtain prey that is almost exclusively benthic-dwelling invertebrates. In much of their range this means that they are concentrated between the shore line and the 50 m depth contour, usually less than 1–3 km from shore. Because they occupy a relatively narrow band of nearshore habitat their populations can be considered essentially linear in most areas. As a consequence, their numbers are relatively easily quantified, either from the shore or from small skiffs or aircraft. Initially, due to their near extirpation, and subsequently because of their role as a "keystone predator" (Chapter 2) and their social and cultural value to humans, there has been a concerted effort to conserve the sea otter. Since populations began recovery following the end of the maritime fur trade in 1910 we have a comprehensive and geographically diverse record of the abundance of sea otters that documents periods of increase, decline, and, in some cases, relative stability.

In this chapter I explore the conservation of sea otters over the past century, primarily in numerical terms, but also in a spatial context imposed by a life history that results in relatively small home ranges and population structuring that occur at correspondingly small spatial scales. From this I will convey the lessons learned from studying the processes contributing to recovery and in some cases stability or declines in this species that may be instructive to the conservation of other species and their habitats.

Historic Distribution and Abundance

The archeological, biological, and fur harvest records (Barabash-Nikiforov, 1947; Lensink, 1962; Kenyon, 1969) provide a fairly concise record of the latitudinal distribution of sea otters on both sides of the Pacific (Figure 3.1). Some uncertainty concerning the northern range limit may have been influenced by effects of seasonal southern ice extent (Schneider and Faro, 1975) and it might be expected that future shifts in sea ice extent corresponding to climate change may affect future sea otter distribution. It is reported that the southern distribution of the sea otter may be limited to the extent of cool, nutrient-enriched temperate waters (Riedman and Estes, 1990), represented by Hokkaido, Japan, in the western Pacific and central Baja California, Mexico, in the east. It seems likely that the physiological adaptations that allow sea otters to reside near the ice edge in high latitudes may also limit the extent of their range into the lower latitudes (Bodkin, 2003).

The abundance of sea otters in the Pacific prior to the commercial fur trade is not well known, with estimates ranging from 150,000 to 300,000

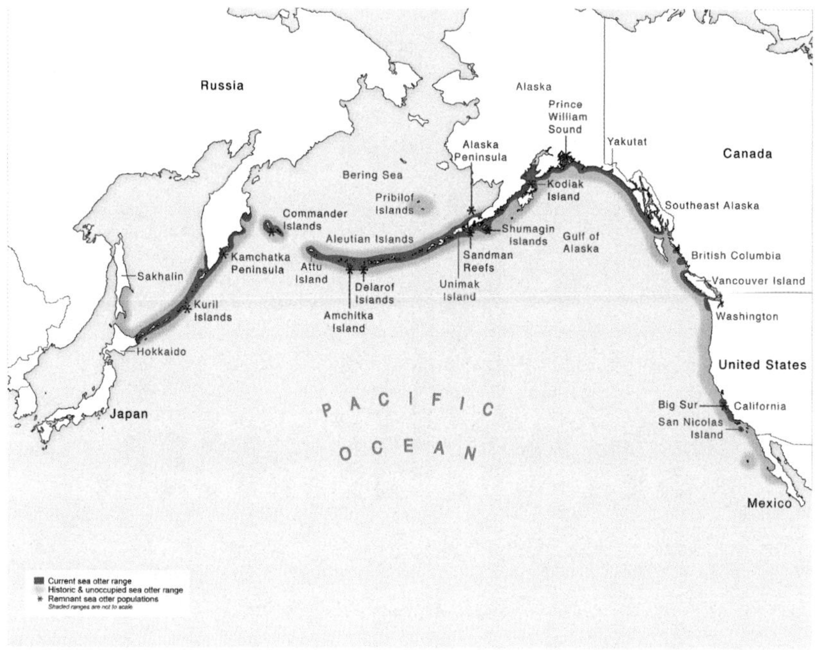

FIGURE 3.1 Historic and contemporary distribution of sea otters in the North Pacific.

individuals (Lensink, 1962; Kenyon, 1969; Johnson, 1982). In some cases the relative abundance of sea otters over millennial time scales has been inferred from midden sites (Simenstad et al., 1978; Lech et al., 2011). However, beginning about 1750 and at the spatial scale of the North Pacific, a fairly clear picture of change in sea otter abundance begins to emerge (Lensink, 1962; Kenyon, 1969; Bodkin, 2003).

The Maritime Fur Trade

The recent history of sea otter populations begins with a rapidly escalating reduction in abundance as the commercial maritime fur trade developed in the western Pacific (Figure 3.2) following the explorations by Vitus Bering that ended in 1742. The harvest of sea otters peaked at the turn of the nineteenth century when nearly 15,000 per year on average were being harvested in Alaska and populations rapidly declined as they neared extinction. The first documented efforts to restore and sustainably manage the harvest were made by the Russian American Co. early in the nineteenth century and included a limited male-only harvest. These efforts eventually led to increasing populations and harvest levels until about 1850. Following the sale of Alaska by Russia in 1867 to the United States, and despite protections

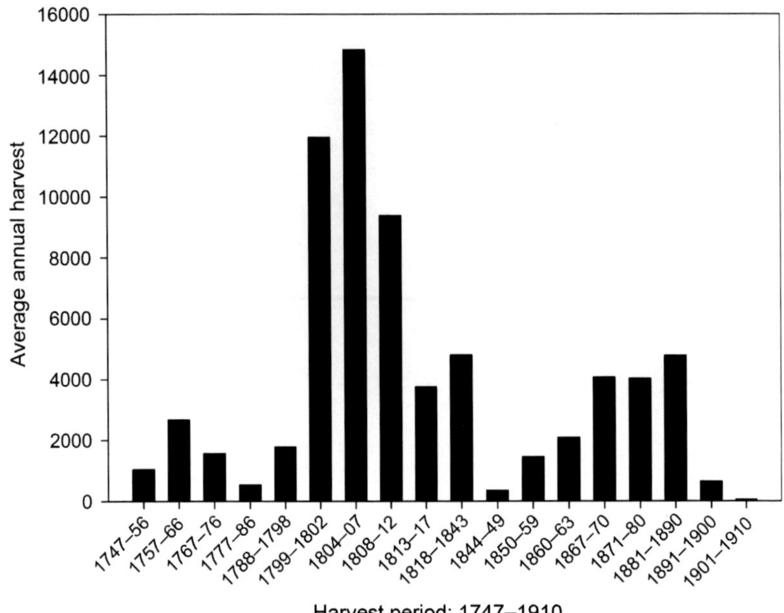

FIGURE 3.2 Average annual commercial harvest of sea otters from Alaska, 1750–1910. Note that not all time increments are equal. *(Data summarized from Lensink (1962).)*

afforded within territorial waters of Alaska, a multinational harvest expanded, once again resulting in a collapse of the harvest as populations became depleted. By late in the 1890s the harvest of sea otters was effectively ended as a consequence of their scarcity (Lensink, 1962). In 1910, the final recognized year of the maritime fur trade, 40 Aleut hunters in the only two vessels still hunting obtained a total of 16 pelts (Lensink, 1962), signaling the near demise of the species.

The commercial harvest that began in earnest about 1750 generally proceeded east from the Asian mainland and Kurile Islands, across the Aleutian Archipelago and Alaska Peninsula and then southward along the west coast of North America. As noted by Gorbics and Bodkin (2001), the annual harvest rate during the commercial fur trade averaged only about 1.5% per year of the global sea otter population. Range-wide reductions and extirpations occurred not simply because of excessive harvest, but because the harvest was not allocated proportional to the abundance and distribution of sea otters. This resulted in the serial depletion of otters beginning in the western Pacific that systematically expanded eastward across the Aleutian Archipelago and southward along North America, as harvested populations became either reduced to unprofitable densities or, more often, became locally extinct. This process of serial depletion was facilitated by the relatively sedentary nature of sea otters.

Annual home range sizes of adult sea otters are relatively small, with male territories ranging from <1 to 11 km^2 and adult female home ranges from <1 to 24 km^2 (Loughlin, 1980, Garshelis and Garshelis, 1984, Ralls et al., 1988; Jameson, 1989). When additional mortality is spatially concentrated in areas equal to or smaller than the cumulative home range of the population being managed, local depletion, potentially leading to serial depletion, may occur. Therefore, it is essential to consider the spatial scales at which sea otter mortality is managed (Bodkin and Ballachey, 2010). The same may be true for other species of conservation concern.

As the maritime fur trade came to a close late in the nineteenth century, sea otters had been reduced to 13 remnant colonies (11 that eventually persisted) distributed almost exclusively near the northern limits of their habitat, with the exception of a single small colony on the remote coast of central California. These colonies were geographically isolated (Figure 3.1) and possibly numbered from a few tens individually to perhaps 2000 collectively (Kenyon, 1969). Relatively little information is available on the initial rate of recovery of these colonies following the end of the maritime fur trade, as surveys were not usually conducted until populations became conspicuous as their numbers increased.

Early Twentieth Century Recovery and Conservation

Sea otters occupy habitat restricted by land masses on one side and ocean depth on the other. Because they forage nearly exclusively on large benthic invertebrates such as clams, crab, mussels, and urchins (Riedman and Estes, 1990), their seaward distribution is limited by their ability to dive to the sea floor. Using archival time depth recorders, we have accumulated a large volume of data that identifies that while some sea otters are capable of diving to >100 m, they rarely do so (Figure 3.3). Rather, most of their foraging takes place in waters less than about 40–50 m deep, and some individuals rarely dive to depths >20 m (Bodkin et al., 2004, 2007). In some instances, where depth contours are widely spaced or where shoals occur far from land, sea otters can be found many kilometers from shore. Examples of offshore populations occur north of the Alaska Peninsula in the Bering Sea and, at least historically, on the Fairweather grounds south of Yakutat (Lensink, 1962). Thus, sea otters find themselves relegated to relatively shallow habitats that are generally adjacent to shore lines. As a consequence of this restricted distribution and proximity to land, sea otter populations are relatively easy to survey, as well as readily accessible for harvest.

Various methods are employed to survey sea otters, including counts from shore, skiffs, or larger vessels, and both rotor and fixed-wing aircraft. Because sea otters must dive for food and spend up to 50% of their time foraging, they are frequently unavailable for detection and thus all survey methods produce results that are biased low to varying degrees, depending on the survey

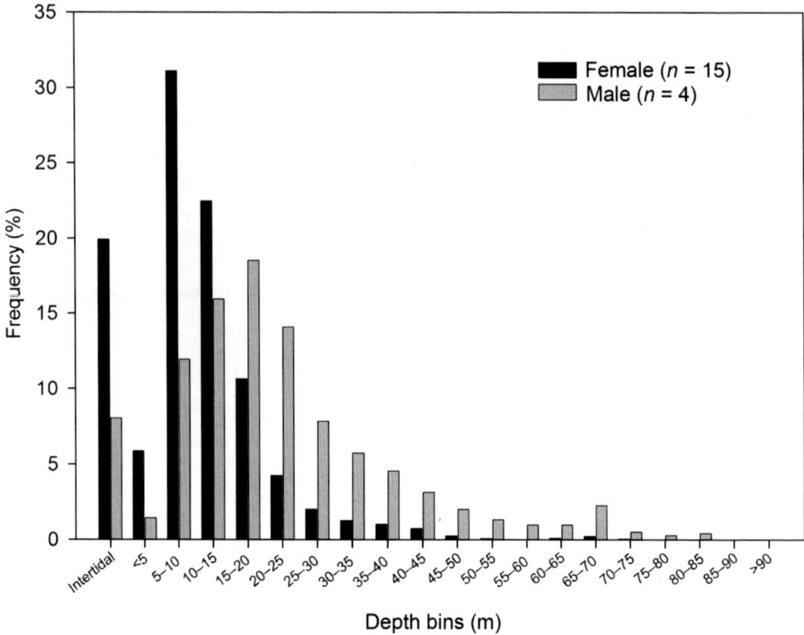

FIGURE 3.3 Forage dive depth distribution of 15 female and 4 male adult sea otters in PWS, Alaska. Less than 0.04% of all dives were >50 m depth. *(From Bodkin et al. (2012).)*

platform (Udevitz et al., 1995). When diving animals are not accounted for it is generally assumed that all individuals are detected (or that detection remains constant over time) and these types of surveys are considered censuses. In some instances, where unbiased estimates of abundance are required (e.g., harvested populations), methods have been developed to correct for this detection bias (Bodkin and Udevitz, 1999) and results for such surveys are generally considered population estimates. The distinction between censuses and unbiased estimates of abundance is critical. Census data can be useful in documenting the distribution of animals and rates of change over time, but cannot estimate actual abundance (unless all animals are truly observed). In some cases, where shore access is good and depth contours are close to shore, such as central California, nearly all animals can be accounted for by well-trained ground observers (Estes and Jameson, 1988). The absence of unbiased estimates of abundance can preclude documenting change in the numbers of individuals, which can be important where populations may be subjected to catastrophic events such as oil spills (Ballachey et al., 1994), or where they are harvested (Bodkin and Ballachey, 2010). Where available, I will present population estimates in evaluating population trends over time. In many cases, routine censuses provide the best available information and those will be included in analyses as appropriate.

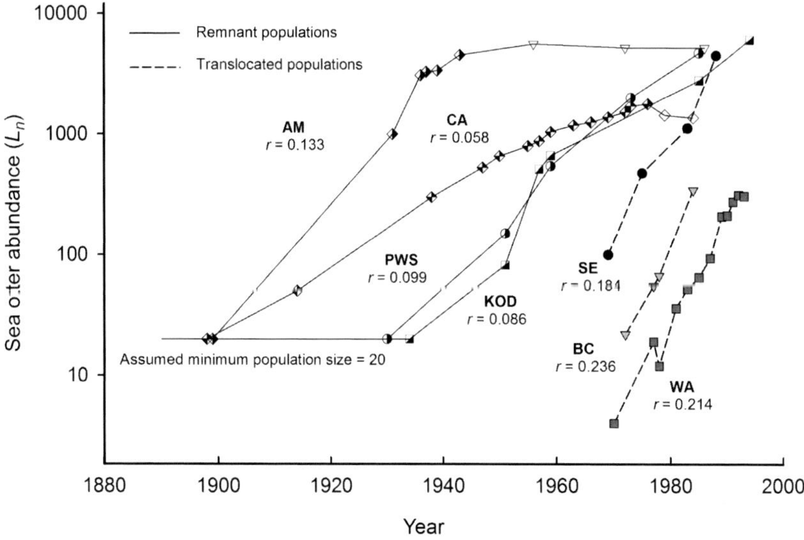

FIGURE 3.4 Population growth trajectories for four remnant and three translocated sea otter populations. Unfilled symbols indicate data points not included in rate (*r*) calculations. *(From Bodkin et al. (1999).)*

Based on maximum estimated reproductive and survival rates, sea otter populations are theoretically capable of achieving annual rates of increase of up to about 22% (Estes, 1990), assuming immigration is not a contributing factor. However, long-term recovery rates of four sea otter populations following protection in 1911 averaged less than one-half of this rate (9%), and ranged from about 6% in California to 13% at Amchitka (Figure 3.4; Bodkin et al., 1999). By the middle of the twentieth century it appeared that at Amchitka Island, in the central Aleutian Archipelago, sea otters had recovered all available habitat and recovered to the point of approaching an equilibrium density (Estes, 1990). The reasons that remnant sea otter population recovery did not attain higher rates are unclear, but may have been related to continued human harvests following widespread protection (Bodkin et al., 1999, Chapters 13 and 14). It has also been suggested that prey populations shortly after the end of the fur harvest were not suitably large for supporting maximum growth rates (Bodkin et al., 1999). However, given suitable habitat and continued protection, remnant populations in Russia, Alaska, and California continued to recover at least until late into the twentieth century.

Translocations Mid-Twentieth Century

Translocations are a widely recognized tool to aid in the recovery of depleted populations (Griffith et al., 1989) as well as the restoration of ecological processes (Moritz, 1999). Translocation success depends on various factors, including an appropriate number of founding individuals, suitable habitat, food resources, and reproductive potential (Griffith et al., 1989). The initial translocation of sea otters was conducted by Russians in 1937, when nine sea otters were captured at Medny Island, in the Commander Islands, for transport to the Murman coast in the southern Barents Sea, far from their natural distribution (Barabash-Nikiforov, 1947). Only two males survived the journey, but they successfully resided in captivity and in the wild for at least 5 years, thus demonstrating the feasibility of translocating this species.

Initial efforts to restore sea otter populations along the coast of North America began in 1951 (Kenyon, 1969). Between 1951 and 1959 attempts to translocate sea otters from Amchitka Island in the central Aleutians to the Pribilof Islands in the Bering Sea and Attu Island in the western Aleutians failed, most likely due to poor husbandry and inadequate numbers (Kenyon, 1969, Chapter 8). In 1965 translocations to St. George, in the Pribilof Islands, from Amchitka were successful and followed by translocations to southeast Alaska (SE AK), British Columbia (BC), Washington (WA), and Oregon. A total of 708 animals were moved from Amchitka and Prince William Sound (PWS) in 13 different translocations (Jameson et al., 1982). The only translocation in this group that failed was to Oregon. In 1987, the most recent translocation of sea otters was initiated when 139 animals were taken from the mainland California coast to San Nicolas Island, off the coast of Southern California (Rathbun et al., 2000).

Numbers of sea otters generally decline rapidly following translocations and in most cases appeared to soon stabilize at about 10% of the number actually moved. We have a poor understanding of the causes behind this rapid diminishment following introductions that were hampered by limited follow-up surveys after most translocations. Post-translocation surveys and marking of individuals moved to San Nicolas Island allowed insights into this phenomenon. Intensive post-translocation surveys throughout California documented that 26% (36 of 139) of the animals that were moved swam back to an area near their original capture location (Rathbun et al., 2000). This finding demonstrates a strong affinity in this species for their home range, despite the selection of San Nicolas Island as their release site due to its suitability of habitat and the abundance of prey present (Rathbun et al., 2000). These observations indicate that factors other than prey can be important in determining the behavior of species being translocated and suggest that social or cultural attributes should be considered in translocations. It is also possible that animals that are translocated simply are unwilling to remain where they are released, despite the suitability of habitat and abundance of prey.

Surveys of translocated populations generally demonstrated growth rates that were at or near the maximum rates feasible for sea otters, averaging 21% per annum (Bodkin et al., 1999, Figure 3.4). One recent exception to this was the 1987 translocation to San Nicolas Island. In this case, following 3 years of consecutive translocations the founding population quickly declined to 16 individuals and remained essentially unchanged for nearly a decade, despite the birth of 50 pups. Clearly, animals were being lost, either through immigration or, more likely, fishery-related mortality (Hatfield et al., 2011). 2014 survey results from San Nicolas indicate a population of at least 68 otters, including eight pups, with a total of at least 201 pups produced at the Island since 1987 (B. Hatfield USGS). Since 2007, when 41 otters were present, the San Nicolas population has increased at an average annual rate of about 7% (USGS unpublished data). Although only about one-third of the average growth rate of other translocated populations, and despite the recent USFWS declaration of this effort as failed, it seems increasingly likely that this population may persist.

Thus several lessons can be gained from experiences with translocating sea otters to aid conservation. First, biology of the species is important. Early translocations were largely unsuccessful due to the lack of understanding of the basic physiology of sea otters and their dependence on maintaining a thermal balance through their pelage. Second, given suitable habitat, prey resources, and protection from human mortality, translocations were an important tool in sea otter conservation. Today, about 35% of the global sea otter abundance can be attributed to the translocations to SE AK, BC, and WA (Table 3.1). Third, a variety of factors will tend to reduce the founding population to a small fraction of the number actually translocated. It appears likely that with sea otters, behavior may have been more important than food in determining the retention rate at the site of introduction. Despite the presence of suitable habitat and abundant prey resources, all translocations suffered high initial losses (Bodkin et al., 1999; Rathbun et al., 2000). A careful consideration of the behavior and social structure within parent populations that may affect the probability of retention at a translocation site may aid in forecasting the success of translocation as a conservation tool.

Late Twentieth Century

As the twentieth century came to a close, the outlook for sea otter conservation and recovery was fairly positive. Increases in abundance and range expansion of remnant populations had resulted in reoccupation of most of the species' northern range, including most of Russia, the Aleutian Archipelago, and much of the Gulf of Alaska, and growth rates suggested continuing recovery. Translocations to SE AK, BC, and WA resulted in established populations exhibiting rather phenomenal rates of increase, and the translocation to San Nicolas Island in California held promise of relieving at least some of the threats to the mainland population. Although there were subtle signs that all was not well with the sea otter, the new millennium

TABLE 3.1 Recent Estimates of the Abundance of Sea Otters at Locations from California to Japan

	Year	Number	Rate Change	% of Total	Source
California[1]	2013	2943	0.03	2.4	USGS
Washington	2011	1154	0.08	0.9	FWS, WDFW
British Columbia[2]	2008	4700	0.08	3.8	DFO, Canada
SE Alaska	2012	25712	0.05	20.9	FWS/USGS
Gulf of Alaska[3]	2004	46703	0.05	38.0	USGS/FWS
AK Peninsula S[4]	2001	4993	−0.18	4.1	FWS
AK Peninsula N[5]	2000	11353	−0.04	9.2	FWS
Aleutian Islands	2000	8742	−0.29	7.1	USGS/FWS
Commander Islands	2007	7010	0.08	5.7	A. Burdin
Kamchatka Peninsula	2008	518	NA	0.4	V. Nikulin
Kuril Islands	2000	9047	NA	7.4	S. Kornev
Japan	2010	present	NA		
Total		**122875**			

[1]*Includes 62 sea otters at San Nicolas Island.*
[2]*Includes Vancouver Island and mainland (Chapter 13).*
[3]*Includes Kodiak Is, Katmai and Kenai Fjords National Parks, Cook Inlet, PWS and south to Cape Spencer.*
[4]*Includes False Pass to Cape Aklek and islands south of Alaska Peninsula.*
[5]*Includes Unimak Islands to Cape Seniavin, north of Alaska Peninsula.*
Rate of change is based on data over approximately 10 years prior to the survey year. Rate of change is calculated by regressing the log of survey data over time. Italicized locations represent translocated populations. See Figure 3.5 for locations referenced.

generally looked positive, particularly considering the precipice of extinction on which the sea otter had rested a century earlier.

Although extremely remote, the Aleutian Islands, and Amchitka Island in particular, in many regards might be considered hallowed habitat as far as sea otters are concerned. Sea otters were an important resource to the Aleuts who originally occupied the islands for several centuries prior to the maritime fur trade (Simenstad et al., 1978). Amchitka was also the site of perhaps the largest population to survive the fur trade and Karl Kenyon conducted much of the research behind his seminal sea otter monograph (1969) at this island. Some of our first evidence of the role of food in structuring sea otter populations was revealed as Amchitka became fully occupied by the otter shortly after WWII (Kenyon, 1969). In the following decades, as the world entered the nuclear age, Amchitka became the site of extensive below-ground testing of nuclear warheads by the United States. As a consequence of the testing program, far-reaching research on the biology and ecology of the island, and on marine systems more generally, was supported by the Nuclear Regulatory Commission (Merritt and Fuller, 1977). It is through these studies and comparisons across islands where sea otters did not occur that we first find documentation of the keystone role that sea otters play in the structuring and functioning of nearshore marine ecosystems (McLean, 1962; Estes and Palmisano, 1974, Chapter 2). But also important in this work was the continuation of the time series of sea otter population surveys that were reported by Kenyon (1969) and continued by researchers at irregular intervals through to the present. It was these long time series of data at individual islands that led to the discovery late in the twentieth century that sea otters across the Aleutian Archipelago and the Alaska Peninsula (Figure 3.1) were unexpectedly once again in dramatic decline (Estes et al., 1998; Doroff et al., 2003; Burn and Doroff, 2005, Chapter 4).

Doroff et al. (2003) chronicle perhaps one of the greatest declines in the abundance of a large mammalian carnivore in recent history. Survey data collected in the 1990s indicate that sometime prior to 1990 sea otter populations in the central Aleutian Archipelago begin to demonstrate significant declines. Over a distance of about 2000 km, sea otter populations declined at about 18% per year, eventually reaching a few percent of their prior abundance. Burn and Doroff (2005) estimate that more than 62,000 sea otters, and perhaps as many as 90,000, were lost across the Aleutian Archipelago by the close of the twentieth century. More recent survey data indicate the decline continued at least through 2008 (USFWS 2008 stock assessment) with no evidence of recovery through at least 2010 (USGS unpublished data), and that the decline extends eastward at least to the central Alaska Peninsula (USGS unpublished data). The decline has been attributed to predation by the killer whale (*Orcinus orca*) (Estes et al., 1998, Chapter 4) possibly precipitated by commercial whaling in the North Pacific that resulted in dramatic declines in killer whale prey. The decline in great whales may have led to killer whales switching to pinnipeds and eventually sea otters (Springer et al., 2003), although this theory is not without controversy (Wade et al., 2007; Estes et al., 2009).

Other than the large temporal variation in the abundance of sea otters and their prey represented in Aleutian Island middens (Simenstad et al., 1978; Lech et al., 2011), we have little insight into how sea otter populations may have varied over time prior to the fur trade. However, the recent decline in sea otters across the Aleutians and Alaska Peninsula, combined with midden records, suggests that large-scale variations in future sea otter abundance should not necessarily be unexpected.

Recent Population Abundance

Although sea otters occupy several thousand kilometers of habitat from Southern California across the Pacific Rim to northern Japan, because of the

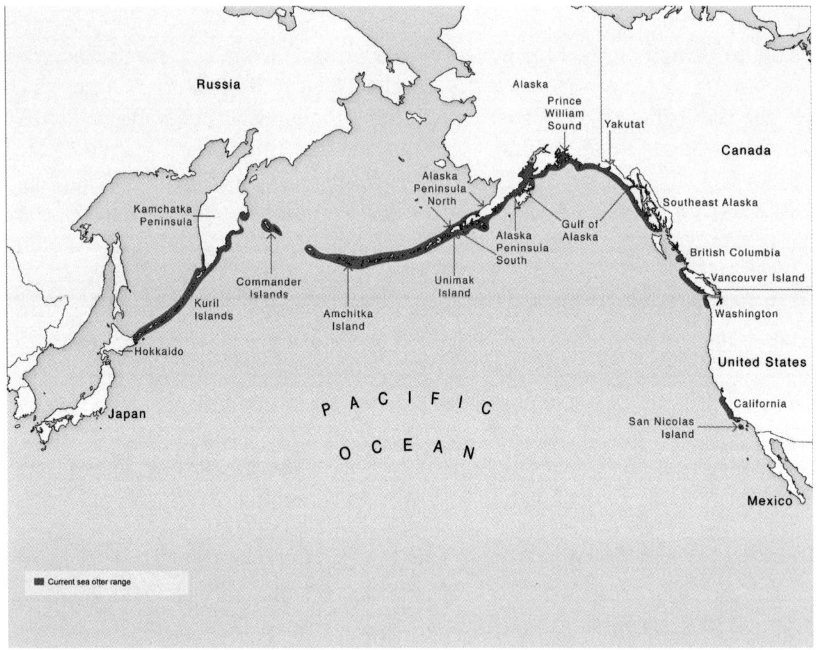

FIGURE 3.5 Contemporary sea otter distribution and locations of aggregated recent sea otter population counts and estimates of abundance referenced in Table 3.1.

widespread interest in sea otter conservation and the relative ease with which sea otters can be enumerated, there is relatively recent survey data over most of the species' range (Table 3.1). Data include results from censuses and surveys and include previously published and unpublished data. The rate of change (Table 3.1) reflects the recent estimated rate of change in each region (Figure 3.5). Rate of change was obtained from published values or calculated by regressing recent counts or estimates over time. The actual number of sea otters today may be more or less than 122,875 as most populations are either increasing or decreasing. The previous published estimate of the abundance of sea otters throughout their range was 93,200 and included survey data from 1994 to 2003 (Bodkin, 2003). Modest increases in many geographic areas have been offset by declines in the Aleutian Islands and Alaska Peninsula, resulting in a modest average annual increase of less than 2% over the past 15 years in the global sea otter population.

The approximately 125,000 sea otters present early in the twenty-first century reflects a remarkable success story in efforts to remedy prior poor management of a natural resource. Conservation was achieved through protection of the sea otter from human harvest and reintroductions into habitats from which they were extirpated. Data presented in Table 3.1 also illustrate mechanisms of population regulation that might not have been previously expected, yet should be considered in future conservation and management efforts. These include the widespread decline of sea otters across a large portion of their northern range, precipitated through predation by killer whales, the persistently low growth rate in the California population, and the reduced growth rates that are now evident in all of the surviving translocated populations. The rate of growth in the mainland California population has remained low (about 5%) and variable (including periods of decline) since regular surveys began in the 1970s (Figure 3.4), and several sources of human- and non-human-related mortality apparently contribute to this (Chapter 4). These include fisheries-related mortality, predation by sharks, and disease (Estes et al., 2003). The California population serves to illustrate the role of cumulative sources of mortality affecting rates of population change. It will be important to consider the role of cumulative mortality in other sea otter populations that demonstrate reduced rates of increase or declines.

Changes in the growth rates within translocated populations provide suggestion of other processes that may be influencing growth. Where two decades ago population growth rates in WA, BC, and SE AK averaged >20% (Figure 3.4), today that average is 7.4%, nearly a third of the prior rate, and we find no evidence of growth exceeding 10% annually in any sea otter population at these spatial scales (Table 3.1). The causes behind these reduced rates of growth are not known and should be cause for additional research. It is possible that behavioral processes that result in spatial and demographic structuring within individual populations are contributing to these reduced growth rates. For example, it is possible that as populations increase in spatial extent and abundance, animals that occur far from the ends of the occupied range may not have access to the abundant prey resources that occur there and that might fuel high growth rates. While such processes are not evident in Table 3.1, the analysis of population trend data over different spatial scales within populations may be informative.

Spatial Scale of Population Structuring

One of the fundamental requirements for management and conservation of species is an understanding of the spatial scale at which populations are structured, specifically in terms of the demographic rates that ultimately determine how and why populations change. For many species movement data acquired by radio or satellite telemetry can provide information useful in identifying spatial aspects of species. However, in the case of sea otters, satellite transmitters are not yet feasible and traditional VHF radio telemetry introduces potential distance biases in movement data that can limit inference regarding movements and eventual population structuring. An alternative approach, however, is the analysis of population abundance data at various spatial scales, where variation in rates of change might indicate the spatial scales at which sea otter populations are demographically structured.

Each of the geographic regions identified in Table 3.1 potentially represents one or more demographically distinct populations that may demonstrate population rates of change that differ from one another. In the following paragraphs I present results of survey data from PWS in the Gulf of Alaska and SE AK that will serve to illustrate variation in population growth rates, and thus demographic rates at various spatial scales.

Prince William Sound

As a consequence of the 1989 *Exxon Valdez* oil spill in PWS, Alaska, frequent surveys of sea otter abundance have been conducted to estimate both acute and chronic effects of the spill and the process of recovery over the past two decades (Bodkin et al., 2002; Bodkin et al., 2011). The resulting data allow us to view population trends at various spatial scales. At the largest scale of PWS as a whole (~5000 km^2 of sea otter habitat), survey data suggest a relatively stable population (Figure 3.6a), although data at this scale are relatively sparse. At the scale of western PWS (~2500 km^2), a relatively consistent growth rate has been evident since 1993 at about 4.5% annually, although variation among years was evident (Figure 3.6a, Bodkin et al., 2011). At the spatial scale of northern Knight Island, within WPWS (168 km^2), there was a relatively long period of stability from 1993–2001, followed by a significant rate of increase of nearly 25% annually through 2011 (Figure 3.6b, Bodkin et al., 2011). At the smallest spatial scale of the north side of Montague Is. (~90 km^2), within WPWS but outside the spill area, there has been no significant trend over this same time period (Figure 3.6b, Bodkin et al., 2011), although population estimates vary annually, likely in response to annual commercial fisheries.

Southeast Alaska

The SE AK translocation was numerically the largest, at 413 individuals over several years, and in 2003 numbered nearly 9000 animals (Esslinger and Bodkin, 2009). In the initial decades following translocation the average annual rate of increase was more than 18%. Late in the twentieth century the rate of increase diminished and divergent trends became evident. In southern SE AK (5320 km^2) the annual rate of change declined to about 7%, while in northern SE AK (3821 km^2) the rate declined to less than 3%. Concurrently, Glacier Bay in northern SE AK (549 km^2) was colonized in 1995 and has demonstrated annual growth of 44%, requiring both intrinsic reproduction and immigration (Esslinger and Bodkin, 2009). The strongly divergent trends among the three geographic areas in SE AK provide further evidence that demographic rates as well as movements differ among areas and at spatial

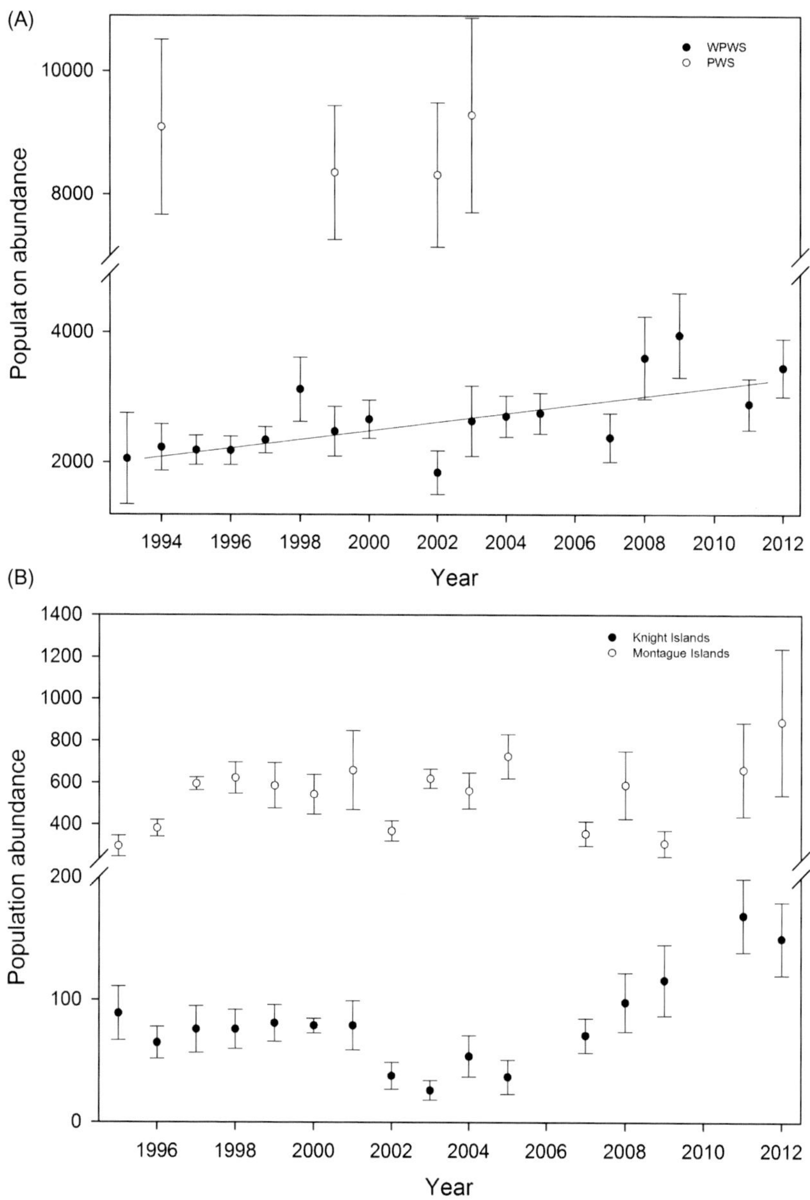

FIGURE 3.6 Results of sea otter population surveys in PWS 1994–2011. Regression lines represent significant trends in abundance. No line indicates no significant trend in abundance. (A) represents data from PWS and western PWS and (B) from Montague and northern Knight Islands in WPWS. *(From Bodkin et al. (2011).*

scales smaller than the total area surveyed. Other examples of divergent rates of change in adjacent sea otter populations that provide further evidence of small scale population structuring in Alaska can be found in Burn and Doroff (2005).

The serial depletion of sea otters noted during the maritime fur trade and the variation in population growth rates noted above suggest that sea otter populations are demographically structured at spatial scales as small as a few hundred square kilometers. This is generally consistent with the relatively small annual home ranges estimated from VHF radio telemetry studies noted above, on the order of a few to a few dozen square kilometers (Garshelis et al., 1984; Ralls et al., 1988; Jameson, 1989, Bodkin unpublished data). Demographic structuring at these small scales becomes important when considering conservation and management strategies and actions, particularly when managing human harvests of sea otters (Bodkin and Ballachey, 2010). It may be equally important when considering conservation and management of other species, including those without human harvest.

CONCLUSION

Unregulated harvest of sea otters over 170 years resulted in near extirpation of the species. Protection from human harvest early in the twentieth century and translocations later in that century brought the sea otter back from the brink of extinction. Today, sea otters occupy much of their historic habitat and their total abundance is approximately 125,000 individuals.

The study of the sea otter decline from over-harvest, subsequent protection and conservation efforts, and trends in abundance over time provide several lessons that may be more broadly applicable to conservation. First, frequent and consistent estimates of abundance are essential to understand the status and trend of populations. Particularly when human sources of mortality are present, accurate abundance estimates are necessary to evaluate the effects of this mortality on populations (Bodkin and Udevitz, 1999). Accurate estimates of population size are also essential to determine the effects of catastrophic mortality on populations, for example in the case of oil spills (Ballachey et al., 1994). Second, translocations can be an important tool in the conservation of diminished species. Careful consideration of a species' biology, behavior, habitat, and social needs may be important in determining the potential success of translocations. Although food resources and suitable habitat are essential, it is apparent in the case of sea otters that other factors, likely related to behavior, resulted in the dispersal of up to 90% of the translocated individuals. As a result, founding population size was a small fraction of the number actually moved. Third, unanticipated events or processes can have large effects on species and populations. Examples from recovering sea otter populations include the emergence of the killer whale as a significant predator, and oil spills (Chapter 4). Although predation by killer whales on sea otters was known to occur (Barabash-Nikiforov, 1947; Kenyon, 1969), it was not expected that predation would cause such widespread and catastrophic decline. Last, spatial scale can be important when considering the status and trend of populations. In the case of sea otters, small increases in mortality relative to the total population size, applied sequentially to small segments of the population, resulted in a process of serial depletion that eventually led to near extirpation (Gorbics and Bodkin, 2001).

There remains some uncertainty in the prospects for the sea otters' future. The catastrophic declines across the Aleutian Archipelago and much of the Alaska Peninsula were not anticipated and provide cause for caution in forecasting the species' future. Presently, population abundance across this vast area remains at a few percentage points of historical abundance, with no evidence of recovery. The present populations are geographically separated throughout the archipelago and remain at risk of local extinction and population bottlenecks (Chapter 5), with potential adverse effects. Across much of the species' range in North America, annual rates of population growth that were at levels from 10% to >20% a few decades ago are now mostly in single digits. Causes for the reduced growth rates across many isolated populations are unknown and provide additional cause for caution regarding future trends and justification for additional research. Evidence of divergent population trend data over various spatial scales within large geographic regions, e.g., PWS and SE AK, provide evidence of fine-scale demographic differences. The spatial scales at which populations structure themselves through behavioral or other processes provide new research questions that remain largely unaddressed. Results from such research may aid in the understanding of current conservation issues of declining populations and diminished rates of increase recently observed in sea otter populations.

ACKNOWLEDGMENTS

A large number of biologists, too numerous to mention individually, from across the North Pacific contributed to the survey data presented in Table 3.1 and I recognize those contributions. This work was supported by the Alaska Science Center, US Geological Survey. Shawn Larson, Glenn VanBlaricom, and Anthony DeGange provided constructive review for this contribution.

REFERENCES

Ballachey, B.E., Bodkin, J.L., DeGange, A.R., 1994. An overview of sea otter studies. In: Loughlin, T. (Ed.), Marine Mammals and the Exxon Valdez. Academic Press, San Diego, CA, pp. 47–59.

Barabash-Nikiforov, I.I., 1947. The sea otter. Translated from Russian by A. Birron and Z.S. Cole for the National Science Foundation by the Israel program for scientific translations, Jeruselam, 1962, 227 pp.

Bodkin, J.L., 2003. Sea otter. In: Feldham, G.A., Thompson, B.C., Chapman, J.A. (Eds.), Wild Mammals of North America, second ed. Johns Hopkins University Press, Baltimore, MD, pp. 735–743.

Bodkin, J.L., Ballachey, B.E., 2010. Modeling the effects of mortality on sea otter populations. US Geological Survey Scientific Investigations Report, 2010–5096, pp. 12.

Bodkin, J.L., Udevitz, M.S., 1999. An aerial survey method to estimate sea otter abundance. In: Garner, G.W., Amstrup, S.C., Laake, J.L., Manly, B.F.J., McDonald, L.L., Robertson, D.G. (Eds.), Marine Mammal Survey and Assessment Methods. Balkema Press, Netherlands, pp. 13–26.

Bodkin, J.L., Ballachey, B.E., Cronin, M.A., Scribner, K.T., 1999. Population demographics and genetic diversity in remnant and re-established populations of sea otters. Conserv. Biol. 13 (6), 1278–1385.

Bodkin, J.L., Ballachey, B.E., Dean, T.A., Fukuyama, A.K., Jewett, S.C., McDonald, L.M., et al., 2002. Sea otter population status and the process of recovery from the *Exxon Valdez* oil spill. Mar. Ecol. Prog. Ser. 241, 237–253.

Bodkin, J.L., Esslinger, G.G., Monson, D.H., 2004. Foraging depths of sea otters and implications to coastal marine communities. Mar. Mammal Sci. 20 (2), 305–321.

Bodkin, J.L., Monson, D.H., Esslinger, G.G., 2007. Population status and activity budgets derived from time-depth recorders in a diving mammal. J. Wildl. Manage. 71 (6), 2034–2044.

Bodkin, J.L., Ballachey, B.E., Esslinger, G.G., 2011. Trends in sea otter population abundance in western Prince William Sound, Alaska: progress toward recovery following the 1989 *Exxon Valdez* oil spill. US Geological Survey Scientific Investigations Report, 2011-5213, pp. 14.

Bodkin, J.L., Ballachey, B.E., Coletti, H.A., Esslinger, G.G., Kloecker, K.A., Rice, S.D., et al., 2012. Long-term effects of the Exxon Valdez oil spill: sea otter foraging in the intertidal as a pathway of exposure to lingering oil. Mar. Ecol. Prog. Ser. 447, 273–287.

Burn, D.M., Doroff, A.M., 2005. Decline in sea otter (*Enhydra lutris*) populations along the Alaska Peninsula, 1986–2001. Fishery Bull. 103 (2), 270–279.

Doroff, A.M., Estes, J.A., Tinker, M.T., Burn, D.M., Evans, T.J., 2003. Sea otter population declines in the Aleutian archipelago. J. Mammal 84 (1), 55–64.

Esslinger, G.G., Bodkin, J.L., 2009. Trends in Southeast Alaska sea otter populations; 1969–2003. USGS Scientific Investigations Report 2009–5045, pp. 18.

Estes, J.A., 1990. Growth and equilibrium in sea otter populations. J. Anim. Ecol. 59, 385–401.

Estes, J.A., Jameson, R.J., 1988. A double-survey estimate for sighting probability of sea otters in California. J. Wildl. Manage. 52 (1), 70–76.

Estes, J.A., Palmisano, J.F., 1974. Sea otters: their role in structuring nearshore communities. Science 185, 1058–1060.

Estes, J.A., Tinker, M.T., Williams, T.M., Doak, D.F., 1998. Killer whale predation on sea otters linking oceanic and nearshore ecosystems. Science 282, 473–476.

Estes, J.A., Hatfield, B.B., Ralls, K., Ames, J.A., 2003. Causes of mortality in California sea otters during periods of population growth and decline. Mar. Mammal Sci. 19 (1), 198–216.

Estes, J.A., Doak, D.F., Springfer, A.M., Williams, T.M., Van Vliet, G.B., 2009. Trend data do support the sequential nature of pinnipeds and sea otter declines in the North Pacific, but does it really matter? Mar. Mammal Sci. 25 (3), 748–754.

Garshelis, D.L., Garshelis, J.A., 1984. Movements and management of sea otters in Alaska. J. Wildl. Manage 48, 665–678.

Garshelis, D.L., Johnson, A.M., Garshelis, J.A., 1984. Social organization of sea otters in Prince William Sound, Alaska. Can. J. Zool. 62, 648–2658.

Gorbics, C., Bodkin, J.L., 2001. Stock identity of sea otters in Alaska. Mar. Mammal Sci. 17 (3), 632–647.

Griffith, B., Scott, J.M., Carpenter, J.W., Reed, C., 1989. Translocations as a species conservation tool: status and strategy. Science 245, 477–480.

Hatfield, B.B., Ames, J.A., Estes, J.A., Tinker, M.T., Johnson, A.B., Staedler, M.M., et al., 2011. Sea otter mortality in fish and shellfish traps: estimating potential impacts and exploring possible solutions. Endangered Species Res. 13, 219–229.

Jameson, R.J., 1989. Movements, home ranges, and territories of male sea otters off central California. Mar. Mammal Sci. 5, 159–172.

Jameson, R.J., Kenyon, K.W., Johnson, A.M., Wight, H.M., 1982. History and status of translocated sea otter populations in North America. Wildl. Soc. Bull. 10, 100–107.

Johnson, A.M., 1982. The sea otter: *Enhydra lutris*. In: Mammals of the Sea. FAO Fisheries Series 5, vol. IV, pp. 521–525.

Kenyon, K.W., 1969. The sea otter in the eastern Pacific Ocean. North Am. Fauna 68, 352.

Lech, V., Betts, M.W., Maschner, H.D.G., 2011. An analysis of seal, sea lion and sea otter consumption patterns on Sanak Island, Alaska: an 1800 year record on Aleut consumer behavior. In: Braje, T., Rick, T.C. (Eds.), Human Impacts on Seals, Sea Lions, and Sea Otters: Integrating Archeology and Ecology in the Northeast Pacific. University of California Press, LA, pp. 111–128.

Lensink, C.J., 1962. The History and Status of Sea Otters in Alaska. Ph.D. dissertation, Purdue University, Indiana, pp. 165.

Loughlin, T.R., 1980. Home range and territoriality of sea otters near Monterey, California. J. Wildl. Manage. 44, 576–582.

McLean, J.H., 1962. Sublittoral ecology of kelp beds of the open coast near Carmel, California. Biol. Bull. 122, 213–219.

Merritt, M.L., Fuller, R.G., 1977. The environment of Amchitka Island. TID-26712, National Technical Information Service, United States Department of Commerce, Springfield, VA.

Moritz, C., 1999. Conservation units and translocations: strategies for conserving evolutionary processes. Hereditas 130, 217–228.

Ralls, K., Eagle, T., Siniff, D.B., 1988. Movement patterns and spatial use of California sea otters. In: Siniff, D.B., Ralls, K. (Eds.), Population Status of California Sea Otters. Minerals Management Service, LA, pp. 33–63. Final Report on Contract No. 14-12-001-3003.

Rathbun, G.B., Hatfield, B.B., Murphey, T.G., 2000. Status of translocated sea otters at San Nicolas Island, California. Southwest. Nat. 45 (3), 322–375.

Riedman, M.L., Estes, J.A., 1990. The sea otter (*Enhydra lutris*): behavior, ecology and natural history. U.S. Fish and Wildlife Service. Biol. Rep. 90 (14), 126.

Schneider, K.B., Faro, J.B., 1975. Effects of sea ice on sea otters (*Enhydra lutris*). J. Mammal 56, 91–101.

Simenstad, C.A., Estes, J.A., Kenyon, K.W., 1978. Aleuts, sea otters, and alternate stable state communities. Science 200, 403–411.

Springer, A.M., Estes, J.A., Williams, T.M., Doak, D.F., Danner, E.M., Forney, K.A., et al., 2003. Sequential megafaunal collapse in the North Pacific Ocean: an ongoing legacy of industrial whaling? Proc. Natl. Acad. Sci. USA 100, 12223–12228.

Udevitz, M.S., Bodkin, J.L., Costa, D.P., 1995. Detection of sea otters in boat-based surveys of Prince William Sound, Alaska. Mar. Mammal Sci. 11 (1), 59–71.

Wade, P.R., Barrett-Lennard, L.G., Black, N.A., Burkanov, V.N., Burdin, A.M., Calambokidis, J., et al., 2007. Killer whales and marine mammal trends in the North Pacific—a re-examination of evidence for sequential megafauna collapse and the prey-switching hypothesis. Mar. Mammal Sci. 23, 766–802.

Chapter 4

Challenges to Sea Otter Recovery and Conservation

Brenda E. Ballachey and James L. Bodkin
US Geological Survey, Alaska Science Center, Anchorage, AK, USA

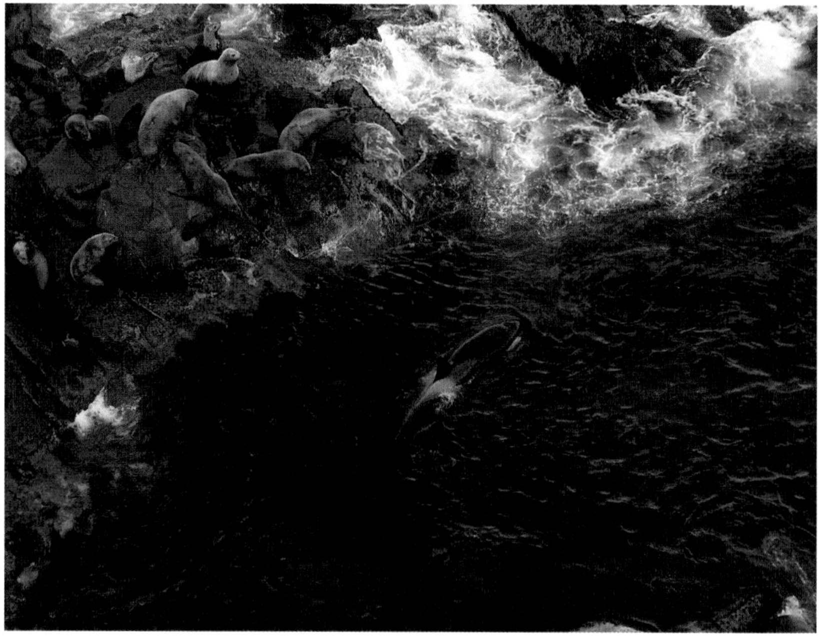

INTRODUCTION

The history of sea otter abundance over the last 270 years demonstrates the vulnerability of this species, with large fluctuations in population size and variation in population growth rates over that period. Hunting for fur in the eighteenth and nineteenth centuries brought sea otters very close to extinction, and recovery depended on legal protections and conservation action

Sea Otter Conservation. DOI: http://dx.doi.org/10.1016/B978-0-12-801402-8.00004-4

(Chapters 3, 13, and 14). More recently, in southwest Alaska, predation pressures have reduced sea otter numbers by close to 90% over a more than 2000 km range (Doroff et al., 2003; Estes et al., 2005), and as yet, there has been no indication of recovery (USFWS, 2013). In California, sea otter abundance has failed to reach conservation goals, and for many decades growth rates have been lower than expected, relative to other remnant populations (Bodkin et al., 1999), with periods of modest growth and decline despite focused efforts to protect sea otters and measures to enhance population growth. With mounting concerns over the stability of coastal marine ecosystems, sea otters continue to be at risk.

We recognize that many species across the globe share a suite of emerging pressures that include habitat degradation and loss, competition from invasive and introduced species, and mortality from a variety of sources, including in some cases incidental or directed mortality caused by humans, often at rates that exceed the ability of the species to replace lost individuals. Perhaps more importantly, threats to survival can be cumulative, with the collective impact of multiple threats difficult to distinguish. Sea otters are one of those species at risk, and face increasing challenges as continuing human development alters coastal marine habitats (Crain et al., 2009), compounded by the consequences of a changing global climate. Marine systems are anticipated to suffer some of the most deleterious effects from climate change, including ocean acidification, increasing temperatures, and rising sea levels (Harley et al., 2006; Doney et al., 2012).

Sea otters have adapted to northern marine ecosystems and hold considerable value for humans (Chapters 2, 11, and 12). Residing exclusively in cold marine waters, they have developed an exceptionally dense fur (Williams et al., 1992). That fur traps air and provides an insulating layer for their skin, allowing sea otters to exist up to the southern extent of sea ice in the North Pacific (Chapter 3). It was for this fur that the first people arriving in the North Pacific developed methods to hunt the sea otter, which facilitated the expansion of human settlement across the Pacific some 15,000 years ago (Chapter 11). The sophisticated communal hunting methods and evidence of sea otter remains in middens left by Native Americans speak to the value and use of the sea otter for the North Pacific's first human residents. Although the sea otter was widely used by humans throughout its range, there is little evidence that native harvests threatened the existence of sea otters at any but the local level, if at all (Chapter 11).

Following the European explorations of the North Pacific in the 1740s, an expanding commercial harvest of sea otters for their pelts demonstrated the value of otter fur to humans on a wider scale. The threat of the harvest was not fully understood until early in the nineteenth century, when Russian managers recognized diminishing sea otter abundance and imposed restrictions on the harvest (Chapter 3). Despite these protections, the emergence of an escalating international harvest later in the nineteenth century eventually extirpated the sea otter throughout nearly all of its range. As the abundance of sea otters diminished to the point of near extinction, the harvest became unprofitable and declined, allowing the process of recovery and conservation to begin. Since the end of the commercial fur trade around 1900, sea otter conservation has largely been a success, with protections and reintroductions resulting in a current population of about 123,000 individuals that occupies an estimated 70–80% of the species' historic range (Chapter 3). However, although sea otters can be considered to have successfully recovered, there are nevertheless large areas of unoccupied habitat along the west coast of the United States and Canada and a number of threats that again could lead to declining sea otter populations.

Our objective in this chapter is to review past threats and then consider present and potential future threats to sea otter populations, to provide a foundation for development of future conservation plans. The many examples of reductions and extinctions of a variety of species, as a direct or indirect consequence of human activity, substantiate the need for ongoing vigilance. The story of the sea otter provides important lessons that may aid not only in their continued welfare but in the conservation of other threatened species as well.

Defining Threats: Scale and Scope

When considering threats to the recovery or growth of a population, the spatial and numerical scales at which threats may be imposed must be considered. Threats may be classified as "global," potentially affecting a species throughout its range, or they may operate at smaller scales, affecting less than the full range of the species. Similarly, threats may affect all individuals in a population or only a subset of the individuals. In some situations, the affected proportion of a population will be so small that changes in demography or emigration are not discernible, whereas in other cases the number of affected individuals will be so high that the viability of the population is jeopardized. Where appropriate for threats discussed in this chapter, we will consider the geographic scale at which the threat may occur, or at which it has been previously observed, as well as the observed and potential numerical extent of the threat.

Discussion of the potential spatial and numerical scale of threats requires some understanding of the spatial scale at which a population is demographically structured. For sea otters, population structure is recognized to occur at scales below subspecies and stocks, at least within Alaska (Gorbics and Bodkin, 2001). In Chapter 3, Bodkin uses divergent population growth rates to describe the concept that sea otter populations are demographically structured at spatial scales as small as perhaps a few hundred square kilometers, and in Chapter 10, Tinker provides further discussion on population structuring and the implications for conservation of the species. Consideration of the spatial structuring of populations may be a critical aspect when assessing the potential effects of threats to many managed populations, including the sea otter.

In this chapter we broadly define threats as either demographic (resulting in either reduced reproduction, increased mortality, or modified rates of immigration/emigration), or habitat related (resulting in reduced environmental quality or habitat suitability), and define mortality as a threat when the magnitude of the mortality results in either a decline in the abundance or a reduction in the growth rate of a demographically distinct population. Excessive human harvest, predation, and incidental fisheries mortality all provide good examples of actual demographic threats. Habitat-related threats may or may not induce mortality, but can diminish the quality or quantity of habitat such that the equilibrium density or carrying capacity is diminished. The construction of harbors disturbing the nearshore environment and the extraction of prey in shellfish fisheries represent actual threats to habitat. In some instances threats may be both demographic and habitat related, such as oil spills where mortality may be direct and habitat may be degraded through clean-up or residual oil persisting in the environment. We choose not to predict the consequences of threats that may meet our defined criteria in the future. Rather, we identify those conditions or situations that, if they occur, could produce either demographic or habitat threats. In Table 4.1, we outline the nature of threats that we discuss in this chapter, recognizing that this list should be considered predominantly retrospective in nature. Finally, we caution the reader to be aware that threats will often be cumulative in nature and that recognizing interactions and complexities of various issues and concerns, and discriminating among them in terms of

TABLE 4.1 List of Threats to Sea Otter Conservation

Threat	Type
• Human take (legal, illegal, incidental, and accidental)	D
• Fisheries (incidental take)	D/H
• Predation	D
• Competition and resource limitation (food)	D
• Oil and other contaminants	D/H
• Disease (direct and indirect)	D
• Genetic diversity	D
• Natural hazards	D/H
• Habitat loss and degradation, disturbance	H
• Density-dependent effects	D/H
• Climate change (potential)	D/H

"D" represents demographic and "H" represents habitat.

management capacity to respond, will present a major challenge for continued conservation efforts.

THREATS: PAST, PRESENT AND FUTURE

Human Take

The beautiful thick fur of the sea otter has made it a target for human harvests for thousands of years, demonstrated by the presence of skeletal remains of sea otters found in midden sites throughout their range. Based on the relative abundance of skeletal remains and invertebrate shells over time, exploitation of sea otters may have been a tool used by native hunters to manage sea otter abundance at local levels, perhaps to limit the predation impact of sea otters on shellfish fisheries (Simenstad et al., 1978; Erlandson et al., 2005; Jones et al., 2011; Chapter 11). Nevertheless, it appears that human impacts on sea otter numbers were limited, as populations generally thrived until the arrival of Europeans in the mid-1700s and the subsequent initiation of intensive hunting to acquire furs for commercial trade. Prior to the maritime fur trade the number of sea otters in the Pacific was thought to be about 300,000, ranging from the northern islands of Japan to Baja California (Figure 4.1). Over the next 170 years, sea otters were decimated, with less than 1% of the original

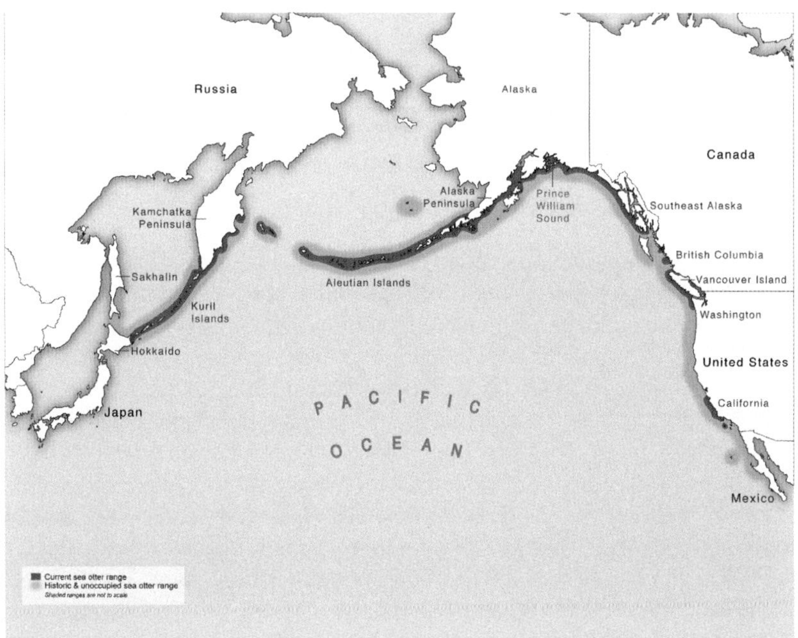

FIGURE 4.1 North Pacific Ocean showing current and historic distribution of sea otters.

number left by 1911, when they were afforded limited legal protection by the International Fur Seal Treaty (Lensink, 1960; Kenyon, 1969; Bodkin et al., 1999; Chapters 3 and 14). The surviving population was severely fragmented, with 13 scattered remnant groups, each probably numbering less than a few hundred individuals (Bodkin, 2003; Bodkin et al., 1999).

Following the end of the commercial harvest, sea otter populations gradually increased, although rates of recovery varied. Recovery was aided by a series of translocations in the 1960s and 1970s (Jameson et al., 1982). A loss of genetic variation likely hindered population growth in remnant populations and in those translocated from a single source (Bodkin et al., 1999; Larson et al., 2002a,b). Although perhaps 20–30% of their former range currently remains unoccupied, sea otters now are found throughout most of the areas they previously occupied and, barring unanticipated impediments to recovery, should further repopulate portions of their former range over the next decades.

Currently, the only legal take of sea otters is in Alaska, and this only for native Alaskans for subsistence purposes, including the creation of qualified handicrafts, under exemption to the US Marine Mammal Protection Act (MMPA) of 1972 (Chapter 14). Products made from pelts by Alaska natives can be traded or sold without restriction. This harvest is largely unregulated, and has the potential to depress otter abundance, particularly on local scales (Bodkin and Ballachey, 2010). Based on US Fish and Wildlife Service records for Alaska, between 1989 and 2012, 18,308 sea otters were reported as harvested by native hunters in Alaska, averaging 763 per year, with an increasing trend over the past decade. Most otters were harvested in relatively small areas of southeast and south-central Alaska, with very few taken from the southwest region of Alaska, particularly in recent years. The 2013 Alaska harvest (reported as of April 11, 2014, V. Gill, US Fish and Wildlife Service, personal communication) was 2037, almost triple the average of the past two decades, perhaps signaling the onset of intensifying harvests as populations in some areas increase (see also Chapter 12).

In consideration of present-day regulations and restrictions, it seems unlikely that subsistence harvests will have large-scale consequences for sea otter conservation. Rather, it is more likely that human harvests could either serve to moderate rates of population growth and expansion or, at localized scales, lead to reduced abundance and distribution of sea otters.

In addition to the reported subsistence harvest, some undetermined number of sea otters are "struck and lost," which refers to an animal that was shot and injured, perhaps fatally, but the carcass was not recovered by the hunter and thus not reported as a harvested animal. There may also be additional sea otters that are harvested by native Alaskans and are not reported to the US Fish and Wildlife Service, but the extent of this is not known.

Illegal harvest of sea otters (by non-native hunters in Alaska, and all hunting outside of Alaska) may also occur throughout the range. Lensink

(1960) suggested that some amount of illegal harvest may have continued after protection early in the twentieth century, constraining recovery of the species following the end of the fur trade. Today, in addition to those taken legally for their pelts, others may be killed illegally to minimize their competition in various fisheries, particularly where otters have increased in abundance and present direct competition in commercial and recreational fisheries (see below; see also Chapter 12).

Conflicts between sea otters and coastal fisheries are well recognized. In California, evidence of illegal efforts to control adverse effects of sea otters on fisheries comes from the recovery of carcasses of sea otters that have died of gunshot wounds (Estes et al., 2003). As sea otters continue to recolonize former habitats, they will face continued pressures from humans over these conflicts, as well as potential incidental mortality from existing fisheries such as gill/trammel nets and trap fisheries.

The extent of mortality from illegal take is unknown and consequently, the effect on populations cannot be ascertained. Generally, illegal harvests are not thought to be a major factor influencing populations. Overall, however, it is likely that losses to the sea otter population from hunting are greater than the number actually recorded, and, when combined with other sources of human-induced mortality, may be contributing to diminished growth in many sea otter populations in recent decades (Chapters 3 and 12; Esslinger and Bodkin, 2009; Estes et al., 2003).

Due to conflicts between sea otters and commercial fisheries, there have been discussions in Alaska about the potential for modifying hunting restrictions, but changes in regulations do not appear to be likely in the near future. South of Alaska, in British Columbia and Washington State, there is also the potential that Native Americans and other hunters may push for the right to legally harvest sea otters (Chapter 12). In Russia and Japan, sea otters are fully protected from subsistence harvest.

Fisheries

Fisheries by-catch is a widely recognized problem in the conservation and management of a variety of marine mammals, birds, and turtles, and sea otters are included among the species suffering deleterious effects. Several commercial fisheries are known to impact sea otters, with documented mortalities and local population-level effects spanning several decades (Wendell et al., 1986; Estes et al., 2003; Hatfield et al., 2011; Chapter 12). Early evidence of incidental mortality between sea otters and human fisheries is found in Newby (1975), who reported a sea otter carcass recovered from a crab trap set at a depth of 97 m in the Aleutian Islands. Hatfield et al. (2011) report on 21 cases of incidental sea otter mortality resulting from entrapment and subsequent drowning in crab, lobster, and various fish traps from California to the Aleutian Archipelago. They suggest that incidental trap mortality is contributing to the lower than expected rate of range expansion in California, at both southern and northern ends of the population, and constraining population growth.

Declining numbers observed in the California sea otter population in the mid-1970s and early 1980s were attributed in part to loss of animals through entrapment and drowning in fisheries gear, primarily in the set-net fishery for California halibut (*Paralichthys californicus*) that developed within the foraging habitat of sea otters (Estes et al., 2003). Eventual restrictions on the fishery, limiting it to waters beyond the forage dive depth of sea otters, apparently alleviated losses. However, continuing interactions with fisheries throughout the range of sea otters should be expected as sea otters move into new areas, and as emerging fisheries develop to take advantage of markets for new species when traditional fish and shellfish stocks become depleted.

In the early 1980s, the US Fish and Wildlife Service permitted a translocation of sea otters to San Nicolas Island, a remote naval facility off the coast of southern California, in exchange for an agreement to limit sea otter range expansion south of Point Conception (Figure 4.1, Chapter 1). A further impetus for the translocation was to establish a second sea otter population in California, in recognition of the vulnerability of the mainland population to a large oil spill. Slow growth in the San Nicolas population over several decades and an inability to contain the mainland population in the "otter free" zone eventually led to classifying the translocation as a failure in 2013 (Chapter 12).

Fishermen may be reluctant to report incidental sea otter mortality (or that of other species) in fishing gear, thus impeding identification and estimation of fishery-related mortality that may be constraining recovery. Thus, managers and groups interested in the conservation and restoration of marine vertebrates, including sea otters, should be cognizant of potential conflicts between depleted species and emerging fisheries. The evolving nature of fisheries and expanding sea otter populations suggests that potential demographic effects at regional scales will result from incidental fisheries interactions. In some cases incidental mortality may be adequate to retard sea otter range expansion into areas where trap or net fisheries exist, as has been observed in California (Hatfield et al., 2011). Conflicts with fisheries should be anticipated as sea otters recolonize suitable habitats in California and elsewhere in their historic range, as understanding mechanisms of impacts on otters may offer approaches, such as gear modification, to ameliorate losses.

Other Accidental Mortality

Other sources of mortality attributed to accidents have been reported. These include entanglement in debris and vessel strikes (Estes et al., 2003). Accidental mortality due to boat strikes is becoming an issue of some concern, particularly in areas of higher sea otter density. In Alaska, sea otters generally occupy areas that have not seen a lot of boat traffic, but in recent years there have been increases in the number of small recreational boats and larger tourism vessels, which are capable of traveling at relatively high speeds. Further south, sea otters are expanding into areas where there may be greater overlap with boat traffic. Boat strikes are already a recognized source of mortality for sea otters in Alaska and California. At this time it would seem that most sources of accidental mortality associated with human activities are numerically minor and localized near human and sea otter population centers. However, with increased human presence in some areas, the potential for elevated rates of accidental mortality to contribute to overall mortality and affect population growth requires continued diligence.

Mortality from Nuclear Testing (Amchitka Island, Aleutian Chain)

From 1965 to 1971, the US government conducted a series of underground nuclear tests at Amchitka Island in the Aleutian chain (Fuller and Kirkwood, 1977; Benning et al., 2009). The last and largest explosion was the Cannikin, a 5 Mt device detonated at a depth of nearly 1800 m. Widespread biological damage, including direct mortality of several hundred sea otters, was observed (Estes and Smith, 1972; Rausch, 1973; Fuller and Kirkwood, 1977) and there was some consideration that the number of sea otter deaths actually was closer to 1000 animals (J. Estes, personal communication).

The greater concern now, for sea otters and other species in the Aleutian Archipelago, is the threat of the release of nuclear contamination that resides deep beneath Amchitka Island as a result of the Cannikin test. Baskaran et al. (2003) evaluated the potential for leakage of nuclear wastes around Amchitka and the Aleutians through assessment of radionuclides in sea otter bone obtained prior to and following the nuclear testing at Amchitka. They found that although there was a signature of earlier atmospheric nuclear testing, there was no evidence of leakage from the Amchitka underground testing. Assays of benthic invertebrates have also indicated that contamination with radionuclides is not a present concern in the area (Burger et al., 2007). However, due to the high level of tectonic activity along the Pacific "rim of fire," disruption of the formations encasing nuclear contamination at Amchitka is a continuing threat that could result in widespread leakage of radiation with ecosystem level consequences (Burger et al., 2005; Benning et al., 2009).

Predation (Other than Human)

As human populations expanded globally, nearly all terrestrial ecosystems were impacted by human predation that effectively reduced or removed many, if not most, large predator populations (Estes et al., 2011). In marine ecosystems, large-scale effects of human predation largely occurred over the past few centuries as humans developed the technology to explore and exploit the world's oceans. The reduction and elimination of large predators through human activity detracted from our ability to understand the role of predation in both ecological and demographic terms. Only recently, and in some part as a consequence of conservation efforts, has the role of predation become appreciated as a dominant factor in structuring the abundance of species and the function of ecosystems (Estes et al., 2011). The literature describing the keystone role of sea otter predation (summarized by Estes, Chapter 2) provides an understanding of the potential role of predation in structuring ecosystems. Thus, it may be a bit surprising that the events that unfolded in southwest Alaska late in the twentieth century, described below, were unanticipated and only somewhat fortuitously detected, through surveys and during the conduct of related research efforts.

The remote Aleutian Archipelago and Alaska Peninsula (Figure 4.1) provided prime habitat for sea otters surviving in that region after the maritime fur trade ended early in the twentieth century (Kenyon, 1969; Chapter 3). Over the following decades, those remnant populations extended their range and exhibited substantial increases in abundance (Kenyon, 1969; Bodkin et al., 1999). This pattern was expected to continue, particularly in those island groups where unoccupied habitat and little human presence, and thus little disturbance, allowed for further expansion. However, in the late twentieth century, perhaps the most catastrophic decline of a mammalian carnivore in recorded history was identified and eventually described (Estes et al., 1998; Doroff et al., 2003; Burn and Doroff, 2005). Because of the remoteness of this area, the precise timing of the event is unclear, but by early in the twenty-first century nearly 80,000 sea otters (>90% of pre-decline abundance) were unexpectedly lost and unaccounted for across the Aleutian Islands and the southern Alaska Peninsula (Chapter 3; Figure 4.2).

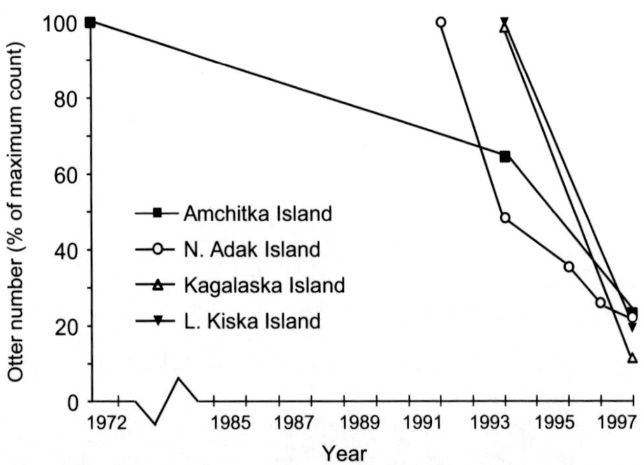

FIGURE 4.2 Declines in sea otter abundance at four locations in the Aleutian Islands. *(From Estes* et al.*, 1998. Reprinted with permission from AAAS.)*

The most recent surveys indicate that by 2008, the southwest Alaska sea otter population, occupying more than 2000 km of prime sea otter habitat, had uniformly declined to a point where current abundance was perhaps 7–8% of historic numbers, resulting in the listing of this stock in 2005 as "threatened" under the US Endangered Species Act (ESA) (USFWS, 2013). Predictably, following the discovery of the sea otter decline in southwest Alaska, the question of cause was raised. Also predictably, and for a variety of reasons often common to species conservation issues, there were relatively few data available to definitively assign cause of the decline. However, following accumulation of various sources of evidence and careful consideration, Estes et al. (1998) identified predation by killer whales (*Orcinus orca*) as the most likely factor behind the decline (see Box 4.1 for a summary of the evidence).

While killer whales are clearly recognized in the literature as potential predators of sea otters (Barabash-Nikiforov, 1962; Kenyon, 1969), prior to this decline, predation in general was not considered to be a significant factor in regulating sea otter population growth or abundance. Historically,

Box 4.1 Evidence for Strong Impacts of Killer Whale Predation on Sea Otters in Southwest Alaska (Summarized from US Fish and Wildlife Service, 2013)

1. An increased rate of observed killer whale attacks consistent with the time the decline began (Estes et al., 1998; Hatfield et al., 1998). The actual number of attacks observed between 1992 and 1997 corresponds with the expected number of attacks that would have been seen given the intensity of observation, and is consistent with the estimated number of otters that had to have died in the central/western Aleutians to produce the observed population decline (Estes et al., 1998).
2. Based on energy requirements of killer whales, a feasibility analysis indicates that predation by as few as five individuals would have been sufficient to account for the decline in sea otter abundance (Williams et al., 2004).
3. There were relatively few beach-cast carcasses recovered during the decline, despite dedicated carcass collection efforts and a history of carcass deposition in the Aleutians (Kenyon, 1969). Throughout the sea otters' range, beach-cast carcasses are a common occurrence (Bodkin and Jameson, 1991; Bodkin et al., 2000; Monson et al., 2000a). A large number of carcasses would be expected in a population declining at the rate estimated if animals were dying from disease, starvation, or other typical causes of sea otter mortality. In the case of predation, few if any carcasses would be expected.
4. The premature disappearance rate of animals that were radio-tagged as part of telemetry-based studies at Amchitka Island (1992–1994) and Adak Island (1995–1996; Estes et al., 1998; Monson and DeGange, 1995; Tinker and Estes, 1996) greatly exceeded that recorded during other radio-telemetry studies that used identical techniques and instrumentation (Siniff and Ralls, 1991). In Clam Lagoon at Adak Island (an enclosed embayment that may

(*Continued*)

Box 4.1 (Continued)

serve as a refuge from predators), the loss rate of tagged otters was significantly lower (Estes et al., 1998) and was consistent with the disappearance rate seen in previous studies (Siniff and Ralls, 1991).

5. Between 1995 and 1999 the population in Clam Lagoon, a largely predator-free habitat, declined only slightly while numbers dropped by approximately 80% outside the lagoon in Kuluk Bay (Estes et al., 1998).
6. The current distribution pattern of sea otters throughout the region of southwest Alaska affected by the decline has shifted from that prior to the decline such that most otters are now closer to shore and tend to be concentrated around protected embayments or areas with extensive shallow reefs. This likely affords some measure of protection from both killer whales and sharks.
7. There have been conspicuous modifications of behavioral responses of sea otters to disturbances in the decline area that are both atypical and suggestive of avoidance of marine predators. These include animals feeding and resting in very shallow (<2 m) water, and hauling out on land in response to both shore-based and water-based disturbances (J. Bodkin, personal observation).

observations of interactions between killer whales and sea otters are rare, although this may reflect the recently reduced abundance of sea otters and the relatively low number of observers in remote areas. One aggressive interaction in the Kuril Islands (Nikolaev, 1965) and a few presumed attacks in the Commander Islands are reported in Hatfield et al. (1998). However, most killer whale–sea otter interactions in Alaska and elsewhere were assumed to be non-aggressive (Kenyon, 1969). In the early 1990s this pattern appeared to change, based on nine documented observations of killer whales attacking sea otters (three incidents in Prince William Sound (PWS), Alaska and six in the Aleutian Islands; Hatfield et al., 1998). It was speculated that these interactions reflected an increase in killer whale predation as a result of "transient" (mammal-eating) killer whales adjusting their diet to include sea otters, possibly in response to declines in other marine mammal prey, including harbor seals and Steller sea lions (Estes et al., 1998; Hatfield et al., 1998; Springer et al., 2003). Additional recent evidence that killer whale predation includes sea otters includes the skeletal remains of at least five sea otters recovered from the stomach of a beach-cast killer whale in PWS, confirming that otters are, at least sometimes, eaten by killer whales (Vos et al., 2006).

Using observed interactions between sea otters and killer whales described by Hatfield et al. (1998), Estes et al. (1998) conducted a feasibility analysis to evaluate the possibility that predation by killer whales could be a contributing factor to the sea otter decline. They showed that extrapolating the frequency of observed interactions across the entire Aleutian archipelago for the period from the late 1980s to 1997 could potentially account for an increase in sea otter mortality sufficient to cause the observed rate of decline (Estes et al., 1998).

The feasibility analysis alone does not constitute definitive evidence that killer whale predation was a contributing factor in the decline. However, the analysis does demonstrate the potential and justifies the consideration of killer whale predation as the significant factor in the decline, despite the relatively few direct observations reported. The cumulative support for predation as a primary cause of the decline consists of a number of independent pieces of information, representing both direct and indirect evidence, and is summarized in Box 4.1.

Sea otters are typically viewed as apex predators in nearshore marine communities, yet there have been many reported instances in which sea otters have been killed and/or consumed by other high-level predators, in addition to killer whales. Documented instances of mortality related to other predators, including bald eagles (*Haliaeetus leucocephalus*), white sharks (*Carcharodon carcharias*), and terrestrial carnivores such as brown bears (*Ursus arctos*), coyotes (*Canis latrans*) and arctic foxes (*Alopex lagopus*), are summarized below.

Bald eagles are fairly common predators of sea otter pups in the Aleutian Islands, where pup remains account for up to 20% of the prey items found in some eagle nests (Sherrod et al., 1975). However, because this predation is limited to very young pups (<5 kg), a stage with naturally high mortality, it generally has been considered that eagle predation has a small demographic impact, at least at high sea otter densities (Riedman and Estes, 1990; Sherrod et al., 1975).

White shark attacks are a source of mortality for sea otters in California, particularly at the north end of the otter's range (Ames and Morejohn, 1980; Estes et al., 2003; Kreuder et al., 2003). It appears that white shark attacks do not represent true predation—that is, sharks are not actually consuming the sea otters but rather killing them incidentally as a by-product of mistaking sea otters for pinnipeds, their preferred prey (Ames and Morejohn, 1980). This interpretation is consistent with the lack of sea otter remains found in white shark stomachs (Riedman and Estes, 1990). Regardless of whether or not fatal shark bites represent true predation or just a case of "mistaken identity," they have a relatively high potential for limiting population growth in parts of the California sea otter range (Gerber et al., 2004) and elsewhere where the two species co-occur.

Shark predation is also a potential source of mortality elsewhere in the sea otters' range, given that white sharks, salmon sharks (*Lamna ditropis*), broadnose sevengill sharks (*Notorynchus cepedianus*), and Pacific sleeper sharks (*Somniosus pacificus*) all occur in coastal Pacific waters and are known to prey on other marine mammals (Hulbert et al., 2006; Klimley, 1985). However, other than the relatively high occurrence of white shark predation in California, there is little evidence that shark predation, at least to date, has played a significant role in sea otter mortality.

Terrestrial predators can prey on sea otters in areas where they haul out frequently, although reports of such predation are remarkably few. In the

Commander Islands, arctic foxes frequently scavenge sea otter carcasses and also occasionally kill live otters, although such predation is largely limited to young or already compromised and moribund animals (Zagrebel'nyi, 2004). In PWS, Alaska, newly weaned sea otters occasionally may be killed and eaten by coyotes (Monnett and Rotterman, 1988), and brown bears have been reported to prey on sea otters along the Kamchatka Peninsula in the late winter/early spring (Riedman and Estes, 1990).

In summary, predation by species other than humans has historically been thought to be a rather minor component of sea otter mortality. However, the recent decline of sea otters in the North Pacific, representing more than 70% of the present global sea otter population, is largely attributable to killer whales and indicates that predation can be a major factor driving sea otter population dynamics. In recent times, only the near extirpation of sea otters by humans for their fur exceeds the magnitude of mortality experienced through predation in the Aleutian Islands and Alaska Peninsula. We now recognize that predation patterns can change relatively rapidly, with profound effects on populations at large spatial scales. In California, where recovery of sea otters has been protracted, the effects of predation-related mortality on sea otter abundance warrant careful monitoring. More generally, the Aleutian decline illustrates the need for and benefit of routine population monitoring, although in that situation there was little that humans could have done to mitigate mortality caused by killer whales. In general, however, the availability of reliable data on abundance and trends may allow earlier recognition of factors affecting mortality, preferably during the initial phase rather than subsequent to a population decline.

Competition and Resource Limitation

As noted above, when sea otters move into unoccupied habitat as part of their recovery, or when new fisheries develop within an existing sea otter range, the potential for adverse interactions is present and can manifest as increased incidental or intentional mortality. Here we consider the potential for adverse effects on sea otter populations mediated through competition with humans for prey resources.

In the absence of sea otters throughout most of their range in the twentieth century, their prey populations, such as clams, crabs, urchins, and abalones, expanded both numerically and in terms of average prey size. In response, humans frequently developed fisheries on these resources, based on shellfish populations that were largely free of this important predator. As the sea otter populations expanded their range following conservation measures, they came into direct competition with humans over these fisheries (Chapters 12 and 13).

The impact and effect on sea otter conservation of the removals and reductions of prey by human fisheries is largely unexplored. In concept, human fisheries are managed to optimize harvest, but often management fails and harvested populations fail, perhaps to some degree from factors other than harvest. Examples of once-prosperous but now failed or failing west coast shellfish fisheries include several species of abalone (*Haliotis* spp.) and urchins (*Strongylocentrotus* spp.). In some cases declining fisheries may be attributed in part to competition with sea otters (Hines and Pearse, 1982; Garshelis et al., 1986; Kvitek et al., 1992; Watson, 2000; Laidre and Jameson, 2006), but in others, they certainly cannot (Pearse et al., 1977; Neuman et al., 2010).

We suggest that the high population growth rates observed in many areas where sea otters were translocated (Bodkin et al., 2000) can be attributed in part to abundant prey populations not exploited by human fishers. Further, diminished growth rates evident today in some of those same populations result in part from diminished prey populations as a consequence of human fisheries. Conservation of sea otters and other species may benefit from consideration of the abundance of prey populations available and the potential effects of human harvests or other effects on those prey.

Oil and other Contaminants

Environmental contaminants are of concern in the conservation and management of sea otter populations throughout the North Pacific, and likely will continue to pose a risk with increasing human exploitation and development of coastal areas (Crain et al., 2009). Some contaminants result from local sources of pollution, whereas others are transported globally. Broad classes of contaminants of concern in coastal environments include persistent organic pollutants (POPs, including polycyclic aromatic hydrocarbons [PAHs], polyhalogenated biphenyls [PCBs and PBBs], brominated flame retardants, organometals, and pharmaceuticals), heavy metals, and plastics (Crain et al., 2009; Cole et al., 2011). Contaminants have the potential to affect the health and survival of sea otters directly (Jessup et al., 2007; Bodkin et al., 2012), as well as indirectly through influences on the community structure and abundance of prey species (Johnston and Roberts, 2009).

Studies of a range of contaminants (organic and inorganic compounds) in sea otters have focused on the California population, which generally exhibits higher burdens than have been measured in northern populations (Estes et al., 1997; Bacon et al., 1999; Kannan et al., 2008; Brancato et al., 2009; Hart et al., 2009; Jessup et al., 2010). Levels of various contaminants measured in California sea otters are sufficiently high to be of concern and have been associated with risk of infectious disease and morbidity (Kannan et al., 2006a,b, 2007; Nakata et al., 1998; Murata et al., 2008). Changes in stable lead isotope compositions from pre-industrial and modern sea otters reflect changes in the sources of lead in coastal marine food webs. In pre-industrial samples, lead was from natural deposits, whereas in contemporary sea otters, lead is primarily from Asian and North American industrial sources (Smith et al., 1990, 1992).

PAHs are a class of POPs that have received greater attention over the past two decades, largely due to recognition of potential damage from petroleum spills. Oil spills from ships or other sources into the marine environment often occur in close proximity to coastlines, and oil frequently accumulates in coastal habitats resulting in potential threats to marine life (Bodkin et al., 2014). An estimated total of 700,000 barrels of oil are spilled into North American waters annually, with only a small proportion of this amount (less than 10%) resulting from marine oil spills (NRC, 2003). The acute impacts of large spills are generally recognized, due to obvious and dramatic direct effects. In contrast, an estimated 625,000 barrels of oil are released each year as chronic, non-point pollution, resulting from oil entering the coastal ocean as run-off at a more consistent, but much less conspicuous, rate. Despite the relatively minor contribution of oil spilled from ships, we have a good understanding of the acute effects of oil spills because of their immediate consequences and the directed response and research activities. The susceptibility of sea otters to oil is well known from early research (Costa and Kooyman, 1982; Siniff et al., 1982), and in part contributed to the listing of the California sea otter population as "threatened," under the ESA.

The effects of the 1989 *Exxon Valdez* oil spill on sea otters were significant, in terms of acute losses within the first weeks and months following the spill (Ballachey et al., 1994), and over the next decades, because of chronic exposure to lingering oil (Bodkin et al., 2002, 2012; Monson et al., 2011; Ballachey et al., 2014). The lessons learned from post-spill studies spanning over two decades implicate chronic oil exposure as a threat perhaps greater than that from the acute effects from oil spills, and demonstrate the need to consider the less conspicuous but clear potential for chronic exposure to petroleum hydrocarbons and other contaminants to affect marine species, including the sea otter. A summary of the sea otter studies following the spill is presented in Box 4.2.

Pollution of the oceans with plastic debris is now widely recognized as an emerging issue of tremendous concern for the health of oceans worldwide. Plastic debris is ubiquitous throughout marine environments, and is found in a range of sizes including small pieces of debris (<5 mm) classified as microplastics (Moore, 2008; Andrady, 2011; Cole et al., 2011; Desforges et al., 2014). Although the adverse effects of microplastics on wildlife are not fully established, they present a threat in multiple ways and are estimated to have caused the demise of millions of higher-level vertebrate organisms (Moore, 2008). Plastics are known to absorb POPs, which are hydrophobic, and thus can concentrate levels of these contaminants, making them available in the food web as plastic items are consumed (Teuten et al., 2009; Hirai et al., 2011). Further, many plastics are manufactured with additives to extend their life, and these additives are toxic and can leach from the plastic

Box 4.2 Sea Otters and the 1989 *Exxon Valdez* Oil Spill

Background

On 24 March 1989, the T/V *Exxon Valdez* went aground in northwest Prince William Sound, AK (Figure 4.1). The ship eventually spilled an estimated 264,000 barrels of crude oil, contaminating more than 2000 km of shoreline (Bragg et al., 1994), with more than half of that washed ashore in PWS (Wolfe et al., 1994).

The acute effects of the spill were dramatic and relatively well documented, at least among birds and mammals. Acute mortality estimates included approximately 250,000 seabirds (Piatt and Ford, 1996), several thousand sea otters (Ballachey et al., 1994), and hundreds of harbor seals (*Phoca vitulina*; Frost et al., 1994). Based on initial loss rates of oil following the spill, it was assumed that lingering oil would be negligible, spill effects would rapidly diminish, and recovery of affected populations would be evident within a few years (Neff et al., 1995). However, continued monitoring identified delayed recovery of several species of nearshore birds and mammals, including sea otters (Peterson et al., 2003, Figure 4.3).

In the decade following the spill, several lines of evidence suggested that *Exxon Valdez* oil persisted in unexpected volumes in nearshore habitats and was related to the protracted period of recovery evident for some species.

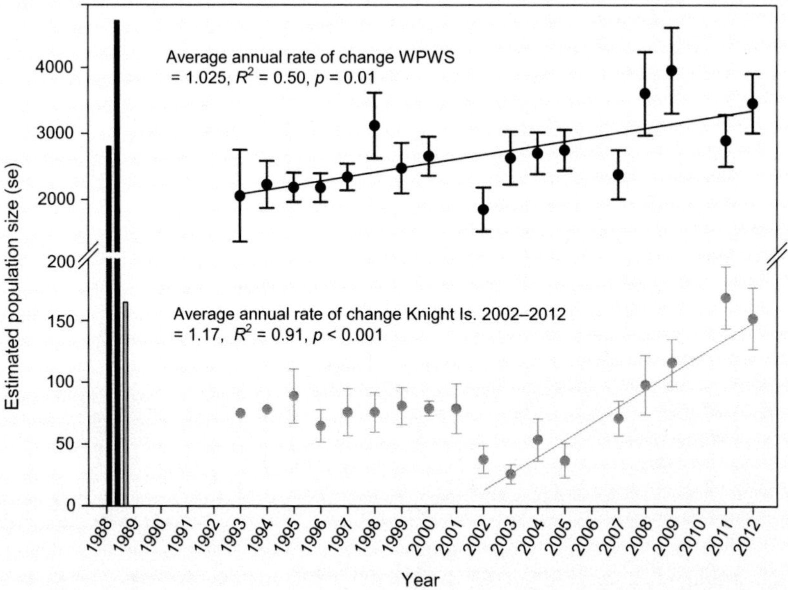

FIGURE 4.3 Estimated sea otter abundance in western Prince William Sound (WPWS) and at the heavily oiled area of northern Knight Island in PWS, after the 1989 Exxon Valdez oil spill, and average annual rate of change for each area. Black bars represent a range of estimated prespill abundance for WPWS, and the gray bar the estimated prespill abundance at Knight Island.

(Continued)

Box 4.2 (Continued)

The evidence included direct observations of oil in the intertidal (Hayes and Michel, 1999), the presence of hydrocarbons sequestered in the tissues of bivalves (Babcock et al., 1996; Fukuyama et al., 2000; Carls et al., 2001), and biological responses in fishes (Jewett et al., 2002) and birds (Trust et al., 2000; Esler et al., 2002, 2010, 2011; Golet et al., 2002).

Lingering oil from other catastrophic spills has been observed in a number of other environments (Corredor et al., 1990; Burns et al., 1993; Vandermeulen et al., 1982; Vandermeulen and Singh, 1994; Reddy et al., 2002; Bernabeu et al., 2009), suggesting that persistence of oil is not an issue that is specific to the *Exxon Valdez* spill. However, in the case of the *Exxon Valdez* spill, the extent of research directed at understanding the persistence and distribution of lingering oil, as well as its effects on wildlife, was unprecedented.

Evidence of Injury and Recovery of Sea Otters from the* Exxon Valdez *Spill

1. Estimates of sea otter mortality due to acute effects of the oil spill in western PWS ranged from 750 to 2650 animals (Garrott et al., 1993; Garshelis, 1997). The disparity among acute mortality estimates largely reflects the lack of accurate pre-spill estimates of sea otter population size.
2. Using population models, Udevitz et al. (1996) predicted recovery of the western PWS sea otter population at 10–23 years, with maximum annual growth rates from 10% to 14%.
3. Elevated mortality was evident in sea otters residing in spill affected areas in PWS for up to two decades following the spill (Ballachey et al., 2003; Monson et al., 2000a, 2011). Mortality associated with chronic exposure or long-term effects of acute exposure was estimated at 900 animals, suggesting that chronic mortality was similar in magnitude to the known immediate, acute mortality represented by the number of carcasses recovered in 1989 following the spill.
4. Gene transcription studies in 2008 showed differences consistent with exposure to oil between sea otters in oiled and unoiled areas (Miles et al., 2012). Gene transcripts, assessed by quantifying mRNA, provide the earliest observable signs of health impairment (Bowen et al., 2011).
5. In the first decade post-spill, numbers of sea otters remained depressed at the heavily oiled northern Knight Island region while recovery was evident to some extent through an increase in abundance in areas less severely affected by oil (Figure 4.3). By 2011, sea otter abundance at Knight Island approached the estimated abundance at the time of the spill, signaling recovery (Figure 4.3).

In summary, sea otter populations residing in the path of oil from the T/V *Exxon Valdez* suffered high rates of mortality through acute exposure, resulting in large-scale declines in abundance. Oil in intertidal sediments unexpectedly has persisted for more than two decades (Short et al., 2004, 2006, 2007; Li and Boufadel, 2010). However, pathways of exposure to the relatively small amounts of lingering oil in the decades following the spill were not readily discerned.

Sea otters excavate sediments when they forage for some prey, including clams, and routinely disturb sediments at depths below which lingering oil occurs

(*Continued*)

Box 4.2 (Continued)

(Short et al., 2004), and as their diet is dominated by clams that burrow in shallow sediments (Dean et al., 2002), this provided a direct pathway to lingering oil. Thus sea otters foraging at sites with patches of lingering oil were likely to be exposed, either through consumption of prey with assimilated hydrocarbons or by disturbing oiled sediments and releasing oil, which could adhere to their fur and subsequently be ingested upon grooming (Bodkin et al., 2012; Ballachey et al., 2013). Chronic exposure to oil over two decades post-spill, and possibly latent acute exposure effects, apparently led to elevated mortality and contributed to a delayed recovery. Most recent evidence suggests that by 2011, effects of chronic exposure had abated and sea otters had recovered (Ballachey et al., 2014).

debris into the environment and the animals that eat them (Moore, 2008; Cole et al., 2011), increasing the overall risk of exposure to POPs. Finally, there is the high potential for mechanical damage to organisms that are consuming plastic debris; this danger is reflected in the mortalities of fishes, birds, amphibians, and marine mammals observed to have plastic debris in their stomachs at necropsy (Moore, 2008).

Although there are no reports to date of sea otters consuming plastics, ingestion of microplastics has been observed in a variety of invertebrate species, including bivalves and echinoderms that could be typical sea otter prey items (reviewed by Cole et al., 2011). Thus there is a potential route for a variety of contaminants associated with plastics to enter the food chain and be incorporated by sea otters, generally increasing the risk of bioaccumulation of harmful contaminants.

Algal toxins, while not considered a pollutant in the same sense as those discussed above, are known to affect sea otter populations and could present a threat of increasing concern, in light of a global increase in the occurrence of algal blooms over the past decades (Van Dolah, 2000). One class of algal toxins are the saxitoxins associated with paralytic shellfish poisoning (PSP) which are common in bivalves in Alaskan waters, particularly *Saxidomus gigantea*, a favored prey item of sea otters (Kvitek and Bretz, 2004). Sea otters are able to detect bivalves with higher levels of saxitoxin and thus can protect themselves from harmful effects by avoiding consumption of toxic prey (Kvitek and Bretz, 2004). However, if more frequent algal blooms lead to more widespread contamination of bivalves, this could decrease the prey base available for sea otters, affecting health and survival. We can speculate that if alternate prey resources were not available, sea otters may have to consume prey with elevated levels of saxitoxin, which would have a direct detrimental effect on health.

In California, there have been reports of other harmful algal blooms contributing to sea otter mortality. Specifically, Miller et al. (2010a) implicated freshwater blooms of cyanobacteria as a source of cyanotoxins that were transferred to coastal marine areas and accumulated in bivalve prey as the probable cause of sea otter deaths from the microcystin hepatotoxin. Further, domoic acid, produced by a toxic algae and the cause of amnesic shellfish poisoning (Van Dolah, 2000; Jessup et al., 2007), was identified as a probable contributing factor to sea otter mortalities in California (Miller et al., 2010b). These incidents are worrisome, particularly as they may present an increasing problem for sea otters in consideration of potential increases in the overall incidence of algal blooms in coastal areas (Van Dolah, 2000).

Disease

Although sea otters are susceptible to a wide range of infectious and noninfectious diseases (Chapters 7 and 9), there is little evidence to date for consequences at a population level, outside of California. Nevertheless, given sea otters' role as apex predators inhabiting coastal areas, often in proximity to human populations and thus with increased potential risk of contaminant and pathogen exposures, disease potentially may affect population status. Disease may be a particular concern in light of other problems faced by a population, including prey limitation, habitat quality, contaminants, and limited genetic diversity (Chapter 5), which could compound losses from disease. Perhaps of greatest concern is the possibility for pathogen range changes due to changing climates (Daszak et al., 2001; Harvell et al., 2002; Goertz et al., 2013; Burge et al., 2014), but at this time, we can only speculate that there is significant potential for effects from novel pathogens.

Habitat Loss or Degradation

A continuing threat to the welfare of sea otters throughout their range is degradation of coastal habitats resulting primarily from increased human development (Crain et al., 2009). Currently, as much of the area occupied by sea otters is relatively remote, particularly in Alaska, British Columbia, and Russia, widespread change does not seem likely and habitat loss due to harbors or other industrial developments is not considered to be an imminent concern. However, habitat quality in specific locations may be diminished by development, which may negatively affect survival of localized populations. For southern sea otters, the extent to which high density coastal human population centers and their impacts on coastal ecosystems may deter continued sea otter range expansion (into southern California and north to the San Francisco Bay area) is generally unknown.

The loss of habitat associated with development of industrial use areas, including log transfer facilities, dredging, and fish processing operations, does present some concerns for sea otters, although effects will likely be localized. For example, a large seafood production facility located in Akutan Harbor in the Aleutian chain was found to be making unauthorized discharges of seafood processing waste in several locations, in violation of the US Clean Water Act. One seafood waste pile in Akutan Harbor was more than 50 acres in size. The waste piles create areas with anoxic conditions that destroy habitat for invertebrates, fish, and other living things. As a result of the release of seafood processing waste, Akutan Harbor has been listed as an "Impaired Water Body" (http://www.dec.state.ak.us/water/wqsar/Docs/2010impairedwaters.pdf). For sea otters, consequences likely will be most evident in terms of reduced sea otter density and abundance in the vicinity of the harbor.

Mariculture operations also have increased in number over the last decades, and although these operations may not directly degrade habitat, if there are conflicts and the presence of sea otters is discouraged, this will likely limit adjacent sea otter populations.

Introduction of invasive species and the expansion of species ranges in coastal ecosystems also may have adverse effects on the overall community structure and prey base available for sea otters (Wonham and Carlton, 2005; Sorte et al., 2010), but the magnitude of this threat is not clear. Sea otters may be at an advantage relative to many species, as they are able to consume a wide variety of prey items (Riedman and Estes, 1990), which perhaps provides some ability to adapt to changing prey structures.

Habitat quality and suitability for sea otters also may be affected by disturbance-related effects. Potential effects include acoustic, visual, or olfactory disturbances as well as physical disturbances caused by vessels or vessel wakes (Thrush and Dayton, 2002; Tyack, 2008). The potential role of non-lethal disturbance to have population-level effects is considered under some protective legislation including the MMPA, which defines disturbance as an effect that modifies behavior. Clearly in the case of the sea otter there appears to be a relatively high level of disturbance that is tolerated, reflected in their adaptation to living in close proximity to chronic disturbance in habitats such as the Monterey Peninsula in California. However, one must consider what the abundance of sea otters might be in the absence of such disturbance before discounting disturbance in general. Our experience suggests that sea otters are fairly tolerant of the presence of humans and adapt fairly easily provided disturbances are minimized and do not result in injury or mortality.

Natural Hazards

Natural disasters that may impact sea otter habitat throughout their range include volcanoes, earthquakes, and tsunamis. The frequency and magnitude of these events are unpredictable, and in most cases, effects on sea otters are at a limited spatial scale and of little consequence. However, there is risk of larger events, such as the 1912 Katmai Novarupta volcanic eruption and the

1964 Alaska earthquake, which have the potential to cause significant mortality of sea otters (through presumably still at only a limited scale).

Sea otters were not present along Katmai coast in 1912 as their numbers had been greatly reduced by the fur trade, and thus there was no reported loss of otters following the Novarupta eruption. However, the resulting deposition of ash in coastal areas (estimated to be several feet deep in the near vicinity of the volcano) caused great devastation, with massive numbers of dead fish, seabirds, and other marine mammals described by eyewitnesses (Schaaf, 2004).

The 1964 Alaska earthquake was centered in south-central Alaska and caused tremendous devastation in coastal areas. Kenyon (1969) reported a decline of over 40% in estimated sea otter abundance in PWS between 1959 and 1964, before and after the 1964 Alaska earthquake, and speculates that the earthquake may have been a factor in mortalities, as intertidal areas in much of the Sound were drastically altered by the earthquake. Kenyon (1969) also refers to Russian reports of dramatic effects of earthquakes on sea otter prey species, and consequent reductions and redistributions of sea otter populations.

Genetic Diversity

Sea otters already have experienced at least one severe genetic bottleneck, during the commercial fur harvest (Chapter 5), and there is ample evidence of reduced diversity in current populations (Bodkin et al., 1999; Larson et al., 2002a,b; Aguilar et al., 2008; Larson et al., 2012). Nevertheless, there have been large increases in otter abundance over the last century, particularly within translocated populations founded by two distinct populations (Bodkin et al., 1999). However, the effects of low gene diversity and continued loss of diversity in small, isolated populations is a real threat which may affect a population's ability to adapt to diseases, prey shifts, and changing environmental conditions. For example, in the Aleutian Islands, where otters now exist in small, isolated groups with reduced chance for genetic exchange, there is a probable continued loss of diversity, and the vulnerability of these small groups may be elevated with combined threats of continuing predation pressure and decreased genetic diversity.

Density-Dependent Threats

Our view of the role of sea otters in coastal marine communities has been formed to a large degree by what we observed during the twentieth century, as sea otter numbers grew from a few hundred animals distributed over thousands of miles of shoreline to many tens of thousands of individuals occupying more than half of their historic range. Although density-dependent factors traditionally have not been considered a threat to populations, for sea otters these factors may become important in restricting population and individual growth and overall abundance at local to regional scales. The situation created by multiple remnant populations surviving the fur trade and the subsequent multiple translocations to aid recovery resulted in nearly 20 geographically distinct and isolated sea otter populations late in the twentieth century. As these populations increase in numbers and range, it seems likely that density-dependent factors will come into play in limiting population growth rates, particularly as the distance from the center to the ends of the range of that population exceed dispersal distance.

Annual adult female sea otter reproductive rates are nearly invariant (85–90%) among populations studied, with most females producing a single offspring each year (Monson et al., 2000b). In contrast to generally high and constant reproductive output, survival appears to be the life history character leading to variation in population size (Estes and Bodkin, 2002). Young sea otters are subject to elevated mortality immediately after birth and again at about six months of age, following weaning. Survival of young sea otters likely is influenced by the age and experience of the mother and by availability of food resources (Bodkin, 2003). Where food and space are abundant (e.g., recently recolonized areas), females will be in better condition and their offspring will benefit through increased survival (Chapter 6). Conversely, in long-established populations where food may be a limiting resource, females may be in relatively poor condition and survival of their offspring may suffer. If a juvenile survives through its second year of life, survival probability remains high until about the age of 10 (Monson et al., 2000b). Like harbor seals and sea lions, sea otter population sizes usually appear to be regulated primarily by mortality (principally by starvation) of young during the first year of life (Estes and Bodkin, 2002). In California and the Aleutian Islands, while at or near equilibrium densities, only about half of the pups reached weaning age, and fewer than half of those survived to their first birthday. At Kodiak Island during early recolonization, when females were in relatively good condition, post-weaning survival was nearly 90% (Monson et al., 2000b), demonstrating the importance of food resources as a factor contributing to population regulation.

We know little about how sea otter prey populations respond to variability in primary production in the Gulf of Alaska. Diets of sea otters in the soft sediment habitats at PWS and Kodiak Island consist predominantly of clams (Dean et al., 2002), although in PWS in the past, Dungeness crabs were important prey until depleted by otters (Garshelis et al., 1986). On Kodiak Island, mussels became more abundant in otter diets in areas where clams were depleted by otters (Doroff and DeGange, 1994). Among rocky reef habitats throughout the sea otter's range, urchins, crabs, snails, and mussels are common prey (Estes and Bodkin, 2002). However, the nature of community structure, the survival and reproductive strategies of benthic invertebrate prey species, and the relatively small magnitude of interannual variability in kelp production tend to dampen fluctuations in biomass yield compared to that common in pelagic food webs. Over longer intervals, increases in prey available to sea otters should eventually have positive effects on population sizes. Conversely, as prey items become less available overall, there is evidence that individuals respond through declining mean weights and reduced survival (Bodkin et al., 2000).

As local population densities reach equilibrium with food and space resources, large, abrupt declines in those populations might occur under stressful conditions, such as a harsh winter. For example, at Bering Island (Commander Islands, Russia) sea otters experienced a major winter die-off and the population declined from about 5000 animals to 3000 in a single year (Bodkin et al., 2000). The principal cause of death was starvation, and nearly 80% of the recovered carcasses were male, resulting in a surviving population with an increased proportion of females, a situation conducive to optimizing recovery potential. Similar phenomena might be expected as recovery continues in regions where sea otters have not declined recently, particularly at individual islands where emigration might be limited, or in local mainland areas where the distance to unoccupied habitat exceeds the dispersal distance of sea otters. For example, in southeast Alaska as unoccupied habitat shrinks and growing populations come into equilibrium with their food resources, it seems unlikely that the high growth rates observed in the twentieth century will continue. Thus an understanding of processes that contribute to spatial variation in population structuring, including dispersal distance, can be essential in conservation and population management.

Climate Change

Major concerns for the continuing well-being of sea otters include anthropogenic factors that will place not only sea otters but countless other species at risk. Harvest by humans has been the greatest threat in the past, and still presents a risk for some populations in some areas. However, going forward, the gravest concerns appear to be associated with overall habitat degradation as a result of changing global climate, including effects on marine systems. Ocean acidification may be the greatest danger, as this could lead to pervasive changes in the overall ecology of marine ecosystems (Harley et al., 2006; Hoegh-Guldberg and Bruno, 2010; Ainsworth et al., 2011; Helmuth et al., 2013; Bijma et al., 2013), including the invertebrate populations upon which sea otters depend for food. As nearshore systems become more acidic, one anticipated response is that many benthic invertebrates—including clams, oysters, mussels, and snails—will have greater difficulties forming and maintaining their shells, resulting in increased mortality. However, sea otters are dietary generalists and able to consume a variety of soft prey and fishes in addition to shelled invertebrates, which may confer some degree of resiliency.

Further, if there are changes in the southern extent of overwinter pack ice, it may be that sea otters will extend their range northward, into coastal areas of the Bering Sea, as there is abundant shallow habitat in that region with benthic prey populations, and winter ice conditions are the only apparent limitation. However, given the huge uncertainties that will come with a warming ocean, anticipating consequences for sea otters is highly speculative.

CONCLUSION

Although, range-wide, sea otters have faced major challenges over the past several centuries, they nevertheless represent a conservation success story, as they rebounded from near extirpation and now reoccupy the majority of their former range. Their conservation has been facilitated through management and modification of human behavior, and has included translocations from remnant populations to reoccupy former habitat. In some areas, they now represent a competitor to human use of marine resources and expanding numbers are not welcomed. In other locations, they are appreciated as a charismatic wild species that plays an important role in structuring diverse marine communities, and have strong public support. As is the case for many large wild mammal species today, their welfare is not assured and they continue to be at risk from a variety of threats, both present and potential.

Human harvests clearly represent the greatest past threat to sea otter populations, and where unmanaged harvests continue, they present a persistent threat (Esslinger and Bodkin, 2009). Mortality incidental to human activities (primarily fishing) also can lead to population-level consequences. However, human harvests, fisheries conflicts, and predation pressures may not present as great a challenge to sea otters in the coming decades compared to the growing threat from climate change and degraded marine environments.

In our view, the potential for cumulative effects represents the greatest and as yet largely unrealized threat to conservation of sea otters, as well as other species. Cumulative effects can be considered as the sum of sources of mortality (including reduced fecundity) in addition to the natural or baseline mortality that results in observed rates of population change. For each individual source of mortality, effects on a population may be minor. Each source of mortality may have a relatively low impact (e.g., reduced genetic diversity, reduced diet diversity or energy intake, or habitat modification), and if cumulative mortality remains low, changes in population size may be indiscernible. However, as sources and magnitude of mortality increase, population level responses become apparent. This phenomenon currently is most evident in California, where various sources of mortality, generally related to human activities (e.g., shooting, drowning in fishing gear, boat strikes, contaminants), may combine with natural mortality (e.g., disease, predation) such that all contribute to growth rates that fluctuate annually but consistently fall below what is widely observed elsewhere, and at times fall to the point of population decline (Estes et al., 2003). We further suggest that unanticipated sources of mortality, perhaps unrelated to human endeavors (e.g., increasing predation, competition, or density dependence), will also be factors in the complex array of conditions that ultimately control abundance of sea otter populations throughout their range.

REFERENCES

Aguilar, A., Jessup, D.A., Estes, J., Garza, J.C., 2008. The distribution of nuclear genetic variation and historical demography of sea otters. Anim. Conserv. 11, 35–45.

Ainsworth, C.H., Samhouri, J.F., Busch, D.S., Cheung, W.W., Dunne, J., Okey, T.A., 2011. Potential impacts of climate change on Northeast Pacific marine foodwebs and fisheries. ICES J. Mar. Sci.: J. Du Conseil 68, 1217–1229.

Ames, J.A., Morejohn, G.V., 1980. Evidence of white shark, *Carcharodon carcharias*, attacks on sea otters, *Enhydra lutris*. Calif. Fish Game 66, 196–209.

Andrady, A.L., 2011. Microplastics in the marine environment. Mar. Pollut. Bull. 62, 1596–1605.

Babcock, M.M., Irvine, G.V., Harris, P.M., Cusick, J.A., Rice, S.D., 1996. Persistence of oiling in mussel beds three and four years after the Exxon Valdez oil spill. In: Rice, S.D., Spies, R.B., Wolfe, D.A., Wright, B.A. (Eds.), Proceedings of the *Exxon Valdez* Oil Spill Symposium. American Fisheries Society, Bethesda, pp. 286–297.

Bacon, C.E., Jarman, W.M., Estes, J.A., Simon, M., Norstrom, R.J., 1999. Comparison of organochlorine contaminants among sea otter (*Enhydra lutris*) populations in California and Alaska. Environ. Toxicol. Chem. 18, 452–458.

Ballachey, B.E., Bodkin, J.L., DeGange, A.R., 1994. An overview of sea otter studies. In: Loughlin, T.R. (Ed.), Marine Mammals and the *Exxon Valdez*. Academic Press, San Diego, CA, pp. 47–59.

Ballachey, B.E., Bodkin, J.L., Howlin, S., Doroff, A.M., Rebar, A.H., 2003. Survival of juvenile sea otters in Prince William Sound, Alaska, 1992–93. Can. J. Zool. 81, 1494–1510.

Ballachey, B.E., Bodkin, J.L., Monson, D.H., 2013. Quantifying long-term risks to sea otters from the 1989 "Exxon Valdez" oil spill: reply to Harwell & Gentile (2013). Mar. Ecol. Prog. Ser. 488, 297–301.

Ballachey, B.E., Monson, D.H., Esslinger, G.G., Kloecker, K., Bodkin, J., Bowen, L., et al., 2014. 2013 update on sea otter studies to assess recovery from the 1989 *Exxon Valdez* oil spill, Prince William Sound, Alaska: US Geological Survey Open-File Report 2014-1030, 40p. Avaliable from <http://dx.doi.org/10.3133/ofr20141030>.

Barabash-Nikiforov, I.I., 1962. The Sea Otter: Kalan. National Science Foundation and the Department of Interior, Washington, DC.

Baskaran, M., Hong, G.-H., Dayton, S., Bodkin, J.L., Kelly., J.J., 2003. Temporal variations of natural and anthropogenic radionuclides in sea otter skull tissue in the North Pacific Ocean. J. Environ. Radioact. 64, 1–18.

Benning, J.L., Barnes, D.L., Burger, J., Kelley, J.J., 2009. Amchitka Island, Alaska: moving towards long term stewardship. Polar Rec. 45, 133–146.

Bernabeu, A.M., Rey, D., Rubio, B., Vilas, F., Domínguez, C., Bayona, J.M., et al., 2009. Assessment of cleanup needs of oiled sandy beaches: lessons from the Prestige oil spill. Environ. Sci. Technol. 43, 2470–2475.

Bijma, J., Pörtner, H.O., Yesson, C., Rogers, A.D., 2013. Climate change and the oceans—What does the future hold? Mar. Pollut. Bull. 74, 495—505.

Bodkin, J.L., 2003. Sea otter *Enhydra lutris*. In: Feldhamer, G.A., Thompson, B.C., Chapman, J. A. (Eds.), Wild Mammals of North America. Biology, Management and Conservation, second ed. The Johns Hopkins University Press, Baltimore and London, pp. 735—743.

Bodkin, J.L., Ballachey, B.E., 2010. Modeling the effects of mortality on sea otter populations. USGS Scientific Investigations Report 2010-5096, 12p.

Bodkin, J.L., Jameson, R., 1991. Patterns of seabird and marine mammal carcass deposition along the central California coast, 1980—1986. Can. J. Zool. 69, 1149—1155.

Bodkin, J.L., Ballachey, B.E., Cronin, M.A., Scribner, K.T., 1999. Population demographics and genetic diversity in remnant and re-established populations of sea otters. Conserv. Biol. 13, 1278—1385.

Bodkin, J.L., Burdin, A.M., Ryzanov, D.A., 2000. Age and sex specific mortality and population structure in sea otters. Mar. Mammal Sci. 16, 201—219.

Bodkin, J.L., Ballachey, B.E., Coletti, H.A., Esslinger, G.G., Kloecker, K.A., Rice, S.D., et al., 2012. Long-term effects of the "Exxon Valdez" oil spill: sea otter foraging in the intertidal as a pathway of exposure to lingering oil. Mar. Ecol. Prog. Ser. 447, 273—287.

Bodkin, J.L., Esler, D., Rice, S.D., Matkin, C.O., Ballachey, B.E., 2014. The effects of spilled oil on coastal ecosystems: lessons from the *Exxon Valdez* spill. In: Maslo, B., Lockwood, J. (Eds.), Coastal Conservation. Cambridge University Press, Cambridge, UK, pp. 311—346.

Bowen, L., Miles, A.K., Murray, M., Haulena, M., Tuttle, J., Van Bonn, W., et al., 2011. Gene transcription in sea otters (*Enhydra lutris*); development of a diagnostic tool for sea otter and ecosystem health. Mol. Ecol. Resour. 12, 67—74.

Bragg, J.R., Prince, R.C., Harner, E.J., Atlas, R.M., 1994. Effectiveness of bioremediation for the *Exxon Valdez* oil spill. Nature 368 (6470), 413—418.

Brancato, M.S., Milonas, L, Bowlby, C.E., Jameson, R., Davis, J.W., 2009. Chemical contaminants, pathogen exposure and general health status of live and Beach-Cast Washington Sea Otters (*Enhydra lutris kenyoni*). Marine Sanctuaries Conservation Series ONMS-08-08. US Department of Commerce, National Oceanic and Atmospheric Administration, Office of National Marine Sanctuaries, Silver Spring, MD, 181 pp.

Burge, C.A., Eakin, C.M., Friedman, C.S., Froelich, B., Hershberger, P.K., Hofmann, E.E., et al., 2014. Climate change influences on marine infectious diseases: implications for management and society. Annu. Rev. Mar. Sci. 6, 249—277.

Burger, J., Gochfeld, M., Kosson, D., Powers, C.W., Friedlander, B., Eichelberger, J., et al., 2005. Science, policy, and stakeholders: developing a consensus science plan for Amchitka Island, Aleutians, Alaska. Environ. Manage. 35, 557—568.

Burger, J., Gochfeld, M., Jewett, S.C., 2007. Radionuclide concentrations in benthic invertebrates from Amchitka and Kiska islands in the Aleutian chain, Alaska. Environ. Monit. Assess. 128, 329—341.

Burn, D.M., Doroff, A.M., 2005. Decline in sea otter (*Enhydra lutris*) populations along the Alaska Peninsula, 1986—2001. Fishery Bull. 103, 270—279.

Burns, K.A., Garrity, S.D., Levings, S.C., 1993. How many years until mangrove ecosystems recover from catastrophic oil spills? Mar. Pollut. Bull. 26, 239—248.

Carls, M.G., Babcock, M.M., Harris, P.M., Irvine, G.V., Cusick, J.A., Rice, S.D., 2001. Persistence of oiling in mussel beds after the *Exxon Valdez* oil spill. Mar. Environ. Res. 51, 167—190.

Cole, M., Lindeque, P., Halsband, C., Galloway, T.S., 2011. Microplastics as contaminants in the marine environment: a review. Mar. Pollut. Bull. 62, 2588—2597.

Corredor, J.E., Morell, J.M., Del Castillo, C.E., 1990. Persistence of spilled crude oil in a tropical intertidal environment. Mar. Pollut. Bull. 21, 385–388.

Costa, D.P., Kooyman, G.L., 1982. Oxygen consumption, thermoregulation, and the effect of fur oiling and washing on the sea otter, *Enhydra lutris*. Can. J. Zool. 60, 2761–2767.

Crain, C.M., Halpern, B.S., Beck, M.W., Kappel, C.V., 2009. Understanding and managing human threats to the coastal marine environment. Ann. NY Acad. Sci. 1162, 39–62.

Daszak, P., Cunningham, A.A., Hyatt, A.D., 2001. Anthropogenic environmental change and the emergence of infectious diseases in wildlife. Acta Trop. 78, 103–116.

Dean, T.A., Bodkin, J.L., Fukuyama, A.K., Jewett, S.C., Monson, D.H., O'Clair, C.E., et al., 2002. Food limitation and the recovery of sea otters following the "Exxon Valdez" oil spill. Mar. Ecol. Prog. Ser. 241, 255–270.

Desforges, J.P.W., Galbraith, M., Dangerfield, N., Ross, P.S., 2014. Widespread distribution of microplastics in subsurface seawater in the NE Pacific Ocean. Mar. Pollut. Bull. 79, 94–99.

Doney, S.C., Ruckelshaus, M., Duffy, J.E., Barry, J.P., Chan, F., English, C.A., et al., 2012. Climate change impacts on marine ecosystems. Annu. Rev. Mar. Sci. 4, 11–37.

Doroff, A.M., DeGange, A.R., 1994. Sea otter, Enhydra lutris, prey composition and foraging success in the northern Kodiak Archipelago. US. Fish. Bull. 92, 704–710.

Doroff, A.M., Estes, J.A., Tinker, M.T., Burn, D.M., Evans, T.J., 2003. Sea otter population declines in the Aleutian archipelago. J. Mammal. 84, 55–64.

Erlandson, J.M., Rick, T.C., Estes, J.A., Graham, M.H., Braje, T.J., Vellanoweth, R.L., 2005. Sea otters, shellfish, and humans: 10,000 years of ecological interactions on San Miguel Island, California. In: Garcelon, D.K., Schwemm, C.A. (Eds.), Proceedings of the Sixth California Islands Symposium. Institute for Wildlife Studies and National Park Service, Arcata, CA, pp. 58–69.

Esler, D., Bowman, T.D., Trust, K.A., Ballachey, B.E., Dean, T.A., Jewett, S.C., et al., 2002. Harlequin duck population recovery following the *Exxon Valdez* oil spill: progress, process and constraints. Mar. Ecol. Prog. Ser. 241, 271–286.

Esler, D., Trust, K.A., Ballachey, B.E., Iverson, S.A., Lewis, T.L., Rizzolo, D.J., et al., 2010. Cytochrome P4501A biomarker indication of oil exposure in harlequin ducks up to 20 years after the Exxon Valdez oil spill. Environ. Toxicol. Chem. 29, 1138–1145.

Esler, D., Ballachey, B.E., Trust, K.A., Iverson, S.A., Reed, J.A., Miles, A.K., et al., 2011. Cytochrome P4501A biomarker indication of the timeline of chronic exposure of Barrow's goldeneyes to residual *Exxon Valdez* oil. Mar. Pollut. Bull. 62, 609–614.

Esslinger, G.G., Bodkin, J.L., 2009. Status and trends of sea otter populations in Southeast Alaska, 1969–2003. US Geological Survey Scientific Investigations Report 2009-5045.

Estes, J.A., Bacon, C.E., Jarman, W.M., Norstrom, R.J., Anthony, R.G., Miles, A.K., 1997. Organochlorines in sea otters and bald eagles from the Aleutian Archipelago. Mar. Pollut. Bull. 34, 486–490.

Estes, J.A., Bodkin, J.L., 2002. Otters. In: Perrin, W.F., Wursig, B., Thewissen, J.G.M. (Eds.), Encyclopedia of Marine Mammals. Academic Press, San Diego, CA, pp. 842–858.

Estes, J.A., Smith, N.S., Amchitka Bioenvironmental Program, Research on the Sea Otter, Amchitka Island, Alaska. Final Report, 1 October 1970–31 December 1972, US AEC Report No. NVO 520-1, 68 pp. and appendices.

Estes, J.A., Tinker, M.T., Williams, T.M., Doak, D.F., 1998. Killer whale predation on sea otters linking oceanic and nearshore ecosystems. Science 282 (5388), 473–476.

Estes, J.A., Hatfield, B.B., Ralls, K., Ames, J., 2003. Causes of mortality in California sea otters during periods of population growth and decline. Mar. Mammal Sci. 19, 198–216.

Estes, J.A., Tinker, M.T., Doroff, A.M., Burn, D.M., 2005. Continuing sea otter population declines in the Aleutian archipelago. Mar. Mammal Sci. 21, 169–172.

Estes, J.A., Terborgh, J., Brashares, J.S., Power, M.E., Berger, J., 19 others, 2011. Trophic downgrading of planet earth. Science 333, 301–306.

Frost, K.J., Lowry, L.L., Sinclair, E.H., Ver Hoef, J., McAllister, D.C., 1994. Impacts on distribution, abundance and productivity of harbor seals. In: Loughlin, T.R. (Ed.), Marine Mammals and the *Exxon Valdez*. Academic Press, San Diego, CA, pp. 97–118.

Fukuyama, A.K., Shigenaka, G., Hoff, R.Z., 2000. Effects of residual *Exxon Valdez* oil on intertidal *Protothaca staminea*—mortality, growth, and bioaccumulation of hydrocarbons in transplanted clams. Mar. Pollut. Bull. 40, 1042–1050.

Fuller, R.G., Kirkwood, J.B., 1977. Ecological consequences of nuclear testing. In: Merrit, M.L., Fuller, R.G. (Eds.), The Environment of Amchitka Island. National Technical Information Service, Alexandria, VA, pp. 627–649, TID-267612.

Garrott, R.A., Eberhardt, L.L., Burn, D.M., 1993. Mortality of sea otters in Prince William Sound following the *Exxon Valdez* oil spill. Mar. Mammal Sci. 9, 343–359.

Garshelis, D.L., 1997. Sea otter mortality estimated from carcasses collected after the Exxon Valdez oil. Spill. Cons. Biol. 11, 905–916.

Garshelis, D.L., Garshelis, J.A., Kimker, A.T., 1986. Sea otter time budgets and prey relationships in Alaska. J. Wildl. Manage. 50 (4), 637–647.

Gerber, L.R., Tinker, M.T., Doak, D.F., Estes, J.A., Jessup, D.A., 2004. Mortality sensitivity in life-stage simulation analysis: a case study of southern sea otters. Ecol. Appl. 14 (5), 1554–1565.

Goertz, C.E.C., Walton, R., Rouse, N., Belovarac, J., Burek-Huntington, K., Gill, V., et al., 2013. *Vibrio parahaemolyticus*, a climate change indicator in Alaska marine mammals. In: Mueter, F.J., Dickson, D.M.S., Huntington, H.P., Irvine, J.R., Logerwell, E.A., MacLean, S.A., et al. (Eds.), Responses of Arctic Marine Ecosystems to Climate Change. Alaska Sea Grant, University of Alaska Fairbanks, Fairbanks, AK, USA. Available from: http://dx.doi.org/10.4027/ramecc.2013.03.

Golet, G.H., Seiser, P.E., McGuire, A.D., Roby, D.D., Fischer, J.B., Kuletz, K.J., et al., 2002. Long-term direct and indirect effects of the "Exxon Valdez" oil spill on pigeon guillemots in Prince William Sound, Alaska. Mar. Ecol. Prog. Ser. 241, 287–304.

Gorbics, C.S., Bodkin, J.L., 2001. Stock structure of sea otters (*Enhydra lutris*) in Alaska. Mar. Mammal Sci. 17 (3), 632–647.

Harley, C.D., Randall Hughes, A., Hultgren, K.M., Miner, B.G., Sorte, C.J., Thornber, C.S., et al., 2006. The impacts of climate change in coastal marine systems. Ecol. Lett. 9 (2), 228–241.

Hart, K., Gill, V.A., Kannan, K., 2009. Temporal trends (1992–2007) of perfluorinated chemicals in northern sea otters (*Enhydra lutris kenyoni*) from south-central Alaska. Arch. Environ. Contam. Toxicol. 56 (3), 607–614.

Harvell, C.D., Mitchell, C.E., Ward, J.R., Altizer, S., Dobson, A.P., Ostfeld, R.S., et al., 2002. Climate warming and disease risks for terrestrial and marine biota. Science 296, 2158–2162.

Hatfield, B.B., Marks, D., Tinker, M.T., Nolan, K., Peirce, J., 1998. Attacks on sea otters by killer whales. Mar. Mammal Sci. 14 (4), 888–894.

Hatfield, B.B., Ames, J.A., Estes, J.A., Tinker, M.T., Johnson, A.B., Staedler, M.M., et al., 2011. Sea otter mortality in fish and shellfish traps: estimating potential impacts and exploring possible solutions. Endangered Species Res. 13, 219–229.

Hayes, M.O., Michel, J., 1999. Factors determining the long-term persistence of *Exxon Valdez* oil in gravel beaches. Mar. Pollut. Bull. 38 (2), 92–101.

Helmuth, B., Babij, E., Duffy, E., Fauquier, D., Graham, M., Hollowed, A., et al., 2013. Impacts of climate change on marine organisms. Oceans and Marine Resources in a Changing Climate. Island Press/Center for Resource Economics, Washington, DC, pp. 35–63.

Hines, A.H., Pearse, J.S., 1982. Abalones, shells, and sea otters: dynamics of prey populations in central California. Ecology 63 (5), 1547–1560.

Hirai, H., Takada, H., Ogata, Y., Yamashita, R., Mizukawa, K., Saha, M., et al., 2011. Organic micropollutants in marine plastics debris from the open ocean and remote and urban beaches. Mar. Pollut. Bull. 62 (8), 1683–1692.

Hoegh-Guldberg, O., Bruno, J.F., 2010. The impact of climate change on the world's marine ecosystems. Science 328, 1523–1528.

Hulbert, L.B., Sigler, M.F., Lunsford, C.R., 2006. Depth and movement behaviour of the Pacific sleeper shark in the north-east Pacific Ocean. J. Fish. Biol. 69 (2), 406–425.

Jameson, R.J., Kenyon, K.W., Johnson, A.M., Wight, H.M., 1982. History and status of translocated sea otter populations in North America. Wildl. Soc. Bull. 10, 100–107.

Jessup, D.A., Miller, M.A., Kreuder-Johnson, C., Conrad, P.A., Tinker, M.T., Estes, J., et al., 2007. Sea otters in a dirty ocean. J. Am. Vet. Med. Assoc. 231 (11), 1648–1652.

Jessup, D.A., Johnson, C.K., Estes, J., Carlson-Bremer, D., Jarman, W.M., Reese, S., et al., 2010. Persistent organic pollutants in the blood of free-ranging sea otters (*Enhydra lutris* ssp.) in Alaska and California. J. Wildl. Dis. 46 (4), 1214–1233.

Jewett, S.C., Dean, T.A., Woodin, B.R., Hoberg, M.K., Stegeman, J.J., 2002. Exposure to hydrocarbons ten years after the *Exxon Valdez*—evidence from cytochrome P4501A expression and biliary FACs in nearshore demersal fishes. Mar. Environ. Res. 54, 21–48.

Johnston, E.L., Roberts, D.A., 2009. Contaminants reduce the richness and evenness of marine communities: a review and meta-analysis. Environ. Pollut. 157, 1745–1752.

Jones, T.L., Culleton, B.J., Larson, S., Mellinger, S., Porcasi, J.F., 2011. Toward a prehistory of the southern sea otter (*Enhydra lutris nereis*). In: Braje, T.J., Rick, T.C. (Eds.), Human Impacts on Seals, Sea Lions, and Sea Otters: Integrating Archaeology and Ecology in the Northeast Pacific. University of California, Oakland, CA, pp. 243–271.

Kannan, K., Agusa, T., Perrotta, E., Thomas, N.J., Tanabe, S., 2006a. Comparison of trace element concentrations in livers of diseased, emaciated and non-diseased southern sea otters from the California coast. Chemosphere 65 (11), 2160–2167.

Kannan, K., Perrotta, E., Thomas, N.J., 2006b. Association between perfluorinated compounds and pathological conditions in southern sea otters. Environ. Sci. Technol. 40 (16), 4943–4948.

Kannan, K., Perrotta, E., Thomas, N.J., Aldous, K.M., 2007. A comparative analysis of polybrominated diphenyl ethers and polychlorinated biphenyls in southern sea otters that died of infectious diseases and noninfectious causes. Arch. Environ. Contam. Toxicol. 53 (2), 293–302.

Kannan, K., Moon, H.B., Yun, S.H., Agusa, T., Thomas, N.J., Tanabe, S., 2008. Chlorinated, brominated, and perfluorinated compounds, polycyclic aromatic hydrocarbons and trace elements in livers of sea otters from California, Washington, and Alaska (USA), and Kamchatka (Russia). J. Environ. Monit. 10 (4), 552–558.

Kenyon, K.W., 1969. The sea otter in the eastern Pacific Ocean. North Am. Fauna, 1–352.

Klimley, A.P., 1985. The aerial distribution and autecology of the white shark (*Carcharodon carcharias*), off the west coast of North America. Mem. South. Calif. Acad. Sci. 9, 15–40.

Kreuder, C., Miller, M.A., Jessup, D.A., Lowenstine, L.J., Harris, M.D., Ames, J.A., et al., 2003. Patterns of mortality in southern sea otters (*Enhydra lutris nereis*) from 1998–2001. J. Wildl. Dis. 39 (3), 495–509.

Kvitek, R.G., Bretz, C., 2004. Harmful algal bloom toxins protect bivalve populations from sea otter predation. Mar. Ecol. Prog. Ser. 271, 233–243.

Kvitek, R.G., Oliver, J.S., DeGange, A.R., Anderson, B.S., 1992. Changes in Alaskan soft-bottom prey communities along a gradient in sea otter predation. Ecology 73 (2), 413–428.

Laidre, K.L., Jameson, R.J., 2006. Foraging patterns and prey selection in an increasing and expanding sea otter population. J. Mammal. 87 (4), 799–807.

Larson, S., Jameson, R., Bodkin, J., Staedler, M., Bentzen, P., 2002a. Microsatellite DNA and mtDNA variation in remnant and translocated sea otter (*Enhydra lutris*) populations. J. Mammal. 83 (3), 893–906.

Larson, S., Jameson, R., Etnier, M., Fleming, M., Bentzen, P., 2002b. Loss of genetic diversity in sea otters (*Enhydra lutris*) associated with the fur trade of the 18th and 19th centuries. Mol. Ecol. 11, 1899–1903.

Larson, S., Jameson, R., Etnier, M., Jones, T., Hall, R., 2012. Genetic diversity and population parameters of sea otters, Enhydra lutris, before fur trade extirpation from 1741–1911. PLoS One 73, e32205.

Lensink, C.J., 1960. Status and distribution of sea otters in Alaska. J. Mammal. 41, 172–182.

Li, H.L., Boufadel, M.C., 2010. Long-term persistence of oil from the *Exxon Valdez* spill in two-layer beaches. Nat. Geosci. 3, 96–99.

Miles, A.K., Bowen, L., Ballachey, B., Bodkin, J.L., Murray, M., Estes, J.L., et al., 2012. Variation in transcript profiles in sea otters (Enhydra lutris) from Prince William Sound, Alaska and clinically normal reference otters. Mar. Ecol. Prog. Ser. 451, 201–212.

Miller, M.A., Kudela, R.M., Mekebri, A., Crane, D., Oates, S.C., Tinker, M.T., et al., 2010a. Evidence for a novel marine harmful algal bloom: cyanotoxin (microcystin) transfer from land to sea otters. PLoS One 5 (9), e12576.

Miller, M.A., Conrad, P.A., Harris, M., Hatfield, B., Langlois, G., Jessup, D.A., et al., 2010b. A protozoal-associated epizootic impacting marine wildlife: mass-mortality of southern sea otters (*Enhydra lutris nereis*) due to *Sarcocystis neurona* infection. Vet. Parasitol. 172 (3), 183–194.

Monnett, C., Rotterman, L., 1988. Sex-related patterns in the post-natal development and survival of sea otters in Prince William Sound, Alaska. In: Siniff, D.B., Ralls, K. (Eds.), Population Status of California Sea Otters, Final Report to the Minerals Management Service, US Department of the Interior 14-12-001-3003, pp. 162–190.

Monson, D.H., DeGange, A.R., 1995. Reproduction, preweaning survival, and survival of adult sea otters at Kodiak Island, Alaska. Can. J. Zool. 73, 1161–1169.

Monson, D.H., Doak, D.F., Ballachey, B.E., Bodkin, J.L., 2011. Could residual oil from the Exxon Valdez spill create a long-term population "sink" for sea otters in Alaska? Ecol. Applic. 21, 2917–2932.

Monson, D.H., Doak, D.F., Ballachey, B.E., Johnson, A., Bodkin, J.L., 2000a. Long-term impacts of the *Exxon Valdez* oil spill on sea otters, assessed through age-dependent mortality patterns. Proc. Natl. Acad. Sci. USA 97 (12), 6562–6567.

Monson, D.H., Estes, J.A., Bodkin, J.L., Siniff, D.B., 2000b. Life history plasticity and population regulation in sea otters. Oikos 90 (3), 457–468.

Moore, C.J., 2008. Synthetic polymers in the marine environment: a rapidly increasing, long-term threat. Environ. Res. 108 (2), 131–139.

Murata, S., Takahashi, S., Agusa, T., Thomas, N.J., Kannan, K., Tanabe, S., 2008. Contamination status and accumulation profiles of organotins in sea otters (*Enhydra lutris*) found dead along the coasts of California, Washington, Alaska (USA) and Kamchatka (Russia). Mar. Pollut. Bull. 56, 641–649.

Nakata, H., Kannan, K., Jing, L., Thomas, N., Tanabe, S., Giesy, J.P., 1998. Accumulation pattern of organochlorine pesticides and polychlorinated biphenyls in southern sea otters (*Enhydra lutris nereis*) found stranded along coastal California, USA. Environ. Pollut. 103 (1), 45–53.

National Research Council, 2003. Oil in the Sea III: Inputs, Fates, and Effects. National Academies Press, Washington, DC, 265 pp.

Neff, J.M., Owens, E.H., Stoker, S.W., McCormick, D.M., 1995. Shoreline oiling conditions in Prince William Sound following the Exxon Valdez oil spill. ASTM Special Technical Publication No. 1219.

Neuman, M., Tissot, B., Vanblaricom, G., 2010. Overall status and threats assessment of black abalone (*Haliotis cracherodii* Leach, 1814) populations in California. J. Shellfish Res. 29 (3), 577–586.

Newby, T.C., 1975. A sea otter (*Enhydra lutris*) food dive record. Murrelet 56, 19.

Nikolaev, A.M., 1965. On the feeding of the Kuril sea otter and some aspects of their behavior during the period of ice. In: Pavlovskii, E.N., Zenkovich, B.A. (Eds.), Marine Mammals. Akademiya Nauk SSR, Moscow, pp. 231–236 (translated from Russian by Nancy McRoy, April 1966).

Pearse, J.S., Costa, D.P., Yellin, M.B., Agegian, C.R., 1977. Localized mass mortality of red sea urchin, *Strongylocentrotus franciscanus*, near Santa Cruz, California. Fishery Bull. 75, 3.

Peterson, C.H., Rice, S.D., Short, J.W., Esler, D., Bodkin, J.L., Ballachey, B.E., et al., 2003. Long-term ecosystem response to the *Exxon Valdez* oil spill. Science 302 (5653), 2082–2086.

Piatt, J.F., Ford, R.G., 1996. How many seabirds were killed by the Exxon Valdez oil spill? In: Rice, S.D., Spies, R.B., Wolfe, D.A., Wright, B.A. (Eds.), Proceedings of the Exxon Valdez Oil Spill Symposium. American Fisheries Society Symposium 18, Bethesda, Maryland, pp. 712–719.

Rausch, R.L., 1973. Post Mortem Findings in Some Marine Mammals and Birds Following the Cannikin Test on Amchitka Island (No. NVO—130). Arctic Health Research Center, Fairbanks, AK.

Reddy, C.M., Eglinton, T.I., Hounshell, A., White, H.K., Xu, L., Gaines, R.B., et al., 2002. The West Falmouth oil spill after thirty years: the persistence of petroleum hydrocarbons in marsh sediments. Environ. Sci. Technol. 36 (22), 4754–4760.

Riedman, M., Estes, J.A., 1990. The sea otter (*Enhydra lutris*): behavior, ecology, and natural history. US Fish and Wildlife Service. Biolog. Rep. (USA) 90 (14), 126.

Schaaf, J.M., 2004. Witness: Firsthand Accounts of the Largest Volcanic Eruption in the Twentieth Century. National Park Service, p. 36.

Sherrod, S.K., Estes, J.A., White, C.M., 1975. Depredation of sea otter pups by bald eagles at Amchitka Island, Alaska. J. Mammal. 56 (3), 701–703.

Short, J.W., Lindeberg, M.R., Harris, P.A., et al., 2004. Estimate of oil persisting on beaches of Prince William Sound, 12 years after the *Exxon Valdez* oil spill. Environ. Sci. Technol. 38, 19–25.

Short, J.W., Maselko, J.M., Lindeberg, M.R., Harris, P.M., Rice, S.D., 2006. Vertical distribution and probability of encountering intertidal *Exxon Valdez* oil on shorelines of three embayments within Prince William Sound, Alaska. Environ. Sci. Technol. 40 (12), 3723–3729.

Short, J.W., Irvine, G.V., Mann, D.H., et al., 2007. Slightly weathered *Exxon Valdez* oil persists in Gulf of Alaska beach sediments after 16 years. Environ. Sci. Technol. 41 (4), 1245–1250.

Simenstad, C.A., Estes, J.A., Kenyon, K.W., 1978. Aleuts, sea otters, and alternate stable-state communities. Science 200 (4340), 403–411.

Siniff, D.B., Ralls, K., 1991. Reproduction, survival and tag loss in California sea otters. Mar. Mammal Sci. 7 (3), 211–299.

Siniff, D.B., Williams, T.D., Johnson, A.N., Garshelis, D.L., 1982. Experiments on the response of sea otters, *Enhydra lutris*, to oil. Biol. Conserv. 23, 261–272.

Smith, D.R., Flegal, A.R., Niemeyer, S., Estes, J.A., 1990. Stable lead isotopes evidence anthropogenic contamination in Alaskan sea otters. Environ. Sci. Technol. 24 (10), 1517–1521.

Smith, D.R., Niemeyer, S., Flegal, A.R., 1992. Lead sources to California sea otters: industrial inputs circumvent natural lead biodepletion mechanisms. Environ. Res. 57 (2), 163–174.

Sorte, C.J., Williams, S.L., Carlton, J.T., 2010. Marine range shifts and species introductions: comparative spread rates and community impacts. Global Ecol. Biogeogr. 19 (3), 303–316.

Springer, A.M., Estes, J.A., Van Vliet, G.B., Williams, T.M., Doak, D.F., Danner, E.M., et al., 2003. Sequential megafaunal collapse in the North Pacific Ocean: an ongoing legacy of industrial whaling? Proc. Natl. Acad. Sci. USA 100 (21), 12223–12228.

Teuten, E.L., Saquing, J.M., Knappe, D.R., Barlaz, M.A., Jonsson, S., Björn, A., et al., 2009. Transport and release of chemicals from plastics to the environment and to wildlife. Philos. Trans. R. Soc. B: Biol. Sci. 364 (1526), 2027–2045.

Thrush, S.F., Dayton, P.K., 2002. Disturbance to marine benthic habitats by trawling and dredging: implications for marine biodiversity. Annu. Rev. Ecol. Syst. 33, 449–473.

Tinker, M.T., Estes, J.A., 1996. The population ecology of sea otters at Adak Island, Alaska. Final Report to the Navy, Contract # N68711-94-LT-4026, Santa Cruz, CA.

Trust, K.A., Esler, D., Woodin, B.R., Stegeman, J.J., 2000. Cytochrome P4501A induction in sea ducks inhabiting nearshore areas of Prince William Sound, Alaska. Mar. Pollut. Bull. 40, 397–403.

Tyack, P.L., 2008. Implications for marine mammals of large-scale changes in the marine acoustic environment. J. Mammal. 89 (3), 549–558.

Udevitz, M.S., Ballachey, B.E., Bruden, D.L., 1996. A population model for sea otters in western Prince William Sound. Exxon Valdez Oil Spill State/Federal Restoration Final Report (Restoration Study 93043-3), National Biological Service, Anchorage, AK. pp. 34.

US Fish and Wildlife Service, 2013. Southwest Alaska Distinct Population Segment of the Northern Sea Otter (*Enhydra lutris kenyonii*)—Recovery Plan. US Fish and Wildlife Service, Region 7, Alaska, 171 pp.

Vandermeulen, J.H., Singh, J.G., 1994. *Arrow* oil spill, 1970–90: persistence of 20-yr weathered bunker C fuel oil. Can. J. Fisheries Aquat. Sci. 51 (4), 845–855. Available from: http://dx.doi.org/10.1139/f94-083.

Vandermeulen, J.H., Platt, H.M., Baker, J.M., Southward, J.Y., 1982. Some conclusions regarding the long-term biological effects of some major spills. Philos. Trans. R. Soc. Lond. 297 (1087), 335–351.

Van Dolah, F.M., 2000. Marine algal toxins: origins, health effects, and their increased occurrence. Environ. Health Perspect. 108 (Suppl. 1), 133–141.

Vos, D.J., Quakenbush, L.T., Mahoney, B.A., 2006. Documentation of sea otters and birds as prey for killer whales. Mar. Mammal Sci. 22 (1), 201–205.

Watson, J., 2000. The effects of sea otters (*Enhydra lutris*) on abalone (*Haliotis* spp.) populations. Can. Spec. Publ. Fisheries Aquat. Sci., 123–132.

Wendell, F.E., Hardy, R.A., Ames, J.A., 1986. An Assessment of the accidental take of sea otters, *Enhydra lutris*, in gill and trammel nets. California Fish and Game, Marine Resources Tech. Report No. 54.

Williams, T.D., Allen, D.D., Groff, J.M., Glass, R.L., 1992. An analysis of California sea otter (*Enhydra lutris*) pelage and integument. Mar. Mammal Sci. 8, 1–18.

Williams, T.M., Estes, J.A., Doak, D.F., Springer, A.M., 2004. Killer appetites: assessing the role of predators in ecological communities. Ecology 85 (12), 3373–3384.

Wolfe, D.A., Hameedi, M.J., Galt, J.A., Watabayashi, G., Short, J., O'Clair, C., et al., 1994. The fate of the oil spilled from the *Exxon Valdez*. Environ. Sci. Technol. 28 (13), 561A–568A.

Wonham, M.J., Carlton, J.T., 2005. Trends in marine biological invasions at local and regional scales: the Northeast Pacific Ocean as a model system. Biol. Invasions 7 (3), 369–392.

Zagrebel'nyi, S.V., 2004. Sea otters (*Enhydra lutris* L.) of Bering Island, the Commander Islands: age and sex composition of dead animals and spatial distribution. Russ. J. Ecol. 35 (6), 395–402.

Chapter 5

Sea Otter Conservation Genetics

Shawn E. Larson[1], Katherine Ralls[2] and Holly Ernest[3]
[1]Department of Life Sciences, The Seattle Aquarium, Seattle, WA, USA, [2]Smithsonian Institution, Washington DC, USA, [3]Department of Veterinary Sciences, University of Wyoming, Laramie, WY, USA

INTRODUCTION

A primary goal of wildlife conservation is to maintain functioning ecosystems and biodiversity (Frankham, 1995; Lynch, 1996). The scientific discipline of conservation genetics applies evolutionary, ecological, and

Sea Otter Conservation. DOI: http://dx.doi.org/10.1016/B978-0-12-801402-8.00005-6

molecular genetics principles to understand and conserve hereditary variation that underpins ecosystem function and biodiversity (Allendorf et al., 2013; Frankham, 2010; Frankham et al., 2011). Conservation geneticists strive to describe genetic diversity and gene flow within and among populations, resolve taxonomic uncertainties, and delineate genetically unique populations. Objectives include maximizing genetic diversity, minimizing inbreeding, and identifying individuals and species for forensic purposes (Frankham et al., 2011). Conservation genetic techniques are typically applied to species believed to be compromised or vulnerable to loss of genetic diversity and are commonly listed as threatened, endangered, or species of concern. The sea otter, *Enhydra lutris*, because of its dramatic population reduction and fragmentation during the fur trade (Chapter 3) and the listing of two populations (the southern sea otter, *E.l. nereis*, and the southwestern stock of the northern sea otter, *E.l. kenyoni*) as threatened under the Endangered Species Act (ESA), has been widely studied over the past several decades as conservation genetics techniques have advanced.

The genetic variation that is present in individuals, populations, and species is thought to be critical for population adaptability and survival in response to changing environmental conditions (such as climate change and degraded habitats). Genetic diversity is typically described using polymorphism, heterozygosity, and allelic diversity (Frankham et al., 2011). Polymorphism literally translates to "many forms" of a gene or locus. Polymorphic genes or loci have different forms or versions of the genetic code called alleles. Polymorphism is defined as the proportion of the sampled population that has at least two alleles within the genes of study. Heterozygotes are defined as individuals that carry two different alleles or versions of the gene being studied, one from the sire and one from the dam, while homozygotes carry two copies of the same allele. This case assumes the species is diploid (as is the case with sea otters) and has two copies of most chromosomes except for the sex chromosomes in the male where there is only one copy of the Y chromosome and one copy of the X chromosome. Heterozygosity is calculated as a proportion of heterozygotes in the sampled population. Allelic diversity or richness is the measurement of the mean number of alleles per locus (genetic location or gene) across the sampled population.

Genetic variation is created within individuals by the two basic processes of mutation and recombination, and can be divided into two types, neutral and adaptive (Kirk and Freeland, 2011). *Adaptive genetic variation* describes genetic variants that affect fitness (survival and/or reproduction), thus subjecting individuals to the forces of natural selection. Conversely, variation within the DNA that has no discernible effect on the fitness of an individual is termed *neutral genetic variation*, and is assumed not to subject an individual to forces of natural selection. Neutral and adaptive categorizations should not be assumed to be constant: one mutation might be effectively *neutral* to selection in one population, place, and time, while incurring a selective evolutionary (*adaptive*) cost or benefit in a different population, place, and time. Analyses of both types of genetic variation are important to provide key information for monitoring and conservation of sea otter and other wildlife populations.

Levels of neutral genetic variation in populations are influenced by mutation (addition or deletion in the genome), gene flow (transfer of alleles from one population to another), and genetic drift (the change in allele frequency due to random sampling). Measurements of these inherited DNA variants, because they are not subject to selection, tend to be maintained in the population and can reflect the demographic history of a population, including past population reductions, expansions, immigration, and emigration. Typically neutral genetic variation is measured within variable and non-protein coding regions of the mitochondrial genome (mtDNA) and/or from DNA within the cell's nucleus (nuclear DNA), which houses the rest of the genome. Genetic diversity is a critical measure in population genetics because it can tell us about the current and likely future adaptive genetic plasticity and possibly future health of a population. For example, low levels of genetic diversity over the long term may reduce population fitness and evolutionary potential because individuals may not have the genetic flexibility to respond to changing conditions or disease. Neutral markers provide geneticists with insights into the overall levels of genetic variation within populations and, by inference, the diversity at gene coding regions or adaptive genetic variation.

Adaptive genetic variation pertains to the differences in genes that are important for survival or reproduction of an organism. Assessment of nuclear DNA and mtDNA gene sequence, as well as studies of gene expression, are some of the ways to measure adaptive variation. Historically, adaptive variation was much more difficult to quantify by molecular methods than neutral variation; therefore, relatively few early conservation genetic studies included such analyses. Early adaptive variation research focused largely on the major histocompatibility complex (MHC), which contains the genes that control immune function and response. More recently, because of the development of higher-throughput parallel DNA sequencing technologies, studies have focused on specific candidate genes to measure adaptive genetic variation that may have potential importance for fitness traits (Allendorf et al., 2013; Frankham, 2010; Funk et al., 2012; Kirk and Freeland, 2011).

Another common goal in conservation genetic analysis is determining population differences. Population structure may be measured (1) by determining whether the sampled individuals represent one or more genetically distinct populations and (2) by quantifying the magnitude of genetic differences among populations. For the first of these goals, one method is to use Bayesian, model-based algorithm analysis on multiple locus data sets to test for presence of genetic clustering indicative of genetically similar individuals (Pritchard et al., 2000). One such algorithm is principal components analysis (PCA), which often provides useful insights into population structure. Once the number and identities of genetically distinct populations are determined, there are a number of statistical genetic indices to provide measures of genetic difference and divergence among individuals and populations. In general the larger the value of the genetic difference indices, the greater the differences among populations.

Commonly used indices to determine genetic differences among populations are the F_{ST} family of measurements. These measure genetic divergence and genetic exchange (gene flow) among populations, and assume that only two alleles occur in each gene within an individual. F_{ST} represents the proportion of the total heterozygosity between subpopulations and has values that vary between zero and one. When there are small differences in allelic variation between populations, F_{ST} is close to zero, and when there are very large differences, F_{ST} is closer to one (Weir and Cockerham, 1984). For a genetic marker, or region of interest, that has more than two alleles, G_{ST}, a modification of F_{ST}, is a useful metric with the same general interpretation: values near zero indicate populations of similar genetic composition, while higher values indicate genetically more distinct groups (Culley et al., 2002).

Other metrics commonly used to determine population differences are gene flow and genetic distance. Gene flow between groups is typically determined by estimating number of migrants per generation (Nm). This metric describes the average number of individuals per generation migrating between populations of equal size (Avise, 1994). Genetic distance measures use allelic frequency differences to determine structure between populations with smaller distances indicating closely related groups. Most genetic distance indices, such as Nei's D, assume that the differences between groups arise due to mutations and genetic drift (Nei, 1972).

THE SEA OTTER STORY AND CONSERVATION GENETICS

Sea otters have been hunted for their fur and meat for thousands of years. Before the fur trade extirpation in the eighteenth and nineteenth centuries, sea otters were routinely harvested by native people (Simenstead et al., 1978; Chapter 11). Most populations were thought to have been able to survive subsistence hunting with some, but not significant, loss of genetic diversity (Aguilar et al., 2008; Larson et al., 2012). However, the maritime fur trade drove the species to near extinction throughout the range (Chapter 3), with a clear reduction in genetic diversity (Aguilar et al., 2008; Larson et al., 2002a,b).

In an effort to re-establish sea otter populations, state and federal agencies made several translocations from the 1950s through the 1970s (Jameson et al., 1982; Chapter 3), resulting in viable populations in Washington, British Columbia, and southeast Alaska (Jameson et al., 1982; Bodkin et al., 1999). In spite of the recovery of the remnant populations and the successful translocation efforts, many sea otter populations have remained isolated, with limited gene flow or migration possible only between adjacent groups (Figure 5.1), thus limiting the exchange of genetic material between populations

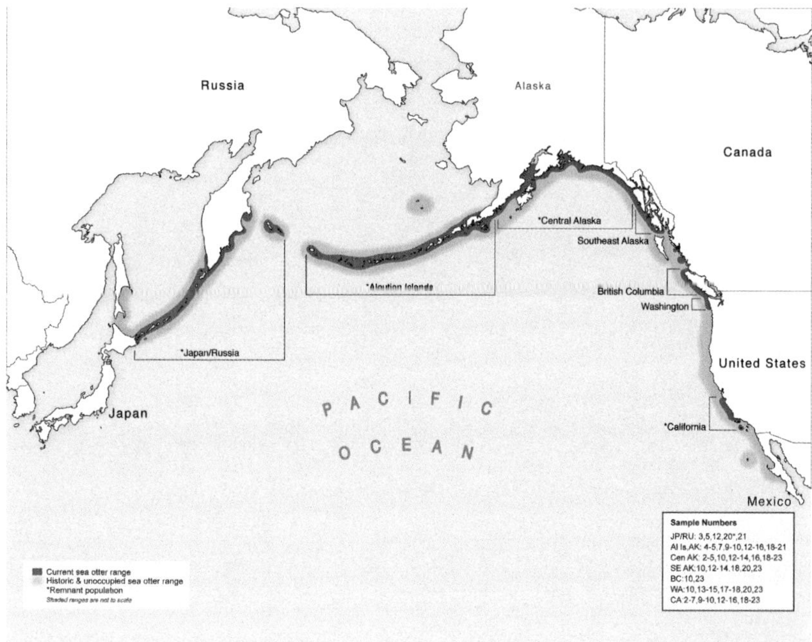

FIGURE 5.1 Map of historical and current sea otter range with brackets indicating where samples were taken from studies listed in Table 5.1.

(Bodkin et al., 1999; Larson et al., 2012). This barrier to gene flow results from both long geographic distances between current populations and long stretches of deep water between island groups that sea otters find difficult to cross because of their high metabolic rates and relatively large food requirements (Chapter 2). In general, the sea otter is capable of moving no more than a few hundred kilometers in a single generation (Rathbun et al., 2000) and most display annual movements of less than a few tens of kilometers (Chapter 2). This isolation of populations is thought to have changed the historical genetic relationships among populations, through factors such as loss of gene flow and small founder population sizes. It is this history of sea otter population extirpation, as well as fragmentation and the resultant population bottlenecks, that has inspired many studies of sea otter population genetics (see Table 5.1 for a complete list of sea otter genetics studies to date).

CONSERVATION GENETIC APPROACHES

Samples

Conservation genetics studies of sea otters have been conducted since the early 1980s using various techniques and samples (Tables 5.1 and 5.2)

TABLE 5.1 Genetic Studies of Sea Otters to Date in Chronological Order

Study	Reference	Year	Location								N	Marker	Objectives
			CA	OR	WA	BC	SEAK	CEAK	ALEU	JN/RU			
1	Ralls et al., 1983	1983									–	–	Variation
2	Rotterman, 1992	1992	X					X			208	Allozymes	Variation
3	Sanchez, 1992	1992	X					X		X	86	MtDNA	Variation
4	Cheney, 1995	1995	X					X	X		79	MtDNA	Phylogenetics
5	Cronin et al., 1996	1996	X					X	X	X	167	MtDNA	Variation
6	King et al., 1996	1996	X								1	Interleukin-6	Phylogenetics
7	Lidicker and Mccollum, 1997	1997	X						X		83	Allozymes	Genetic variation
8	Scribner et al., 1997	1997									–	–	Interpretation
9	Koepfli and Wayne, 1998	1998	X						X		2	MtDNA	Phylogenetics

10	Bodkin et al., 1999	1999	X		X	X	X	X	X		144	MtDNA	Variation
11	Gorbics and Bodkin, 2001	2001					X	X	X			Microsatellites, mtDNA	Stocks
12	Cronin et al., 2002	2002	X				X	X	X	X	480	Microsatellites, mtDNA	Stocks
13	Larson et al., 2002a	2002	X		X		X	X	X		174	Microsatellites, mtDNA	Variation
14	Larson et al., 2002b[a]	2002	X		X		X	X	X		186	Microsatellites, mtDNA	Variation
15	Bowen et al., 2006	2006	X		X				X		3	MHC class II genes	Variation
16	Aguilar et al., 2008	2007	X					X	X		396	Microsatellites, MHC	Variation
17	Valentine et al., 2008[a]	2008		X	X						18	MtDNA	Phylogenetics
18	Larson et al., 2009	2009	X		X		X	X	X		174	Microsatellites	Variation
19	Bowen et al., 2012	2012	X					X	X		42	10 immune function genes	Reference
20	Larson et al., 2012[a]	2012	X	X	X		X	X	X	X	337	Microsatellites	Variation
21	Miles et al., 2012	2012	X					X	X		90	10 immune function genes	Oil spill

[a]*Includes ancient DNA*
SEAK = Southeast Alaska; CEAK = Central Alaska (Prince William Sound, Kodiak, Kenai, Alaska Peninsula); ALEU = Aleutian chain; JN/RU = Japan and Russia

TABLE 5.2 Genetic Methods Employed in Sea Otter Studies

Genetics Method	What	Specifically	Tissues	Method	Co-D[a]	Neu[b]	Uses	Studies[c]
Allozymes	Protein	Protein variants	Blood	Staining	Yes	No	Variability, pop structure and systematics	2,7
Restriction fragment-length polymorphism (RFLP)	Genomic DNA	Variable length cut fragments due to variable gene sequences	Skin or muscle	Enzyme	Yes	No	Variability, pop structure and systematics	3,4,5
Microsatellites	Genomic DNA	Repeating units of 1–6 base pairs	Skin or muscle, blood, hair, teeth or bone	PCR	Yes	Yes	Variability, pop structure and systematics	11,12,13,14,16,18,20
D-loop Mitochondrial DNA	Organelle DNA	Variable region of mtDNA	Skin or muscle, blood, bone	PCR	No	Yes	Variability, pop structure and systematics	9,10,11,12,13,14,17

Interleukins	Protein	Variable immune response	Blood	PCR	Yes	No	Variability, pop structure and systematics	6
Major Histocompatibility Complex (MHC)	Genomic DNA	Variable immune response	Skin or muscle, blood, hair, teeth or bone	PCR	Yes	No	Health, variability, pop structure and systematics	15,16
Gene expression	RNA and cDNA	Variable gene response	Blood	PCR	Yes	No	Health	19,21
Next-generation sequencing	Genomic DNA	Regional variants	Skin or muscle, blood, hair, teeth or bone	PCR	Yes	No	Health, variability pop structure and systematics	Na
Whole genome sequencing	Genomic DNA	Regional variants	Skin or muscle, blood, hair, teeth or bone	PCR	Yes	No	Health	Na

[a]Co-D = co-dominant meaning that each genome has two copies of the region of interest, one from the sire and one from the dam.
[b]Neu = neutral meaning that the region is not in a protein coding region thus mutations persist and pass on to the next generation because they do not affect viability of the individual.
[c]Studies = Numbers refer to studies listed in Table 5.1 and Figure 5.1.

spanning the sea otters' range (Figure 5.1). Sample types used to date include blood, small skin biopsies or tissue plugs from live caught animals, muscle or skin from stranded dead animals, teeth, and bone (Table 5.2). Fresh tissue samples from skin or muscle preserved in alcohol or any other commercially available DNA preservative, or samples frozen immediately after removal from live or freshly dead animals, provide the greatest quantity of intact DNA. Skin or muscle samples from animals that are not freshly collected and preserved may still yield good quality DNA, although the best quality DNA is from those that are freshly sampled and preserved. Following centrifugation of blood samples, high-quality DNA can be extracted from white blood cells ("buffy coat" layer in centrifuged blood samples), which are the only blood cells in mammals that have nuclear DNA. Fecal DNA, commonly employed in other wildlife genetics studies because it can be sampled non-invasively, has not yet been employed in sea otter genetics studies because of the difficulty in collecting fecal samples from free-ranging marine mammals living in water. Finally, DNA can also be harvested from remains of sea otters hundreds of years old by drilling for tissue samples from teeth and bones. The quality of this DNA is generally degraded and amplification of many larger and desired regions in the genome is generally poor with the methods used to date. However, these difficulties are being overcome with newer techniques that may stimulate further work using ancient sea otter samples (Dabney et al., 2013).

METHODS

Nuclear

Both proteins (the product of DNA via gene expression) and DNA have been used to measure population genetics in sea otters. Protein variation was studied early in the history of the discipline of conservation genetics. Allozymes are proteins used in the early 1990s to study genetic variation and genetic differences between sea otter populations, specifically between northern and southern sea otters (Lidicker and McCollum, 1997; Rotterman, 1992).

DNA-based methods quickly became the preferred approach for studying population genetics in the mid-1990s. Those that have been employed in sea otter studies include restriction fragment-length polymorphism (RFLP), microsatellites, mtDNA, and gene expression. RFLP works by cutting DNA with a restriction enzyme, resulting in different lengths of DNA fragments based on their different sequences. Most commonly, the most variable region within the mtDNA, the D-loop, is cut with restriction enzymes to determine genetic variation and differences between northern and southern sea otters (Cheney, 1995; Cronin et al., 1996; Sanchez, 1992).

Microsatellites are nuclear markers that gained popularity after restriction enzymes during the 1990s because they are highly variable and abundant within the genome (Bentzen et al., 1991; Park and Moran, 1994; Wright and Bentzen, 1994). Microsatellites are stretches of DNA with repeating units of 1–6 base pairs (bp), are assumed to be genetically neutral and variable, and are easily amplified and assayed using polymerase chain reaction (PCR), a procedure performed easily in any modern molecular laboratory. Several sea otter population genetics studies have employed this molecular marker and it remains the most popular to measure genetic diversity and differences within and among populations (Aguilar et al., 2008; Cronin et al., 2002; Larson et al., 2002a,b, 2009, 2012; Table 5.1).

Gene expression involves conversion of the information encoded in a gene into messenger ribonucleic acid (mRNA) and then into a protein. Variability is measured by the amounts of mRNA and, indirectly, the protein being expressed. This is a relatively new and growing technology with only a few recent studies published to date (Bowen et al., 2012; Miles et al., 2012; Chapter 6).

Mitochondrial DNA

Mitochondrial DNA is a useful tool for surveying genetic variation and structure within and among populations (Avise, 1994; O'Reilly and Wright, 1995). MtDNA is a small organelle genome, composed of dozens of genes in a tightly organized structure with few non-coding regions, and, like microsatellites, can be used as a neutral marker. Unlike the single set of two copies of the genomic DNA in the nucleus (one from the sire and one from the dam), mtDNA exists in thousands of copies per cell within the mitochondria organelles (Wiesner et al., 1992). One non-coding region that evolves rapidly and is useful for measuring genetic variation and population structure is the D-loop or control region (Avise, 1994; Cheney, 1995). MtDNA is often used to track maternal lineages because, in most mammalian cases, it is passed on directly from mother to offspring in the fertilized egg and thus represents maternal genetics only (Avise, 1994). Several genetic studies have used mtDNA to describe sea otter diversity, population differences, and evolutionary relationships (Bodkin et al., 1999; Gorbics and Bodkin, 2001; Larson et al., 2002a; Koepfli and Wayne, 1998; Valentine et al., 2008; Table 5.1).

Adaptive Genetics

Genetic studies can not only infer indirect fitness through the measurement of genetic diversity but also provide valuable insights on how well sea otters and other species may adapt to environmental challenges, such as exposure to pathogens and toxins. One type of adaptive genetics focuses on the examination of sections of the genome associated with immune function and how sea otter genes respond to fight threats to their health. Immunocompetence is an organism's ability to resist infection and comprises two main functions: innate immunity and adaptive immunity (Hughes, 2002). Each of these functions of the sea otter's immunity is genetically regulated. Innate immunity is the first line of defense against attacking pathogens or exposure to toxins, and includes pre-existing nonspecific defense mechanisms that are ready to work without prior activation (Kaufmann and Kabelitz, 2010). One example of innate immunity is the inside of the mouth which contains enzymes to degrade and kill bacteria and viruses (Sugawara et al., 2002). Interleukins are a group of molecules that function within the innate immune response. They may be useful indicators of acute inflammation due to invasion of a pathogen or toxin and can be measured using adaptive genetics techniques. For example, King et al. (1996) described the genetic sequence and inferred protein of interleukin-6 in southern sea otters and how it related to and differed from that sequence found in other marine mammals.

Adaptive immunity is engaged if the invading agent is not inactivated or killed by innate mechanisms. This type of immunity can be specific to the invading agent and can have memory. Specificity is enabled through recognition of particular proteins or "antigens" and memory is developed so that a second exposure to the same invading agent will elicit a rapid response of antigen-specific cells to destroy the invader. Development of antibodies via vaccination against a disease organism is one example of the adaptive immune system. Examination of genes of the MHC was one of the first measurements of adaptive genetics in sea otters and provides a valuable tool for evaluating the immune vigor and potential plasticity of populations. The MHC genes are generally categorized into three groups: Class I, II, and III (Avise, 1994; Bowen et al., 2006). Expression of MHC genes is influenced by environmental change and exposure to invaders or disease vectors. Evaluation of MHC genes is through direct sequencing of genomic DNA and through indirect measurement of mRNA. Variability within this region has been measured in sea otters and may provide a relevant measure of immune potential and gene expression within individuals (Aguilar et al., 2008; Bowen et al., 2006).

SEA OTTER GENETIC RESEARCH FINDINGS

Genetic Diversity

Long before many population genetics methods were commonly available to study genetic diversity, conservation biologists were already engaging in theoretical work to examine the potential loss of genetic variation within sea otters due to the fur trade. Ralls et al. (1983) used mathematical models based on effective population size and natural history data (sex ratio, generation length, etc.) to estimate the loss of genetic diversity within the southern sea otter population following the fur trade. Their models assumed that the effective breeding population size, or N_E, was 27% of the total population size, there was a breeding ratio of one male to five females, and there were 50% juveniles within the population. They concluded that, based on these parameters, sea otters would have retained about 77% of their original diversity, measured as genetic heterozygosity, following the fur trade. This estimate was corroborated almost 30 years later when Larson et al. (2012) measured genetic diversity within southern sea otters before and after the fur trade and found that modern southern sea otters had retained 72% of their original genetic diversity.

Genetic diversity within sea otters has been measured using many methods and was first measured using allozymes, RFLP, and mtDNA; all methods revealed low genetic diversity (Cheney, 1995; Cronin et al., 1996; Lidicker and McCollum, 1997; Rotterman, 1992; Sanchez, 1992; Table 5.1). Later, microsatellites were used in repeated studies to demonstrate low genetic variation in both remnant and translocated populations, with expected heterozygosities ranging between 41% and 57% (Aguilar et al., 2008; Cronin et al., 2002; Larson et al., 2002a). These levels were comparable to other mammalian species that had experienced severe population bottlenecks or reductions (Garner et al., 2005; DiBattista, 2008). Aguilar et al. (2008), using microsatellite data, found relatively low genetic diversity within sea otters and suggested that there might have been a genetic bottleneck that pre-dated the fur trade. However, this result remains in question, as a recent re-evaluation of the statistical methods used to determine whether or not a population bottleneck occurred questioned the ability of those methods to detect true bottlenecks (Peery et al., 2012).

The finding of low genetic diversity within modern sea otters, using multiple methods, stimulated further work to determine the potential cause and answer the following question: Did sea otters historically have low genetic diversity from multiple bottlenecks as suggested by Aguilar et al. (2008) or was this a result of fur trade extirpations? In an attempt to answer this, Larson et al. (2002b) used microsatellite data taken from 40 sea otter bone fragments approximately 400 years old from the now-extinct Washington sea otter population found in the Makah tribe midden remains. The genetic diversity found within this one pre-fur trade population was compared to the microsatellite diversity measured within five modern sea otter populations throughout the range sampled in the 1990s. The extinct Washington population was found to have 82% heterozygosity, twice that of any measured modern sea otter population and well within the range expected for mammalian populations that have not experienced severe bottlenecks (DiBattista, 2008; Frankham et al., 2011; Garner et al., 2005). Larson et al. (2012) later expanded the work to include five pre-fur trade sea otter populations throughout the sea otter's historical range. They found similar high levels of diversity, averaging 77% heterozygosity, among the five sampled pre-fur trade populations.

Many studies on sea otter genetic diversity to date agree that large amounts of diversity have been lost within sea otter populations. However, much has been retained, up to 72% heterozygosity in California alone and an average of 58% retained over all populations sampled before and after the fur trade (Larson et al., 2012). It should be noted that population bottlenecks generally are believed to have a greater effect on allelic diversity than on heterozygosity (Allendorf, 1986). To that end it has also been demonstrated that modern sea otter populations contain significantly fewer alleles (average of 4.4) within all loci than that found prior to the fur trade (average of 19.8 alleles per locus, Larson et al., 2012).

Maintaining and recovering genetic diversity, specifically adaptive genetic diversity, such as variation found in the MHC, is thought to be important for the long-term health of sea otter populations. The first analysis of MHC diversity in sea otters, published by Bowen et al. (2006), involved examination of several MHC class II genes. Their findings of geographic variation in MHC genes arose from a gene expression analysis using mRNA taken from white blood cells from three individual sea otters. While the study was limited in sample size, the analysis of one of the MHC genes suggested the possible existence of geographically distinct groups. Aguilar et al. (2008) also examined genomic DNA from 70 sea otters to determine variation in three of the same MHC genes examined by Bowen et al. (2006). They employed direct sequence analysis of genomic DNA rather than the mRNA techniques that Bowen et al. (2006) used and found low levels of genetic variation within sea otter MHC. However, this genomic sequence analysis method did not provide information on gene expression and would not have detected the MHC variants that Bowen et al. (2006) found due to gene duplication (one of the important MHC mechanisms to generate allelic diversity in addition to mutational differences between alleles). Thus additional gene expression research was conducted by Bowen et al. (2012) and Miles et al. (2012), who found levels of MHC gene expression similar to that found in other species such as dogs and humans (see Chapter 6 for more information about gene expression). Thus even though sea otters have low levels of MHC variation, the variants that are there perform as expected based on the gene expression found in similar species with more genetic variation.

Population Structure

The initial sea otter population genetic studies conducted in the 1990s focused on evaluating population structure and taxonomic (specifically subspecific) differences (Bodkin et al., 1999; Cheney, 1995; Cronin et al., 1996, 2002; Gorbics and Bodkin, 2001; King et al., 1996; Lidicker and McCollum, 1997; Rotterman, 1992; Sanchez, 1992; Valentine et al., 2008). Lidicker and McCollum (1997) found low genetic variability within sea otters with only five of 30 allozymes variable and no allozymes specific to subspecies, thus rejecting valid subspecific differences using this marker. Rotterman's (1992) allozyme analysis also found low genetic variability (only three of 41 allozymes variable) among northern and southern sea otter populations, again with little difference between subspecies, and came to the same conclusion as Lidicker and McCollum. However, because of low variability at these allozyme loci, this method had little discriminating power to determine differences between groups.

At around the same time as these allozyme studies, restriction fragment-length polymorphism (RFLP) analysis of mtDNA became widely available and was used to further investigate genetic differences between northern and southern sea otters (Cronin et al., 1996; Sanchez, 1992). Unlike the allozyme analyses, the distributions of the mtDNA variations observed in these studies supported the subspecies earlier suggested by skull morphometrics (Wilson et al., 1991), in that unique versions of the mtDNA restriction fragments were found in each population. Later, in the mid-1990s, D-loop mtDNA sequences were tested for subspecific differences between southern and northern sea otters (Cheney, 1995). This study found relatively small genetic differences between areas, providing only weak support for these two subspecies. Clearly more genetic work could be done to determine if genetic differences support the current taxonomy.

Over 100 years after the end of the fur trade, after both international and national protection, after multiple translocations to repopulate areas and other conservation efforts, most sea otter populations today remain too geographically isolated, preventing substantial genetic mixing. Statistical analyses to measure potential gene flow and genetic differences among sampled populations have documented both similarities and differences among groups. As expected, translocated populations that were founded with animals from the same populations are genetically similar to and less differentiated from each other than the isolated remnant populations. For example, sea otters translocated to sites in southeast Alaska, Vancouver Island, and Washington all were founded in part or wholly from the Amchitka population and thus share Amchitka genetic signatures (Larson et al., 2002a). Within modern sea otters, mean genetic differences or microsatellite G_{ST} were found to be 0.05 (relatively low) for pairs of populations related by translocation and 0.18 (intermediate) for remnant groups. Among the three measured remnant populations (California, Prince William Sound, and Amchitka), California was the most divergent, followed by Prince William Sound with larger pairwise G_{ST} values above 0.20 (Larson et al., 2002a and Larson, unpublished data).

Most modern sea otter populations are thought to have moderate genetic structuring. To determine if these differences are artifacts of the fur trade, Larson et al. (2012) investigated population differences and gene flow estimates among five pre-fur trade populations and compared them to those found in five modern populations. They found genetic differences among pre-fur trade populations comparable to and not statistically different from those found today. For example, the G_{ST} between OLDCA and OLDAK was 0.23, while the G_{ST} estimate between California and Prince William Sound was 0.29. However, caution should be employed when comparing the genetic differences between pre-fur trade and modern sea otter populations. The numbers of alleles found in the pre-fur trade populations are higher than the modern population but these are still likely an underestimate of the true number due to the degraded quality of the DNA found in ancient bone samples. Thus, some alleles may not have been measured and the data may underestimate genetic diversity and allelic richness, which may result in increased genetic differentiation values. Thus, genetic differentiation values from degraded DNA for pre-fur trade otters, such as F_{ST} and G_{ST}, are likely inflated and should be regarded with care when interpreting population structure.

Microsatellites and mtDNA have been widely used in modern sea otters to determine population structuring and significant stocks for management purposes. Bodkin et al. (1999), Gorbics and Bodkin (2001), and Cronin et al. (2002) used both microsatellite markers and the D-loop of mtDNA to examine stock structure within the northern sea otter. Gorbics and Bodkin (2001) and Cronin et al. (2002) defined three distinct northern sea otter stocks: a southwestern stock from Cook Inlet to Attu Island, a central stock from Cook Inletto Cape Yakataga, and a southeastern stock from Cape Yakataga to Cape Yakataga. The designation of these stocks proved to be critical for the management of the species in Alaska because of the decline of the southwestern stock in the 1990s, which led it to be listed as a threatened population under the ESA in 2005 (Burn and Doroff, 2005; Doroff et al., 2003; Estes et al., 1998; Gorbics and Bodkin, 2001).

Inbreeding and Genetic Problems in Small, Isolated Populations

What was the effect, if any, of the fur trade on sea otter population viability? There is a growing body of evidence linking fitness variables with genetic diversity. Small isolated populations can suffer from three genetic threats: (1) inbreeding or the breeding of genetically related individuals, (2) the gradual accumulation of new deleterious mutations, known as "mutational meltdown," and (3) reduced ability to adapt to environmental changes due to reduced genetic variation (Frankham, 2010; Frankham et al., 2011). Inbreeding results in a decrease in reproductive fitness, known as inbreeding depression, in most species of plants and animals (Frankham, 2010; Frankham et al., 2011; Ralls et al., 2013). Although inbreeding depression is likely to be the most serious genetic problem faced by small populations in the short term, diminished capacity to evolve in response to environmental change may be a serious problem in the long term. Unfortunately, inbreeding is unavoidable in small, closed populations such as the isolated remnant populations of sea otters following the fur trade that numbered between 10 and 100 individuals, and within translocated populations, where some animals may have died or dispersed, resulting in very small founding population sizes estimated to be as low as 10 in some areas (Bodkin et al., 1999).

The extent of inbreeding depression often depends upon environmental conditions (Bijlsma and Loescheke, 2011; Reed et al., 2012). Inbreeding depression in mammals is approximately seven times higher in presumably more stressful wild environments than under captive conditions (Crnokrak and Roff, 1999). A meta-analysis of 58 data sets on stress and inbreeding depression indicated that the magnitude of inbreeding depression increases with the magnitude of stress in a linear fashion (Fox and Reed, 2011). That is, there appears to be little or no effect on the magnitude of inbreeding depression when organisms are subjected to a mild stress, but when the stress increases, inbreeding depression becomes increasingly apparent. The genetic mechanisms responsible for inbreeding/stress interactions are poorly known. If the same genes are responsible in a variety of species and environments, general stress-coping mechanisms are likely involved, whereas if different groups of genes are responsible in various species and environments, different mechanisms are likely involved in different species (Fox and Reed, 2010). Recent work on fruit flies has found detectable molecular fingerprints of inbreeding on gene expression (Kristensen et al., 2006), but it is not known yet if these patterns are seen only in inbred animals.

The finding that inbreeding depression is more serious under stressful conditions has important implications for conservation biology, as natural environments are probably becoming increasingly stressful for many wild populations due to climate change, habitat degradation and loss, and pollution (Bijlsma and Loescheke, 2011; Frankham, 2005). Most studies on the relationship between stress and inbreeding depression have been conducted under laboratory conditions and may not be representative of more complicated natural environments where organisms are exposed to multiple stresses that vary over time. Thus, studies that measure both stress levels and inbreeding depression under natural conditions are critically important.

Stress response can be measured in several ways. Cortisol and corticosterone are adrenal stress-related hormones that become elevated in response to physical and psychological stressors (Gorbman et al., 1983). Inbreeding has been linked to increased corticosterone in humans and fruit flies (Dahlgard and Hoffman, 2000; Zlotogora, 1995). It is thought that inbred groups may suffer from a breakdown in the negative feedback control of circulating stress levels, maintaining rather than breaking down elevated levels of the stress hormone (Brodkin et al., 1998). Within sea otters, Larson et al. (2009) found a significant negative relationship between the stress hormone corticosterone and genetic diversity within both individuals and populations. They found that the populations with the lowest genetic diversity also had the highest circulating stress levels (California and Prince William Sound) and the populations with the highest genetic diversity had the lowest stress hormone levels (southeast Alaska, Washington, and Amchitka). New research on stress levels in sea otter populations with higher genetic diversity suggests that this significant correlation may have a threshold effect, in that above a certain level of genetic diversity the correlation between stress levels is no longer significant (Larson, unpublished data).

More work could be done to better understand the link between genetic diversity and stressful environments and their importance for sea otter population viability. Reed et al. (2012) proposed several hypotheses linking inbred population viability and environmental stress. Interactions of inbreeding and stressful environment may be reflected by measures at the cellular level such as increased heat shock protein gene expression in sea otters (Bowen et al., 2012; Chapter 6).

FUTURE WORK

Nuclear Variability and Genetic Fingerprinting

Sea otter conservation genetics have been profoundly influenced by the fur trade, which resulted in significant loss of genetic diversity. However, more work needs to be done with pre-fur trade samples using recently developed methods that make it possible to get higher-quality DNA and a broader sampling of the genome as well as a broader sampling of extinct populations (Dabney et al., 2013). A study based on high numbers of loci, such as with large microsatellite panels and thousands of single nucleotide polymorphisms revealed by next-generation sequencing, might give a different picture than the existing studies (Larson et al., 2002b, 2012).

The maintenance of genetic diversity remaining within sea otter populations is thought to be crucial for their long-term survival. The translocations have been very successful from the standpoint of repopulating previously occupied areas and making up a large portion of the sea otters alive today (Chapter 3) but more work is needed on the potential effects on diversity and fitness from mixing genetically distinct stocks to recover genetic variation. Evidence that the mixing of genetically distinct populations resulted in healthy translocated groups of sea otters is provided by the successful translocations within southeast Alaska, British Columbia, and Washington during the 1960s and the 1970s (Bodkin et al., 1999; Jameson et al., 1982). These translocated populations today have some of the highest genetic diversity and growth rates of any measured modern sea otter population (Bodkin et al., 1999; Larson et al., 2009). In addition, there is some evidence that the translocated populations in Washington and British Columbia may be mixing, resulting in some of the highest genetic diversity measured in modern sea otters (Larson, unpublished data). It is important to note that genetic diversity is only one of the many factors, such as habitat quality and dispersal of translocated animals, that affect the ultimate success of translocated populations.

Unlike humans and domestic animals with documented genealogies, wildlife do not provide us with easy access to information on their relatives. Genetic fingerprinting data, such as that generated by microsatellites, may allow reconstruction of sea otter family trees. Genetic data can be used to assess the validity of such relationships through calculations such as probability of identity (Waits et al., 2001), identity by descent kinship indices (Blouin, 2003), and similar measures. These metrics inform researchers how many molecular markers (typically microsatellites) are necessary for each study and the statistical reliability of particular kinship relations. Once relatedness information is assembled through genetic analyses, familial associations with disease, behavior, and other attributes can be assessed (DeWoody, 2005).

Sea otter dispersal or movement can also be evaluated using population genetic assignment techniques (Paetkau et al., 2004). For example, the DNA samples of particular sea otters can be tested to determine whether they are genetically similar to other sea otters in the region where they were captured, or whether they have the DNA signature of a more distant population, suggestive of a long-distance movement. This work has already been performed using microsatellites and assignment tests on lone sea otters found on the Oregon coast, where there is no current sea otter population, to determine where a particular animal came from—California or Washington—with two found to be from Washington and one from California (Larson, unpublished data). As sea otter populations continue to grow and individuals continue to disperse and migrate over long distances, construction of DNA databases for large numbers of sea otters across their range would greatly enhance this work.

In an analogous application, future work involving development of DNA data sets may facilitate forensic evaluation of illegal activities involving sea otters, including poaching and the selling of illegally acquired artifacts (Alacs et al., 2010). Forensic population assignment tests could compare evidence material (such as a confiscated sea otter pelt) and the genetics from potential source populations. Analogous cases in wildlife and domestic animals involve animal abuse cases, poaching, and dog attacks on endangered species (Ernest, unpublished data).

Sequencing

Whole genome analysis is an exciting next step for sea otter conservation genetics. At the time of this publication, no complete sequence of the sea otter genome has been accomplished. The number of species that have whole genome sequences has increased rapidly (Genomes Online Database, US Department of Energy Joint Genome Institute). Whole genome, next-generation DNA sequence, and expressed sequence (mRNA sequence) data will provide numerous tools for evaluating sea otter health (Ryder, 2005; Primer, 2009). Whole genome sequence can provide avenues for development of much larger numbers of molecular markers and expand abilities (such as with ancient samples that previously did not yield adequate DNA levels). Perhaps most importantly, whole genome analyses can allow identification and conservation of functionally relevant or adaptive genetic variations.

Next-generation sequencing (NGS) methods have opened vast new avenues of genomic discovery for DNA and RNA sequencing. These methods typically involve breaking DNA into millions of smaller, more manageable fragments. Adaptors and "DNA barcodes" as molecular labels are bound to each of those fragments. The adaptors allow attachment to a surface that looks like a microscope slide. Millions or more of sequencing reactions are performed microscopically at each of those fragments simultaneously (Metzker, 2010). Data analysis on these huge data sets is conducted using specialized bioinformatics software and supercomputing facilities (Casci, 2012).

RNA sequencing is a procedure that uses next-generation sequencing technologies to sequence the RNA molecules within a biological sample in order to determine the sequence and expression (indicated by relative abundance) of each gene and thus RNA type (Romero et al., 2012). Many traits of importance for sea otter conservation, including disease susceptibility, involve many interacting genes rather than single genes (Bowen et al., 2012; Miles et al., 2012). Thus, the high-throughput capabilities of these newer technologies allow cost-effective assessment of expression at many genes of interest.

CONCLUSION

Sea otters provide an excellent example of how the principles of conservation genetics can inform conservation management decisions. They also provide a classic case study for the implementation of conservation genetics tools after severe and multiple population bottlenecks, which is sadly the case within many targeted, threatened, and endangered wildlife populations today. The lessons learned from sea otter population genetics may be adapted to any similar wildlife population. Key lessons are the necessity to use multiple methods, both neutral and adaptive, to assess population status and to determine the effects that historical patterns of abundance may have on current population viability. The genetic tools described here may be used to determine genetic health and to inform managers when diversity is recovered through mixing of populations, genetic rescue, or natural emigration.

New methods such as whole genome sequencing and analysis of gene expression will be critical to the understanding of the importance of certain areas of the genome for future sea otter and wildlife population viability in the face of changing environments. Our understanding of sea otter genetics and its importance for long-term conservation is impressive but still rudimentary. Work focusing on understanding the whole sea otter genome, the role of adaptive genetics in a changing environment, and the potential need to restore lost genetic diversity through future translocations may inform appropriate strategies for ensuring long-term sea otter conservation.

REFERENCES

Aguilar, A., Jessup, D.A., Estes, J., Garza, J.C., 2008. The distribution of nuclear genetic variation and historical demography of sea otters. Anim. Conserv. 11, 35–45. Available from: http://dx.doi.org/doi:10.1111/j.1469-1795.2007.00144.x.

Alacs, E.A., Georges, A., FitzSimmons, N.N., Robertson, J., 2010. DNA detective: a review of molecular approaches to wildlife forensics. Forensic Sci. Med. Pathol. 6, 180–194.

Allendorf, F., 1986. Genetic drift and the loss of alleles versus heterozygosity. Zoo. Biol. 5, 181–190.

Allendorf, F.W., Luikart, G., Aitken, S.N., 2013. Conservation and the Genetics of Populations. Wiley-Blackwell, Hoboken, NJ.

Avise, J.C., 1994. Molecular Markers, Natural History and Evolution. Chapman and Hall, New York, NY.

Bentzen, P., Harris, A.S., Wright, J.M., 1991. Cloning of hypervariable minisatellite and simple sequence microsatellite repeats for DNA fingerprinting of important aquacultural species and salmonids and tilapia. In: Burke, T., Dolf, G., Jeffreys, A.L., Wolf, R. (Eds.), DNA Fingerprinting: Approaches and Applications. Basel, Switzerland, pp. 243–262.

Bijlsma, R., Loescheke, V., 2011. Genetic erosion impedes adaptive responses to stressful environments. Evol. Appl. 5, 117–129.

Blouin, M.S., 2003. DNA-based methods for pedigree reconstruction and kinship analysis in natural populations. Trends Ecol. Evol. 18, 503–511.

Bodkin, J.L., Ballachey, B.E., Cronin, M.A., Scribner, K.T., 1999. Population demographics and genetic diversity in remnant and translocated populations of sea otters. Conserv. Biol. 13, 1378–1385.

Bowen, L., Aldridge, B., Miles, A.K., Stott, J.L., 2006. Partial characterization of MHC genes in geographically disparate populations of sea otters (*Enhydra lutris*). Tissue Antigens 67 (5), 402–408.

Bowen, L., Miles, A.K., Murray, M., Haulena, M., Tuttle, J., Van Bonn, W., et al., 2012. Gene transcription in sea otters (*Enhydra lutris*); development of a diagnostic tool for sea otter and ecosystem health. Mol. Ecol. Resour. 12, 67–74.

Brodkin, E.S., Carlezon Jr., W.A., Haile, C.N., Kosten, T.A., Heninger, G.R., Nestler, E.J., 1998. Genetic analysis of behavioral, neuroendocrine, and biochemical parameters in inbred rodents: initial studies in Lewis and Fisher 344 rats and in A/J and C57BL/6J mice. Brain Res. 805 (1–2), 55–68.

Burn, D., Doroff, A., 2005. Endangered and threatened wildlife and plants; determination of threatened status for the southwest Alaska distinct population segment of the northern sea otter (*Enhydra lutris kenyoni*). Fed. Regist. 70, 46366.

Casci, T., 2012. Bioinformatics: next-generation genomics. Nat. Rev. Genet. 13, 378–379. Available from: http://dx.doi.org/doi:10.1038/nrg3250.

Cheney, L.C., 1995. An Assessment of Genetic Variation within and between Sea Otter (*Enhydra lutris*) Populations Off Alaska and California. Masters thesis, Moss Landing Marine Laboratories. California State University, San Jose, CA.

Crnokrak, P., Roff, D.A., 1999. Inbreeding depression in the wild. Heredity 83, 260–270.

Cronin, M.A., Bodkin, J.L., Ballachey, B.E., Estes, J.A., Patton, J.C., 1996. Mitochondrial DNA variation among subspecies and populations of sea otters (*Enhydra lutris*). J. Mammal 77, 546–557.

Cronin, M.A., Jack, L., Buchholz, W.G., 2002. Microsatellite DNA and mitochondrial DNA variation in Alaskan sea otters. Final report prepared for US Fish and Wildlife Service by LGL Alaska Research Associates.

Culley, T.M., Wallace, L.E., Gengler-Nowak, K.M., Crawford, D.J., 2002. A comparison of two methods of calculating GST, a genetic measure of population differentiation. Am. J. Bot. 89 (3), 460–465.

Dabney, J., Knapp, M., Glocke, I., Gansauge, M.T., Weihmann, A., Nickel, B., et al., 2013. Complete mitochondrial genome sequence of a Middle Pleistocene cave bear reconstructed from ultrashort DNA fragments. PNAS, 110 (39), 15758–15763. Available from: http://dx.doi.org/10.1073/pnas.1314445110.

Dahlgard, J., Hoffman, A.A., 2000. Stress resistance and environmental dependency of inbreeding depression in *Drosophila melanogaster*. Conserv. Biol. 14, 1187–1193.

DeWoody, J.A., 2005. Molecular approaches to the study of parentage, relatedness, and fitness: practical applications for wild animals. J. Wildlife Manage. 69, 1400–1418.

DiBattista, J.D., 2008. Patterns of genetic variation in anthropogenically impacted populations. Conserv. Genet. 9, 141–156.

Doroff, A.M., Estes, J.A., Tinker, M.T., Burn, D.M., Evans, T.J., 2003. Sea otter population declines in the Aleutian Archipelago. J. Mammal 84, 55–64.

Estes, J.A., Tinker, M.T., Williams, T.M., Doak, D.F., 1998. Killer whale predation on sea otters linking oceanic and nearshore ecosystems. Science 282, 473–476.

Fox, C., Reed, D.H., 2010. Inbreeding depression increases with environmental stress: an experimental study and meta-analysis. Evolution 65, 246–258.

Fox, C.W., Reed, D.H., 2011. Inbreeding depression increases with environmental stress: an experimental study and meta-analysis. Evolution 65, 246–258.

Frankham, R., 1995. Conservation genetics. Annu. Rev. Genet. 29, 305–327.

Frankham, R., 2005. Stress and adaptation in conservation genetics. J. Evol. Biol. 18, 750–755.

Frankham, R., 2010. Challenges and opportunities of genetic approaches to biological conservation. Biol. Conserv. 143, 1919–1927.

Frankham, R., Ballou, J.D., Briscoe, D.A., 2011. Introduction to Conservation Genetics, second ed. Cambridge University Press, Cambridge, UK.

Funk, W.C., McKay, J.K., Hohenlohe, P.A., Allendorf, F.W., 2012. Harnessing genomics for delineating conservation units. Trends Ecol. Evol. 27, 489–495.

Garner, A., Rachlow, J.L., Hicks, J.E., 2005. Patterns of genetic diversity and its loss in mammalian populations. Conserv. Biol. 19, 1215–1221.

Gorbics, C.S., Bodkin, J.L., 2001. Stock structure of sea otters (*Enhydra lutris kenyoni*) in Alaska. Mar. Mammal Sci. 17 (3), 632–647. Available from: http://dx.doi.org/doi:10.1111/j.1748-7692.2001.tb01009.x.

Gorbman, A., Dickhoff, W.W., Vigna, S.R., Clark, N.B., Ralph, C.L., 1983. Comparative Endocrinology, second ed. John Wiley and Sons, Inc., New York, NY.

Hughes, A., 2002. Evolution of the host defense system. In: Kaufmann, S.H.E., Sher, A., Ahmed, R. (Eds.), Immunology of Infectious Diseases. ASM Press, Washington, DC, pp. 67–78.

Jameson, R.J., Kenyon, K.W., Johnson, A.M., Wright, H.M., 1982. History and status of translocated sea otter populations in North America. Wildl. Soc. Bull. 10, 100–107.

Kaufmann, S.H.E., Kabelitz, D., 2010. 5. Proteomic approaches to study immunity in infection. In: Immunology of Infection, Academic Press, London, pp. 116–130.

King, D.P., Schrenzel, M.D., McKnight, M.L., Reidarson, T.H., Hanni, K.D., Stott, J.L., et al., 1996. Molecular cloning and sequencing of interleukin 6 cDNA fragments from the harbor seal (*Phoca vitulina*), killer whale (*Orcinus orca*), and Southern sea otter (*Enhydra lutris nereis*). Immunogenetics 43 (4), 190–195. Available from: http://dx.doi.org/doi:10.1007/s002510050045.

Kirk, H., Freeland, J.R., 2011. Applications and implications of neutral versus non-neutral markers in molecular ecology. Int. J. Mol. Sci. 12, 3966–3988.

Koepfli, K.P., Wayne, R.K., 1998. Phylogenetic relationships of otters (Carnivora: Mustelidae) based on mitochondrial cytochrome b sequences. J. Zool. 246 (4), 401–416.

Kristensen, T.N., Sorensen, P., Kruhoffer, M., Loeschcke, V.T., 2006. Inbreeding by environmental interactions affect gene expression in *Drosophila melanogaster*. Genetics 173, 1329–1336.

Larson, S.E., Jameson, R., Bodkin, J., Staedler, M., Bentzen, P., 2002a. Microsatellite and mitochondrial DNA variation in remnant and translocated sea otter (*Enhydra lutris*) populations. J. Mammal 83, 893–906.

Larson, S.E., Jameson, R., Etnier, M., Fleming, M., Bentzen, P., 2002b. Loss of genetic diversity in sea otters (*Enhydra lutris*) associated with the fur trade of the 18th and 19th centuries. Mol. Ecol. 11, 1899–1903.

Larson, S.E., Monson, D., Ballachey, B., Jameson, R., Wasser, S.K., 2009. Stress-related hormones and genetic diversity in sea otters (*Enhydra lutris*). Mar. Mammal Sci. 25 (2), 351–372.

Larson, S.E., Jameson, R., Etnier, M., Jones, T., Hall, R., 2012. Genetic diversity and population parameters of sea otters, *Enhydra lutris*, before Fur Trade Extirpation from 1741–1911. PLoS ONE 7 (3), e32205. Available from: http://dx.doi.org/doi:10.1371/journal.pone.0032205.

Lidicker, W.Z., McCollum, F.C., 1997. Genetic variation in California sea otters. J. Mammal 78, 417–425.

Lynch, M., 1996. A quantitative-genetic perspective on conservation issues. In: Avise, J.C., Hamrick, J.L. (Eds.), Conservation Genetics. Case Histories from Nature. Chapman and Hall, New York, NY, pp. 471–501.

Metzker, M.L., 2010. Sequencing technologies—the next generation. Nat. Rev. Genet. 11, 31–46. Available from: http://dx.doi.org/doi:10.1038/nrg2626.

Miles, A.K., Bowen, L., Ballachey, B., Bodkin, J.L., Murray, M., Keister, R.A., et al., 2012. Variations of transcript profiles between sea otters (*Enhydra lutris*) from Prince William Sound, Alaska, and clinically normal reference otters. Mar. Ecol. Prog. Ser. 451, 201–212.

Nei, M., 1972. Genetic distance between populations. Am. Nat. 106, 283–292.

O'Reilly, P., Wright, J.M., 1995. The evolving technology of DNA fingerprinting and its application to fisheries and aquaculture. J. Fish. Biol. 47, 29–55.

Paetkau, D., Slade, R., Burden, M., Estoup, A., 2004. Genetic assignment methods for the direct, real-time estimation of migration rate: a simulation-based exploration of accuracy and power. Mol. Ecol. 13, 55–65.

Park, L.K., Moran, P., 1994. Developments in molecular genetic techniques in fisheries. Rev. Fish Biol. Fisheries 4, 272–299.

Peery, M.Z., Kirby, R., Reid, B.N., Stoelting, R., Doucet-Beer, E., Robinson, S., et al., 2012. Reliability of genetic bottleneck tests for detecting recent population declines. Mol. Ecol. 21, 3403–3418.

Primer, C.R., 2009. From conservation genetics to conservation genomics. Ann. NY Acad. Sci. 1162, 357–368.

Pritchard, J.K., Stephens, M., Donnelly, P., 2000. Inference of population structure using multilocus genotype data. Genetics 155, 945–995.

Ralls, K., Ballou, J., Brownell Jr., R.L., 1983. Genetic diversity in California sea otters: theoretical considerations and management implications. Biol. Conserv. 25, 209–232.

Ralls, K., Frankham, R., Ballou, J.D., 2013. Inbreeding and outbreeding. In: second ed. Levin, S.A. (Ed.), Encyclopedia of Biodiversity, vol. 4. Academic Press, Waltham, MA, pp. 245–252.

Rathbun, G.B., Hatfield, B.B., Murphey, T.G., 2000. Status of translocated sea otters at San Nicolas Island. Southwest. Nat. 45 (3), 322–328.

Reed, D.H., Fox, C.W., Enders, L.S., Kristensen, T.N., 2012. Inbreeding-stress interactions: evolutionary and conservation consequences. Ann. NY Acad. Sci. 1256, 33–48.

Romero, I.G., Ruvinsky, I., Gilad, Y., 2012. Comparative studies of gene expression and the evolution of gene regulation. Nat. Rev. Genet. 13, 505–516.

Rotterman, L.M., 1992. Patterns of genetic variability in sea otters after severe population subdivision and reduction. Ph.D. dissertation, University of Minnesota, Minneapolis, Minnesota, 227 pp.

Ryder, O.A., 2005. Conservation genomics: applying whole genome studies to species conservation efforts. Cytogenet. Genome Res. 108, 6–15.

Sanchez, M.S., 1992. Differentiation and Variability of Mitochondrial DNA in Three Sea Otter, *Enhydra lutris*, Populations. M.S. thesis, University of California, Santa Cruz, CA, 101 pp.

Scribner, K.T., Bodkin, J., Ballachey, B., Fain, S.R., Cronin, M.A., Sanchez, M., 1997. Population genetic studies of the sea otter (*Enhydra lutris*): a review and interpretation of available data. In: Dizon, A.E., Chivers, S.J., Perrin, W.F. (Eds.), Molecular Genetics of Marine Mammals. Society for Marine Mammalogy Special Publication 3. Allen Press, Lawrence, KS, pp. 197–208.

Simenstead, C.A., Estes, J.A., Kenyon, K.W., 1978. Aleuts, sea otters, and alternate stable state communities. Science 200, 403–411.

Sugawara, S., Uehara, A., Tamai, R., Haruhiko, T., 2002. Innate immune responses in oral mucosa. J. Endotoxin Res. 8, 465–468.

Valentine, K., Duffield, D.A., Patrick, L.E., Hatch, D.R., Butler, V.L., 2008. Ancient DNA reveals genotypic relationships among Oregon populations of the sea otter (*Enhydra lutris*). Conserv. Genet. (4), 933–938.

Waits, L.P., Luikart, G., Taberlet, P., 2001. Estimating the probability of identity among genotypes in natural populations: cautions and guidelines. Mol. Ecol. 10, 249–256.

Weir, B.S., Cockerham, C.C., 1984. Estimating F-statistics for the analysis of population structure. Evolution 38 (6), 1358–1370.

Wiesner, R.J., Rüegg, J.C., Morano, I., 1992. Counting target molecules by exponential polymerase chain reaction: copy number of mitochondrial DNA in rat tissues. Biochem. Biophys. Res. Commun. 183, 553–559.

Wilson, D.E., Bogan, M.A., Brownell Jr., R.L., Burdin, A.M., Maminov, M.K., 1991. Geographic variation in sea otters, *Enhydra lutris*. J. Mammal 72, 22–36.

Wright, J.M., Bentzen, P., 1994. Microsatellites: genetic markers for the future. Rev. Fish Biol. Fisheries 4, 384–388.

Zlotogora, J., 1995. Hereditary disorders among Iranian Jews. Am. J. Med. Genet. 58, 32–37.

Chapter 6

Evaluating the Status of Individuals and Populations: Advantages of Multiple Approaches and Time Scales

Daniel H. Monson[1] and Lizabeth Bowen[2]
[1]*US Geological Survey, Alaska Science Center, Anchorage, AK, USA,* [2]*US Geological Survey, Western Ecological Research Center, Davis Field Station, UC Davis, Davis, CA, USA*

Sea Otter Conservation. DOI: http://dx.doi.org/10.1016/B978-0-12-801402-8.00006-8

INTRODUCTION

What is "population status"? In conservation, "population status" has a number of definitions including (1) the current trajectory of the population—that is, growing, stable, or declining; (2) the legal status of the population—for example, "threatened" or "endangered" under the US Endangered Species Act; or (3) the population's probability of persistence—that is, extinction risk within a population viability analysis framework (PVA; e.g., Morris and Doak, 2002). These three definitions require a general sense of a population's abundance and trend but not necessarily an understanding of the factors controlling a population's trajectory. The population's trajectory is more precisely defined as a population's intrinsic rate of growth (*r*), or per capita births—per capita deaths. Intuitively, *r* appears to be the most relevant indicator of population status because it integrates all the regulator mechanisms controlling birth and death rates. However, we generally do not measure birth and death rates directly; instead, we estimate *r* from population trend. Because birth and death rates are influenced by a multitude of processes working at the level of the individual, it is not always obvious what factors are driving changes in *r* because population trend generally lags behind factors controlling demographic rates. Moreover, population trend by itself reveals little about the actual mechanisms controlling a population's dynamics, which is critical information to management and conservation (e.g., is birth rate or death rate the primary factor in a decline?). Finally, it can be difficult to estimate population size with the precision required to detect changes in population trends; thus, estimates of *r* may be highly uncertain.

A fourth definition of population status is based on the per capita availability of resources (e.g., food or space) to individuals in a given population. More precisely, population status in the ecological sense is the status relative to the carrying capacity of the environment (*K*) and is traditionally thought of as the amount of bottom-up pressure exerted on a population. In contrast to the population abundance and trend-based definitions listed above, the ecological-based definition of population status explicitly provides information on a major driver of a population's dynamics. Specifically, per capita resource availability is the primary mechanism behind density-dependent changes in population demographic rates and thus *r* (Caughley, 1970; Fowler, 1981, 1987; Skogland, 1985).

We expect that when food is plentiful individuals will (1) obtain resources quickly and efficiently, (2) be able to acquire surplus energy stores as a buffer against leaner times, and (3) grow and mature at their maximum potential rate. As a result, individuals will (4) expend relatively little effort on foraging, (5) have excellent body condition (defined as their relative amount of on-board energy stores), (6) have short offspring dependency periods and/or high weaning mass, (7) mature at a young age, and (8) attain their maximum potential adult body size at a young age. In addition, large average body size and excellent average body condition will lead to (9) high population level reproductive rates and success, (10) low mortality rates especially in young and prime-age animals. Finally, all these positive influences on demographic rates will result in (11) the population growing at its maximum intrinsic rate (r_{max}). As the population approaches K and food becomes scarce and harder to find, the intrinsic rate of population growth (r) declines as density-dependent changes in demographic rates occur via the cascade of effects beginning with the decline of an individual's foraging efficiency. The above predictions led to the recommendations that life history and demographic parameters be used as indicators of a population's status relative to K (Eberhardt, 1977a, b; Fowler, 1987). Likewise, the status of populations has more recently been used to examine the state of the environment itself (Boyd and Murray, 2001; Cairns, 1987; Piatt and Sydeman, 2007).

THE NORTH PACIFIC EXAMPLE

Estes (Chapter 2) and Bodkin (Chapter 3) provide details of the sea otter story in the North Pacific. In short, following the near extirpation of sea otters in the eighteenth and early nineteenth centuries, the population recovered in the twentieth century throughout much of its range via natural and translocation-aided population expansion (Kenyon, 1969; Lensink, 1960). This scenario provided a natural experiment that demonstrated the sea otter's role as a keystone species (sensu Paine (1969)) in the nearshore ecosystems that they occupy (Estes and Palmisano, 1974). Sea otters began a second precipitous decline in the western Aleutians in the late 1980s (Doroff et al., 2003), resulting in the listing of the southwestern Alaska stock of sea otters as threatened in 2005. Ecologists studying the decline found no evidence that lack of food resources was involved in the recent decline (Estes et al., 1998, 2004; Laidre et al., 2006) and birth rates were sufficient for population stability or growth (Monson et al., 2000b). Mortality factors such as disease, contaminants, and human harvest could not explain observations such as divergent population trends in open coast versus protected habitats, lack of carcasses on beaches, and changes in the distribution and behavior of surviving animals. Ultimately, a top-down driver was found to be the most parsimonious explanation for the decline with killer whale predation the most likely specific cause (Estes et al., 1998, 2004, 2009; Springer et al., 2003, 2008; Williams et al., 2004).

Two challenges presented by the second sea otter population decline are common in conservation. Specifically, (1) because of the lack of survey data and/or the imprecision of survey data, declines are often not detected until they are well under way, (2) resulting in difficulty understanding cause of declines because the responsible factors may no longer be acting by the time research to understand them begins. Consequently, controversy over the cause of the decline can arise (e.g., was it a top-down or bottom-up process?), delaying necessary and appropriate conservation actions. In these situations, measures of the post-decline population status and ideally some retrospective measure of pre-decline population status can be valuable in understanding the population's past dynamics. While we use an example of a population decline for illustration, population increases can also pose conservation challenges (e.g., current conflicts between the growing sea otter population in southeast Alaska and shellfish fisheries that developed during the period of sea otter absence), and measures of population status can also be useful in these situations.

To understand the change in sea otter abundance in the Aleutian Islands noted above, as well as the historic recovery of remnant and translocated populations of sea otter populations in the twentieth century, researchers needed to develop a suite of approaches to assess population status. Remnant and translocated populations remained geographically and demographically isolated for long periods because of the limited movement of sea otters. Thus, early in the recovery period populations were almost certainly not food limited, but became so as populations increased in abundance and fully reoccupied available habitat. This situation enabled the testing of various ways to evaluate the status of populations within and across populations where status was known with confidence.

In this chapter, we will describe and review the "tools" used in sea otter research to understand the status of populations. Most tools will focus on the ecological definition of population status (i.e., relative to *K*), and we refer to this as "PS." We will use the case study of sea otters in the North Pacific to illustrate the utility of various PS indices for understanding population dynamics. Specifically, we will discuss how understanding of PS informed our understanding of sea otter population trends including situations where a change in an index of PS could have foreshadowed a change in population trend. In addition, we will describe tools that can indicate the presence of top-down pressure or other challenges populations can face. The tools described fit into several broad categories including (1) energetic, (2) morphological, and (3) demographic. Most of these are traditional tools that have been used in many systems, although we also describe novel approaches to using some of these metrics. In addition, we will discuss a fourth category of emerging tools not yet employed in many other situations.

TOOLS OF THE TRADE

A variety of tools have been developed and tested with sea otters (and other species) to assist in evaluating the mechanisms underlying changes in abundance. As referenced above, the ecological definition of population status reflects the relative availability of food resources, and most tools are based on this ecological relation (Table 6.1). In addition to indices of PS that respond to food availability, other indices may reflect environmental

TABLE 6.1 Metrics Used to Assess Population Status in Sea Otters. Temporal Scale Will Depend on the Life History Parameters of a Species and for Sea Otters May Be Days or Weeks (i.e., Immediate to Short) or Months to Several Years (i.e., Mid to Long)

Metric	Temporal Scale of Integration	Sensitivity to Short-Term Fluctuations	
		↑ food	↓ food
Energy intake rate	Immediate	High	High
Total foraging time	Immediate	High	High
Body condition	Short	Mod	Mod
Growth rate	Long	Low	Low
Structural body size	Long "recorded"	Low	Low
Demographic rates			
1. Juvenile survival	Mid	High	High
2. Age of first reproduction	Mid	Mod	Mod
3. Age at weaning/weaning mass	Mid	Mod	Mod
4. Female reproductive rate	Mid	Low	Mod
5. Adult survival	Mid	Low	Mod
6. Female reproductive success	Mid	Mod	High
Reproductive synchrony	Mid	Low	High
Dietary diversity	Long	High	Mod
Community structure	Long	Low	Low
Spatial distribution	Immediate—Long "top-down"	Low	Mod
Gene transcription	Short—Long "tunable"	Variable	Variable

conditions independent of food. These include the emerging field of gene transcription, which can provide insight into challenges from the physical environment such as temperature, the chemical environment (e.g., contaminants), or the biological environment in terms of disease or parasites. Many of the ecologically based tools are not new and were first promoted by Eberhardt (1977a,b) and Fowler (1987). However, we will expand on the concepts behind each tool by focusing on the different temporal scales each index reflects. In addition, each index has its own functional response to changes in food availability (i.e., linear, nonlinear, threshold, e.g., Costa, 2007). Further, each index may respond symmetrically or asymmetrically to positive versus negative changes in food availability, or exhibit lag times between when a change in food resources occurs and when that change is manifest in a particular PS index.

When employing the various tools described below, researchers should also consider the spatial scale on which the population is acting as a unit. Sea otter populations can be demographically structured (i.e., consistent birth and death schedules) at relatively small spatial scales, perhaps as small as a few hundred square kilometers (Chapter 3) because of the limited movements and high fidelity to small home ranges or territories exhibited by sea otters. As a result, estimating PS at spatial scales larger than the scale of consistent demographic structuring could lead to false inference. Finally, we examine the spatial distribution of sea otters across their habitat and the community structure of the rocky reef systems they inhabit that reflect processes not necessarily related to food.

Energetic Tools

Energy Intake Rate

Food availability acts on individuals by modulating the rate at which they acquire caloric resources. Energy intake rates will constantly fluctuate at various time scales, but for an individual to remain healthy, it must maintain a positive or at least equilibrium energy balance over the long term. Apex predators in particular may experience boom and bust intake rates, with extended periods of negative energy balance followed by periods of extremely positive energy gain when prey are successfully captured (e.g., wolves and African lions that cannot capture prey at will) or available (e.g., whales and elephant seals that migrate between feeding and breeding grounds). Because energy intake can be so variable and sensitive to environmental fluctuations, analysis of data on energy intake rates must be considered over appropriate time scales to reflect biologically meaningful intake rates. For example, sea otters have high metabolic rates with little ability to store excess resources, so they must feed often (Costa and Kooyman, 1982; Yeates et al., 2007). Thus, energy intake scaled on a daily and even hourly basis may be appropriate. In contrast, a weekly energy intake rate may be more appropriate for predators such as the wolf or African lion, both of which can go days between successful kills. Regardless of scale, interpretation must consider seasonal or episodic fluctuations in environmental productivity and the current physiological demands on the individual. In particular, the demographic consequences of energy intake rates depend on the sex and reproductive status of individuals. For example, when resources are limited, seasonal peaks in prey quantity/quality and/or degree of other environmental challenges (e.g., thermal challenges) can be important for female reproductive success as mothers face the added energetic demands of lactation. In contrast, males may have higher energy intake rates in some areas during some seasons, but this will have minimal demographic

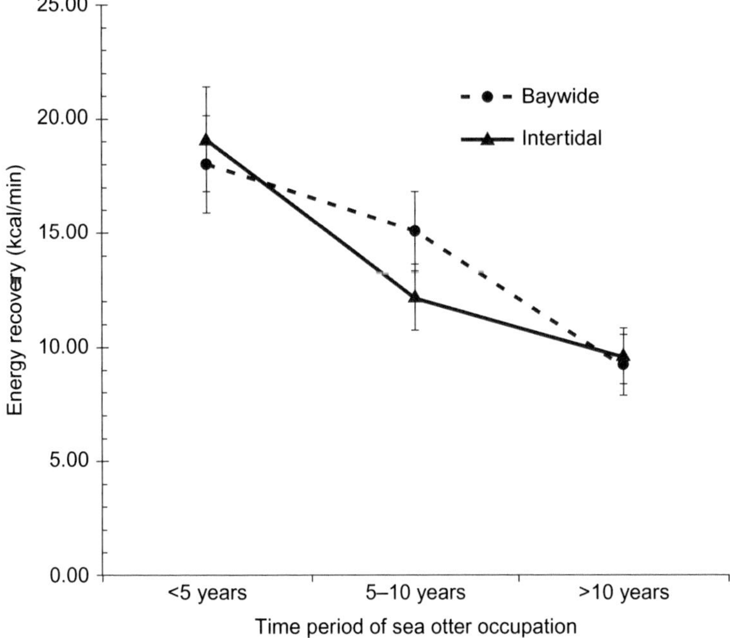

FIGURE 6.1 Energy intake rates relative to years of occupation by sea otters in Glacier Bay, Alaska. *(Figure from Weitzman (2013).)*

effects if reproductive females are energetically limited. In addition, limitations due to prey discovery, capture, handling, and processing time set the maximum energy intake rate, and increased food availability beyond a certain level will have no effect on the rate of energy acquisition.

Sea otters are unique among predators because not only are they relatively easy to watch forage, but their captured prey items are visible to observers for identification and size estimation (Calkins, 1978; Dean et al., 2002; Doroff and Bodkin, 1994; Doroff and DeGange, 1994; Tinker et al., 2008, 2012; Watt et al., 2000). Thus, we are able to make relatively precise estimates of energy intake rates from foraging characteristics and prey sizes (Dean et al., 2002; Chapter 10), and using energy intake as an index of PS appears meaningful (Figure 6.1). For more cryptic species, other technology-based metrics of energy intake rates have been developed such as stomach temperature recorders in pinnipeds (Kuhn et al., 2009) or acoustic surveys for killer whales near fur seal rookeries (Newman and Springer, 2008).

Total Foraging Effort

If energy intake rate changes with the availability of resources, then so should the total amount of time an individual needs to forage to meet its metabolic needs. We define this as total foraging effort, which is the component of an activity–time budget dedicated to finding, recovering, physically handling, and consuming food. Isotopically labeled water techniques (Costa, 1988; Costa and Kooyman, 1982; Reilly and Fedak, 1990) measure foraging effort at the physiological level by directly measuring field metabolic rates. When both field metabolic rates and energy intake rates are known, simply dividing total energy requirement by energy intake rate at some appropriate time scale will provide an estimate of the minimum required foraging effort. Alternatively, a researcher could determine the actual foraging effort by observation or by using technological tools such as radio telemetry and/or instruments equipped with sensors (e.g., accelerometers and inclinometers) that allow a researcher to discern time spent doing various behaviors. Foraging effort will also be sensitive to short-term food availability, and limitations set by the maximum energy intake rates imply there will be a corresponding minimum total foraging effort.

Within the sea otter literature there is a long history of using foraging effort as a metric of PS specifically because of the near-constant energetic demands of the species. However, the methods to estimate foraging effort have evolved and provide increasingly more detailed information. Initially, the proportion of animals feeding during diurnal visual scans of sea otter habitat provided an index of foraging effort (Estes et al., 1982). Methods to track sea otters with radio telemetry developed in the 1980s (Ralls et al., 1989; Williams and Siniff, 1983), allowing researchers to record 24 h activity budgets (Garshelis et al., 1986; Gelatt et al., 2002; Loughlin, 1979; Ralls and Siniff, 1990; Ribic, 1982; Tinker et al., 2008). Foraging effort has also been estimated by dividing sea otter metabolic rate by energy intake rates (Dean et al., 2002). Development of archival time-depth recorders (TDRs) in the 1990s added a new tool for measuring foraging effort (Bodkin et al., 2004, 2007). TDRs provided yearlong continuous records of activity data that allowed detailed partitioning of foraging effort into age-sex classes and reproductive states as well as estimation of the effect of environmental variables on foraging effort (e.g., Figures 6.2 and 6.3; Esslinger et al., 2014), all of which increase our power to detect differences between groups or populations.

Morphometric Tools

Body Condition

Body condition, defined here as the amount of energy reserves stored within the individual, measures PS over the relatively short term. That is, an individual's energy reserves integrate energy intake over a period, reflecting each species' tolerance for extended periods of negative energy balance, with "normal" body condition values varying on seasonal or other temporal

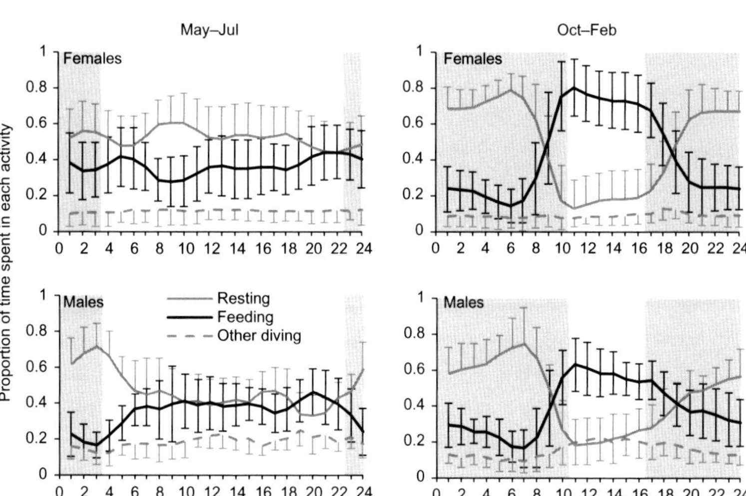

FIGURE 6.2 Diurnal activity patterns for female (top row) and male (bottom row) sea otters inhabiting Prince William Sound, Alaska, during spring/summer (left column) and autumn/winter (right column). *(Figure from Esslinger et al. (2014).)*

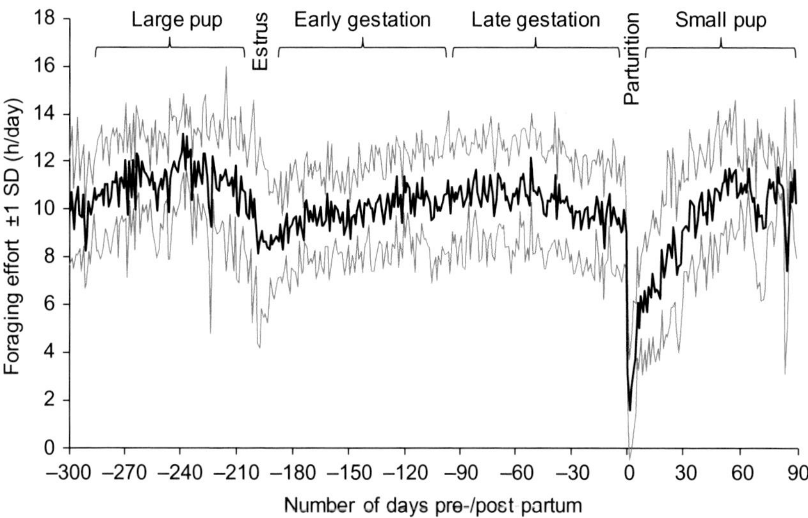

FIGURE 6.3 Foraging effort as a function of reproductive state for female sea otters occupying Prince William Sound, Alaska. *(Figure from Esslinger et al. (2014).)*

scales. Some species, including sea otters, need to maintain a nearly constant energy balance and thus a relatively stable body condition throughout the year, while other species may experience regular (e.g., seasonal) fluctuations in body condition, for example, some pinnipeds and cetaceans. Thus, any assessments of PS based on average body condition must first establish normal annual variation in body condition for the species.

Individuals generally do not put on unlimited energy stores (bears being an exception; Erlenbach et al., 2014) even when living with an abundance of food resources because trade-offs occur between energy stores and other life functions (e.g., locomotion and maneuverability). As a result, body condition will not necessarily increase linearly with food availability. For example, individuals may maintain some minimum body condition over a range of lower food availability values, foraging as long as needed to maintain this minimum value. At the other end of the spectrum, body condition may reach a plateau above which carrying additional energy stores is not useful at best and detrimental at worst when food is abundant. In addition, there may be a range of body condition values where individuals are perfectly "fit" and capable of surviving and reproducing at a maximum rate and higher body condition values provide no additional fitness advantage, which makes interpreting body condition as an index of PS difficult. In contrast, survival and reproductive rates may change relatively slowly as body condition drops below optimal levels (though specific demographic parameters may be more sensitive to changes in body condition than others—see demographic indices section below) until it hits a minimum body condition when severe reproductive and survival effects kick in. In part, this lower threshold occurs because poor body condition may compromise an individual's physiological and immunological response capabilities, increasing its susceptibility to other environmental stressors (e.g., parasite loads and exposure to toxins).

Given these caveats, body condition is an important metric in the conservation of species, as the population average should reflect how well a population is "positioned" to handle novel changes to its environment. That is, when the average body condition of the population is close to the optimal value for the species, most individuals are buffered against short-term environmental fluctuations. In contrast, when the average body condition is near to the minimum requisite level for the species, most individuals have little buffer against reductions in food availability due to environmental perturbation. However, when interpreting body condition values relative to PS, one must account for the potentially confounding factors that affect body condition independent of PS including season, age, reproductive status, or other indicators of physiological state. In addition, one must be aware of current short-term environmental fluctuations that may give a biased view of population-level body condition that does not represent the true long-term status of the population. That said, a short-term environmental fluctuation that is so severe that it causes significant mortality can itself change the population's status even after "normal" conditions resume.

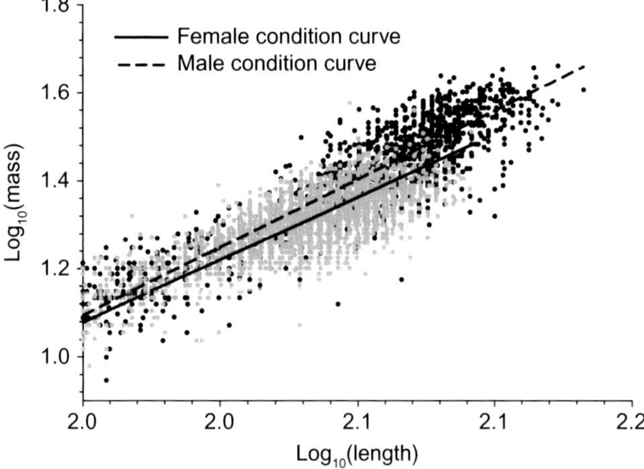

FIGURE 6.4 Sex-specific regressions of $\log_{10}$(mass)/$\log_{10}$(standard length) for sea otters fit to data collected throughout their range. Residuals from these regressions became the basis for a length independent RCI of body condition. *(Figure from Monson (2009).)*

Researchers have found many ways to measure body condition (Speakman, 2001) including morphometric-based indices, which generally fall into two categories: (1) ratios of mass to some measure of size (e.g., mass/length) or (2) residuals of mass-to-size relationships. Ratio-based indices can be difficult to interpret because they are correlated with size. For example, based on a regression of log(mass)/log(length), two female sea otters with standard lengths of 120 and 130 cm and having body masses of 20.1 and 25.2 kg, respectively, would both be in comparable condition (i.e., their log(mass)/log(length) ratios fall on the regression line; Figure 6.4), but their actual log(mass)/log(length) ratios would be 0.63 and 0.66 (biological equivalent of 168 and 194 g/cm). Thus, size (e.g., length) should be considered in any analysis of simple ratio-based body condition indices.

Residual-based indices (e.g., the difference between an individual's actual and predicted log(mass)/log(length) ratio; Figure 6.5) are size independent; however, the regression model on which the residuals are based can influence the outcome (Hayes and Shonkwiler, 2001). In addition, when interpreting residuals, one must consider the environmental conditions acting on the individuals used in building the regression model. For example, if most animals in "excellent" body condition originate from a population with no food limitation, they may also have a larger average body size (see body size section). Similarly, if all "poor" condition animals originate from populations under significant food limitation, they may have smaller average body sizes. Under this scenario the "average" body condition represented by the regression line will be biased high at the upper end of the curve (i.e., average for a group of

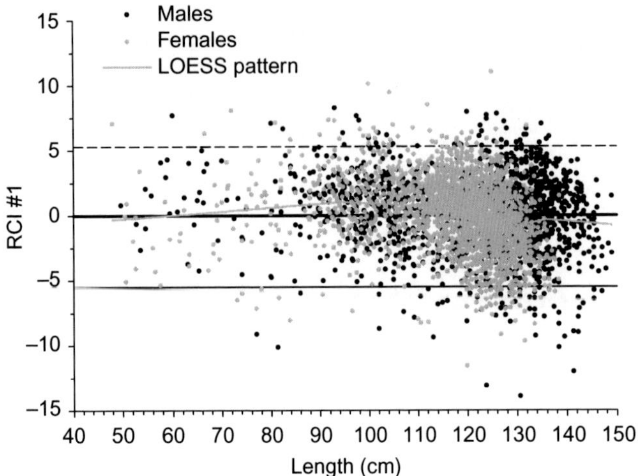

FIGURE 6.5 Length-independent body condition index (RCI #1) based on residual of $\log_{10}$(mass)/$\log_{10}$(standard length) relation developed for sea otters. Note: The LOESS pattern (gray line) is consistent with on average high body condition values for nursing juveniles followed by on average low body condition values for juveniles soon after weaning. *(Figure from Monson (2009).)*

structurally larger excellent condition animals) and the lower end of the curve will be biased low (i.e., average for a group of structurally smaller poor condition animals). In effect, this increases the slope of the regression line, which may complicate comparisons of residual body condition values between food-limited and non-limited populations. However, comparisons within populations of similar PS should not be affected by this bias.

Finally, regardless of the condition index used, the biological meaning of the index should be specified (e.g., energy reserves) and the index verified to determine if it reflects the desired meaning (Hayes and Shonkwiler, 2001). Not all body condition indices measure true energy stores with demographically meaningful effects, which is likely why inconsistent results can often be found in the literature where various body condition indices have little correlation with one another (Bowen et al., 1998; Gales and Renouf, 1994; Hall and McConnell, 2007; Pitcher et al., 2000; Ryg et al., 1990; Tierney et al., 2001).

Several morphologically based measures of body condition have been used in the study of sea otters. Monson et al. (2000b) found a relationship between the simple ratio of mass to standard length and reproductive success (Figure 6.6), which suggests mass/length is an informative indication of body condition for sea otters. Intuitively this makes sense, as female sea otters must put on enough energy reserves during pregnancy to allow for restricted energy intake rates following parturition (Gelatt et al., 2002; Monson et al., 2000b). However, while Monson et al. (2000b) did demonstrate that sea otters with more food availability grew at a faster rate and

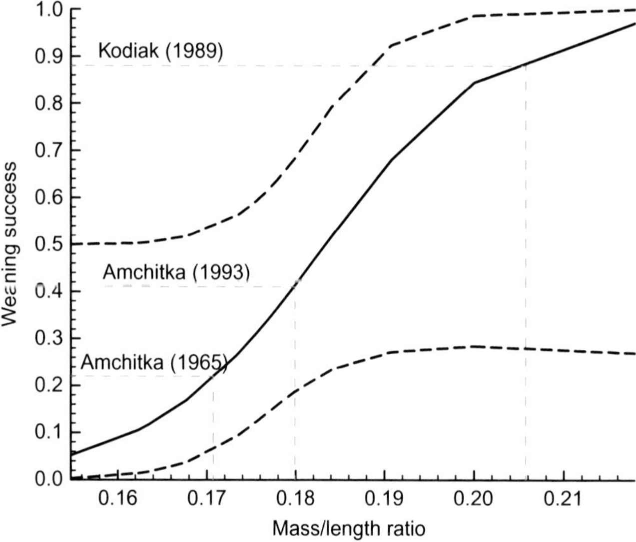

FIGURE 6.6 Logistic regression of female sea otter body mass to length ratio and the probability of successfully weaning their pups. Dotted gray lines indicate average mass/length ratios and predicted success at Amchitka (during the 1960s with the population at *K* and the 1990s after a top-down driven population decline), and at Kodiak during a time when the population was well below *K*. Note: Measured weaning success was near predicted values for 1980s Kodiak and 1990s Amchitka (0.83 and 0.47, respectively) but not measured for 1960s Amchitka. *(Figure from Monson et al. (2000b).)*

reproduced at a younger age than otters living with less food, they could not definitively demonstrate that higher reproductive success was because animals were in superior body condition or simply because they were larger. We reanalyzed reproductive success with a size-independent residual condition index (RCI1; Figure 6.5; Monson, 2009) and found a nearly identical relation (logistic regression, Wald $X^2 = 2.4$, $P = 0.1$), suggesting that body condition did increase reproductive success in sea otters. In addition, it suggests that in this case, the average length differences between the two populations examined did not confound the use of simple mass/length ratios as an index of body condition. Similarly, Tinker et al. (2008) found a large difference in mass/length ratios between the sparsely populated San Nicolas Island population in Southern California and its food-limited founding population in the central range along the Big Sur coast.

Dean et al. (2002) also used a simple mass/length ratio to demonstrate that sea otters living in the *Exxon Valdez* Oil Spill-affected area of Prince William Sound (where reduced sea otter numbers presumably allowed an increase in prey resources) had higher body condition values than sea otters in an unaffected area (where higher sea otter densities kept prey levels low). This analysis

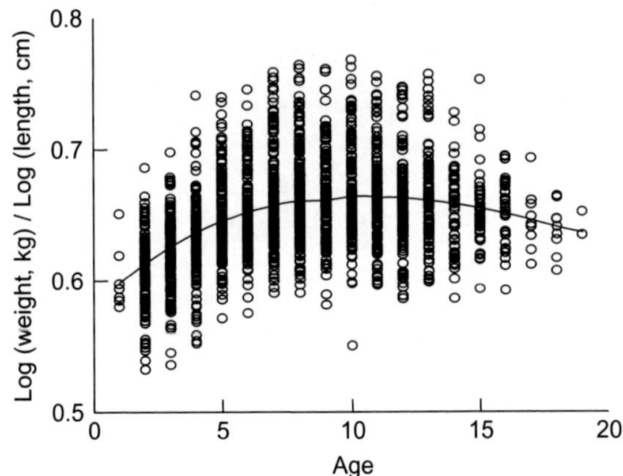

FIGURE 6.7 Log-transformed mass/length ratios for sea otters plotted against age. The LOESS-smoothed, second-order polynomial regression curve is plotted over the raw data, and the residuals used to produce an age-independent body condition index for the Aleutian Islands. *(Figure from Laidre et al. (2006).)*

was limited to young, non-reproductive females (≤4 years old) and mean length differences between the two areas was <3 cm, thus minimizing size-dependent influences, and demonstrates one way of controlling for age, sex, reproductive status, and season in an analysis of body condition. However, in many situations these restrictions would leave few animals to compare.

Laidre et al. (2006) created an age-independent body condition index for sea otters by fitting a locally weighted, second-order polynomial regression (LOESS) to log-transformed mass/length ratios (Figure 6.7). Results suggested that sea otters living in the Aleutian Islands in the 1960s and 1970s (a time when the population was believed to be at equilibrium densities) were shorter and in poorer condition than sea otters living in the same area 30–40 years later after an approximately 80–90% density-independent decline (Figure 6.8). However, the log(mass)/log(length) relationship is size dependent, and because average length-at-age was greater during the later period, at least some of the difference in these residual condition values (Figure 6.8) may be explained by the increase in size over time. However, the general conclusions of Laidre et al. (2006) were confirmed with the length independent condition index RCI1 (Monson, 2009).

Monson (2009) developed a length-independent body condition index (RCI1) to do a range-wide comparison of body condition in sea otters and explored the factors that affect body condition and its value as a PS index. He found that reproductive state was the most influential factor on female sea otter body condition. Specifically, late-term pregnant females could attain particularly high condition values due to both the addition of the

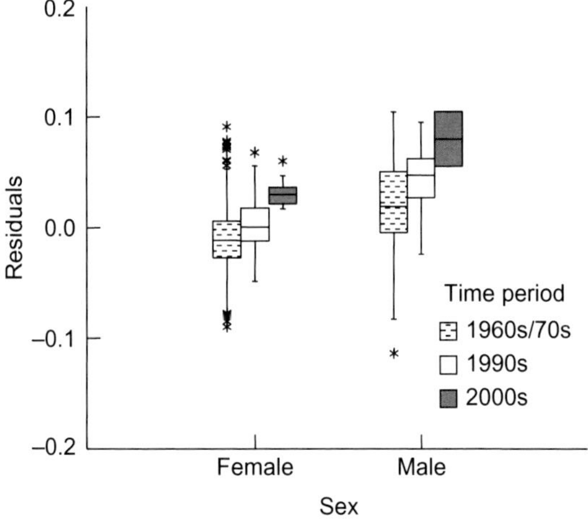

FIGURE 6.8 Box-and-whisker plots of age-independent residual distributions are shown grouped by sex and by time period. Differences in body condition in the Aleutian Islands due to sex and time period were significant. *(Figure from Laidre et al. (2006).)*

developing fetal mass and being in a physiological state that appears to allow them to put on more fat reserves than generally attainable (Figure 6.9). RCI1 also indicated sex- and age-specific patterns in body condition that can aid in differentiating populations near *K* versus populations below *K* (Figure 6.10). However, RCI1 was based on standard length (i.e., including tail length), and Monson (2009) found unexpectedly high body condition values in Prince William Sound, Alaska (a population recognized to be near *K*), to be the result of proportionally shorter tail lengths in that population. As a result, a second residual condition index (RCI2) based on body length alone eliminated variation caused by tail length, and highlights the importance of using a measure of body size that reflects a biologically meaningful measure of structural body size in morphometric indices of body condition. The final conclusion of Monson (2009) was that body condition is not a simple PS index, and required a full understanding of the pertinent physiological and environmental covariates acting on sea otters throughout their range.

The number of methods and studies that examine body condition of wildlife species reflects the great interest biologists and ecologists have in this metric of PS. This interest is no doubt generated by the obvious logical connection between resource availability and an individual's presumed ability to acquire energy reserves. However, similar to energy intake rates and total foraging effort, body condition can be highly variable and sensitive to not only short-term environmental fluctuations but also changes in an individual's physiological and reproductive state. Because of these various factors,

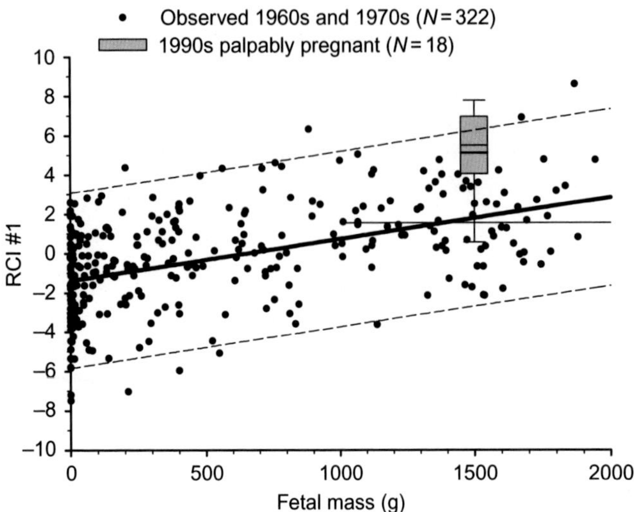

FIGURE 6.9 Length independent body condition values (RCI #1) for pregnant females at Amchitka Island, Alaska. The box-and-whisker plot denotes range and average RCI #1 values for near-term females in the 1990s (i.e., fetuses of palpably pregnant animals assumed to average ~1500 g). The filled dots represent measured fetal masses in the 1960s with the average for females with fetal mass >1000 g denoted by the long horizontal line. Note: The difference in mass between average late term females translates to 4.4 kg for a typical 125 cm adult female. *(Figure from Monson (2009).)*

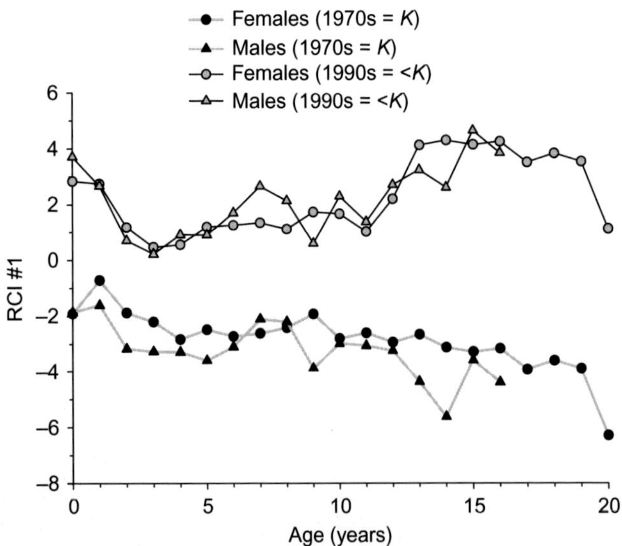

FIGURE 6.10 Age- and sex-specific patterns in sea otter length independent body condition (RCI #1) between the 1970s and 1990s at Amchitka Island, Alaska. Note: The difference in mass between an RCI value of −4 and +4 for a typical adult female or male is approximately 8 and 10 kg, respectively (palpably pregnant individuals not included in calculation of age-specific female patterns). *(From Monson (2009).)*

using body condition as an index of PS is not as simple as intuition would suggest and requires a thorough understanding of all the factors that can influence body condition independent of PS.

Growth Rates and Structural Body Size

Assuming that high energy intake rates allow individuals to achieve high body condition with relatively little foraging effort, presumably newly independent juveniles with little experience feeding on their own may be the most sensitive to this cascade of effects. When food availability is abundant such that inexperienced juveniles can achieve high body condition with little foraging effort, then they should grow at their maximum possible rate, allowing them to attain their genetically pre-determined maximum adult structural body size early in life. Conversely, as food resources become limited, inexperienced juveniles will likely be the first to experience reduced energy intake rates, potentially leading to reduced growth rates and adult body size. The effect of food availability on eventual adult structural size is a classic example of phenotypic plasticity whereby a genotype–environment interaction modifies the expressed phenotype (e.g., adult body size) of an individual during the developmental stages of its life (Reznick, 1990; Stearns, 1989; Tanaka, 2011). However, it should be noted that adult body size can be affected by environmental factors other than food availability (e.g., predation risk, Reznick, 1982; Rodd and Reznick, 1997), and the degree and predictability of environmental fluctuations regularly experienced by a species will influence how responsive growth and adult body size are to environmental fluctuations (Gotthard and Nylin, 1995). With these complications in mind, phenotypic plasticity in growth rate and adult body size due to PS-related changes in food availability have been noted in mammals (Fowler, 1987, 1990) and birds (Teplitsky et al., 2008), and can provide a unique longer-term perspective on PS.

The origin of environment-induced variability in growth and adult body size implies that these measures of PS reflect food availability over the moderately long period of growth and development of organisms. For sea otters, this is the first 2–3 years of life as they reach adult body size in 4 or 5 years. The unique aspect of structural body size (e.g., body length) is that for determinate growers, such as most birds and mammals, it does not change after maturity. As a result, the average adult structural size of a cohort reflects the abundance of resources available during that cohort's development (Monson, 2009). Further, this PS information is "recorded" in the structural body size distributions of all the living cohorts (i.e., 15–20 years for sea otters) and even farther back in time in the skeletal dimensions of the dead cohorts (Monson, 2009). Thus, tracking changes in cohort-specific structural body size allows us to identify how structural size is changing through time, which can indicate whether the population is moving toward or away from carrying capacity (Monson, 2009).

Sea otters provided a good example of the utility of this approach. For example, in the 1960s and 1970s, sea otters were expanding their range in the Andreanof Islands of the Aleutian chain (Figure 6.11). The expansion originated from a remnant colony in the Delarof Islands that survived the era of commercial harvests (Kenyon, 1969). By the 1970s, when collections of animals occurred, the expansion had reached Adak. These animals were descendants from the Delarof population, but the standard lengths of otters living in the recently occupied habitats of Adak were on average 6–7 cm (~6%) longer than the standard lengths of those living in the long-occupied habitats of Delarof and Tanaga Islands (Figure 6.12). Kanaga Island had been occupied for an intermediate length of time and its otters were of intermediate length (Figure 6.12). Further, by the mid-1990s the standard lengths of the oldest cohorts at Adak had declined relative to the 1970s, signifying that the population had moved toward *K* over the intervening 20 years. However, at the same time, the younger cohorts had higher growth rates and appeared to be reaching adult lengths comparable to the days of early occupation (Figure 6.12), indicating that the population was again well below *K*.

The spatial and temporal contrasts in body size trends in the Andreanofs make two points. First, when building a growth curve from cross-sectional data (i.e., from size-at-age data from a single sample) collected from a population approaching *K*, the slow reduction in cohort-specific body size can easily be misinterpreted as evidence of indeterminate growth (Monson, 2009). Second, older cohorts can have smaller adult structural body size than younger cohorts when the opposite has occurred and the population has been released from food limitations. Further, this second pattern of change in body size will not be evident in traditional monotonic growth curves that do not allow for a decline in size with age. The only way to see these patterns is to include individual cohort identifiers (e.g., birth-year) in the analysis (Monson, 2009).

Similarly, standard lengths of sea otters at Amchitka Island, Alaska, declined ~8% from the 1940s to the 1960s, as the population appeared to reach *K*. In contrast, standard length was on the rise again by the mid-1990s after an ~90% density-independent decline in otter numbers (Figure 6.13) and a corresponding increase in energy intake rates (Estes et al., 2010a). The temporal body size trends at Amchitka make two additional points. First, a moderately long-lasting environmental fluctuation can generate a signal in structural body size. In this case an apparent influx of lumpsuckers (*Aptocyclus ventricosus*), which are easily captured and commonly eaten when available (Watt et al., 2000), appeared to supplement food availability over a 1- or 2-year period in the 1950s (note short-term peak in standard length; Figure 6.13). Second, adult body size begins to decline well before the population reaches *K*—that is, the population trend

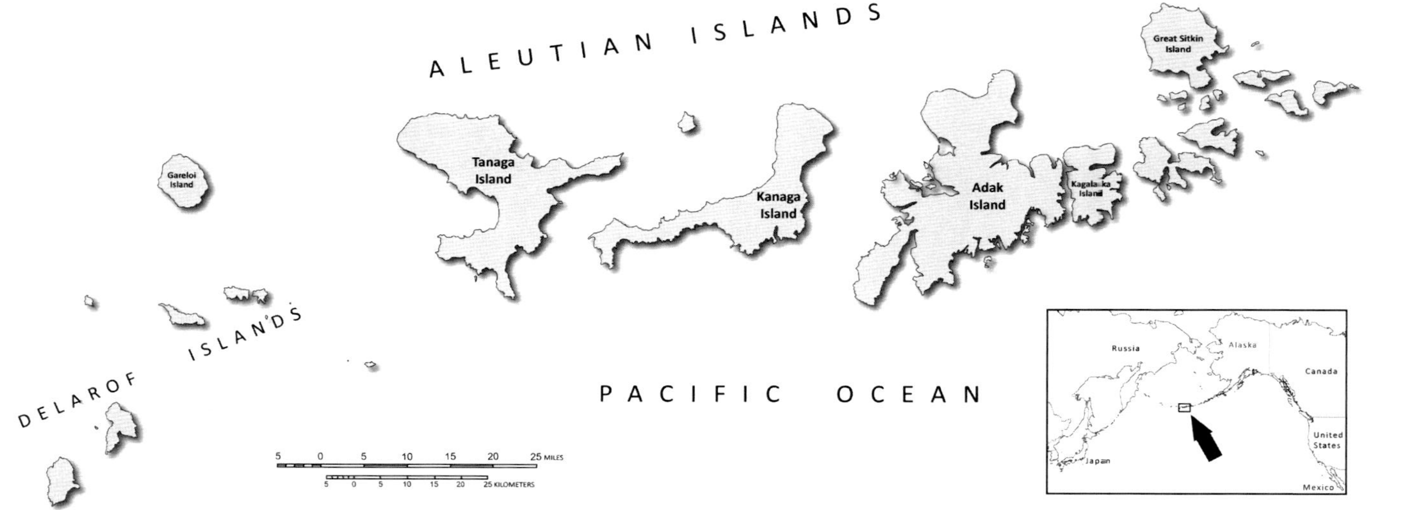

FIGURE 6.11 Map of the Andreanof Islands in the Aleutian Island chain. Note: A remnant sea otter population survived eighteenth and early nineteenth century commercial harvests in the Delarof Islands and then spread east to Tanaga, Kanaga and reached Adak Island in the early 1950s (Lensink, 1960). The population then collapsed in the 1990s presumably due to killer whale predation (Estes et al., 1998, 2005).

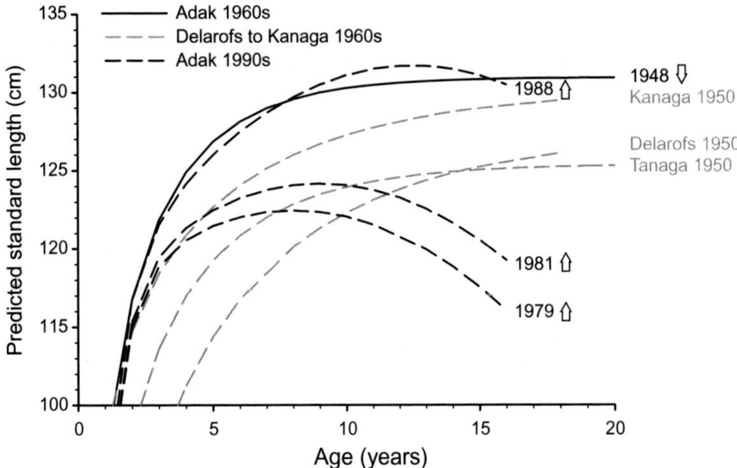

FIGURE 6.12 Standard length growth curves including birth-year affects (Monson, 2009) for female sea otters from the Andreanof Islands of the Aleutian chain (Delarof Islands to Adak Islands) fit to data collected in the 1960s to the 2000s. Note: Birth-year for the oldest cohorts represented in the curve noted at the end of each curve, and the trend in body lengths going forward in time noted by arrows. *(Figure from Monson (2009).)*

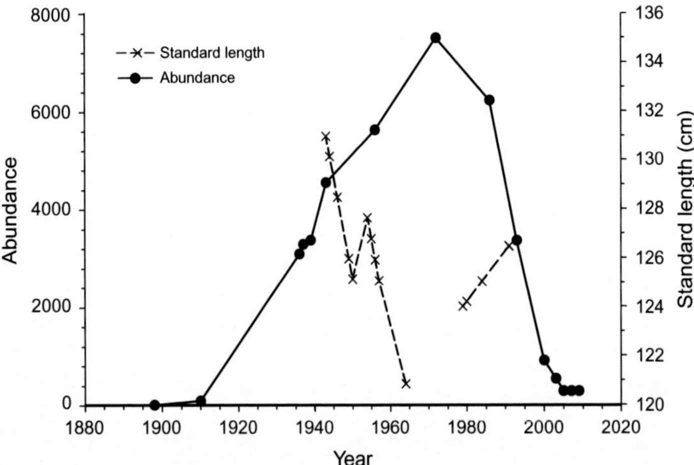

FIGURE 6.13 Historical change in abundance from 1890 to 2005 and trends in adult female standard length from the 1940s to 1990s. Note: drop in sea otter abundance after 1980 presumably due to top-down forcing (i.e., killer whale predation; Estes et al., 1998, 2005). *(Figure from Monson (2009) at Amchitka Island, Alaska.)*

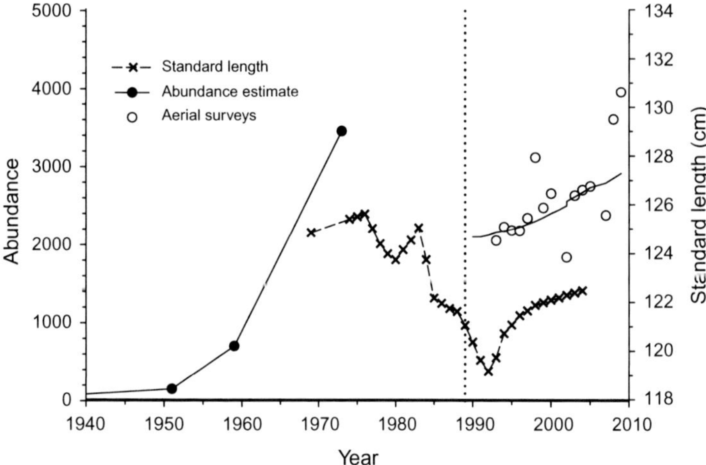

FIGURE 6.14 Historical sea otter abundance, and trends in female sea otter standard length. Note: The *Exxon Valdez* Oil Spill of 1989 (vertical dotted line) reduced the western Prince William Sound population with recovery delayed in the most heavily oiled areas until at least 2005 (Bodkin et al., 2002; Dean et al., 2002; Monson et al., 2000a, 2011). *(Figure from Monson (2009) in western Prince William Sound, Alaska.)*

lagged well behind the change in body size. Moreover, when both population abundance and structural size trends are known, inferences regarding the forces acting on the population (bottom-up vs. top-down) can be made. Specifically, a growing population moving toward food limitation should exhibit a declining trend in structural body size well before the population reaches carrying capacity, with the eventual leveling of both body size and population trend a strong indication of the population reaching *K*. In contrast, a declining population trend followed by an increasing body size trend suggests density independent or top-down forces controlling the population, which allowed prey resources to recover (e.g., predation, Figure 6.13, or an oil spill, Figure 6.14). Finally, a declining population trend and unchanging body size suggests bottom-up forces at work resulting from decreased environmental productivity. However, just as declines in population growth may lag behind reductions in body size when a population approaches *K*, the opposite lag occurs during a top-down reduction of the population. That is, the ability of a predator's prey resources to recover following a density-independent driven decline of the predator will determine the lag time between population decline and when body size begins to increase in the predator population.

Monson (2009) further suggests that measures of structural body size should not include mass. Clearly, mass is dynamic and increases or decreases with current environmental conditions or the physiological state of an individual. Mass does co-vary with body size and in some circumstances it may be an acceptable measure of structural body size, but we do not consider a population level change in body mass without a concurrent change in structural body size as phenotypic plasticity. That is, our concept of phenotypic plasticity concerns environmentally induced alterations of the phenotype of an individual that, once established, do not change (e.g., color morphs of *Biston betularia* moths or plant forms of *Hieracium umbellatum*).

Overall, structural body size integrates the PS signal over relatively long periods and may be one of the better measures of long-term average PS. The body size signal will lag well behind the immediate indicators of food availability (i.e., energy intake, foraging effort, and body condition) but this longer response time also dampens its response to short-term environmental fluctuations. Thus, when interpreting shorter-term PS indices, an understanding of body size trends can be advantageous. Body size also has the advantage of being recorded in the size distribution of both living and dead cohorts and can be useful in retrospective analyses of PS, possibly providing information prior to the initiation of specific research into a population's status.

Demographic Tools

Demographic Rates

The use of demographic rates as indices of PS originates from the hypothesis that a sequence of changes in demographic rates occurs that controls density-dependent growth as a population of long-lived animals approaches *K* (Eberhardt, 1977a, 2002). Specifically, as a population approaches *K*, the population first exhibits a decline in juvenile survival rates, followed by an increase in the average age of first reproduction, reductions in adult female reproductive rates, and finally a decline in adult survival rates. Thus, monitoring these demographic parameters can be informative as to a population's status. In addition, demographic rates can be incorporated into population models that can be used to predict population trends (Chivers, 1999). This paradigm appears to be generally applicable to a wide variety of long-lived species (Bonenfant et al., 2009; Eberhardt, 2002), though different species may show differences in the sensitivity of each demographic rate to food limitation.

The logic behind the sequence of demographic rate changes begins with the assumption that inexperienced juvenile animals will likely be the first to experience the cascade of effects produced by reduced energy intake as food becomes less available. In addition, females of a species living in relatively unpredictable environments are expected to invest less in offspring when environmental conditions are poor to ensure their own survival (Goodman, 1979; Schaffer, 1974; Stearns, 1976; Wilbur et al., 1974), leading to lowered female reproductive success as populations approach *K*. Consistent with this

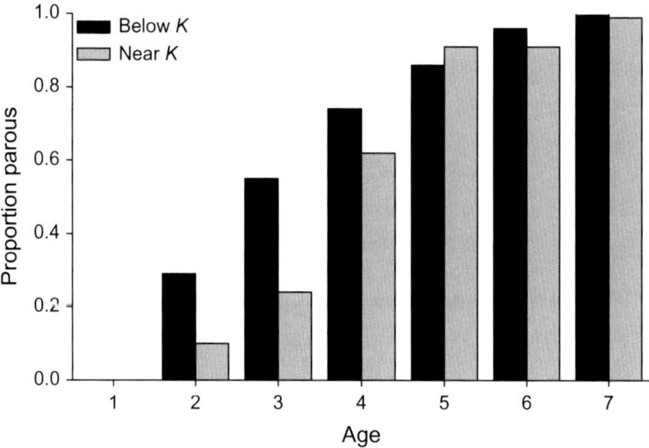

FIGURE 6.15 Proportion of parous females in sea otter populations below and near *K*. *(Figure from von Biela et al. (2009).)*

hypothesis, juvenile survival and female reproductive success appear to vary with PS and appear to be the primary mechanisms controlling density-dependent growth in sea otters (Monson et al., 2000b). In addition, sea otters are rather unique in their capability to reproduce at any time of year. Therefore, when food is abundant there is no seasonal advantage to when a pup is born and reproduction is asynchronous with broad or no peak in annual pupping (Monson et al., 2000b). In contrast, as sea otter populations approach *K* and female reproductive success declines, reproductive synchrony increases as seasonal advantages to successful pup rearing become evident (Monson et al., 2000b). Thus, the degree of female reproductive synchrony can be an indicator of PS and the timing of pupping peak(s) informative as to when environmental conditions are most conducive to pup survival.

In addition, lower juvenile energy intake rates leading to slow juvenile growth are consistent with a delayed onset of maturity and thus increased age of first reproduction as resources become limiting (Bengtson and Laws, 1985; Choquenot, 1991; Festa-Bianchet et al., 1995; Riedman, 1990; Skogland, 1985). For sea otters, average age of first reproduction appears to vary with PS and differs by over 1 year between newly occupied and long-established sea otter populations at or near *K* (Figure 6.15; von Biela et al., 2009). As predicted, age at first reproduction appears to be less sensitive to changes in PS than juvenile survival and female reproductive success. For example, Monson et al. (2000b) found two below-*K* sea otter populations with documented differences in juvenile survival and female reproductive success that did not differ in age of first reproduction. Also consistent with the predicted pattern, female reproductive rates appear mostly invariant across sea otter populations of different PS (Monson et al., 2000b), with reductions occurring only under extreme food limitation—that is, the lowest documented reproductive rates are from Amchitka Island in the 1950s and 1960s when the population was believed to be at *K* (Kenyon, 1969). Similarly, compared to juvenile survival, adult survival rates do not appear nearly as sensitive to PS although adult survival can decline substantially when populations are under extreme food stress (Bodkin et al., 2000; Kenyon, 1969). In particular, when food limitation is extreme, females in the late stages of the lactation may experience significant mortality due to the energetic demands of pup rearing. This seems to be the case for sea otters occupying the central portion of the range in California (Staedler, 2011; Tinker et al., 2013), suggesting this population may be one of the most severely food limited of contemporary populations.

Overall, demographic signals of PS are generally operating over moderate time scales and integrating food availability over several months to a year in large mammals. However, they may be slow to respond to significant increases in environmental productivity due to lags between the increase in resources and the production of offspring, but possibly quick to respond to extreme declines in productivity as survival of dependent young and juveniles in particular may be very sensitive to reductions in prey, especially below some minimum threshold level.

Emerging Tools

Dietary Diversity

Dietary diversity may increase as populations approach *K* because as preferred prey populations decline due to predation pressure, a predator may switch to or supplement their diet with less preferred prey species. Increased diet diversity can occur in populations of generalists, who non-selectively take what prey types are available (Taillon et al., 2006) and it can occur in populations of specialists where multiple foraging strategies can develop and coexist within a population (Paez-Rosas and Aurioles-Gamboa, 2010; Paez-Rosas et al., 2012). However, the dietary diversity of generalist predators should respond quickly to changes in prey specifically because they simply take what is available. In contrast, specialist predators may be less sensitive to changes in prey availability because prey selection includes an element of behavioral inertia. That is, specialist predators often employ specialized hunting skills in order to exploit various prey types efficiently. These foraging skills are often learned and require time to develop, making prey switching difficult. Thus, at the individual level, the dietary diversity of specialists may not respond until the density of a preferred prey reaches some minimum threshold. In addition, the alternate prey available to an individual may be a subset of total prey types available to the population, as it will include only prey types that require foraging skills similar to those already learned.

However, as the population approaches *K* and new foraging strategies develop within the population, new niches open up, thus increasing dietary diversity of the population and increasing the carrying capacity of the habitat.

Sea otters are specialist predators with learned foraging skills passed down from mothers to offspring (Estes et al., 2003; Newsome et al., 2009; Tinker et al., 2007, 2008). As such, sea otters are a good example of how behavioral inertia affects the prey-switching process. Historically, sea urchins were the major prey for the recovering sea otter population at Amchitka Island, which increased until *K* was reached by the late 1940s (Kenyon, 1969). Following a density-dependent decline in sea otter abundance, the population began consuming fish (Estes, 1977). The inclusion of fish increased the carrying capacity of the now kelp forest-dominated habitat and the population increased to higher levels than attainable before fish were incorporated into their diet (Estes, 1977). Similarly, the California sea otter population demonstrates strong prey diversification in populations existing at carrying capacity in contrast to populations existing below *K* (Newsome et al., 2009; Tinker et al., 2008, 2012).

In contrast, sea otter diet diversity can also respond quickly to the influx of an easily accessible preferred prey. For example, Pacific smooth lumpsuckers are a pelagic fish species (Orlov and Tokranov, 2008) not normally available to sea otters and thus are rarely part of a sea otter's diet. However, large numbers of adult lumpsuckers episodically occur in nearshore habitats of the Aleutian Islands to spawn. Lumpsuckers are poor swimmers and easily captured by sea otters; thus, essentially all age classes begin consuming them when present in the western Aleutians (Watt et al., 2000). Because all male lumpsuckers and many spawning females remain in the nearshore after spawning (Orlov and Tokranov, 2008), they become a supplemental resource that lasts many months (Watt et al., 2000).

Community Structure

Sea otters are one example of a species that has large effects on the structure of communities they occupy including the overall species composition of the community, and the abundances and interactions of the species present. Such species are typically considered "keystone" in that their presence can have an influence on both the structure and function of the community that is disproportionate to their own abundance (Paine, 1969; Power et al., 1996). Where the effects of keystone species are known and predictable, the state of the community should reflect PS (Eberhardt, 1977b). Estes et al. (2010b) used the well-documented role of sea otters in nearshore marine ecosystems as a novel approach to aid in the assessment of the "threatened" southwestern stock of sea otters in Alaska. They found that the state of nearshore rocky reefs in the Aleutian Archipelago would accurately reflect sea otter abundance. Specifically, below a threshold density of approximately six otters/km, the green sea urchin consistently dominated rocky reefs and grazing resulted in easily defined urchin barrens. At densities above six otters/km, predation by otters on urchins limited grazing and predictably resulted in reefs dominated by canopy-forming kelps. Thus, by simply measuring the abundance of kelps and urchins at a small number of islands, Estes et al. (2010b) were able to accurately infer the PS of sea otter populations and define recovery endpoints in the absence of costly sea otter population survey data. In addition, their approach provided inference regarding the potential role of a preferred prey (the urchin in this case) as a limiting resource. While such approaches may be suitable for evaluating PS in other keystone species, assessing PS from community structure does require a quantitative understanding of the relation between the status of a population and the phase state of the community. Moreover, although the urchin barren/kelp forest phase states in this system were largely exclusive, other systems may have intermediate phase states or alternate predators that complicate this approach for evaluating PS.

Spatial Distribution

An animal's spatial distribution is often examined with the goal of determining its habitat preferences, which, in the absence of predation risk, often correlate with where its preferred food resources occur. Conversely, predation pressure can influence the distribution of prey by restricting them to less preferred habitats where predation risk is lower (Heithaus and Dill, 2006). Thus, spatial distribution is unique among the tools presented here in that it reflects population status relative to an important top-down force. For example, sea urchins are an epibenthic species that in the absence of sea otters conspicuously occurs at high densities on the surface of nearly all subtidal substrates including open soft-sediment habitats and rocky reefs, where they graze with impunity. In the presence of sea otters, urchins revert to a cryptic lifestyle as they disappear from open habitats and are restricted to the cracks and crevices of rocky reefs where they are out of reach of foraging otters.

Sea otters themselves are no different in this respect. That is, historically when subjected to top-down pressure (i.e., commercial harvest) their remnant populations survived where hunting pressure was lowest. For example, historically sea otters exhibited extremely close associations with shore, but following extensive commercial harvest initiated by Russian fur traders in the 1700s, sea otters began to take refuge in offshore waters (Hooper, 1897). An extreme example of this is a remnant population described as "pelagic" that survived well offshore (2–50 miles) in the shallow waters of the Bering Sea north of Unimak Island (Lensink, 1960). Presumably, once most of the otters were gone, finding the few remaining animals out in the open expanse of the Bering

Sea was difficult and inefficient for the hunters, which allowed a low-density population to survive. On Kodiak Island, the remnant population survived at the north end of the Archipelago at a location called Latex Rocks (Lensink, 1960). The currents and wind that create consistently inhospitable boating conditions at this location (Monson, personal observation) suggest that hunters rarely attempted to hunt there. Presumably, all remnant sea otter populations survived in similarly inaccessible or inhospitable locations (Lensink, 1960). Thus, in the face of intense hunting pressure, sea otters persisted where hunting pressure was least.

Following the end of commercial harvests, sea otters reoccupied much of their former habitat in the western Aleutian Islands and they occurred over essentially all suitable foraging habitats with animals commonly foraging >1 km from shore, including documented sightings >50 km from shore when sufficiently shallow habitat was available (Kenyon, 1969; Lensink, 1960). However, by the early 2000s otters in the western Aleutians had shifted to a distribution similar to that of their urchin prey—that is, they shifted their distribution inshore to the nooks and crannies of the rocky coast where they are presumably less vulnerable to killer whale predation and are now rarely observed >100 m from shore (Estes et al., 2010a). Moreover, the only habitats in the western Aleutians that did not experience sea otter declines (e.g., Clam Lagoon on Adak Island) are inaccessible to killer whales (Estes et al., 1998). Thus, this change in distribution is a strong indication of top-down pressure, with their distribution patterns reflecting the level of protection from predation given by specific habitats.

Gene Transcription

Wildlife populations face both physiological and environmental pressures that will ultimately affect their survival and reproduction. Gene transcription is an emerging tool that can be used to assess the physiological consequences of environmental perturbations (i.e., changes in climate, pathogen prevalence, contaminant exposure) within individuals and populations. One of the earliest observable signs of physiological impairment is altered levels of gene transcripts, evident prior to clinical manifestation (McLoughlin et al., 2006). Gene transcription is the process by which information from the DNA template of a particular gene is transcribed into mRNA and eventually translated into a functional protein. Both intrinsic and extrinsic factors, including stimuli such as temperature change, infectious agents, toxin exposure, or trauma, dictate the amount of a particular gene that is expressed. Thus, analysis of mRNA can provide information not only about genetic potential but also about dynamic changes in the functional state of an organism. The costs associated with mounting a transcript response create a delicate balance between a protective phenotype (transcription pattern, in this case) and a potential misallocation of resources. Immune defense exists to impede infections, but other ecological demands (i.e., nutrition, weather, and predation) can supersede this, causing immune defenses to be compromised (Martin et al., 2011). The utility of gene transcription methodology relies on the assumption that sub-lethal pathologies are accompanied by predictable and specific patterns and changes in gene transcription.

Thus, gene transcription can be used to identify potentially vulnerable or susceptible populations prior to the onset of clinical signs and changes in population trajectory. In fact, through the process of targeted gene panel selection, gene transcription can actually be adjusted or tuned to address questions specific to particular ecosystems or populations, and at varying temporal scales. Many studies have examined the effects of xenobiotics on gene transcription in wildlife. Additionally, gene transcription has been successfully validated and applied to the study and diagnosis of disease and susceptibility in human medicine (Deirmengian et al., 2005; McLoughlin et al., 2006), setting the stage for parallel studies in wildlife (Bowen et al., 2007, 2012; Connon et al., 2012; Mancia et al., 2012; Miles et al., 2012; Sitt et al., 2010; Stott and McBain, 2012).

Disease is not believed to have played a significant role in the sea otter decline in the western Aleutians because it is inconsistent with a number of observations made by researchers conducting fieldwork during the decline (Estes et al., 2009; Laidre et al., 2006; Springer et al., 2003, 2008). Specifically, researchers found no sick or moribund animals and few fresh carcasses, and the broad geographical range of the decline included widely spaced island groups with little if any exchange of sea otters (Estes et al., 1998). However, serologic evidence from sea otters living at Kodiak Island, Alaska, shows evidence of recent exposure to phocine distemper virus (Goldstein et al., 2009). No active disease or sea otter population decline has been linked to this exposure, but its presence is still a concern (Goldstein et al., 2009, 2011). Given scientific inconsistencies such as these, a measure of susceptibility of this population via gene transcription would be informative.

In Prince William Sound, Alaska, an anthropogenic catastrophe (the *Exxon Valdez* oil spill of 1989) created a situation where sea otters faced acute oil effects during and soon after the spill that caused significant mortality (Garrott et al., 1993). In addition, residual oil left in the environment acted as a chronic stressor to sea otters and other nearshore invertebrate predators that survived the spill (Bodkin et al., 2002; Dean et al., 2002; Monson et al., 2000a, 2011; Peterson et al., 2003), which can serve as an example of the value of gene transcription for identifying populations under stress. Specifically, Miles et al. (2012) compared sea otters from Knight and Montague Islands and Prince of Wales Passage in Prince William Sound (varying levels of historic oil exposure) with "clinically normal" sea otters, both from the Alaska Peninsula (little to no known overt environmental perturbations) and from aquaria (Bowen et al., 2012). The "reference" range of gene transcription values provided by the clinically normal sea otters facilitated interpretation of natural variation among healthy subjects from variation created by compromised immune function, which is a key requirement for application of gene transcript technology at the population level (McLoughlin et al., 2006). In this case, genes were selected based on known functions and pathways most likely involved in immune or environmental effects on sea otters identified in studies of captive mink exposed to hydrocarbons (Bowen et al., 2007). These gene transcripts were representative of activities involving multiple physiological systems that play roles in immune modulation, inflammation, cell protection, tumor progression, cellular stress response, xenobiotic metabolizing enzymes, and antioxidant enzymes. Genes were also responsive at different temporal scales, the more ephemeral transcripts suggesting recent insults while enduring transcripts potentially reflected long-term perturbations. Variation in transcript levels among genes and across populations was apparent. When analyzed without *a priori* structure (e.g., location) using non-metric multidimensional scaling, sea otters separated into significantly different, well-defined groups with non-oiled locations distinct from oiled locations (cluster analysis (SIMPROF, $p < 0.001$ to 0.03; Figure 6.16)), indicating that otters from oil-exposed areas had distinct gene expression patterns from otters from non-oiled areas.

One of the more striking results was the association of transcript profiles of sea otters from the Alaska Peninsula (the eastern boundary of the western sea otter stock) with captive sea otters. This suggests that susceptibility in the western stock of sea otters is relatively low as transcript profiles indicated limited pathogenic or environmental impacts, which again implicates other pressures such as predation in the decline of the western sea otter stock (Estes et al., 1998). In contrast, a majority of sea otters from the three different locations in Prince William Sound were delineated by higher transcription of certain genes compared to clinically normal, reference sea otters. In general, transcription patterns in the Prince William Sound sea otters were indicative of molecular reactions to organic exposure, tumor formation, inflammation, and viral infection that seemed consistent with chronic, low-grade exposure to an organic substance, and are consistent with the decreased life expectancy documented for sea otters inhabiting oiled areas (Monson et al., 2000a, 2011).

CONCLUSION

Assessing a population's status can be challenging and often requires researchers to develop innovative metrics that will work for the species they study. Moreover, no single metric for assessing PS is likely to give a complete picture of factors controlling a population's dynamics. Instead, we recommend using several indices of PS operating at different temporal scales (Table 6.1). Short-term indices such as energy intake rates, total foraging

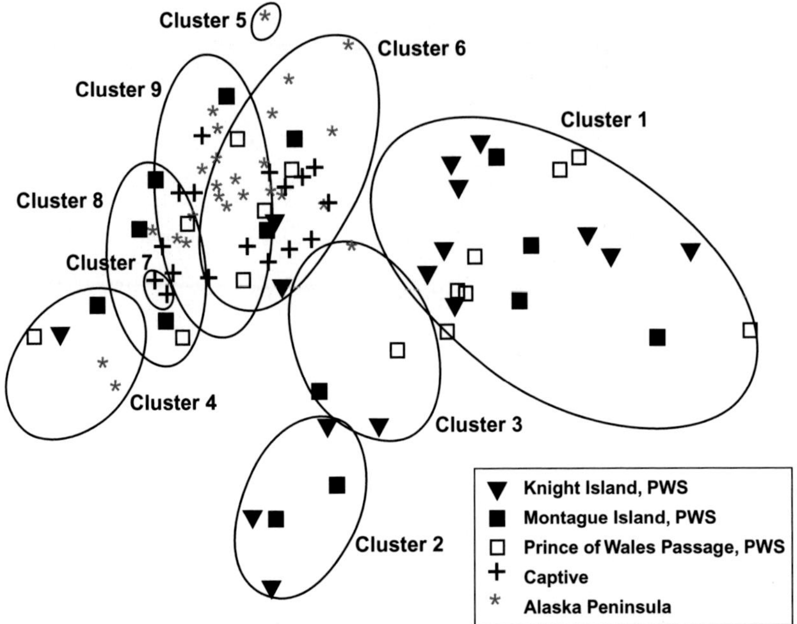

FIGURE 6.16 Multivariate, non-parametric, multidimensional scaling of gene transcription profiles of sea otters sampled at three locations in Prince William Sound, Alaska, at the Alaska Peninsula, and in captivity at aquaria. Interpretive cluster analysis and SIMPROF (similarity profile permutation test; Primer, v6, Plymouth, UK) indicated significant ($p < 0.001-0.034$) separation between all clusters depicted by circles. The figure is two-dimensional (Stress = 0.11); however, three-dimensional (Stress = 0.07) representation depicted distinct separation among Clusters 6–9. *(Figure from Miles et al. (2012).)*

effort, and body condition can give early clues to changes in food availability but only if normal seasonal, sex, and age-specific covariates are included in the analyses. In addition, changes in these indices due to unexpected short-term environmental fluctuation can complicate interpretation and provide a view of PS that may not reflect the true long-term status of the population. Mid-term indices of PS including most demographic rates are also useful as inputs into population models that predict population trend, and they can indicate the severity of food limitation because the various demographic rates have different sensitivity to food restrictions. However, there may be lags between when environmental conditions change and when these changes are reflected in demographic rates. Long-term indices of PS, including growth rates, structural body size, dietary diversity, and community structure, will generally be the least sensitive to short-term environmental fluctuations except in special cases (e.g., dietary diversity reaction to an influx of preferred prey) and thus should provide the best picture of overall average status of the population. However, the long lags between an environmental change and when that change is reflected in the various PS indices could result in missing the onset of an environmental change if no short-term indices of PS have been monitored. Structural body size is unique among these indices as PS is "recorded" in the body size distribution of each cohort. Thus, PS can be tracked back through time with the trajectory providing insight into the direction a population is moving (i.e., away from or toward *K*). Gene transcription is another unique index of PS as (1) it is "tunable" in that the choice of gene transcripts monitored can be geared toward the types of environmental challenges a population is expected to face, and (2) it can examine both short-acting and chronic stressors depending upon the genes monitored.

Spatial distribution can be insightful because top-down pressure can influence space use. That is, predators can limit their prey directly by consuming them and/or by limiting their spatial distribution. However, prey generally adopt antipredator strategies that come at a cost (Creel, 2011). For example, when a predator has limited access to its preferred foraging habitats due to predation risk from its own predators, it may also experience concurrent bottom-up pressure (Frid et al., 2007). In the case of sea otters, all other metrics of PS suggest food is not a limiting factor, indicating this particular scenario does not apply in the most recent western Aleutian decline. However, for a diving mammal in particular, predation risk can have other non-lethal effects (Wirsing et al., 2008), including modifying diving behavior (Wirsing et al., 2011). In the western Aleutians, sea otters changed their "escape" behavior, indicative of such a "fear" response. In the early 1990s, sea otters dove and swam for deep water when disturbed. However, by the mid-2000s, following intense predation, most animals occurred nearshore and climbed onto the nearest rock when alarmed (Estes et al., 2010a). It is interesting to speculate whether or not the current distribution and behavior of sea otters in southwestern Alaska is at all similar to the pre-harvest distribution and behavior described by Hooper (1897), and if so, whether it had a common origin. At present, we do not know if these types of non-lethal predator-driven physiological or behavioral effects are influencing the western Aleutian sea otter population. Future research should investigate this possibility as non-lethal effects could still accelerate loss of genetic diversity within local sea otter populations (e.g., predation-induced limitations to sea otter movement and dispersal that result in reduced reproductive rates and gene flow), which may increase extinction risk at many island groups.

Overall, the various indices of PS measured in sea otter populations throughout their range have been valuable in understanding the trajectory of various sea otter populations. We believe this approach (i.e., measuring PS at multiple spatial and temporal scales) can serve as a model for how to understand populations of other species with conservation concerns.

ACKNOWLEDGMENTS

The work covered in this chapter represents the collective wisdom of an army of colleagues, collaborators, and energetic field assistants. Karl Kenyon, Karl Schneider, Cal Lensink, and Ancel Johnson pioneered sea otter studies beginning in the 1950s and 1960s, which became the basis for all the work that followed. Much of the work done since their time was done under the intellectual direction of James Estes and James Bodkin as leaders of their respective USGS projects in California and Alaska. Don Siniff of the University of Minnesota and Kathy Ralls of the Smithsonian laid the groundwork for many ideas presented here. They also mentored students who made important contributions including Dave Garshelis, Charles Monnett, Lisa Rotterman, and Tom Gelatt. Jack Ames with the California Department of Fish and Wildlife contributed in ways both tangible and intangible through his many years of experience in California, and as an always-welcome crewmember of many Alaska field trips. Glenn VanBlaricom, Marianne Riedman, and Ron Jameson all made significant historical contributions. Alaska Science Center project members who contributed both intellectually and as valued participants of many field adventures include Brenda Ballachey, George Esslinger, Kim Kloecker, Heather Coletti, Jennifer DeGroot, Angie Doroff, Dana Bruden, Mike Fedorko, and Kelly Modla. Esteemed members of the California sea otter crew include Tim Tinker (current USGS project lead), Brian Hatfield, Mike Kenner, Ben Weitzman (currently shared with AK), and Joe Tomoleoni. Revered colleagues from the Monterey Bay Aquarium include Mike Murray, Michelle Staedler, and Gena Bentall. Seth Newsome of the University of New Mexico provided valuable insights through the study of stable isotopes. Many other names could be added but they would fill a whole volume, and we thank them all just the same. Most of the work discussed here was supported as basic research projects under the auspices of the US Fish and Wildlife Service and the USGS. The University of California, Santa Cruz, and the University of Minnesota supported many students and studies often through grants from the National Science Foundation and the North Pacific Research Board. All work was conducted under authority of marine mammal research permits issued from the US Fish and Wildlife Service's Office of Management Authority. Any use of trade, product, or firm names is for descriptive purposes only and does not imply endorsement by the US government.

REFERENCES

Bengtson, J.L., Laws, R.M., 1985. Trends in crabeater seal age at maturity: an insight into Antarctic marine interactions. In: Siegfried, W.R., Condy, P.R., Laws, R.M. (Eds.), Antarctic Nutrient Cycles and Food Webs. Springer-Verlag, Berlin, pp. 669–675.

Bodkin, J.L., Burdin, A.M., Ryazanov, D.A., 2000. Age- and sex-specific mortality and population structure in sea otters. Mar. Mammal Sci. 16, 201–219.

Bodkin, J.L., Ballachey, B.E., Dean, T.A., Fukuyama, A.K., Jewett, S.C., McDonald, L., et al., 2002. Sea otter population status and the process of recovery from the 1989 "Exxon Valdez" oil spill. Mar. Ecol. Prog. Ser. 241, 237–253.

Bodkin, J.L., Esslinger, G.G., Monson, D.H., 2004. Foraging depths of sea otters and implications to coastal marine communities. Mar. Mammal Sci. 20, 305–321.

Bodkin, J.L., Monson, D.H., Esslinger, G.G., 2007. Activity budgets derived from time-depth recorders in a diving mammal. J. Wildl. Manage. 71, 2034–2044.

Bonenfant, C., Gaillard, J.M., Coulson, T., Festa-Bianchet, M., Loison, A., Garel, M., et al., 2009. Empirical evidence of density-dependence in populations of large herbivores. Advances in Ecological Research, Vol 41. Elsevier Academic Press Inc, San Diego, CA, pp. 313–357.

Bowen, L., Riva, F., Mohr, C., Aldridge, B., Schwartz, J., Miles, A.K., et al., 2007. Differential gene expression induced by exposure of captive mink to fuel oil: a model for the sea otter. EcoHealth 4, 298–309.

Bowen, L., Miles, A.K., Murray, M., Haulena, M., Tuttle, J., Van Bonn, W., et al., 2012. Gene transcription in sea otters (*Enhydra lutris*); development of a diagnostic tool for sea otter and ecosystem health. Mol. Ecol. Resour. 12, 67–74.

Bowen, W.D., Boness, D.J., Iverson, S.J., 1998. Estimation of total body water in harbor seals: how useful is bioelectrical impedance analysis? Mar. Mammal Sci. 14, 765–777.

Boyd, I.L., Murray, A.W.A., 2001. Monitoring a marine ecosystem using responses of upper trophic level predators. J. Anim. Ecol. 70, 747–760.

Cairns, D.K., 1987. Seabirds as indicators of marine food supplies. Biol. Oceanogr. 5, 261–271.

Calkins, D.G., 1978. Feeding behaviour and major prey species of sea otter, *Enhydra-lutris*, in Montague Strait, Prince William Sound, Alaska. Fish. Bull. 76, 125–131.

Caughley, G., 1970. Eruption of ungulate populations, with emphasis on Himalayan than in New Zealand. Ecology 51, 53–72.

Chivers, S.J., 1999. Biological indices for monitoring population status of walrus evaluated with an individual-based model. In: Garner, G.W., et al., (Eds.), Marine Mammal Survey and Assessment Methods. A. A. Balkema Publishers, Rotterdam, Netherlands, pp. 239–247.

Choquenot, D., 1991. Density-dependent growth, body condition, and demography in feral donkeys: testing the food hypothesis. Ecology 72, 805–813.

Connon, R.E., D'Abronzo, L.S., Hostetter, N.J., Javidmehr, A., Roby, D.D., Evans, A.F., et al., 2012. Transcription profiling in environmental diagnostics: health assessments in Columbia River Basin steelhead (*Oncorhynchus mykiss*). Environ. Sci. Technol. 46, 6081–6087.

Costa, D.P., 1988. Methods for studying the energetics of freely diving animals. Can. J. Zool. 66, 45–52.

Costa, D.P., 2007. A conceptual model of the variation in parental attendance in response to environmental fluctuation: foraging energetics of lactating sea lions and fur seals. Aquat. Conserv. Mar. Freshw. Ecosyst. 17, S44–S52.

Costa, D.P., Kooyman, G.L., 1982. Oxygen consumption, thermoregulation, and the effect of fur oiling and washing on the sea otter, *Enhydra lutris*. Can. J. Zool. 60, 2761–2767.

Creel, S., 2011. Toward a predictive theory of risk effects: hypotheses for prey attributes and compensatory mortality. Ecology 92, 2190–2195.

Dean, T.A., Bodkin, J.L., Fukuyama, A.K., Jewett, S.C., Monson, D.H., O'Clair, C.E., et al., 2002. Food limitation and the recovery of sea otters following the "Exxon Valdez" oil spill. Mar. Ecol. Prog. Ser. 241, 255–270.

Deirmengian, C., Lonner, J.H., Booth Jr, R.E., 2005. The Mark Coventry Award: white blood cell gene expression: a new approach toward the study and diagnosis of infection. Clin. Orthop. Relat. R. 440, 38–44.

Doroff, A.M., Bodkin, J.L., 1994. Sea otter foraging behavior and hydrocarbon levels in prey. In: Loughlin, T.R. (Ed.), Marine Mammals and the *Exxon Valdez*. Academic Press, San Diego, CA, pp. 193–208.

Doroff, A.M., DeGange, A.R., 1994. Sea otter, *Enhydra lutris*, prey composition and foraging success in the northern Kodiak Archipelago. Fish. Bull. 92, 704–710.

Doroff, A.M., Estes, J.A., Tinker, M.T., Burn, D.M., Evans, T.J., 2003. Sea otter population declines in the Aleutian Archipelago. J. Mammal. 84, 55–64.

Eberhardt, L.L., 1977a. Optimal policies for conservation of large mammals with special reference to marine ecosystems. Environ. Conserv. 4, 205–212.

Eberhardt, L.L., 1977b. "Optimal" management policies for marine mammals. Wildl. Soc. Bull. 5, 162–169.

Eberhardt, L.L., 2002. A paradigm for population analysis of long-lived vertebrates. Ecology 83, 2841–2854.

Erlenbach, J.A., Rode, K.A., Raubenheirmer, D., Robbins, C.T., 2014. Macronutrient optimization and energy maximization determine diets of brown bears. J. Mammal. 95, 160–168.

Esslinger, G.G., Bodkin, J.L., Breton, A.R., Burns, J.M., Monson, D.H., 2014. Temporal patterns in the foraging behavior of sea otters in Alaska. J. Wildl. Manage. 78, 689–700.

Estes, J.A., 1977. Population estimates and feeding behavior of sea otters. In: Merritt, M.L., Fuller, R.G. (Eds.), The Environment of Amchitka Island, Alaska. Technical Information Center, US Energy Research and Development Administration, Springfield, VA, pp. 511–526.

Estes, J.A., Palmisano, J.F., 1974. Sea otters: their role in structuring nearshore communities. Science 185, 1058–1060.

Estes, J.A., Jameson, R.J., Rhode, E.B., 1982. Activity and prey election in the sea otter: influence of population status on community structure. Am. Nat. 120, 242–258.

Estes, J.A., Tinker, M.T., Williams, T.M., Doak, D.F., 1998. Killer whale predation on sea otters linking oceanic and nearshore ecosystems. Science 282, 473–476.

Estes, J.A., Riedman, M.L., Staedler, M.M., Tinker, M.T., Lyon, B.E., 2003. Individual variation in prey selection by sea otters: patterns, causes and implications. J. Anim. Ecol. 72, 144–155.

Estes, J.A., Danner, E.M., Doak, D.F., Konar, B., Springer, A.M., Steinberg, P.D., et al., 2004. Complex trophic interactions in kelp forest ecosystems. Bull. Mar. Sci. 74, 621–638.

Estes, J.A., Tinker, M.T., Doroff, A.M., Burn, D.M., 2005. Continuing sea otter population declines in the Aleutian Archipelago. Mar. Mammal Sci. 21, 169–172.

Estes, J.A., Doak, D.F., Springer, A.M., Williams, T.M., 2009. Causes and consequences of marine mammal population declines in southwest Alaska: a food-web perspective. Phil. Trans. R. Soc. B 364, 1647–1658.

Estes, J.A., Bodkin, J.L., Tinker, M.T., 2010a. Threatened southwest Alaska sea otter stock: Delineating the causes and constraints to recovery of a keystone predator in the North Pacific Ocean, North Pacific Research Board Final Report 717. 117 p.

Estes, J.A., Tinker, M.T., Bodkin, J.L., 2010b. Using ecological function to develop recovery criteria for depleted species: sea otters and kelp forests in the Aleutian Archipelago. Conserv. Biol. 24, 852–860.

Festa-Bianchet, M., Jorgenson, J.T., Lucherini, M., Wishart, W.D., 1995. Life history consequences of variation in age of primiparity in bighorn ewes. Ecology 76, 871–881.

Fowler, C.W., 1981. Density dependence as related to life history strategy. Ecology 62, 602–610.

Fowler, C.W., 1987. A review of density dependence in populations of large mammals. In: Genoways, H.H. (Ed.), Current Mammology. Plenum Press, New York, NY, pp. 401–441.

Fowler, C.W., 1990. Density dependence in northern fur seals (*Callorhinus ursinus*). Mar. Mammal Sci. 6, 171–195.

Frid, A., Heithaus, M.R., Dill, L.M., 2007. Dangerous dive cycles and the proverbial ostrich. Oikos 116, 893–902.

Gales, R., Renouf, D., 1994. Assessment of body condition of harp seals. Polar Biol. 14, 381–387.

Garrott, R.A., Eberhardt, L.L., Burn, D.M., 1993. Mortality of sea otters in Prince William Sound following the *Exxon Valdez* oil spill. Mar. Mammal Sci. 9, 343–359.

Garshelis, D.L., Garshelis, J.A., Kimker, A.T., 1986. Sea otter time budgets and prey relationships in Alaska. J. Wildl. Manage. 50, 637–647.

Gelatt, T.S., Siniff, D.B., Estes, J.A., 2002. Activity patterns and time budgets of the declining sea otter population at Amchitka Island, Alaska. J. Wildl. Manage. 66, 29–39.

Goldstein, T., Mazet, J.A.K., Gill, V.A., Doroff, A.M., Burek, K.A., Hammond, J.A., 2009. Phocine distemper virus in northern sea otters in the Pacific Ocean, Alaska, USA. Emerg. Infect. Dis. 15, 925–927.

Goldstein, T., Gill, V.A., Tuomi, P., Monson, D., Burdin, A., Conrad, P.A., et al., 2011. Assessment of clinical pathology and pathogen exposure in sea otters (*Enhydra lutris*) bordering the threatened population in Alaska. J. Wildl. Dis. 47, 579–592.

Goodman, D., 1979. Regulating reproductive effort in a changing environment. Am. Nat. 113, 735–748.

Gotthard, K., Nylin, S., 1995. Adaptive plasticity and plasticity as an adaptation: a selective review of plasticity in animal morphology and life history. Oikos 74, 3–17.

Hall, A.J., McConnell, B.J., 2007. Measuring changes in juvenile gray seal body composition. Mar. Mammal Sci. 23, 650–665.

Hayes, J.P., Shonkwiler, J.S., 2001. Morphometric indicators of body condition: worthwhile or wishful thinking. In: Speakman, J.R. (Ed.), Body Composition Analysis of Animals. Cambridge University Press, Cambridge, UK, pp. 8–38.

Heithaus, M.R., Dill, L.M., 2006. Does tiger shark predation risk influence foraging habitat use by bottlenose dolphins at multiple spatial scales? Oikos 114, 257–264.

Hooper, C.L., 1897. A Report on the Sea-otter Banks of Alaska. US Government Printing Office, Washington, DC.

Kenyon, K.W., 1969. The sea otter in the Eastern Pacific ocean. North Am. Fauna 68, 1–352.

Kuhn, C.E., Crocker, D.E., Tremblay, Y., Costa, D.P., 2009. Time to eat: measurements of feeding behaviour in a large marine predator, the northern elephant seal *Mirounga angustirostris*. J. Anim. Ecol. 78, 513–523.

Laidre, K.L., Estes, J.A., Tinker, M.T., Bodkin, J., Monson, D., Schneider, K., 2006. Patterns of growth and body condition in sea otters from the Aleutian Archipelago before and after the recent population decline. J. Anim. Ecol. 75, 978–989.

Lensink, C.J., 1960. Status and distribution of sea otters in Alaska. J. Mammal. 41, 172–182.

Loughlin, T.R., 1979. Radio telemetric determination of the 24-hour feeding activities of sea otters, *Enhydra lutris*. In: Amalaner Jr., C.J., Macdonald, D.W. (Eds.), Handbook on Biotelemetry and Radio Tracking. Permagon Press, Oxford, UK, pp. 717–724.

Mancia, A., Ryan, J.C., Chapman, R.W., Wu, Q., Warr, G.W., Gulland, F., et al., 2012. Health status, infection and disease in California sea lions (*Zalophus californianus*) studied using a canine microarray platform and machine-learning approaches. Dev. Comp. Immunol. 36, 629–637.

Martin, L.B., Hawley, D.M., Ardia, D.R., 2011. An introduction to ecological immunology. Funct. Ecol. 25, 1–4.

McLoughlin, K., Turteltaub, K., Bankaitis-Davis, D., Gerren, R., Siconolfi, L., Storm, K., et al., 2006. Limited dynamic range of immune response gene expression observed in healthy blood donors using RT-PCR. Mol. Med. 12, 185–195.

Miles, A.K., Bowen, L., Ballachey, B., Bodkin, J.L., Murray, M., Estes, J.L., et al., 2012. Variations of transcript profiles between sea otters *Enhydra lutris* from Prince William Sound, Alaska, and clinically normal reference otters. Mar. Ecol. Prog. Ser. 451, 201–212.

Monson, D.H., 2009. Sea Otters (*Enhydra lutris*) and Steller Sea Lions (*Eumetopias jubatus*) in the North Pacific: Evaluating Mortality Patterns and Assessing Population Status at Multiple Time Scales. Department of Ecology and Evolutionary Biology. University of California, Santa Cruz, Santa Cruz, CA, pp. 207.

Monson, D.H., Doak, D.F., Ballachey, B.E., Johnson, A., Bodkin, J.L., 2000a. Long-term impacts of the *Exxon Valdez* oil spill on sea otters, assessed through age-dependent mortality patterns. Proc. Natl. Acad. Sci. USA 97, 6562–6567.

Monson, D.H., Estes, J.A., Bodkin, J.L., Siniff, D.B., 2000b. Life history plasticity and population regulation in sea otters. Oikos 90, 457–468.

Monson, D.H., Doak, D.F., Ballachey, B.E., Bodkin, J.L., 2011. Could residual oil from the *Exxon Valdez* spill create a long-term population "sink" for sea otters in Alaska? Ecol. Appl. 21, 2917–2932.

Morris, W.F., Doak, D.F., 2002. Quantitative Conservation Biology: Theory and Practice of Population Viability Analysis. Sinaurer Associates, MA.

Newman, K., Springer, A.M., 2008. Nocturnal activity by mammal-eating killer whales at a predation hot spot in the Bering Sea. Mar. Mammal Sci. 24, 990–999.

Newsome, S.D., Tinker, M.T., Monson, D.H., Oftedal, O.T., Ralls, K., Staedler, M.M., et al., 2009. Using stable isotopes to investigate individual diet specialization in California sea otters (*Enhydra lutris nereis*). Ecology 90, 961–974.

Orlov, A.M., Tokranov, A.M., 2008. Specific features of distribution, some features of biology, and the dynamics of catches of smooth lumpsucker Aptocyclus ventricosus (*Cyclopteridae*) in waters of the Pacific Ocean off the Kuril Islands and Kamchatka. J. Ichthyol. 48, 81–95.

Paez-Rosas, D., Aurioles-Gamboa, D., 2010. Alimentary niche partitioning in the Galapagos sea lion. Zalophus Wollebaeki. Mar. Biol. 157, 2769–2781.

Paez-Rosas, D., Aurioles-Gamboa, D., Alava, J.J., Palacios, D.M., 2012. Stable isotopes indicate differing foraging strategies in two sympatric otariids of the Galapagos Islands. J. Exp. Mar. Biol. Ecol. 424, 44–52.

Paine, R.T., 1969. A note on trophic complexity and community stability. Am. Nat. 103, 91–93.

Peterson, C.H., Rice, S.D., Short, J.W., Esler, D., Bodkin, J.L., Ballachey, B.E., et al., 2003. Long-term ecosystem response to the *Exxon Valdez* oil spill. Science 302, 2082–2086.

Piatt, J.F., Sydeman, W.J., 2007. Seabirds as indicators of marine ecosystems. Mar. Ecol. Prog. Ser., 352.

Pitcher, K.W., Calkins, D.G., Pendleton, G.W., 2000. Steller sea lion body condition indices. Mar. Mammal Sci. 16, 427–436.

Power, M.E., Tilman, D., Estes, J.A., Menge, B.A., Bond, W.J., Mills, L.S., et al., 1996. Challenges in the quest for keystones. BioScience 46, 609–620.

Ralls, K., Siniff, D.B., 1990. Time budgets and activity patterns in California sea otters. J. Wildl. Manage. 54, 251–259.

Ralls, K., Siniff, D.B., Williams, T.D., Kuechle, V.B., 1989. An intraperitoneal radio transmitter for sea otters. Mar. Mammal Sci. 5, 376–381.

Reilly, J.J., Fedak, M.A., 1990. Measurement of the body composition of living gray seals by hydrogen isotope dilution. J. App. Psychol. 69, 885–891.

Reznick, D.N., 1982. The impact of predation on life history evolution in trinidadian guppies: genetic basis of observed life history patterns. Evolution 36, 1236–1250.

Reznick, D.N., 1990. Plasticity in age and size at maturity in male guppies (*Poecilia reticulata*): an experimental evaluation of alternative models of development. J. Evol. Biol. 3, 185–203.

Ribic, C.A., 1982. Autumn activity of sea otters in California. J. Mammal. 63, 702–706.

Riedman, M., 1990. The Pinnipeds: Seals, Sea Lions, and Walruses. University of California Press, Berkeley and Los Angeles, CA.

Rodd, F.H., Reznick, D., 1997. Variation in the demography of guppy population: the importance of predation and life histories. Ecology 78, 405–418.

Ryg, M., Lydersen, C., Markussen, N.H., Smith, T.G., Oritsland, N.A., 1990. Estimating the blubber content of phocid seals. Can. J. Fish. Aquat. Sci. 47, 1223–1227.

Schaffer, W.M., 1974. Optimal reproductive effort in fluctuating environments. Am. Nat. 108, 783–790.

Sitt, T., Bowen, L., Blanchard, M.T., Gershwin, L.J., Byrne, B.A., Dold, C., et al., 2010. Cellular immune responses in cetaceans immunized with a porcine erysipelas vaccine. Vet. Immunol. Immunop. 137, 181–189.

Skogland, T., 1985. The effects of density-dependent resource limitations on the demography of wild reindeer. J. Anim. Ecol. 54, 359–374.

Speakman, J.R., 2001. Body Composition Analysis of Animals: A Handbook of Non-Destructive Methods. Cambridge University Press, Cambridge, UK.

Springer, A.M., Estes, J.A., van Vliet, G.B., Williams, T.M., Doak, D.F., Danner, E.M., et al., 2003. Sequential megafaunal collapse in the North Pacific Ocean: an ongoing legacy of industrial whaling? Proc. Natl. Acad. Sci. USA 100, 12223–12228.

Springer, A.M., Estes, J.A., van Vliet, G.B., Williams, T.M., Doak, D.F., Danner, E.M., et al., 2008. Mammal-eating killer whales, industrial whaling, and the sequential megafaunal collapse in the North Pacific Ocean: a reply to critics of Springer *et al.* 2003. Mar. Mammal Sci. 24, 414–442.

Staedler, M.M., 2011. Maternal care and provisioning in the southern sea otter (*Enhydra lutris nereis*): Reproductive consequences of diet specialization in an apex predator. University of California, Santa Cruz, CA, p. 67.

Stearns, S.C., 1976. Life-history tactics: a review of the ideas. Q. Rev. Biol. 51, 3–47.

Stearns, S.C., 1989. The evolutionary significance of phenotypic plasticity. BioScience 39, 436–445.

Stott, J.L., McBain, J.F., 2012. Longitudinal monitoring of immune system parameters of cetaceans and application to their health management. In: Miller, R.E., Fowler, M.E. (Eds.), Fowler's Zoo and Wild Animal Medicine Current Therapy. Saunders, St Louis, MO.

Taillon, J., Sauve, D.G., Cote, S.D., 2006. The effects of decreasing winter diet quality on foraging behavior and life-history traits of white-tailed deer fawns. J. Wildl. Manage. 70, 1445–1454.

Tanaka, K., 2011. Phenotypic plasticity of body size in an insular population of a snake. Herpetologica 67, 46–57.

Teplitsky, C., Mills, J.A., Alho, J.S., Yarrall, J.W., Merila, J., 2008. Bergmann's rule and climate change revisited: disentangling environmental and genetic responses in a wild bird population. Proc. Natl. Acad. Sci. USA 105, 13492–13496.

Tierney, M., Hindell, M., Lea, M.A., Tollit, D., 2001. A comparison of techniques used to estimate body condition of southern elephant seals (*Mirounga leonina*). Wildl. Res. 28, 581–588.

Tinker, M.T., Costa, D.P., Estes, J.A., Wieringa, N., 2007. Individual dietary specialization and dive behaviour in the California sea otter: using archival time-depth data to detect alternative foraging strategies. Deep-Sea Res. Part II-Top. Stud. Oceanogr. 54, 330–342.

Tinker, M.T., Bentall, G., Estes, J.A., 2008. Food limitation leads to behavioral diversification and dietary specialization in sea otters. Proc. Natl. Acad. Sci. USA 105, 560–565.

Tinker, M.T., Guimarães, P.R., Novak, M., Marquitti, F.M.D., Bodkin, J.L., Staedler, M., et al., 2012. Structure and mechanism of diet specialisation: testing models of individual variation in resource use with sea otters. Ecol. Lett. 15, 475–483.

Tinker, M.T., Jessup, D., Staedler, M., Murray, M., Miller, M., Burgess, T., et al., 2013. Sea otter population biology at Big Sur and Monterey California: investigating the consequences of resource abundance and anthropogenic stressors for sea otter recovery. In: Tinker, M.T. (Ed.), Final Report to California Coastal Conservancy and U.S. Fish and Wildlife Service. University of California, Santa Cruz, CA.

von Biela, V.R., Gill, V.A., Bodkin, J.L., Burns, J.M., 2009. Phenotypic plasticity in age at first reproduction of female northern sea otters (*Enhydra lutris kenyoni*). J. Mammal. 90, 1224–1231.

Watt, J., Siniff, D.B., Estes, J.A., 2000. Inter-decadal patterns of population and dietary change in sea otters at Amchitka Island, Alaska. Oecologia 124, 289–298.

Weitzman, B.P., 2013. Effects of sea otter colonization on soft-sediment intertidal prey assemblages. Glacier Bay, Alaska, Department of Ecology and Evolutionary Biology. University of California, Santa Cruz, CA, 78 pp.

Wilbur, H.M., Tinkle, D.W., Collins, J.P., 1974. Environmental certainty, trophic level, and resource availability in life history evolution. Am. Nat. 108, 805–817.

Williams, T.D., Siniff, D.B., 1983. Surgical implantation of radiotelemetry devices in the sea otter. J. Am. Vet. Med. Assoc. 183, 1290.

Williams, T.M., Estes, J.A., Doak, D.F., Springer, A.M., 2004. Killer appetites: assessing the role of predators in ecological communities. Ecology 85, 3373–3384.

Wirsing, A.J., Heithaus, M.R., Frid, A., Dill, L.M., 2008. Seascapes of fear: evaluating sublethal predator effects experienced and generated by marine mammals. Mar. Mammal Sci. 24, 1–15.

Wirsing, A.J., Heithaus, M.R., Dill, L.M., 2011. Predator-induced modifications to diving behavior vary with foraging mode. Oikos 120, 1005–1012.

Yeates, L.C., Williams, T.M., Fink, T.L., 2007. Diving and foraging energetics of the smallest marine mammal, the sea otter (*Enhydra lutris*). J. Exp. Biol. 210, 1960–1970.

Chapter 7

Veterinary Medicine and Sea Otter Conservation

Michael J. Murray
Monterey Bay Aquarium, Monterey, CA, USA

Veterinarians and veterinary medicine, in general, have historically played, and continue to play, an important role in the efforts to understand sea otters and assure the conservation of free-ranging sea otter populations. There are, however, important differences between the "traditional" doctor–patient–client relationship paradigm associated with the health management of individual animals, and the manner in which health and disease are evaluated and interpreted in populations of animals. It comes as no surprise that animals in the wild die from one or combinations of a myriad of causes.

Sea Otter Conservation. DOI: http://dx.doi.org/10.1016/B978-0-12-801402-8.00007-X

Therefore, if one examines a large enough number of animals, you will find some that are, in fact, critically ill and dying. It is important to recognize that the veterinarian's role in conservation medicine is not to "cure every sick sea otter," but rather to provide insight into the etiopathogenesis, the epidemiology, and the implications of these diseases. This information may then be integrated into the bigger population picture involving wildlife ecology, natural history, environmental factors, and demography, all of which are critically important in the understanding, and therefore the preservation, of sea otter populations.

In addition to this role in population-level conservation medicine, there remain important individual animal aspects of the veterinarian's role in sea otter conservation. Examination of adequate numbers of individuals within a population can provide data which can be used to compare and contrast different sea otter populations, and which have contributed to gaining an understanding of the factors involved in either constraining or contributing to the recovery of depleted sea otter populations. In this case, the doctor–patient paradigm is modified: each individual otter provides a data point which contributes to an evaluation of the "patient," a population of individual otters.

The remaining critical role for veterinarians and veterinary medicine is much more closely related to traditional individual animal medicine. By adapting and exporting medical and surgical techniques from the veterinary clinical setting into field settings, veterinarians have been able to dramatically increase the capability of conservation biologists to better understand sea otter ecology. In some cases, pre-existing clinical methodology may be readily transferred from domestic animal models into sea otter settings. In other cases, a more measured, meticulous process to evaluate safety and efficacy is needed. In either case, veterinary medicine has provided tools which may be used to help address the questions posed in sea otter conservation.

UNIQUE FEATURES IN NATURAL HISTORY ARE IMPORTANT

There are several aspects of the sea otter's natural history which make them unique among marine mammals. These not only make sea otters ideal sentinels of ocean health, but also facilitate the use and interpretation of data obtained from these unique carnivores (Table 7.1).

Sea otters are apex predators. For the most part, they are at the top of the food chain with significant "top-down" trophic influence on other aspects of the nearshore food web (Estes and Palmisono, 1974). Beyond these trophic level attributes, however, sea otters, as apex predators, are also bioaccumulators. Over their decade-plus lifespan, fat-soluble contaminants, such as many of the persistent organic pollutants, as well as some infectious agents, such as *Toxoplasma gondii*, accumulate in various tissues with little or no opportunity for depuration over time. These tissues essentially function

TABLE 7.1 Many Consider Sea Otters to Be Ideal Sentinels for Evaluation of the Health of the Nearshore Ecosystem

Trophic level	Apex predator, bioaccumulator
Migratory behavior	Relatively strong site fidelity, no significant migratory behavior
Prey species	Low trophic level, benthic invertebrates
Otter lifespan	Nearshore, generally within site of land
Prey migratory behavior	Relatively sessile species, minimal movement outside of local area.
Size	Small for a marine mammal. Can be handled with relative ease compared to other large species

These six attributes justify such a designation.

as "time capsules" which provide a record of exposure to a variety of contaminants and pathogens throughout their life (Jessup et al., 2010).

Sea otters demonstrate a very high degree of site fidelity to relatively small home ranges, probably unique among marine mammals. While male sea otters may travel tens or even hundreds of miles over time, females tend to have very small home ranges. And, in general, the species does not demonstrate extensive migrations for breeding or foraging, a feature of the natural history of many marine mammals. As a result, the source of exposure and subsequent accumulation of contaminants and pathogens may be more readily determined for sea otters. In most cases, these compounds or organisms are encountered within the sea otter's rather restricted home range.

Not only do the sea otters have relatively small home ranges, but their preferred prey species are sessile, benthic invertebrates with even smaller home ranges. They not only have small longitudinal ranges, but there is also very little, if any, vertical migration in their life history. As a result, interpretation of origin for nutrition, contaminants, and pathogens is simplified and is generally a reflection of the local geographic area of the otter's home range.

Again, unique among marine mammals, the sea otter's entire lifespan is within the shallow nearshore environment integrating the coastal marine system over relatively small spatial scales. Therefore, information gained from studying the sea otter is reflective of this environmentally critical nearshore marine habitat. Since the life of the sea otter occurs relatively close to shore, the sea otter's activities are relatively accessible to shore- or skiff-based observers. This ability to monitor and record sea otter behaviors and foraging activities provides data which may provide insight into risk factors for health and disease.

Lastly, and importantly for the veterinarian, sea otters are small, at least on a marine mammal scale. The largest sea otter recorded was a male weighing 50.8 kg (112 lb) in Washington. Most sea otters are significantly smaller, with adult females weighing 20–30 kg and males weighing 30–40 kg. These weights are within the weight range of many of the large breed dogs encountered in companion animal practice. Therefore traditional drugs and equipment are readily available through traditional resources. Also, the animals' weights are such that the sea otter can be captured, lifted, and moved around without specialized cranes or lifts. This allows the biologists and veterinarians to actually handle the animal, measure it, collect biologic specimens, and apply identifying tags with relative ease, at least when compared to other marine mammal species.

VETERINARY MEDICINE AND THE EVOLUTION OF SEA OTTER CAPTURE AND HANDLING

Despite their relatively small size on the marine mammal scale, sea otters are formidable and potentially dangerous carnivores, capable of delivering serious bites and scratches to those attempting to handle them. These defensive attributes, coupled with their keen sense of environmental awareness associated with their highly developed olfactory capabilities and exceptional mobility in the water, make capture problematic.

Over the years, there have been three basic methods of sea otter capture that have demonstrated both relative safety and efficacy: dip nets, entanglement nets, and diver-propelled Wilson traps. Each method is associated with a series of advantages and disadvantages, indications and contraindications, and varying degrees of efficacy and safety. Modifications and improvements for each of these methods, while not totally devoid of veterinary input, were driven primarily by the cadre of field biologists involved in sea otter research (Ames et al., 1986).

The earliest capture efforts involved the use of long-handled nets. Capture efforts were limited primarily to animals hauled out of the water. While records of success per unit effort are not available, it is likely that success rates were relatively low; there was likely significant bias in sea otters sampled based on haul-out locations and accessibility, and the demographics of the animals captured were unlikely to represent a uniform cross-section of sea otter populations. Dip net methodology was subsequently taken "out into the water" and some nets were modified to break away from the net frame and included a purse string to close the net opening. Dip netters were positioned in the bows of small, maneuverable skiffs and animals were captured either by surprise or by pursuit (following their bubble trails under water). Again, this method was likely associated with geographic and demographic bias. The success of dip netting required open water without kelp canopy or rocks and reefs approaching the water's surface. Water visibility was also important in order

FIGURE 7.1 Young sea otter captured in entanglement net. The large diameter woven line is the float line. Note the presence of a second otter, in this case the mom of the capture juvenile, not captured adjacent to the net. *(Photo credit: MJ Murray.)*

to be able to follow submerged sea otters. Most of the sea otters captured with this method tended to be naïve juveniles and geriatric animals whose endurance and overall health condition may have been suboptimal.

The use of modified gill nets placed at the surface of the water in areas of sea otter aggregation or transit routes eliminated many of the biases associated with dip net methods. Animals were captured much more randomly, providing a better representation of populations sampled. Nets could be allowed to soak for longer time periods, thereby improving the capture rates per unit of effort. And, with entanglement nets, biologists were able to capture animals in geographic regions in which haul-outs either didn't exist or weren't accessible.

As designed in their early days, nets were approximately 100 m in length, 6 m in depth, and had a 23 cm stretch mesh size. The bottom of the net was a lightweight lead line and the top was a self-floating polyform core whose buoyancy was enhanced by a series of floats. The underlying principle for their use was that surface or shallow-diving sea otters would inadvertently swim into the net and efforts to extricate themselves would result in a more entangled animal. Gear was designed lightweight enough to permit the otter to remain at the surface; however, submarine hazards such as rocks, pinnacles, or floating debris might snag the net, preventing an entrapped animal from surfacing (Figure 7.1).

Initially, the use of gill nets proved somewhat problematic. There were numerous reports of entrapped otters being attacked and severely injured by free-swimming or adjacent con-specifics. There were troubling cases of non-target species by-catch in sea otter nets. Initial reports of mortality rates of nearly 13% dead-in-nets were reported in Alaska (Ames et al., 1986). Modifications in methodology, such as more frequent checking of nets even overnight, and increased reliance on daytime net sets, coupled with ever-increasing skill and expertise associated with the use of the methodology, have nearly eliminated sea otter mortality and significantly reduced the incidence of non-target species by-catch. Current entanglement nets are 100–200 m long with a 9-inch (23 cm) stretch and a depth of 12–15 meshes. They are used in snag-free, smooth-bottomed areas as shallow as 15 ft (4.5 meters) or as deep as 120 ft (36.6 meters) (Esslinger, personal communication).

The most advanced, albeit technically difficult, capture method involves the use of diver-delivered Wilson traps. The Wilson trap is a teacup-shaped tubular aluminum frame which supports a net bag. The opening of the net bag can be closed tightly by the diver pulling closed its purse string. The initial design by Ken Wilson had the trap mounted on a cylindrical pole and carried underwater to a position under the sleeping sea otter by a pair of scuba divers. This original method has been vastly improved by mounting the trap on the front of a dive scooter and shifting from traditional, bubble-producing scuba divers to non-bubble-producing rebreather circuits used by specially trained scientific divers (Figure 7.2).

Wilson trap-based captures have proven to be a significant improvement for a number of reasons. Capture-related mortality is nearly non-existent, as there is constant observation of the animal's behavior from before capture until it is transferred into a transport container. Since divers are able to visualize the animal before a capture attempt is made, there is no non-target species by-catch. And, of critical importance to investigators, specific animals can be targeted for capture. Disadvantages are limited. There is a significant and important learning curve for both the use of the rebreathing system and its implementation in the capture of the sea otter. Capture efforts are weather and visibility dependent. Additionally, one cannot overlook the potential danger inherent to situations in which the diver is attempting to capture a large, aggressive marine carnivore in its aquatic environment.

In addition to the improvements in capture-related methodology, there have been significant and important changes in the ways in which sea otters are held and transported, which have significantly decreased the incidence of iatrogenic anthropogenic morbidity and mortality. In the short term, sea otters removed from their natural habitat face four primary threats associated with their short-term holding and transportation: hyperthermia, fouling of their pelage, damage to their teeth, and respiratory compromise. Varying combinations of these threats had significant impact on success rates of early capture-based studies. Recognition and subsequent mitigation by the biologists and wildlife veterinarians of the circumstances that predispose sea

FIGURE 7.2 Use of a scooter-mounted Wilson trap by a diver with a rebreathing apparatus. *(Photo by J.L. Bodkin, with permission.)*

otters to these events remains an ongoing process which continues to minimize the untoward impact that scientific intervention has on the sea otter.

Sea otters are designed to conserve heat; there is little adaptation for off-loading excess body heat when the animal is removed from the relatively cold waters of the northern Pacific Ocean. Therefore, captured sea otters out of water develop hyperthermia (core body temperatures >38.6°C/101.5°F) relatively quickly. This concern was recognized relatively early on in the history of sea otter captures. Initially, efforts were made to hold sea otters in plastic dog kennels which held a shallow pool of water. This proved impractical for a number of reasons and was replaced by using shaved or chipped ice in the holding container. While ice probably doesn't affect core body temperature as a result of conductive loss through the fur, the cooling of the air column and the resulting convective heat loss associated with respiration appear to be critically important. Additionally, sea otters appear to habitually consume ice in their holding areas. Further cooling can occur as this ice is consumed and swallowed.

Experience with field captures has facilitated accurate prediction of risk factors associated with hyperthermia. Aggressive animals that struggle extensively during the capture period tend to produce excessive body heat from muscular exertion. Long transports from capture sites to ship- or shore-based veterinary labs place otters in confined, poorly ventilated spaces, predisposing them to hyperthermia. While significantly more comfortable for field personnel, bright sunny days tend to be associated with more hyperthermic sea otters compared to colder, overcast conditions. This increased awareness of predisposing conditions permits veterinarians and biologists to take prophylactic measures before core body temperature rises become critical. The use of prolonged immersion of otter-bearing transport containers prior to definitive handling has been tremendously effective. Periodic monitoring of rectal body temperature coupled with application of cold packs as needed during veterinary procedures has also helped mitigate hyperthermia.

Whether it is the stress of capture or the relatively short gastro-intestinal transit time, sea otters often defecate in transport or holding containers. The relatively amorphously formed sea otter stool, which normally has a relatively high mucus content, easily fouls the otter's fur within the confined space of a dry holding box or kennel. Failure to either prevent intimate contact or clean the fur before release is likely to compromise the critically important thermoregulatory character of the fur and subject the released otter to heat loss and resultant hypothermia.

Modifications in the form of false bottoms or raised racks upon which the otter sits in transport boxes have been important changes which have decreased the incidence of fecal fouling of fur. In addition, the increased sensitivity of the veterinary team to the importance of "clean fur" results in cleansing efforts as needed to return the animal's fur to its natural state prior to release.

Traumatic injury to the sea otter's teeth, most commonly the canine teeth, during capture or transport, while of minimal concern in the short term, may have a profound impact on the otter's health over time. These fractures tend to be diagonal fractures of the tooth's crown and result in significant exposure of the pulp cavity (Figure 7.3). Ascending bacterial pulpitis, osteomyelitis, and even septicemia are likely sequelae to this capture-related event. Tooth fractures seem to occur most commonly when sea otters struggle within the Wilson trap, biting the padded aluminum frame, or when the otter is being transferred from the trap's net bag into its transport box. The metal doors of dog transport kennels, which were favored for moving captured sea otters, were notorious for causing damage to teeth when otters struggled to escape. The procedural shift from the dog kennel to large, wooden capture boxes has dramatically reduced the incidence of this serious iatrogenic injury.

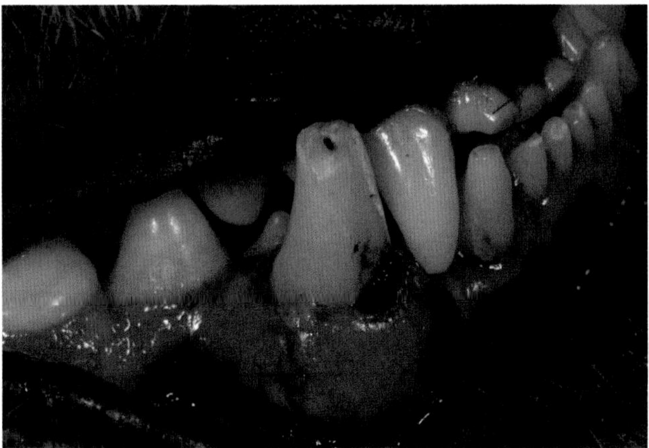

FIGURE 7.3 Slab fracture of the leading edge of the right lower canine tooth. Injury occurred as a result of biting on the metal door of a transport cage. The pulp cavity, the dark spot at the apex of the damaged tooth, is exposed and subject to ascending infection. *(Photo credit: MJ Murray.)*

In an effort to conserve space, reduce the potential for otter escape, and minimize weight combinations, dog kennels and net bags have historically been used for transporting recently captured sea otters. There are concerns about their use beyond those described previously. Sea otters' long, fusiform body shape and relatively large lung volume require a relatively long transport container to facilitate normal respiration. The importance of this is amplified during the struggles and associated oxygen demand associated with capture. Both dog kennels and net bags prevent sea otters from stretching out and assuming their natural dorsally recumbent posture. As a result, they don't breathe efficiently, and the respiratory-based compensatory mechanisms for recovery from oxygen debt and acid–base imbalance are compromised. The negative impacts of these imbalances are then amplified when chemical or physical immobilization occurs. The shift away from the dog kennel and net bag has been critically important in alleviating these concerns.

The most important improvement in sea otter holding and short-term transportation has been the nearly universal adoption of the marine-plywood capture box (Figure 7.4). This box, while relatively heavy and time-consuming to fabricate, has mitigated nearly all of the anthropogenic sources of capture/transport-related morbidity and mortality. Boxes are perforated with numerous holes to allow ventilation and movement of water through the box when soaked in water. The raised PVC grate on the bottom of the box prevents nearly all contact between the otter and its excrement. The wood construction of the box is far more forgiving to teeth when bitten than is the metal door of the dog kennel. In addition, the length of the box is

FIGURE 7.4 Wooden capture box currently used to hold and transport sea otters in field operations. Note the PVC grate on the box's floor which separates otter from excrement; the multiple holes drilled into the box's walls permitting ventilation and water entry when the box is floated in water; and the heavy rubber dog "chew toy" used to give the sea otter an opportunity to chew on something besides the box. *(Photo credit: Colleen Young.)*

adequate to allow the otter to stretch out into its more normal posture. Since the box is opaque and relatively dark, otters tend to relax, and even to start grooming, allowing recovery from the catecholamine-based effects on critical body functions, many of which may be impacted by chemical immobilizing agents.

PHYSICAL AND CHEMICAL IMMOBILIZATION OF SEA OTTERS

While veterinarians have been only tangentially involved in the advancement of sea otter capture and transportation improvements, they have been intimately involved in the development of immobilization protocols. Development of safe and effective chemical immobilization protocols has been challenged by the need to have rapid and relatively complete recovery following a medical or surgical procedure, as the sea otter needs to return to its aquatic environment. As a result, many of the drug regimens utilized in terrestrial carnivores, including mustelids, are inappropriate for use in the sea otter.

Early work with varying combinations of opioids (fentanyl, etorphine), neuroleptics (azaperone), alpha-agonists (xylazine, medetomidine), and benzodiazepine compounds (diazepam) were met with varying degrees of efficacy and safety (Williams and Kocher, 1978, Williams et al., 1981; Monson et al., 2001). Other, more common "terrestrial" immobilizing agents, such as ketamine, were found to be problematic. The most significant improvement in sea otter anesthesia was the work published by Monson et al. in 2001, using the opioid fentanyl citrate in combination with the benzodiazepine diazepam. This combination had the benefit of intramuscular administration, rapid onset, profound immobilization including anesthesia at higher doses, and reversibility. There were two notable drawbacks to its use, namely respiratory suppression and a significant frequency of seizures likely attributable to the erratic absorption of intramuscular diazepam.

In its most current use, the fentanyl is combined with a water-soluble benzodiazepine, midazolam. Midazolam has a much more consistent and predictable absorption when administered intramuscularly, making the incidence of anesthesia-related seizures rare. The improvements and reliability of chemical sedation of sea otters have facilitated dramatic advances in the development of conservation-related studies of *in situ* sea otter populations throughout their range. To date, there have been well over 1500 fentanyl/midazolam events with only 2–3 mortalities, all in compromised animals—a track record unprecedented in chemical immobilization of wildlife (Murray, unpublished data).

The efficacy and safety of current immobilization protocols have resulted in a dramatically increased level of comfort with their use in field settings. This, coupled with the recognition of the need to increase the data and the number/variety of biological samples collected from each captured sea otter, has nearly eliminated the indications for physical immobilization as the sole form of restraint. This trend has not been overlooked by institutional animal care and use oversight committees. Many, if not most, are now mandating both significant veterinary involvement in field research involving animal captures and the use of pharmaceutical methods for the control of pain and anxiety in captured animals.

ROLE OF SEA OTTER ANATOMY AND PHYSIOLOGY ON CONSERVATION

On a geological time scale, sea otters have only recently entered the aquatic environment as marine mammals; the fossil record suggests a time frame of approximately 1–3 million years ago (Berta and Sumich, 1999). As a result, they retain many of the anatomic and physiologic features of their terrestrial relatives. There are, however, three major features of the normal sea otter anatomy and physiology that have profound conservation relevance: thermoregulation (fur), metabolic rates and diet (food), and oxygen use. These factors place limitations on available habitat, and therefore on sea otter range, variability of daily activity budgets, reproductive strategies and limitations, and susceptibility and response to infectious and non-infectious disease.

The most notable external feature of the sea otter, besides its deceivingly cute and cuddly outward appearance, is its fur. With estimated hair densities ranging from 120,000 to 140,000/cm^2 on the body and 60,000/cm^2 on the head (Kuhn et al., 2010), the sea otter has the densest pelage of any mammal. It is this feature which was nearly their undoing; they were hunted almost to extinction for the fur trade in the 1700s and 1800s. The fur consists of a longer guard hair surrounded by a number of under-coat hairs. These hairs have a series of surface scales that permit a series of interlocking sheets which entrap air (Weisel et al., 2005). It is this hydrophobic configuration of the fur with its entrapped air which provides the sea otter with its sole insulation, necessary for survival in the aquatic environment. But, unlike the behaviorally passive nature of insulation by blubber seen in pinnipeds and cetaceans, the sea otter's thermoregulatory "organ" requires active maintenance throughout the day. It is estimated that approximately 10% of the sea otter's daily activity budget is spent grooming (Yeates et al., 2007). Obviously, anything that compromises the otter's ability to maintain its pelage, such as contamination by fouling agents such as oil or concurrent illness, will have a significant impact on the animal's ability to maintain its core body temperature.

Management and protection of the fur is critically important to the sea otter. With a thermal neutral zone of 20–25°C (Morrison et al., 1974; Costa and Kooyman, 1984; Yeates, 2006), yet living in nearshore northern Pacific areas with temperatures ranging from 0°C to 20°C, the sea otter lives continuously in an environment with a temperature gradient favoring net loss of heat. The relatively homogeneous thermal environment in which the sea otter lives is a medium with high thermal conductivity and heat capacity (Denny, 1993). The impact of heat loss is more substantial for relatively small marine mammals, like the sea otter, with larger surface-area-to-volume ratios that provide a greater area from which to lose heat and less volume in which to produce it.

There are a number of behavioral and physiologic mechanisms employed by the sea otter to conserve heat. In addition to the efficient thermal insulation provided by well-maintained fur, the sea otter has an elevated metabolic rate, demonstrates both increased activity and significant periods of inactivity which conserve energy, and makes use of heat produced during the digestion of food (Costa and Kooyman, 1984; Williams et al., 1988; Yeates et al., 2007).

It has been well documented that sea otters are voracious eaters, consuming 20–25% of their body weight daily (Kenyon, 1969; Costa and Kooyman, 1982), which requires devotion of 20–50% of the day to foraging activities (Tinker, 2004). Terrestrial carnivores, on the other hand, average 5–14% of body weight in food consumed daily associated with only 14% of the day spent hunting for prey (Williams et al., 2004). The net gains realized through this increased caloric consumption are partially offset by the energy expense of inefficient locomotion, the energy expense associated with buoyancy compensation while diving, the loss of thermal insulation associated with the pressure-related loss of air in the fur when submerged, hunting effort, food handling and processing, and heat loss associated with the consumption of cold food (Yeates, 2006). There is obviously a close relationship between the density of prey, the nutritional content of these food items, and the foraging expense and the sea otters to meet their high caloric demands.

Sea otters demonstrate a number of behavioral traits that are important for survival as the smallest marine mammal in a cold ocean. Changes in wind, weather, and photoperiod are often associated with alterations in foraging activity (Esslinger et al., 2014). Additionally, sea otter prey selection may change periodically to take advantage of changes in the reproductive status or availability of prey species. Despite the apparent demand for an increased daily energy cost for sea otters, their field metabolic rate is allometrically aligned with predicted values for other marine mammals (Yeates, 2006). This is accomplished behaviorally by taking advantage of the relatively small home range of the sea otter and its site fidelity; otters rest nearly motionless on the water's surface for up to 49% of the day (Yeates et al., 2007).

At rest, there is little energy expenditure; however, there is little energy generation associated with either muscular activity or food ingestion. Therefore a gradual loss of core body temperature is expected. This phenomenon is also mitigated to some degree by sea otter behavior. In order to more efficiently capture heat generated by muscular activity and during the digestion of food, southern sea otters tend to demonstrate a predictable sequence of daily activities: foraging, swimming, grooming, resting, grooming, foraging (Yeates, 2006). The extent to which sea otters in other geographic regions follow similar patterns is not well described yet.

It is easy to appreciate that food availability is of critical importance for sea otter health. Availability of this resource is likely one of the most significant limiting factors for sea otter population growth. Our understanding of the magnitude of the sea otter's metabolic needs is nearly exclusively based on studies in captive animals in "maintenance" settings. It is probably safe to assume that the dramatically elevated nutritional requirements of sea otters for normal physiologic states such as growth, lactation, and reproduction are similar to those of terrestrial carnivores, if not greater.

While the nature and implications of the sea otter's pelage, thermoregulation, metabolism, and diet have been vigorously investigated over the past 50 years, study of the animal's use, storage, distribution, and limits related to oxygen and respiration and how these change throughout a sea otter's development and reproductive lifespan have not been well scrutinized to date. Such studies are currently under way; however, they are limited to short- and long-term captive animals with limited variation in age or physiologic/pathologic status. Nonetheless, an appreciation for oxygen accession, storage, and distribution and how these limit various aspects of the animal's behavior and response to disease and its environment is important in sea otter conservation.

The metabolic oxygen expense for diving and foraging is highest in sea otters when compared to other marine mammals studied. In a recent study, it was found that adult male sea otter performing a one-minute dive had an oxygen expense of 17.1 mL O_2/kg/min (Yeates, 2006). In comparison, a phocid seal has an expense of 4.5–5.7 mL O_2/kg/min (Castellini et al., 1992; Sparling and Fedak, 2004; Yeates, 2006). This high oxygen requirement is exceeded only by costs for surface swimming and intense grooming (Yeates, 2006). Marine mammals have three basic mechanisms to tolerate hypoxia: metabolic adjustments for heat production associated with muscular activity; the animal's oxygen storage capacity and respiratory properties of the blood; and economic use of oxygen associated with cardiovascular and respiratory responses (Lenfant et al., 1970). It is unclear how efficiently, if at all, sea otters employ hypometabolism or a "dive response" to mitigate their high oxygen costs during diving.

Like other marine mammal species studied, sea otters have respiratory characteristics of blood which suggest that they are better adapted for quick recovery from dives than for prolonged or deep dives (Lenfant et al., 1970). They have a slightly increased oxygen–hemoglobin affinity when compared to commensurately sized terrestrial carnivores. Consistent with other marine mammals, they also have significantly increased bicarbonate and buffering capacity in their blood which is protective against potential deleterious shifts in blood pH (Lenfant et al., 1970).

Oxygen is stored, and therefore "available" for metabolic use, in three primary locations within the body: in the lungs, in the blood, and in muscle-based myoglobin. The sea otter's lungs are the largest per unit weight for any of the marine mammals studied, with a volume of 345 mL/kg. Not surprisingly, the lungs account for the greatest percentage, 55%, of oxygen store in the sea otter. This volume may be somewhat deceiving, however, as these oxygen stores may become less accessible during diving due to gas compression and movement of air from the compressible alveoli into the more rigid airways (Lenfant et al., 1970).

Sea otters have a relatively low blood volume, 91 mL/kg, for a diving marine mammal. This low volume accounts for approximately 29% of the otter's oxygen stores (Lenfant et al., 1970). While there have been no published studies describing changes in circulating blood volume associated with sea otter growth and development, changes in the peripheral blood appear to be consistent with those described in terrestrial carnivores (Boyd and Bolon, 2010). All three measured components of the erythron—red blood cell count, hemoglobin levels, and packed cell volume—are lower in dependent pups and gradually rise to adult levels by the time they become independent (Williams et al., 1992).

Oxygen stores found in myoglobin tend to be relatively low when compared to other marine mammal species, 2.6 g/100 g^{-1}, accounting for only 16% of total stored oxygen (Lenfant et al., 1970). Because myoglobin only incompletely desaturates during diving, it is unclear what role this oxygen store plays in sea otters. It may serve to protect active muscle from ischemic necrosis or may aid in the diffusion of oxygen into muscle during exertion (Ponganis, 2011). Given the relatively short duration and shallow nature of sea otter dives, myoglobin-based oxygen stores may not be as important as they are in other, deeper-diving marine mammals.

Based upon metabolic rates and oxygen storage capacities, the aerobic dive limit for an adult male sea otter has been calculated to range from 2.9 to 4.3 min, depending upon the volume of air within the lungs (Yeates, 2006). This observed limit, coupled with the observed data that most foraging dives occur in water less than 100 m in depth (Bodkin et al., 2004; Esslinger, 2011), has a profound impact upon the nature of sea otter habitat and food availability.

There are many other idiosyncratic attributes of the sea otter, such as those associated with the anatomy and physiology of vision, olfaction, fluid/electrolyte balance, and reproduction, but the three described above—fur, food, and oxygen—are arguably the most influential on individual and subsequently on population health. They are directly related to day-to-day behavioral activity budgets, the sea otter's ability to survive in waters outside of its thermal neutral zone, and in the character of nearshore habitat suitable for sea otter habitation.

THE VETERINARY CONTRIBUTION TO SEA OTTER CONSERVATION

One need not look too deeply into the history of sea otter conservation to find evidence for the veterinarian's contribution to conservation of this marine mammal. Unlike many of the *ex situ* programs, however, the veterinarian's role is typically more supportive and collaborative when directly involved in the *in situ* conservation efforts. Many of the methods upon which field biologists and ecologists rely, such as chemical immobilization protocols, surgical methods, and a myriad of diagnostic tools, were tested and developed *ex situ* under the leadership of many innovative zoo and wildlife veterinarians.

The most important clinical contribution made by veterinary medicine to sea otter conservation is the development of safe, effective, and consistent protocols for immobilization and sedation of free-ranging sea otters. The rather unique nature of the field environment in which sea otter conservation work is done, coupled with the idiosyncrasies of sea otter anatomy and physiology, made it necessary to establish some characteristics that any immobilizing protocol must have (Table 7.2).

Once it was demonstrated that sea otters could be safely anesthetized in laboratory and field settings, a variety of advances in the study of free-ranging sea otters became possible. Field biologists were no longer limited to observing and tracking individual sea otters by recognition of inherently labile physical characteristics or the attachment of color-coded plastic tags

TABLE 7.2 While No Chemical Immobilizing Agent May Be Considered Ideal, the Combination of Fentanyl and Midazolam Demonstrates Many of the Characteristics Identified as Ideal

Characteristic	Fentanyl/Midazolam
Single agent	N
Intramuscular injection	Y
Small volume	Y
Few side effects/contraindications	Y
Reversal agent	Y
Rapid recovery	Y
Complete reversal/no re-Narc	Y
Expense	N
Regulatory oversight	Y

Note that "regulatory oversight" in the form of designation as "controlled substances" is not considered ideal.

affixed to their flippers (Ames et al., 1983). Tags, which could be chewed off by recalcitrant sea otters, fade in color as a result of long-term exposure to UV radiation, and could be hard to see in varying lighting and weather conditions. VHF radios could now be surgically implanted into sea otters. This advancement allowed biologists not only to follow individual sea otters with their unique radio frequency, but to more effectively monitor the otters' behavior under a variety of weather and lighting conditions, including at night; to find individual otters by scanning for a radio signal; and to interpret activities based on changes in the radio signal without necessarily seeing the otter (Williams and Siniff, 1983; Ralls et al., 1989, 1995).

This breakthrough was followed by development and application of additional monitoring technologies, such as the time-depth recorder (TDR) (Bodkin et al., 2004, 2007). Like its sister instrument, the VHF transmitter, the TDR is implanted within the otter's abdominal cavity. The TDR records both temperature and pressure, which can be mathematically converted to depth. The frequency and duration of data collection can be set by the researcher by instrument calibration. With this next tool, biologists are able to better understand the otters' behavior beneath the ocean surface. Determination of depth, dive times, and other activities not only permits a better understanding of the daily activity budget of the animal, but also permits calculation of energy recovery rates, a critically important aspect of the evaluation of the relationship between sea otter populations and the food resources upon which they depend (Bodkin et al., 2007; Esslinger et al., 2014).

These instruments, and the veterinarian's ability to safely surgically implant them in field settings, have provided tremendous insight into the status of sea otter populations throughout their range. But the data have given insight not only into daily activities and foraging behavior. Measured variation in core body temperature provides data concerning the sea otter's manipulation of metabolic rate in both the short-term daily scale and in the longer term through a variety of normal and pathological processes. Not only can biologists and veterinarians now identify and follow metabolic changes associated with the thermal energy of feeding, decay of thermoregulatory stability in between grooming events, and the loss of core body heat associated with the consumption of a large meal, but they have also gained insight into the metabolic changes associated with normal reproductive processes, such as the significant down-regulation of metabolic rate noted in the later stages of pregnancy (Esslinger, 2011).

While the advent and wide acceptance and use of these surgically implanted electronic tools have been of critical importance to sea otter conservation, their application is relatively invasive. Admittedly, the technical aspects of the performance of a laparotomy are not beyond the capability of almost any veterinarian; however, the additional layers of performance of an abdominal surgery in a field setting, on a marine mammal destined for return to the ocean within minutes of completion of the last suture, and the presence of a critically important suture line on the belly of one of the most fastidious groomers on the planet make the process significantly more daunting (Figure 7.5). More important, however, is the fact that implantation of the instruments and the need to surgically remove the TDR to access the data are both invasive procedures from the perspective of the sea otter. And, despite the myriad examples of advances in electronic technology and miniaturization of devices, both the VHF transmitter and the TDR remain relatively unchanged from their original design as much as three decades ago (Ralls et al., 1989; Walker et al., 2012).

In the case of abdominally implanted electronic devices, smaller is not necessarily better. Smaller instruments are more likely to become entrapped in one of the numerous fossae, membranes, canals, or folds found in the abdominal cavity. And, when surgical retrieval is necessary, as it is with archival data storage instruments like TDRs, a smaller instrument is much more difficult to find, which potentially subjects the surgical patient to longer procedures and more intraoperative trauma. There is an ongoing and nearly continous collaboration among veterinarians, biologists, and engineers to identify and develop less invasive and more secure locations to implant electronic instruments and to simultaneously reduce instrument size, increase the instrument's energy efficiency, and identify parameters which, when measured, provide insight into the life of the sea otter.

As a greater understanding of the nature of various sea otter populations has grown, so too have the indications for application of a variety of

FIGURE 7.5 Surgical extraction of a TDR in a field setting on Bering Island, Russia. Despite the remote, primitive setting, efforts are made to assure aseptic methodology in the performance of these surgeries. *(Photo credit: Brian Hatfield.)*

clinico-pathological tools for the study of health and disease in sea otters. Initially, veterinarians adapted the suite of diagnostic tests available in conventional domestic animal medicine to use in captive and free-ranging sea otters. These tests, such as complete blood counts (CBC), serum chemistry panels, and culture/sensitivity testing from microbial isolates, while potentially useful for individual animals, were of limited value in population-level health assessments. Even at the individual level, these tests suffered from the absence of validation, lack of reference range, and exceptionally limited clinical experience necessary for interpretation and application. Despite these limitations, their use and importance continue to grow.

Diagnostic tools have become increasingly sophisticated. The spectrum of available test parameters, the greater number of indications for use, and the value of such ancillary testing in the assessment of health and disease in both individual and populations of sea otters have grown exponentially. Many of the more important infectious diseases, such as *Toxoplasmosis*, *Sarcocystosis*, and morbillivirus infections, have serologic tests useful in sea otters. The clinical database of disease pathogens has been extensively fleshed out, facilitating a more comprehensive understanding of the etiopathogenesis and implications of many of the diseases to which sea otters are susceptible.

Molecularly based, technologically intensive laboratory procedures are now capable of answering many of the questions that may be central to sea otter conservation. Historic diets of individual sea otters can be evaluated through the sectioning of whiskers and measurement of levels of isotopes of carbon and nitrogen (Newsome et al., 2009). A variety of organic and inorganic pollutants are now measurable in many sea otter tissues, such as blood and fat, which can be harvested from the living animal with minimal morbidity or mortality. The use of evaluations of a suite of gene expression parameters capable of identifying the sea otter's response to a variety of environmental exposures is just in its infancy. If its promise is realized, biologists and veterinarians will be able to identify how, when, and why sea otters respond to the multiple changes, both natural and anthropogenic, in their environment (Chapter 6). Detection may be possible days, weeks, or even months before external changes are visible or alterations occur in many of the more traditionally evaluated parameters (Bowen et al., 2012; Miles et al., 2012).

The sea otter disease database has been developed and fleshed out as a result of an active veterinary pathology program. Hundreds, if not thousands, of sea otter carcasses have been subjected to a thorough post-mortem examination by veterinary pathologists associated with the US Fish and Wildlife Service in Alaska, the USGS National Wildlife Health Center in Madison, WI, and at the California Department of Fish and Game's Marine Wildlife Veterinary Care and Research Center in Santa Cruz, CA. Complete investigation of sea otter carcasses is providing valuable insight into the reasons sea otters die, and the scientific determination of both proximate and ultimate causes of death are providing insight into the role of anthropogenic and natural sources of pathogens and pollution, as well as the importance of a variety of environmental resource limitations and excesses (Kreuder et al., 2003). The combined impact of the necropsy data with that obtained from the living population is critically important in identifying and mitigating, when possible, those factors which may have a negative impact on the conservation of sea otter populations.

SEA OTTER CONSERVATION, DISEASE, AND VETERINARY MEDICINE

Veterinarians involved in sea otter conservation must make multiple changes in the clinical paradigm traditionally practiced in domestic animal medicine. Much of the veterinarian's training and experience is found in the health management of individual patients. Assessments are made of numerous aspects of the holistic picture, including clinical history, physical findings, and a myriad of laboratory and other ancillary tests of various body organs and systems, as well as specific infectious and non-infectious diseases. These are then accumulated and considered en masse, a health assessment is made, appropriate therapeutic measures are recommended, and a prognosis is given. When dealing with the conservation of free-ranging wildlife, the "patient" becomes the population, and the various "organs and body systems" are individual animals within the population. Health of this "patient" is not simply a statistical analysis of the incidence of health and/or disease within the population, but must also take into consideration a number of external factors, such as reproductive rates, resource availability, habitat use, and the effects of anthropogenic influences.

It is important to recognize that "disease" is not synonymous with infection by virus, bacteria, or parasite. According to the US National Library of Medicine, disease is defined as "an impairment of the normal state of the living animal or plant body or one of its parts that interrupts or modifies the performance of the vital functions, is typically manifested by distinguishing signs and symptoms, and is a response to environmental factors, to specific infective agents, to inherent defects of the organism, or to combinations of these factors" (National Library of Medicine, 2012). Using a more practical method, the vast majority of diseases fall into one of the etiopathogenic categories described by the acronym DAMN IT: degenerative, autoimmune, allergic, metabolic, nutritional, neoplastic, infectious, traumatic, or toxic (Table 7.3). While the genetic disorders are not listed therein, they are not well understood or appreciated in sea otter populations, although the existence of genetic bottlenecks appears to be of significant importance to many sea otter populations (Chapter 5).

One cannot take the traditional terrestrial model for evaluation of the impact of disease on wildlife populations and apply it to sea otters. The land-based models do not provide an analogous situation to the aquatic habitat of the marine mammal, an ecosystem that is subject to influences that may be brought to bear from remote areas from either the oceanic system or from associated freshwater watersheds. The impact and importance of these influences, particularly the role of land–sea transfer of pathogens and pollutants, has only recently been described and is still in need of more thorough study (Miller et al., 2010; Oates et al., 2012; Shapiro et al., 2012).

One aspect of the terrestrially-based system study that does seem to remain applicable to the study of sea otter disease is the use of historic information of the ecology of disease and its impact on wildlife populations; however, it is of limited applicability to current populations (Munson and Karesh, 2002). Having been hunted nearly to extinction for the fur trade, the remnant sea otter populations tended to be found in remote, nearly inaccessible locations. And, since population expansion tends to occur slowly over

TABLE 7.3 By Definition, Disease is "Impairment of the Normal State in the Living Organism"

DAMN IT: Acronym for Disease Etiologies	
D	Degenerative
A	Autoimmune, Allergic
M	Metabolic
N	Nutritional, Neoplastic
I	Infectious
T	Toxic, Traumatic
	Genetic

As such, the causes of disease are not limited to infectious agents, such as viruses, bacteria, and parasites. An acronym, "DAMN IT" has been used to describe the majority of disease etiologies. Note that "genetic causes" of disease are not addressed in the acronym.

relatively small spatial scales, the ability to study sea otter populations was limited. Like terrestrial species, however, the sea otters' populations, and therefore disease ecology, have been subjected to significant anthropogenic impact through habitat loss, competition for food resources, and the introduction of a spectrum of potentially harmful chemical and biological agents.

One of the most important aspects of the study of disease ecology that sea otter and terrestrial systems have in common is the unavoidable and expected inconsistency of data which occur subsequent to advances in technology. There are ever-increasing "samples" available as more people move into and frequent what were once inaccessible areas. And, society is changing, with more and more academic and conservation interest in the study of free-ranging wildlife. This impacts a wide spectrum of specialties—ecology, wildlife management, physiology, and veterinary medicine, to name just a few. With this increased interest comes a natural increase in the sensitivity and specificity of a myriad of diagnostic laboratory tools to increase our understanding of changes in the environment and the individual animal. As this develops, the reliance on the often inaccurate extrapolation of methodology from domestic animals and systems to wildlife species and systems is being replaced by situation- and species-specific tests. For example, the use of the domestic animal-based serologic test for *Toxoplasma gondii* has been replaced by a sea otter-validated assay, thereby increasing both sensitivity and specificity for this relatively common protozoal parasite of the southern sea otter (Miller et al., 2002).

Another important paradigm shift for the wildlife veterinarian involves the interpretation of the importance and role of disease in free-ranging populations. In the traditional setting, disease is a condition in which there is a disorder in a body system, function, or organ, and it needs to be corrected. When evaluating populations of animals, it is not an oxymoron to state that "disease is normal." In sea otter populations, the relationship between births and/or immigration and deaths and/or emigration determines whether the population is growing, shrinking, or remaining the same as births are greater than, less than, or equivalent to deaths, respectively. If there was no disease and subsequent death, populations would continue to grow indefinitely, an obviously impossible situation. Therefore, if one evaluates a large enough sample of any sea otter population, disease will always be found, and that is a normal finding.

The role of disease in the overall health of the sea otter population can be more difficult and often controversial in its interpretation. Ecologic attributes of the various diseases identified, such as transmissibility, impact on the animal (i.e., morbidity versus mortality), frequency, rate of occurrence, and characteristics of the disease causing agent (i.e., primary versus opportunistic) are important characteristics which warrant consideration when evaluating the impact of various diseases on free-ranging wildlife populations. In the cases of contagious or communicable diseases, there is an important "density dependence" which must also be considered, as there is a decreased opportunity for lateral disease transmission when there aren't many animals around. Additionally, as a population is exposed to any particular pathogen over time, there is generally a significant decrease in the incidence of infection, as "herd immunity," the development of pathogen-specific disease resistance through antibody formation, is stimulated.

Beyond the previously described DAMN IT categorization of disease, an ecologically important differentiation in the cause of disease is infectious diseases (those caused by living organisms such as viruses, bacteria, fungi, and parasites) versus non-infectious (those which are not caused by living organisms, which may be of internal or external origin). Infectious diseases may further be ecologically described as either contagious (transferred from one sea otter to another) or non-contagious. While such classifications may appear to be an exercise in semantics, having an understanding of the ecology of various diseases is critically important when attempting to assess the impact of a disease at the population level. As diseases are investigated at either the individual or population level, it is important to recognize that most diseases are not "mutually exclusive," and multiple pathologic conditions, often precipitated by one another, can be present at any given time (Figure 7.6).

Anecdotal reports of the spectrum of diseases seen in sea otters cover nearly every etiologic category, both infectious and non-infectious in nature. There is, however, significantly less understanding of the ecology and impact that this scope of diseases has on free-ranging sea otter populations. There is, however, one aspect of disease in this species that is likely relatively

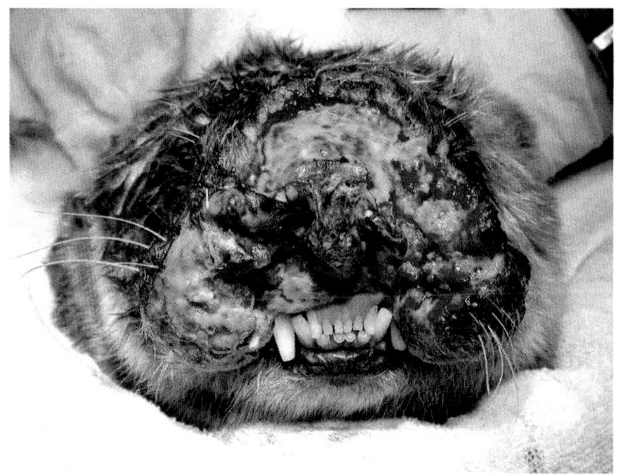

FIGURE 7.6 Extensive nose wound associated with breeding activities in a female sea otter. Multiple diseases are present in this case: secondary bacterial infection of the wounds; loss of blood cells, fluids, proteins, and electrolytes from the wound; significant heat loss through this large thermal window; significant respiratory compromise secondary to damage and swelling of the nasal passages; pre-existing and ongoing weight loss associated with recent pup weaning. This sea otter survived these injuries with medical and nutritional support and was released approximately 3 months post-stranding. *(Photo credit: Andrew Johnson.)*

consistent regardless of the cause: the impact of disease on the animal's energetic demands. As described previously, the lack of thermoneutrality in the sea otter mandates diligent and ongoing maintenance of the pelage and caloric intake through foraging and feeding. Any disease that impacts the sea otter's ability to groom, decreases the quality of the fur, or increases the number or size of the otter's thermal windows may dramatically increase energetic demands. A further increase in caloric demand may be associated with disease-specific pathology, such as pyrexia (Burkholder, 1995; Chan, 2004). If demands aren't met, energy stores are relatively rapidly mobilized and consumed, resulting in weight loss.

Heat (energy) loss through convection and conduction tends to be problematic for sea otters, given their high surface-area-to-mass ratio (Yeates et al., 2007). In the case of an otter that is unable to meet caloric demands, the weight loss decreases the denominator of this ratio. Sea otters that fail to meet energy needs experience a simultaneous increase in surface area, as the loss of body weight and condition is accompanied by increased prominence of numerous bony protuberances on the appendicular and axillary skeleton as a result of muscle atrophy (Figure 7.7). Increased surface area therefore further increases heat loss and caloric requirements. While not well described to date, these increased bony prominences will have a negative impact on the

FIGURE 7.7 Severely emaciated wild sea otter. The prominent ribs are accompanied by commensurately prominent boney projections of the vertebrae, pelvis, and other aspects of the skeleton. All combine to significantly increase body surface area and decrease hydrodynamic efficiency. *(Photo credit: Nicole LaRoche.)*

sea otter's hydrodynamics. Already a relatively inefficient swimmer biomechanically and energetically, loss of the normal body conformation must have a negative energetic impact (Cashman, 2002; Yeates et al., 2007).

Disease has a potential impact on the third critically important aspect of sea otter physiology, oxygen use. Many infectious and non-infectious diseases can negatively impact the quantity of available oxygen in the peripheral blood either by decreasing the rate of erythropoesis in the bone marrow or through direct blood loss (Brockus, 2011). Should the compensatory mechanism be incapable of correcting this deficiency, the resulting anemia and associated decrease in oxygen capacity may impact the sea otter's ability to forage for food by decreasing both dive times and depths.

This vicious cycle of impact on the sea otter through fur, food, and oxygen can occur with both infectious and non-infectious diseases. It would be expected, however, that the importance of this impact would be less in acute or peracute diseases which result in a relatively rapid death. Conversely, it should be significantly greater in those instances in which the pathogenesis is less virulent and the disease more chronic and insidious. In nearshore habitats with abundant food resources, the influence of disease on sea otters should be less significant. In those areas in which resources are limited, however, the impacts may be catastrophic for affected sea otters.

A variety of diseases, both infectious and non-infectious, have been diagnosed in free-ranging sea otters, although the ecology and pathogenesis of most of them are not well understood. As a result, their population conservation impact is speculative at best. There is evidence to support the theory that many of the non-infectious diseases with potential anthropogenic connections may impact sea otters at a population level through direct effects or secondarily by predisposing the otter through indirect actions, such as immune suppression. Most notable among these in free-ranging sea otters are the biotoxins, domoic acid and microcystin, and exposure to many of the anthropogenic persistent organic compounds, such as petroleum-related compounds, polychlorinated biphenyls (PCBs), and polycyclic aromatic hydrocarbons (Brancato et al., 2009; Jessup et al., 2010). In the case of many of these non-infectious diseases of sea otters, there is not only the potential for importance to sea otter populations, but the sea otter may be serving as an important environmental sentinel, as described earlier, as many of these compounds may have undesirable human health impacts.

An array of infectious diseases, representing nearly all of the taxa of pathogens, have also been described in sea otters (Table 7.4). Interestingly, however, the number of viral pathogens recognized remains relatively small. While the ecology of all of these diseases is still not well described, there appear to be both contagious and non-contagious diseases represented, as well as several whose behavior can likely be accurately predicted by their activity in other species. For the most part, infectious diseases in sea otters tend not to be highly virulent with significant mortality, such as seen in the 1988 and 2002 phocine distemper outbreaks involving harbor seals in northern Europe (Harkonen et al., 2006). Epizootics, the animal equivalent of epidemics, have rarely been reported in sea otter populations.

Unusual mortality events (UMEs) may be declared by the US Fish and Wildlife Service under the Marine Mammal Protection Act (MMPA). Defined as "a stranding that is unexpected; involves a significant die-off of any marine mammal population; and demands immediate response," UMEs involving sea otters have been declared only three times, in 2002, 2004, and 2006, since the inception of the program in 1991 (National Oceanic and Atmospheric Administration (NOAA), 2012). The most recent UME involved an undetermined number of northern sea otters in the Kachemak Bay region of Alaska, many dying from cardiac damage caused by infection with *Streptococcus bovis/equinus*. In April 2004, 40 southern sea otters were found sick or dead, suffering from a *Sarcocystis neurona* infection in the area near Morro Bay, CA. The 2002 event was a multispecies event involving small cetaceans, California sea lions, and sea otters that were affected by the marine biotoxin, domoic acid, a non-infectious disease in this case.

While UMEs are quite dramatic events by their nature and the associated bio-politics, they aren't necessarily reflective of population-level events. The three events declared above involving sea otters did not result in the initiation of any preventative or control measures, and there were no significant long-term impacts described on the sea otter population in the areas of these UMEs. While the impact of various sea otter pathogens isn't completely understood, there is evidence from terrestrial models that may be applicable

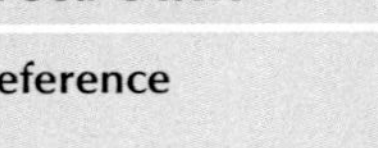

TABLE 7.4 Numerous Infectious Diseases Caused by a Broad Spectrum of Agents Have Been Diagnosed in Sea Otters

Disease	Taxon	Body System Predilection	Contagious (Y/N)	Human Health Impact	Reference
Herpes virus	Virus	Oral cavity	Unknown	No	Tseng et al. (2012)
Papilloma virus	Virus	Oral cavity	Unknown	No	1, 2
Pox virus	Virus	Skin	Unknown	No	1, 2
Phocine distemper virus	Virus	Unknown, immune system?	Unknown	No	Goldstein et al. (2009)
Canine distemper virus	Virus	Systemic	Unknown	No	1, 3
West Nile virus	Virus	Neurologic	Unknown	Yes	4
Bordetella bronchiseptica	Bacteria	Respiratory tract	Yes	No	Staveley et al. (2003)
Bartonella spp.	Bacteria	Systemic	Yes	Yes	1
Salmonella spp.	Bacteria	GI tract, sepsis	Yes	Yes	Miller et al. (2010)
Streptococcus bovis/equinus	Bacteria	Heart, sepsis	Yes	Yes	Brownstein and Miller (2011)
Streptococcus phocae	Bacteria	Wounds, sepsis	Unknown	Unknown	2
Brucella spp.	Bacteria	Wounds, bone	Unknown	Yes	Brancato et al. (2009)
Leptospira spp.	Bacteria	Urinary system	Yes	Yes	Brancato et al. (2009)
Campylobacter spp.	Bacteria	GI tract	Yes	Yes	Miller et al. (2010)

Arcanobacterium phocae	Bacteria	Wound	Unknown		Kreuder et al. (2003)
Vibrio spp.	Bacteria	Multiple	No	Yes	Miller et al. (2010)
Staphylococcus spp.	Bacteria	Multiple	Yes/no	No	Brownstein and Miller (2011)
Enteric bacteria	Bacteria	Multiple	Yes/no	Yes	Miller et al. (2010)
Clostridium spp.	Bacteria	Multiple	No	Yes	Miller et al. (2010)
Coccidiodes immitis	Fungus	Lung, systemic	No	Yes	Cornell et al. (1979)
Histoplasma capsulatum	Fungus	Systemic	Unknown	Yes	Morita et al. (2001)
Cryptococcus spp.	Fungus	Respiratory tract	Unknown	Yes	Cornell et al. (1979)
Sarcocystis neurona	Protozoa	Brain, systemic	No	Unknown	Dubey et al. (2001)
Toxoplasma gondii	Protozoa	Brain, systemic	No	Yes	Thomas and Cole (1996)
Neospora caninum	Protozoa				Dubey et al. (2003)
Giardia spp.	Protozoa	GI tract	Yes	Yes	1
Microphallus spp.	Trematode	GI tract	No	No	Margolis et al. (1997)
Plenosoma minimum	Trematode	GI tract	No	No	Margolis et al. (1997)
Nanophyetes sp.	Trematode	GI tract	Unknown	No	Margolis et al. (1997)
Orthosplanchnus fraterculus	Trematode	Gall bladder	Unknown	No	Margolis et al. (1997)

(*Continued*)

TABLE 7.4 (Continued)

Disease	Taxon	Body System Predilection	Contagious (Y/N)	Human Health Impact	Reference
Corynosoma enhydri	Acanthocephalid	GI tract	No	No	Hennessy and Morejohn (1977)
Profilicollis spp.	Acanthocephalid	GI tract, peritoneal cavity	No	No	Mayer et al. (2003)
Diplogonoporus spp.	Cestode	GI tract	No	No	Margolis et al. (1997)
Anisakis spp.	Nematode	GI tract	No	No	Tuomi and Burek (1999)
Pseudoterranova spp.	Nematode	GI tract	Unknown	No	Tuomi and Burek (1999)
Halarachne spp.	Arthropod	Nasal cavity, upper respiratory tract	Yes	No	Kenyon et al. (1965)

The contagious nature of some is known; in others it is uncertain. The diseases' impact on human health may reflect the otters' ability to transmit the disease directly to humans (zoonotic disease) or the otter and human sharing a common infection (sentinel). Some of the infectious diseases are represented in the literature; some are anecdotal experiences from sea otter veterinarians.
1 = Tuomi, personal communication
2 = Murray, unpublished data
3 = Thomas, personal communication
4 = Takle, personal communication

to sea otter populations. Pathogens that impact populations tend to do so either by affecting individual survival or by impacting reproduction. In the case of diseases that affect survival, pathogens with an intermediate virulence tend to have a greater population-level impact than do those that are highly virulent. In addition, modeling predicts that those infections that reduce reproductive rates should have a greater impact on population size than do those that affect survival (Kilpatrick and Altizer, 2012). This hypothesis may warrant further investigation given the incidence of *Toxoplasma gondii* infection in sea otters and the potential impact of the protozoan on reproduction in some species (Dubey et al., 2003; Dubey, 2008).

There has been a great deal of interest in developing a better understanding of the ecology, pathogenesis, and role in sea otter population dynamics of the infectious diseases of sea otters. While the list provided in Table 7.4 is by no means complete, it does provide insight into the spectrum of pathogens capable of infecting sea otters. But it must be noted that the presence of a potential pathogen, identified either directly or indirectly through serologic testing, is not necessarily an indication that the pathogen caused disease. Many animals, including sea otters, have serologic evidence of current or historic infection, yet no evidence of clinical disease has been detected in the affected animal.

For most of the infectious diseases recognized in sea otters, the pathogenesis is not well described. For example, numerous aspects of the infection and subsequent sequelae for the most thoroughly studied sea otter pathogen, *Toxoplasma gondii*, are not yet understood. As most evidence supports the notion of a land–sea transfer of the pathogen, many questions remain on how it is getting into the sea otter. Additionally, toxoplasmosis is a classic example of a disease in which there are far more cases of infection, diagnosed serologically, than there are clinically diseased animals (Murray, unpublished data). What makes some animals sick, and others not?

While some pathogens are considered primary pathogens, of those capable of causing disease in healthy individuals, many, if not most, are secondary or opportunistic pathogens causing disease in compromised individuals. Numerous predisposing conditions have been recognized as capable of making sea otters susceptible to these opportunistic infections. These include malnutrition, due to either quality or quantity of food; exposure to a variety of man-made or natural contaminants; concurrent infection with immune-suppressive pathogens, such as morbillivirus; and the natural wear and tear to which the sea otter is subjected over time as it senesces. All have been demonstrated or theorized as predisposing sea otters to various infectious diseases (Jessup et al., 2007; Goldstein et al., 2009, 2011; St. Leger et al., 2011). The relatively recent increase in the incidence of fatal shark bite wounds, which are having a significant population-level impact on southern sea otters in parts of its range, may be related to pre-existing disease that impacts the behavior of the sea otter, in turn predisposing it to attack (Kreuder et al., 2003).

As previously described, sea otters are considered to be an ideal sentinel species for the nearshore environment. Their susceptibility to such a wide range of pathogens has provided a means to monitor a variety of infectious diseases which may have potential human health impacts (Jessup et al., 2007). Some of these diseases, such as toxoplasmosis, coccidiomycosis, and cryptosporidiosis, are shared pathogens; both humans and sea otters are susceptible, but there is no evidence of direct transfer between the species. Some of the diseases may be considered zoonotic, transmissible from animals to humans; however, the opportunities for direct contact with the species are limited to marine mammal biologists, veterinarians, zoo/aquarium workers, and rehabilitators who work directly with the animals.

Sea otters can serve as environmental indicators for known pathogens, as well as sentinels for novel or emerging infectious diseases. Epidemiologists and public health organizations are recognizing the importance of free-ranging wildlife populations in the early detection of a variety of infectious diseases. Sea otters may prove to be important sentinels for some of these pathogens which may impact animal and/or human populations. As the incidence and apparent virulence of West Nile Virus has increased, the value of a sentinel marine mammal species may increase. The domestic mustelid cousin of the sea otter, the ferret, has been used as an animal model for influenza research for many years. Since one of the recognized wildlife reservoirs for influenza type A is marine mammals, sea otters may prove to be important sentinels for this important pathogen (Webby et al., 2007).

While the role of disease, both infectious and non-infectious, may be important in sea otter conservation, it must be considered at population levels. As "disease is normal in populations of free-ranging wildlife," not all disease is indicative of a sick population. As one considers the impact of any particular malady, a holistic approach must be taken. Does the disease impact reproductively critical components of the population? Is the disease self-limiting, dissipating after population levels reach a certain point? Is it a novel disease, or one with which the sea otter has evolved over the millennia? Is the disease indicative of other environmental conditions, such as resource limitation, presence of pollutants, or other anthropogenic factors? All of these questions and others must be considered as a part of the evaluation of the role of disease in sea otter conservation.

FUTURE ROLES OF VETERINARY MEDICINE IN SEA OTTER CONSERVATION

As more answers are discovered concerning sea otter conservation, more questions are developed. As research plans evolve to address these questions, veterinarians will most assuredly be part of the collaborative effort involved in the process. Some aspects of the investigation will be continuations of current activities, such as medical and surgical support for

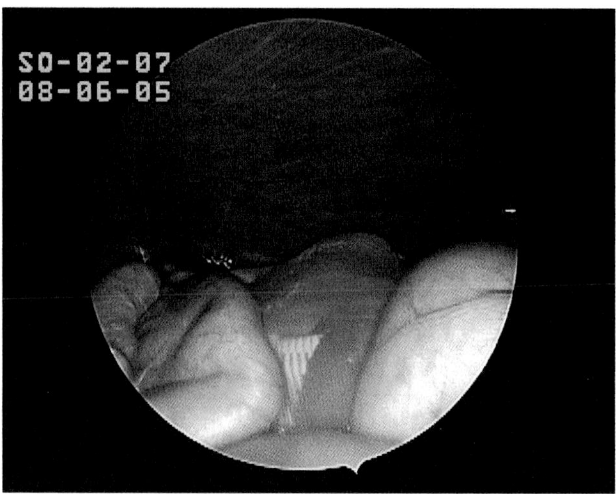

FIGURE 7.8 Laparoscopic image of a VHF radio transmitter free floating in the abdominal cavity of the sea otter. In this case it is nestled among several loops of small intestine. There is no evidence of any inflammatory or untoward reaction to its presence. *(Photo credit: MJ Murray.)*

field research, diagnostic evaluations of both live and dead sea otters, and work with basic research to better understand what makes a sea otter a sea otter. There are other anticipated roles in which veterinary medicine can play an important part.

The tracking and archival data storage instruments that are in use today are essentially identical to the instruments used over two decades ago. Improvements in instrument lifespan, advances in the amount and nature of data collected, and the degree of invasiveness of instrument application are under current development. Newer instruments should be able to take advantage of the technological advances in energy use and storage with advanced generation batteries. Network tags will facilitate data recovery through shared downloading of information to one another or to pre-established buoys, thereby eliminating the need to recapture otters for instrument extraction (Figure 7.8).

While current chemical immobilization protocols have demonstrated unprecedented safety and efficacy, they are not without drawbacks. There is a continual evolution of veterinary anesthetic methods. Wildlife veterinarians will continue to evaluate newer drugs in light of the needs of sea otters both in and ex situ. By doing so, a more balanced approach to sea otter immobilization without the respiratory and neurologic side effects may be possible.

While great strides have been made in the understanding of the role of infectious and non-infectious disease in sea otter conservation, significant gaps in the understanding of disease ecology and pathogenesis remain.

Ongoing efforts involving population-level study of sea otter populations, coupled with ever-increasing understanding of disease in zoo/aquarium sea otters and those encountered through live stranding programs, will be important in gaining an understanding of the diseases which may impact individual, as well as populations of sea otters.

A strong case has been made for the critical importance of fur, food, and oxygen in the survival of sea otters. While many of the basic building blocks of this understanding have been laid, the information is limited in scope to a small number of individuals, clinically healthy individuals, and adults. The database needs to be expanded to include various normal physiologic states in a sea otter's life, such as growth, pregnancy, lactation, and senescence. In addition, the impact of diseases on these three important parameters needs to be studied. Answering questions like "How does an infection alter the sea otter's daily caloric requirement?" will be important in addressing the role that various diseases have on free-ranging sea otter populations, especially those faced with resource limitation. While this work will be led by researchers in the basic sciences, anatomists, and physiologists, there will undoubtedly be a need for veterinary support and participation in the various studies.

An area that has yet to receive any significant work to date, yet may prove to be critically important for sea otter conservation, is the impact of climate change on sea otter populations. At first glance, the concepts of increased ocean temperatures and the rise of sea levels seem to be of little or no consequence to sea otters. But there may be significant impacts on sea otters in the more northern latitudes. Unlike the southern sea otter, whose habitat is found in the more temperate latitudes of California, northern sea otters, particularly those in Alaska and Russia, face harsh winter months. This "winter effect" on sea otter populations tends to impact animals at the "health" margins of the population, margins defined by age and/or physical condition. If winters become more temperate, what impact will be realized on reproduction, resource availability, and the role of disease on sea otter populations? Again, veterinary medicine is likely to be a partner in such an investigation.

Another, more glaring, impact of climate change on sea otter populations is the effect that temperature and pH may have on the sea otter's prey species. Annual recruitment levels may be reduced, or many of the prey may move to deeper water, having a significant impact on accessibility. If there is a shift in preferred or available prey type, will there be an associated increase in the risk for exposure to pathogens or pollutants? Alterations in prey availability and prey type may have a profound impact on sea otter populations, either through resource availability, sea otter daily activity budgets, or exposure to disease.

Lastly, and most unscientifically, yet still important, all of the basic and applied sea otter conservation research and recommendations for conservation action need to be conveyed to the public. Without their support, sea otter conservation can't happen. Veterinarians have a unique position among professionals and scientists. Recent polls have shown that the veterinarian is one of the most trusted professionals in modern society (Gallup News Service, 2006). Veterinarians are viewed as staunch advocates for animal well-being, including that of wildlife. While most scientists have very limited training in communication with the general public (Ecklund et al., 2012), the development and use of these skills in explaining the nuances of complex medical concepts to the animal-owning public is part of the daily routine of the clinical veterinarian, who is well trained for it. As such, veterinarians may be appropriate conduits for the delivery of information regarding the status and importance of sea otter populations and may be important promoters for conservation action.

There is no doubt that sea otters are an important component of the kelp-dominated northern Pacific Ocean's nearshore ecosystem. They are also a valuable sentinel whose study can provide insight into the relatively cryptic nature of ecosystem health. Over time, as conservation research efforts have developed and advanced, veterinarians and veterinary medicine have played important roles in the effort. They have been important not only in application of traditional clinical medicine and surgery, but also in the modification of these technologies to meet the needs of *in situ* conservation. Over time, the contributions of helping describe the nature of the "normal sea otter," the implications of sea otter energetics, and the nature and impact of disease on sea otter populations have been invaluable as science and society struggle to preserve this species, one of the most charismatic on the planet. The multi-disciplinary collaborative efforts continue and will advance into the future with the veterinarian as a valued partner as efforts in sea otter conservation continue.

REFERENCES

Ames, J.A., Hardy, R.A., Wendell, F.E., 1983. Tagging materials and methods for sea otters, *Enhydra lutris*. Calif. Fish Game 69, 243–252.

Ames, J.A., Hardy, R.A., Wendell, F.E., 1986. A simulated translocation of sea otters, *Enhydra lutris*, with a review of capture, transport and holding techniques. Marine Resources Technical Report No. 52. Calif. Dep. Fish Game, pp. 17.

Berta, A., Sumich, J.L., 1999. Marine Mammals: Evolutionary Biology. Academic Press, San Diego, CA.

Bodkin, J.L., Esslinger, G.G., Monson, D.H., 2004. Foraging depths of sea otters and implications to coastal marine communities. Mar. Mammal Sci. 20, 305–321.

Bodkin, J.L., Monson, D.H., Esslinger, G.G., 2007. Activity budgets derived from time-depth recorders in a diving mammal. J. Wildl. Manage. 71, 2034–2044.

Bowen, L., Miles, A.K., et al., 2012. Gene transcription in sea otters (*Enhydra lutris*); development of a diagnostic tool for sea otter and ecosystem health. Mol. Ecol. Resour. 12, 67–74.

Boyd, K.L., Bolon, B., 2010. Embryonic and fetal hematopoiesis. In: Weiss, D.J., Wardrop, K.J. (Eds.), Schalm's Veterinary Hematology, sixth ed. Wiley-Blackwell, Ames, IA, pp. 3–7.

Brancato, M.S., Milonas, L., Bowlby, C.E., Jameson, R., Davis, J.W., 2009. Chemical contaminants, pathogen exposure and general health status of live and beach-cast Washington sea otters (*Enhydra lutris kenyoni*). Marine Sanctuaries Conservation Series (ONMS-09-01).

Brockus, C.W., 2011. Erythrocytes. In: Latimer, K.S. (Ed.), Veterinary Laboratory Medicine: Clinical Pathology, fifth ed. Wiley-Blackwell, Ames, IA, pp. 3–44.

Brownstein, D., Miller, M.A., 2011. Antimicrobial susceptibility of bacterial isolates from sea otters (*Enhydra lutris*). J. Wildl. Dis. 47, 278–292.

Burkholder, W.J., 1995. Metabolic rates and nutrient requirements of sick dogs and cats. J. Am. Vet. Med. Assoc. 206, 614–618.

Cashman, M.E., 2002. Diving in the Southern Sea Otter, *Enhydra lutris nereis*: Morphometrics, Buoyancy, and Locomotion. University of California, Santa Cruz, CA, Master's Thesis.

Castellini, M.A., Kooyman, G.L., Ponganis, P.J., 1992. Metabolic rates of freely diving Weddell seals: correlations with oxygen stores, swim velocity and diving duration. J. Exp. Biol. 165, 181–194.

Chan, D.L., 2004. Nutritional requirements of the critically ill patient. Clin. Tech. Small Anim. Pract. 19, 1–5.

Cornell, L.H., Osborn, K.G., Antrim, J.E., 1979. Coccidioidomycosis in a California sea otter (*Enhydra lutris*). J. Wildl. Dis. 15, 373–378.

Costa, D.P., Kooyman, G.L., 1982. Oxygen consumption, thermoregulation, and the effect of fur oiling and washing on the sea otter, *Enhydra lutris*. Can. J. Zool. 60, 2761–2767.

Costa, D.P., Kooyman, G.L., 1984. Contribution of specific dynamic action to heat balance and thermoregulation in the sea otter, *Enhydra lutris*. Physiol. Zool. 57, 199–203.

Denny, M.W., 1993. Air and Water: The Biology and Physics of Life's Media. Princeton University Press, Princeton, NJ.

Dubey, J.P., 2008. The history of *Toxoplasma gondii*—the first 100 years. J. Eukaryot. Microbiol. 55, 467–475.

Dubey, J.P., Rosypal, A.C., Rosenthal, B.M., Thomas, N.J., Lindsay, D.S., Stanek, J.F., et al., 2001. *Sarcocystis neurona* infections in sea otter (*Enhydra lutris*): evidence for natural infections with Sarcocystis and transmission of infection to opossums (*Didelphis virginiana*). J. Parasitol. 87, 1387–1393.

Dubet, J.P., Zarnke, R., Thomas, N.J., Wong, S.K., Van Bonn, W., Briggs, M., et al., 2003. *Toxoplasma gondii, Neospora caninum, Sarcocystis neurona,* and *Sarcocystis canis*-like infections in marine mammals. Vet. Pathol. 116, 275–296.

Ecklund, E.H., James, S.A., Lincoln, A.E., 2012. How academic biologists and physicists view science outreach. PLoS ONE 7 (5), e36240. Available from: http://dx.doi.org/doi:10.1371/journal.pone.0036240.

Esslinger, G.G., 2011. Temporal Patterns in the Behavior and Body Temperature of Sea Otters in Alaska. Master of Science Thesis. University of Alaska Anchorage, pp. 74.

Esslinger, G.G., Bodkin, J.L., Breton, A.R., Burns, J.M., Monson, D., 2014. Temporal patterns in the foraging behavior of sea otters in Alaska. J. Wildl. Manage. 78, 689–700.

Estes, J.A., Palmisono, J.F., 1974. Sea otters: their role in structuring nearshore communities. Science 185, 1058–1060.

Gallup News Service, 2006. Gallup Poll December 8–10, 2006.

Goldstein, T., Mazet, J.A.K., Gill, V.A., Doroff, A.M., Burek, K.A., Hammond, J.A., et al., 2009. Phocine distemper virus in northern sea otters in the Pacific Ocean, Alaska, USA. Emerging Infect. Dis. 15, 925–927.

Goldstein, T., Gill, V.A., Tuomi, P., Monson, D., Burdin, A., Conrad, P.A., et al., 2011. Assessment of clinical pathology and pathogen exposure in sea otters (*Enhydra lutris*) bordering the threatened population in Alaska. J. Wildl. Dis. 47, 579–592.

Harkonen, T., Dietz, R., Reijnders, P., Teilmann, J., Harding, K., Hall, A., et al., 2006. The 1988 and 2002 phocine distemper virus epidemics in European harbor seals. Dis. Aquat. Org. 68, 115–130.

Hennessy, S.L., Morejohn, V.J., 1977. Acanthocephalan parasites of the sea otter, *Enhydra lutris*, of coastal California. Calif. Fish Game 63, 268–272.

Jessup, D.A., Miller, M.A., Kreuder-Johnson, C., Conrad, P.A., Tinker, M.T., Estes, J., et al., 2007. Sea otters in a dirty ocean. J. Am. Vet. Med. Assoc. 231, 1648–1652.

Jessup, D.A., Johnson, C.K., Estes, J., Carlson-Bremer, D., Jarman, W.M, Reese, S., et al., 2010. Persistent organic pollutants in the blood of free-ranging sea otters (*Enhydra lutris* spp.) in Alaska and California. J. Wildl. Dis. 46, 1214–1233.

Kenyon K.W., 1969. The sea otter in the eastern Pacific Ocean. US Fish and Wildlife Service North American Fauna, 68, pp. 1–352.

Kenyon, K.W., Yunker, C.D., Newell, I.M., 1965. Nasal mites (Halarachnidae) in the sea otter. J. Parasitol. 51, 960.

Kilpatrick, A.M., Altizer, S., 2012. Disease ecology. Nat. Educ. Knowl. 3, 55–65.

Kreuder, C., Miller, M.A., Jessup, D.A., Lowenstines, L.J., Harris, M.D., Ames, J.A., et al., 2003. Patterns of mortality in southern sea otters (*Enhydra lutris nereis*) from 1998–2001. J. Wildl. Dis. 39, 495–509.

Kuhn, R.A., Ansorge, H., Godynicki, S., Meyer, W., 2010. Hair density in the Eurasian otter *Lutra lutra* and the sea otter *Enhydra lutris*. Acta Theriol. 55, 211–222.

Lenfant, C., Johansen, K., Torrance, J.D., 1970. Gas transport and oxygen storage capacity in some pinnipeds and the sea otter. Respir. Physiol. 9, 277–286.

Margolis, L., Groff, J.M., Johnson, S.C., McDonald, T.E., Kent, M.L., Blaylock, R.B., et al., 1997. Helminth parasites of sea otters (*Enhydra lutris*) from Prince William Sound, Alaska: comparisons with other populations of sea otters and comments on the origin of their parasites. J. Helminthol. Soc. Wash. 64, 161–168.

Mayer, K.A., Dailey, M.D., Miller, M.A., 2003. Helminth parasites of the southern sea otter *Enhydra lutris nereis* in central California: abundance, distribution and pathology. Dis. Aquat. Org. 53, 77–88.

Miles, A.K., Bowen, L., Ballachey, B., Bodkin, J.L., Murray, M., Estes, J.L., et al., 2012. Variations of transcript profiles between sea otters *Enhydra lutris* from Prince William Sound, Alaska, and clinically normal reference otters. Mar. Ecol. Prog. Ser. 451, 201–212.

Miller, M.A., Gardner, L.A., Packham, A., Mazet, J.K., Hanni, K.D., et al., 2002. Evaluation of an indirect fluorescent antibody test (IFAT) for demonstration of antibodies to *Toxoplasma gondii* in the sea otter (*Enhydra lutris*). J. Parasitol. 88, 594–599.

Miller, M.A., Byrne, B.A., Jang, S.S., Dodd, E.M., Dorfmeier, E., Harris, M.D., et al., 2010. Enteric bacterial pathogen detection in southern sea otters (*Enhydra lutris nereis*) is associated with coastal urbanization and freshwater runoff. Vet. Res. 41, 01.

Monson, D.H., McCormick, C., Ballachey, B.E., 2001. Chemical anesthesia of northern sea otters (*Enhydra lutris*): results of past field studies. J. Zoo Wildl. Med. 32, 181–189.

Morita, T., Kishimoto, M., Shimada, A., Matsumoto, Y., Shindo, J., et al., 2001. Disseminated Histoplasmosis in a sea otter (*Enhydra lutris*). J. Comp. Pathol. 125, 219–223.

Morrison, P., Rosenmann, M., Estes, J.A., 1974. Metabolism and thermoregulation in the sea otter. Physiol. Zool. 47, 218–228.

Munson, L., Karesh, W.B., 2002. Disease monitoring for the conservation of terrestrial animals. In: Aguirre, A.A., Ostfeld, R.S., Tabor, G.M., House, C., Pearl, M.C. (Eds.), Conservation Medicine: Ecological Health in Practice. Oxford University Press, New York, NY, pp. 95–103.

National Library of Medicine, 2012. <http://www.nlm.nih.gov/medlineplus/mplusdictionary.html>.

National Oceanic and Atmospheric Administration (NOAA), 2012. <http://www.nmfs.noaa.gov/pr/health/mmume>.

Newsome, S.D., Tinker, M.T., Monson, D.H., Oftedal, O.T., Ralls, K., Staedler, M.M., et al., 2009. Using stable isotopes to investigate individual diet specialization in California sea otters (*Enhydra lutris nereis*). Ecology 90, 961–974.

Oates, S.C., Miller, M.A., Byrne, B.A., Choulcha, N., Hardin, D., Jessup, D., et al., 2012. Epidemiology and potential land-sea transfer of enteric bacteria from terrestrial to marine species in the Monterey Bay region of California. J. Wildl. Dis. 48, 654–668.

Ponganis, P.J., 2011. Diving mammals. Compr. Physiol. 1, 517–535.

Ralls, K., Siniff, D.B., Williams, T.D., Kuechle, V.B., 1989. An intraperitoneal radio transmitter for sea otters. Mar. Mammal Sci. 5, 376–381.

Ralls, K., Hatfield, B.B., Siniff, D.B., 1995. Foraging patterns of California sea otters as indicated by telemetry. Can. J. Zool. 73, 523–531.

Shapiro, K., Miller, M., Mazet, J., 2012. Temporal association between land-based runoff events and California sea otter (*Enhydra lutris nereis*) protozoal mortalities. J. Wildl. Dis. 48, 394–404.

Sparling, C.E., Fedak, M.A., 2004. Metabolic rates of captive grey seals during voluntary diving. J. Exp. Biol. 207, 1615–1624.

St. Leger, J.A., Righton, A.L., Nilson, E.M., Fascetti, A.J., Miller, M.A., Tuomi, P.A., et al., 2011. Vitamin A deficiency and hepatic retinol levels in sea otters, *Enhydra lutris*. J. Zoo Wildl. Med. 42, 98–104.

Staveley, C.M., Register, K.B., Miller, M.A., Brockmeier, S.L., Jessup, D.A., Jang, S., et al., 2003. Molecular and antigenic characterization of *Bordetella bronchiseptica* isolated from a wild southern sea otter (*Enhydra lutris nereis*) with severe suppurative bronchopneumonia. J. Vet. Diagn. Invest. 15, 570–574.

Thomas, N.J., Cole, R.A., 1996. The risk of disease and threats to the wild population. Endangered Species Update 13, 23–27.

Tinker, M.T., 2004. Sources of Variation in the Foraging Behavior and Demography of the Sea Otter, *Enhydra lutris*. University of California, Santa Cruz, CA, PhD Thesis, pp. 180.

Tseng, M., Fleetwood, M., Reed, A., Gill, V.A., Harris, R.K., Moeller, R.B., et al., 2012. Mustelid Herpesvirus-2, a novel Herpes infection in northern sea otters (*Enhydra lutris kenyoni*). J. Wildl. Dis. 48, 181–185.

Tuomi P., Burek K., 1999. Septic peritonitis in an adult northern sea otter (*Enhydra lutris*) secondary to perforation of gastric parasitic ulcer. In: Proceedings of the 30th Annual International Association for Aquatic Animal Medicine, vol. 30, pp. 63–64.

Walker, K.A., Trites, A.W., Haulena, M., Weary, D.M., 2012. A review of the effects of different marking and tagging techniques on marine mammals. Wildl. Res. 39, 15–30.

Webby, R.J., Webster, R.G., Richt, J.A., 2007. Influenza viruses in animal wildlife populations. Curr. Top. Microbiol. Immunol. 315, 67–83.

Weisel, J.W., Nagaswami, C., Peterson, R.O., 2005. River otter hair structure facilitates interlocking to impede penetration of water and allow trapping of air. Can. J. Zool. 83, 649–655.

Williams, T.D., Kocher, F.H., 1978. Comparison of anesthetic agents in the sea otter. J. Am. Vet. Med. Assoc. 173, 1127–1130.

Williams, T.D., Siniff, D.B., 1983. Surgical implantation of radiotelemetry devices in the sea otter. J. Am. Vet. Med. Assoc. 183, 1290–1291.

Williams, T.D., Williams, A.L., Siniff, D.B., 1981. Fentanyl and azaperone produced neuroleptanalgesia in the sea otter (*Enhydra lutris*). J. Wildl. Dis. 17, 337–342.

Williams, T.D., Rebar, A.H., Teclaw, R.F., Yoos, P.E., 1992. Influence of age, sex, capture technique, and restraint on hematologic measurements and serum chemistries of wild California sea otters. Vet. Clin. Pathol. 21, 106–110.

Williams, T.M., Kastelein, R.A., Davis, R.W., Thomas, J.A., 1988. The effects of contamination and cleaning on sea otters (*Enhydra lutris*). Thermoregulatory implications based on pelt studies. Can. J. Zool. 66, 2776–2781.

Williams, T.M., Fuiman, L.A., Horning, M., Davis, R.W., 2004. The cost of foraging by a marine predator, the Weddell seal (*Leptonychotes weddellii*): pricing by the stroke. J. Exp. Biol. 207, 973–982.

Yeates, L.C., 2006. Physiological Capabilities and Behavioral Strategies for Marine Living by the Smallest Marine Mammal, the Sea Otter (*Enhydra lutris*). University of California, Santa Cruz, CA, PhD Thesis, pp. 121.

Yeates, L.C., Williams, T.M., Fink, T.L., 2007. Diving and foraging energetic of the smallest marine mammal, the sea otter (*Enhydra lutris*). J. Exp. Biol. 210, 1960–1970.

Chapter 8

Sea Otters in Captivity: Applications and Implications of Husbandry Development, Public Display, Scientific Research and Management, and Rescue and Rehabilitation for Sea Otter Conservation

Glenn R. VanBlaricom[1], Traci F. Belting[2] and Lisa H. Triggs[3]

[1]*Washington Cooperative Fish and Wildlife Research Unit, USGS, and School of Aquatic and Fishery Sciences, College of the Environment, University of Washington, Seattle, WA, USA,* [2]*Department of Life Sciences, The Seattle Aquarium, Seattle, WA, USA,* [3]*Aquatic Animal Department, Point Defiance Zoo & Aquarium, Tacoma, WA, USA*

Sea Otter Conservation. DOI: http://dx.doi.org/10.1016/B978-0-12-801402-8.00008-1

INTRODUCTION

The knowledge gained from captive sea otters has contributed to marine conservation for more than six decades. Captive sea otters have served as sources of research data applicable to conservation and management of wild sea otters and their marine habitats, as subjects for research on methods for husbandry in captive circumstances, and as facilitators of public involvement in conservation education, policy development, and implementation. The impact of captive sea otters on marine conservation results from a number of attributes, four of which are particularly important in our collective view. First, sea otters are adaptable to captivity and public display because they are physically manageable. Sea otters are the smallest of the northern hemisphere marine mammal species in terms of body mass, reaching maxima of ~50 kg in adult males and ~38 kg in adult females. In contrast, sustained high-quality life in captivity remains intractable for many marine mammal species because of extreme body sizes and highly mobile lifestyles at sea. Second, the history of harvest of sea otters for the commercial fur trade from 1741 until at least 1939, and by one interpretation to the present (see Chapter 14), involved the taking of nearly one million pelts range-wide (Coxe, 1780; Berkh, 1823; Cobb, 1906; Lensink, 1960; Kenyon, 1969; Abegglen, 1977). The harvests drove the species to the threshold of extinction (Kenyon, 1969) and stimulated significant public empathy. Third, the public has become acquainted with scientific evidence of the importance of sea otters to the ecological structure and dynamics of coastal marine ecosystems (see Chapter 2). The representation of sea otters as disproportionately important to coastal ecosystems dates from observations in California (McLean, 1962) and Alaska (Estes and Palmisano, 1974). Terms such as "keystone species" seem to resonate with public sentiment, contributing to the premise that conservation effort invested in such species provides multiple positive benefits. Finally, sea otters are among a small group of animal species whose appearance and behavioral repertoire stimulate adoration by the public on a largely emotional basis, particularly when the animals can be viewed at close range in a high-quality zoological setting.

Here we review insights into the ecology and conservation of sea otters resulting from their presence in captivity, in two sections:

1. Small numbers (<10 individuals): Such settings typically involve zoos or aquaria in which the principal purpose of captivity is display and associated provision of information and entertainment to the interested lay public. This category of captivity also includes research programs at institutions such as universities, where the primary purpose of captivity is to provide subjects for research in physiology or methods for captive husbandry.
2. Large numbers (two cases, one involving 150 animals, the other 357): Such circumstances have involved either temporary newly constructed holding facilities or utilization of space and resources in permanent

facilities typically used for purposes other than captive sea otter husbandry. Holding of large numbers of animals has been done on short time scales (days to months). The two subject cases are sequential accumulation of groups of animals for involvement in the California translocation program of 1987–1990, and accumulation of animals for treatment and rehabilitation following rescue in habitats affected by the *Exxon Valdez* oil spill (EVOS) in 1989.

PART ONE: SMALL NUMBERS OF SEA OTTERS IN CAPTIVITY

Background and History

Public exhibits provide a unique opportunity for people to feel a connection to many types of animals (Clayton et al., 2009). Clayton further suggested that if zoos can successfully increase these feelings and encourage discussion of such feelings, they may enhance visitors' support for conservation initiatives. The first public aquarium in the United States was part of Barnum's American Museum which opened in 1856 in New York. The Boston Aquarial Gardens were founded in 1859. America's first zoo, the Lincoln Park Zoological Garden, opened in Chicago in 1868, followed by the Philadelphia Zoo in 1874. Marine mammals were first publicly displayed in the United States in 1938 with the opening of Marineland in Florida. Since 1954, 172 sea otters (as of March 2013) have been housed, and in many cases publicly exhibited in North American facilities (Kenyon, 1969; Gruber and Hogan, 1990; Casson, 2013b; Figure 8.1).

In 1954 sea otters from Alaska were briefly displayed in the United States at the Woodland Park Zoo (WPZ) in Seattle, Washington, then transported to the National Zoological Park (NPZ) in Washington, DC.

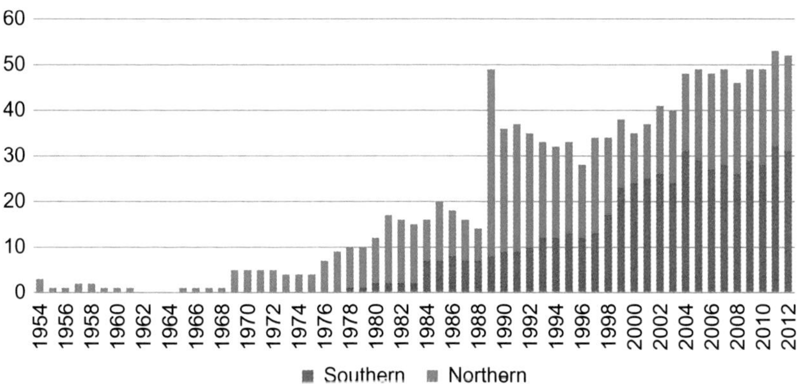

FIGURE 8.1 Number of sea otters in captive facilities in North America (Kenyon, 1969; Gruber and Hogan, 1990; Casson 2013a,b).

Unfortunately, none survived at NZP for more than 10 days (Kirkpatrick et al., 1955; Kenyon, 1969). A female sea otter ("Susie") was housed for 6 years (October 1955 to October 1961) in a freshwater pool at WPZ, joined by a male sea otter ("Dave") from December 1957 to September 1958 (Vincenzi, 1962).

Sea otter numbers in North American zoos and aquaria remained at <20 individuals until the aftermath of EVOS in Prince William Sound, Alaska, in 1989 (Figure 8.1). During and following EVOS the collective efforts of wildlife biologists, aquarium staffs, veterinarians, and volunteers helped to rehabilitate and release 197 animals exposed to spilled oil (Townsend and Heneman, 1989; Bayha and Kormendy, 1990a; Williams and Davis, 1990, 1995; Williams and Williams, 1996; Chapter 4). An additional 37 sea otters were deemed non-releasable and sent to facilities around the world with 22 remaining in North America (Gruber and Hogan, 1990). The last surviving non-releasable sea otter rescued after EVOS and held in North America, a beloved 25-year-old female named Homer, died at the Point Defiance Zoo and Aquarium (PDZA) in June 2013.

In 1978 Sea World in San Diego, California, became the first facility to house sea otters native to California ("southern sea otters," *Enhydra lutris nereis* [Merriam, 1904]) for public display. The Monterey Bay Aquarium (MBA; Monterey, California) followed in 1984. The first southern sea otters displayed publicly outside the state of California were a male and female moved to the New York Aquarium in 1991. Sixteen facilities in North America currently hold resident captive sea otters. Fourteen are zoos or aquaria that display the animals publicly. Two others are research facilities not providing public display of captive otters. As of March 2013 there were 19 northern (*Enhydra lutris kenyoni* [Wilson, 1991]) and 32 southern sea otters in North American facilities living at 16 locations, 14 of which offer public display (Casson, 2013a).

Conservation Value of Captive Animals in Shaping Public Perceptions

Facilities displaying sea otters represent significant potential for effective public outreach in the context of marine conservation by virtue of attendance. For example, in 2013 the combined annual attendance figure for the 14 North American facilities with sea otters on public display was over 20 million,[1] exceeding by ~2.5 million people the combined ticket sales at every professional football game for every team in the entire US National Football League in the same year.[2] In a national survey ("Harris Poll") of

1. http://www.aza.org/azapublications.
2. http://espn.go.com/nfl/attendance/_/year/2013.

more than 1000 adults conducted for the Alliance of Marine Mammal Parks and Aquariums, 91% of those polled agreed that seeing a living marine mammal at an aquarium or zoo fosters a connection to the animal.[3] Ninety-four percent of respondents agreed that children are more likely to be concerned about animals if they learn about them at aquaria and zoos, and that such visits can inspire conservation action helpful to marine mammals and their habitats.

Zoo visits appear to enable experiences of shared identity between humans and displayed animals. In a number of different ways, visitors to zoos show that they are trying to put themselves in the place of the animals or make a behavioral or conceptual connection between themselves and the animals (Clayton et al., 2009). The learning potential of a zoo or aquarium experience was illustrated when 89% of respondents to the Harris Poll agreed that children learn more about marine mammals at an aquarium or zoo than in a school classroom, and 88% agreed that one can learn about animals at marine parks in a way that cannot be replicated by watching film or television programs.

Facility managers with the goal of inspiring an ethic of marine conservation have found it imperative to understand and promote the specific aspects of a zoo or aquarium visit most effectively enhancing learning, thereby potentially encouraging environmentally sustainable behavior. The quality of (zoo and aquarium) exhibits and staffs are most important in influencing outcomes favorable to a conservation perspective (Wagner et al., 2009). Staff enthusiasm and availability to answer questions in an understandable manner, backed by expertise, are the specific experiences particularly important to facilitation of positive conservation attitudes and values. Personal interpretation in zoos and aquaria provides opportunities for connections that can enhance the visitor experience and increase knowledge gained by visitors (Visscher et al., 2009). Such gains are more likely to be realized when visitors (a) are left with the perception that the zoo or aquarium takes good care of its animals; (b) are easily able to spot animals in exhibits; and (c) have opportunities to experience close proximity to some animals (Wagner et al., 2009).

An effective way to combine quality exhibits, professional interpretation, and enhanced animal experience is through formal animal presentations to visitors. Behavioral enrichment, especially animal training, can highlight species-specific natural behaviors and provide visitors with an opportunity to see a wide range of behaviors in a relatively short time span, thereby providing a greater appreciation of an animal's unique abilities. Learning can be defined as any change in behavior due to experience, and operant learning as any behavior that is consequently strengthened or weakened (Chance, 1999). Through the use of operant learning and conditioning, with an emphasis on positive

3. http://www.ammpa.org/doc_harrispoll2012.html.

reinforcement, sea otters can be trained to perform a wide variety of behaviors serving to highlight critical natural history and conservation stories. In this context, sea otter behaviors fall loosely into two categories: natural behaviors and those enhancing the efficacy of animal husbandry. Natural behaviors typical of animals in the wild include surface swimming, diving, rolling, grooming, vocalizing, manipulating objects, and climbing onto and moving about on solid surfaces. Husbandry behaviors can be as simple as training a sea otter to shift between exhibits, enter a transport crate, or stand on a scale to be weighed, or include more complex actions such as holding the mouth open for a dental examination and teeth cleaning, or remaining motionless in a position necessary for an unrestrained blood sample, intramuscular injection, or radiological or ultrasonic examination. Combining both natural and husbandry behaviors into an interpreted presentation can provide the public with a positive zoo or aquarium experience. Conservation-learning outcomes are postulated to be consequences of pleasurable engaging experiences with others in the presence of animals, particularly experiences that promote empathy with the animals (Clayton et al., 2009).

Conservation Value of Research on Resident Animals in Zoos and Aquaria

Although education is a key component of conservation initiatives in zoos and aquaria, the role of research is equally important, linking the management of the captive population with the conservation needs of wild populations (Foose, 1989). Research programs at facilities provide opportunities for scientists to obtain data from individual animals over a period of time and across life history stages, a task which is difficult with free-ranging wild animals. For marine mammals amenable to captivity, successful conservation is most effective when the expertise, efforts, and resources of aquaria, academia, and wildlife biology are combined and coordinated. Carefully planned research can advance an understanding of the basic biology of a species, and therefore improve the scientific foundation of conservation programs. Ninety percent of Harris Poll respondents felt it was essential or somewhat important to help species in the wild by studying their biology in marine life parks, aquaria, and zoos. When linked to education programs, research findings can be used to shape public perceptions about conservation issues, leading to public empathy for the environment and the animals that may translate into positive conservation policies and actions implemented by government agencies.

To determine if zoos and aquaria successfully promote conservation, the Association of Zoos and Aquariums undertook a 3-year, nationwide study of the impacts of visits to US zoos or aquaria (Falk et al., 2007). Most visitors (61% of respondents) indicated that their zoo and aquarium experiences supported and reinforced their values and attitudes toward conservation. Visits to accredited zoos and aquaria[4] prompted many individuals (54%) to reconsider their role in addressing environmental problems and contributing to conservation action, and to seeing themselves as part of the solution. A majority (57%) of responding visitors described time at zoos and aquaria as strengthening their connection to nature, while a higher proportion (66%) of respondents said that zoos and aquaria serve important roles in species preservation and improved public awareness of conservation issues.

Brennan and Houck (1996) evaluated efficacy of strategies for managing the captive sea otter population in North America to maximize conservation benefits. A key identified constraint was lack of necessary space for captive animals in North American zoos and aquaria to effectively contribute to conservation of sea otters without periodic reliance on replacement animals from the wild. Accordingly, Brennan and Houck recommended increasing the carrying capacity of facilities capable of holding resident captive sea otters, increasing the number of captive resident adult females that are allowed to breed, utilizing input of stranded animals to the captive breeding population more effectively, and expanding the international components of management efforts, among other recommendations.

The Seattle Aquarium became the first institution to raise captive sea otter pups from conception to adulthood in 1979 with the successful birth and development of a male, "Tichuk." Ten pups have since been born to resident captive adults at the Seattle Aquarium. Nine of the ten have survived to adulthood. Most sea otters in zoos and aquaria in North America originate from populations listed as threatened pursuant to the US Endangered Species Act of 1973[5] as amended. The California subspecies was listed as threatened in 1977 and the Southwest Alaska Stock of the northern sea otter (*Enhydra lutris kenyoni* [Wilson, 1991]) was similarly listed in 2005 (see also Chapters 3, 13, and 14). In recent years animal managers from facilities housing sea otters in North America, in conjunction with the US Fish and Wildlife Service (USFWS), have agreed informally to a moratorium on breeding of captive sea otters to preserve space required for rescued sea otters deemed non-releasable by USFWS. The new protocol and other measures have addressed some of the concerns noted by Brennan and Houck, including the moratorium on captive breeding and an agreed moratorium on transfer of rehabilitated sea otters out of the United States.

Values and Benefits of Aquarium Staffs as Resources for Sea Otter Conservation

Public zoos and aquaria have contributed historically to *in situ* conservation projects by donating staff, resources, and funding. Zoo and aquarium

4. http://www.aza.org/what-is-accreditation.

5. 87 US Statutes at Large 884-903, Public Law 93-205, 28 December 1973.

personnel and infrastructure have benefited sea otter recovery activities (federal plan for sea otters in California, state plan in Washington) by providing facility access and expertise, and by sharing experiences. Biennial sea otter conservation workshops have been hosted by the Seattle Aquarium, in Seattle, Washington, since 1999. The workshops afford opportunities for scientists studying sea otters to meet and share insights, experiences, and data, and to collaborate among themselves and with aquarium personnel working every day with captive resident sea otters in the realm of husbandry. Zoo and aquarium staffs have assisted in field population surveys of sea otters in California and Washington since 1984 and 2001, respectively. Zoo and aquarium staffs responded in extraordinary numbers to the consequences of EVOS, providing guidance, assistance, expertise, and hard labor on behalf of sea otters in Alaska, and opened their respective facilities for placement of sea otters that could not be rehabilitated and returned to native habitats.

Pioneering Research Projects

In this section we review two ground-breaking research and development programs, one Soviet and the other American. The two programs resolved fundamental issues pertaining to successful captivity of sea otters, and solved informational challenges in sea otter conservation as well. The US program operated without knowledge of the prior existence of the Soviet program as a consequence of language barriers (Kenyon, 1969). The two programs initiated and advanced the cause of sea otter science and conservation and, despite primitive working conditions, disparate goals, and an inability to communicate, came independently to many of the same conclusions.

Soviet Projects, Medney Island and Kola Peninsula, 1932–1940

The first recorded attempts to hold sea otters in captivity for purposes of research or management were at Medney Island in the Commander Islands, Union of Soviet Socialist Republics (USSR), beginning in spring 1932 (Barabash-Nikiforov, 1947). The principal goal was domestication of sea otters for fur production. At the time of the Soviet research sea otters were perceived to be incapable of surviving in captivity (Elliott, 1875). Thus, the first goal of Soviet research at Medney was to determine if sea otters could be maintained in good health in captivity.

The first captive holding facility was an abandoned house at Preobrazhenskoe Village (54.8°N, 167.6°E) with a sand-covered floor and a small concrete pool supplied with freshwater. Two sea otters ("Fomka" and "Bek") were held separately in the enclosure during March and April 1932. Food items provided were primarily sea urchins and fish, all consumed readily. One otter died after 10 days in captivity, the other after 13. Both animals experienced episodes of uneasiness, diminished appetite, diarrhea, and weight loss in the few days prior to their respective deaths. *Post mortem* examinations indicated acute gastrointestinal inflammation in both cases. Deaths were attributed to "the general unsuitability of the building" as a facility for maintenance of captive wild sea otters.

T.A. Mal'kovich studied captive sea otters at Medney Island from 1934 through 1936 (Barabash-Nikiforov, 1947). Research goals involved development of captive breeding protocols. Mal'kovich maintained an outdoor enclosure for captive wild sea otters at Gladkovskaya Bay (54.7°N, 167.8°E), holding 11 different sea otters for observation until completion of the field project in 1936. Significant findings included observation of substantial individual variation in behavior patterns, and the importance of sea urchins in the diet of sea otters, both for nutritional benefit and sustained gastrointestinal health.

In 1937 Soviet managers and biologists initiated an experimental translocation of sea otters from Medney Island to the Kola Peninsula on the southwestern Barents Sea, to a specific area termed the "east Murman shore" (Barabash-Nikiforov, 1947) and located ~5600 km (great circle distance) from the nearest native sea otter habitat in the North Pacific (Kamchatka Peninsula). Goals of the project were to determine the feasibility of translocation, to evaluate the potential for establishment of sea otter populations in areas far removed from native geographic range, and to investigate methods for supporting captive breeding, all with the overarching goal of domestication for purposes of commercial-scale fur production. In 1937 nine sea otters were captured at Medney Island and placed onboard a ship destined for Vladivostok, ~2900 km distant (great circle distance). Transport cages were secured on deck, fully exposed to the weather, and included pools with flowing seawater supplied by shipboard plumbing. Severe storms with frigid temperatures occurred during the vessel crossing, exposing the sea otters to extreme chilling. Seven of the nine otters perished in the first few days after departure from Medney Island. *Post mortem* examinations revealed pulmonary inflammation in all cases, suggesting hypothermia as the primary cause of mortalities. The two surviving sea otters, "Buyan" and "Yashka" (both males), traveled from Vladivostok to the east Murman shore by railroad, arriving November 27, 1937. The excessive time in transit (>30 days) led to the suggestion that use of aircraft in future projects would be a "very likely and expedient method" (Barabash-Nikiforov, 1947). The two translocated sea otters were placed in a captive holding facility extending across the intertidal zone at Yarnyshnaya Bay (69.1°N, 35.8°E; Figure 8.2).

The enclosure was largely flooded by seawater at high tidal levels, and partially drained at low tide. Retention of high concentrations of fecal material and discarded food remains compromised water quality within the enclosure soon after it was occupied by the sea otters, particularly at low tide.

FIGURE 8.2 Holding facilities for sea otters on the Murman shore, USSR, November 1937 to January 1940. Upper image faces seaward and shows the original (at right) and modified (center) holding facility, each extending from shoreline across the intertidal zone. Lower image shows interior of modified holding facility facing shoreward from the seaward end. Otters were moved to the modified enclosure in March 1939. Captive sea otters had free access to steps and decking above as haul-out space. *(Reproduced from Barabash-Nikiforov (1947).)*

A new enclosure was constructed, to which the two sea otters were transferred on March 5, 1939 (Reshetkin and Shidlovskaya, 1947). The new enclosure also spanned the intertidal zone, was larger than the original enclosure, and featured improved water circulation and controllable flushing characteristics that largely eliminated water quality difficulties experienced earlier.

A number of food categories were offered to the sea otters on a trial basis during early phases of captivity at the east Murman shore. A diet of fish alone over an extended period eventually led to diarrhea. In such cases provision of live sea urchins or mussels eliminated the problem within 2–4 days. Experiments demonstrated that a diet of urchins alone facilitated good health and weight gain without ill effects over a period of 35 days (Reshetkin and Shidlovskaya, 1947). Over the period from November 1937 to January 1940, Buyan's body mass increased from 30 to 34 kg, Yashka's from 25 to 27 kg.

Buyan escaped from the east Murman shore facility on January 13, 1940, thereby completing the first known sea otter translocation project that included "release" to free-ranging status. Buyan was seen regularly in or near Yarnyshnaya Bay until February 7, 1940, and at a location 140 km to the east in May 1940. Commercial fishermen reported seeing a sea otter at the Gusintsky Islands near Yarnyshnaya Bay in spring 1942. Thus, Buyan may have survived at liberty for over 2 years. Yashka was euthanized on an unspecified date sometime after Buyan's escape because of imminent encroachment of international military conflict, requiring termination of the project (Reshetkin and Shidlovskaya, 1947).

Major findings of the projects at Medney Island and the east Murman shore were as follows (Shidlovskaya, 1947):

1. Sea otters were sustained for long periods in captivity with a diverse diet, including cod, flatfish, and other fish that may not appear in their natural diet when free-ranging.
2. Captive sea otters required access to hard-shelled invertebrate prey that can be ingested entirely as a source of insoluble fiber, with shells, tests, and spines included, on a regular basis in order to sustain gastrointestinal health and to supply important nutritional values. Sea urchins and mussels served the need most effectively, and crabs were a significant supplement.
3. Gastrointestinal problems progressed rapidly in sea otters, and affected animals were at mortal risk if effective response was not immediate.
4. Scrupulous cleanliness, high water quality, and high water turnover rates were essential attributes of enclosures and pools in which sea otters were held. Any attribute of enclosures promoting retention of fecal matter or discarded food particles required repair or redesign without delay.
5. Condition of the pelage in captive sea otters required continuous monitoring. Indications of fouling of the pelage by food particles or fecal matter, or abnormal wetting such that water was penetrating to the skin, required immediate response to ensure avoidance of hypothermia. As with gastrointestinal problems, hypothermia progressed rapidly, often leading to death within hours of initial signs.
6. Access to clean water was crucial for proper grooming behaviors by the sea otters, allowing retention of the thermoregulatory function of the pelage, and for cooling during unusually warm weather.

US Projects at Amchitka Island, 1951–1957

Initial goals of early US research on sea otters involved husbandry procedures facilitating successful translocation projects in the interest of population restoration. To our knowledge the first published recommendations for translocations as a population restoration or enhancement tool, rather than as support for domestication for commercial fur production, appeared in the early 1930s (Eyerdam, 1933). Eyerdam had observed sea otters to be relatively abundant in the Andreanof Islands of the central Aleutian Archipelago, Territory of Alaska, in 1932, and consequently recommended translocation from the Andreanofs to the Pribilof Islands. Following World War II consideration of sea otter translocation programs emerged on a broad geographic scale in North America. The envisioned purpose was increasing rates of recovery of sea otters from commercial harvests by placing animals in areas previously hunted to regional extinction (Kirkpatrick et al., 1955). Incentives for restoration were commercial as well as ecological, reflecting aspirations "that the sea otter will once again become an article of commerce" (Kenyon, 1969).

The first sea otters in captivity under US jurisdiction were held at Amchitka Island in the Aleutian Islands, Alaska, from 1951 through 1954. US biologists recognized, as Soviet biologists had in the 1930s, that the crucial first steps in implementation of translocation projects were development of methods for successfully holding sea otters in captivity, and for transportation of sea otters quickly over long distances with minimal distress. The first holding facilities at Amchitka were natural or artificial freshwater pools. The initial trials with captivity were done during winter months because wild sea otters could be captured more readily then as compared to other seasons. However, temperatures of freshwater in holding facilities were lower than ambient sea surface temperatures, leading to fatal chilling of the captive animals, with none of the subject animals living more than 11 days while captive (Kenyon, 1969). In 1954 three animals, each ~1 year of age, were held indoors in a small warehouse. Fresh water for ingestion was provided either as snowballs or as liquid in tip-proof pans. The three sea otters survived for ~2.5 months in the warehouse despite the absence of contained water adequate for swimming and proper grooming. Baseline data on respiration and heart rates in air were obtained from the otters (Stulken and Kirkpatrick, 1955). The subject animals were shipped to the WPZ in Seattle, Washington, arriving June 1, 1954 to become the first sea otters placed on public display in the United States. The otters were transported by air to the NPZ in Washington, DC, on June 14, 1954, all three succumbing within 10 days of arrival (Kirkpatrick et al., 1955). Causes of death were thought to be complications from heat stress and fouling of pelage during air transport (Kenyon, 1969).

Early translocation projects consistently failed (Kenyon, 1969). Captive sea otters were accumulated at Amchitka Island in 1951 for translocation purposes, but the effort was thwarted by the nearly immediate deaths of all animals taken into captivity. In 1955, 31 otters were captured at Amchitka and

FIGURE 8.3 Holding facilities for sea otters at Amchitka Island, Alaska, 1957. Upper image is an overview of the facility and environs under construction. Lower image shows concrete pool occupied by sea otters within the completed enclosure. *(Photographs by Karl W. Kenyon, reproduced from Kenyon (1969).)*

translocated in cages on the open deck of a ship to the Pribilof Islands. Otter fur quickly became soiled and mortality of sea otters was high during transport and soon after release, and the project failed to establish a stable new population of sea otters. Seven sea otters were translocated by ship from Amchitka to Attu Island in 1956, but none were seen in subsequent surveys.

Three new experimental enclosures for sea otters were constructed on the Amchitka shore to continue research toward identification of acceptable protocols for facility design and captive husbandry (Kenyon, 1969). Indoor and outdoor facilities built in 1956 both proved inadequate because of problems with excessive humidity and adequacy of water supply, respectively. The third facility, built in 1957 (Figure 8.3), became the first in US territory to hold captive sea otters successfully for indefinite periods.

The enclosure included a concrete pool supplied continuously with fresh water diverted from a nearby natural stream (Figure 8.3). The facility remained effective, with contained animals in good health, as long as water quality was maintained by limiting the number of captive sea otters. At high otter densities water became dangerously contaminated with fecal material and from accumulation of uneaten food fragments in the pool (Kenyon, 1969). A second attempt to relocate sea otters from Amchitka to the Pribilof Islands was made in 1959, and a few animals were observed by island residents up to 2 years after release. However, the effort again failed to establish a resident population at the release site.

The principal implications of the Amchitka studies for sea otter captivity and translocation were as follows:

1. Water: Captive sea otters required abundant clean water in order to groom the pelage effectively and maintain thermoregulatory efficiency. If the pelage became fouled due to poor water quality, otters required access to clean, dry haul-out surfaces in order to maintain internal body temperature. Individuals with such access were in some cases able to restore pelage function to the point that they could return to the water regularly, allowing more effective grooming.
2. Density: Effective captive sea otter facilities had a finite capacity for the number of otters that could be held together humanely. If the capacity was exceeded, water quality problems ensued.
3. Food: Interest in food was variable among sea otters first introduced to captivity, typically stabilizing at healthy levels after a few days. Captive otters generally showed greater interest in food when given access to water for swimming than when held in dry enclosures. Sea otters often groomed while consuming food, apparently to minimize fouling of the pelage by stray food fragments, and did so more effectively in water than on dry surfaces.
4. Disease: Many of the sea otters that failed to survive the Amchitka studies were found to have advanced gastroenteritis on *post mortem* examination. Other factors associated with mortalities were hypothermia due to fouled and improperly groomed pelage, infections in the extremities, refusal of food, peritonitis due to intestinal perforation by the cod worm (*Terranova decipiens* Krabbe, 1878, a parasitic nematode since renamed *Phocanema decipiens*) ingested with fish, and lung congestion caused by nasal mite infestation (*Halarachne miroungae* Ferris, 1925; Kenyon et al., 1965).
5. Implications for translocation: The emerging critical risk factors for survival of translocation attempts were the time required for transportation, ambient air temperature and the associated potential for overheating in transporting vehicles, the tendency for sea otter pelage to become fouled and the age of the animals being transported. Use of aircraft chartered for dedicated transport of sea otters was found effective at delivering relatively healthy sea otters if flight duration was minimized by direct travel to the ultimate

destination for release, and if air temperatures within the aircraft were kept at <10°C. Shipboard transportation was most effective when otters in transit were placed in cages that allowed access to pools provided with continuous seawater flow. In general young sea otters (1–2 years of age) tolerated transportation more readily than older animals.

Insights from the completed Amchitka captivity experiments were crucial in supporting successful translocations to southeastern Alaska, British Columbia, and Washington in the 1960s and 1970s. Experiences at Amchitka also influenced procedures used in the translocation of sea otters from the California mainland coast to San Nicolas Island (SNI), off southern California, from 1987 through 1990 (see Chapters 3, 12, 13, and 14).

Review of the History of Research on Captive Sea Otters in Zoos, Aquaria, Research Facilities, Research Vessels, and Temporary Field Installations

Early Observations in Zoos and Aquaria

The first research on sea otters in a public display facility in North America was done by R.C. Clark at WPZ from March 1956 through December in 1957, involving feeding rates of the captive adult female "Susie" (as reported by Kenyon, 1969). Daily food consumption averaged 4.04 kg, the equivalent of ~23% of Susie's body mass (range 6–42%), during calendar year 1957. On August 29, 1957 and January 21, 1960, K.W. Kenyon and V.B. Scheffer measured activity time budgets for Susie at WPZ. Primary activities were grooming and exercising (Kenyon, 1969). From December 1957 through September 1958 a young adult male sea otter ("Dave") was held in the same enclosure as Susie at WPZ and additional data on sea otter activity and feeding rate were accumulated, with daily feeding rates of the two animals combined averaging the equivalent of 24.6% of summed body mass from January through August 1958.

From September 1966 through August 1967 at PDZA, foraging behaviors and food consumption rates were observed for an adult male sea otter ("Gus;" Kenyon, 1969). Mean daily food consumption rate was the equivalent of ~20% of body mass. Gus's pelage molting rates were monitored by C. Brosseau from October 1967 to October 1968 by collecting molted fur from a screen over the exhibit pool drain. Shedding of fur occurred year round, with maximum rates in late spring and early summer and minimal rates in mid-winter (reported by Kenyon, 1969).

Studies of Sensory Acuity

In March 1967 studies of visual acuity were conducted with Gus at PDZA in Washington (Gentry and Peterson, 1967). Gus was trained to participate in two alternative discrimination tests involving underwater targets differing in surface area. Data suggested that Gus's underwater visual acuity was slightly less than that known for California sea lions and harbor seals.

Sound production and communication were evaluated by recording vocalizations of captive sea otters, and wild sea otters temporarily held in captivity, in California (McShane et al., 1995). Ten basic vocal categories were determined, and vocalizations were thought to be most suitable for short-range communication among individuals familiar with one another.

Olfaction was studied collaboratively with captive resident sea otters in Washington, Oregon, California, and Massachusetts (Hammock, 2005). Discrimination tasks were applied to animals at two facilities to determine whether sea otters have weak olfactory sensitivity as compared with terrestrial mammals. Data from discrimination tasks, together with neuroanatomical information, were inconsistent with the hypothesis of reduced olfactory sensitivity in sea otters.

Captive resident sea otters and wild sea otters in California were utilized to assess sound production patterns as well as sensitivities to received sound levels in air and water (Ghoul and Reichmuth, 2011, 2012). Early results indicated the estimated frequency range of effective hearing in air to be 0.125–32 kHz (Ghoul and Reichmuth, 2011).

Studies of Anatomy, Physiology, and Metabolism

Oxygen storage capacities of sea otters and six species of North Pacific or sub-Arctic pinnipeds were examined onboard a research vessel in the Bering Sea in 1968 (Lenfant et al., 1970). Sea otters were found to utilize air storage in lungs as the major repository of oxygen (O_2) storage during dives, with lesser O_2 quantities stored in association with blood hemoglobin or muscle myoglobin. Data were reported from the same research cruise on resting metabolic rates and ratios of surface area to body mass in sea otters and the six pinniped species (Iversen and Krog, 1973). Resting metabolic rate of sea otters was above expectation based on body mass as compared to the "mammalian standard." The positive deviation from expectation was similar to that for northern fur seals and above deviations for the five other pinniped species evaluated in the study. External body measurements were taken for sea otters and compared to expectation for terrestrial mammals based on a standard allometric function linking surface area to body mass, with results indicating a reduced ratio of surface to volume in sea otters as compared to terrestrial mammals of similar mass.

Thermoregulatory behaviors of live captive sea otters were observed at PDZA in Washington in August 1971 (Tarasoff, 1974).[6] Frequencies of elevation of the flippers into the air by animals resting at the water surface

6. PDZA was incorrectly identified as the Woodland Park Zoo in the paper (Tarasoff, 1974).

were higher in summer than in winter. Flipper elevation was observed only when ambient air temperature exceeded water temperature. Tarasoff concluded that flipper elevation likely was a significant approach for dissipation of surplus internal heat by sea otters at rest on the water surface.

Ten sea otters were used for studies of metabolic rate and tolerance of temperature extremes in captivity at Amchitka Island, Alaska (Morrison et al., 1974). Mean estimated basal metabolic rate for sea otters in air was ~2.7 times the predicted rate for a typical mammal with body mass similar to sea otters (Kleiber, 1975). Observed basal rates were effectively constant across temperature ranges of −19°C to +21°C in air, and <7–25°C in water, defining the thermal neutral zone in each medium for sea otters.

Metabolic and physiological studies were done on five sea otters held captive in California (Costa, 1982). Mean observed food consumption rate was the equivalent of 21.6% of body mass per day or 234 kcal/kg-day of consumed food energy. Caloric intake rates were similar to those reported in pioneering Soviet studies (Reshetkin and Shidlovskaya, 1947). Data indicated that moderately concentrated urine is excreted in relatively large volumes by sea otters. Mean total influx of water was 269 mL/kg-day, of which 68% was derived from food and the remainder from direct ingestion of seawater. The energy assimilation rate from food observed in sea otters was low compared to other mammals, likely a result of high gastrointestinal passage rates.

Three adult female sea otters were subjects in studies of patterns and significance of specific dynamic action (SDA) (Costa and Kooyman, 1984). SDA is a post-feeding increase in resting metabolic rate characteristic of mammals. Variation in SDA correlated with prey type but not water temperature. Elevated O_2 consumption rates associated with SDA persisted for ~250–320 min after completion of feeding. Activity of animals increased as SDA subsided over time, suggesting that heat generated by activity was replacing the decaying heat supply from SDA. It was proposed that heat generated by SDA offsets heat lost due to reduced activity levels, allowing sea otters a longer post-feeding rest period.

A study in California assessed patterns of swimming and associated energy costs in captive sea otters (Williams, 1989). Surface swimming speeds of subject animals were limited to ≤0.80 m/s, while submerged swimming speeds ranged from 0.60 to 1.39 m/s. Oxygen consumption rates were ~41% less during submerged swimming than for surface swimming. Mass-specific swimming transport costs for sea otters were found to be generally larger than those reported for other marine mammals.

Two captive sea otters were subject to metabolic studies in California (Yeates et al., 2007). The study design included a water tower allowing animals to dive to ~10 m depth in a controlled setting. Subject animals were found to experience high metabolic costs associated with diving and foraging as compared to other marine mammal species for which data were available (primarily phocid seals). Contributing factors include swimming style, retention of positive buoyancy during dives, metabolic impacts of food processing (handling, ingestion, digestion, and absorption), and mode of thermoregulation in sea otters.

The anatomical, metabolic, and physiological studies reviewed indicate important general patterns in the biology of sea otters. Clearly the energy cost of life at sea is high. Resting metabolic rates are much higher than in terrestrial mammals of similar body mass, and the added costs of diving and foraging are substantial. Travel on the sea surface is also costly in energy, but can be ameliorated at least temporarily by subsurface swimming. Sea otters have greater lung volume than terrestrial mammals of comparable mass, extending the limits of dive durations. The high metabolic demands of life at sea are supported by high rates of food mass intake, averaging the equivalent of 20–25% of individual body mass per day. As a result of post-feeding elevations of resting metabolic rate, sea otters can extend the duration of post-feeding rest at the sea surface. Sea otters are able to reduce energy consumption by virtue of a ratio of surface to volume lower than that of terrestrial mammals of similar mass. Sea otters are able to supply freshwater needs through metabolically generated water and by drinking sea water. Sea water consumption is enabled by production of highly concentrated urine. Finally, sea otters can tolerate a surprisingly broad range of ambient air and sea surface temperatures without physiological stress.

Reproduction and Pup Development

In a study of reproductive patterns of resident sea otters in Washington (Brosseau et al., 1975), fresh, bleeding nose injuries in adult females were found to be useful indicators of recent copulations. Injuries were caused when male sea otters grasped females during mating by biting into the nasal area of the female. In addition, adult females were observed to come into estrus approximately 30 days following the death of a dependent pup.

Ten pups had been born at PDZA in Washington during the early 1970s, none surviving to weaning (Antrim and Cornell, 1980). All first births by each adult female were stillborn. Survival duration was greatest for third-born pups. Breeding females were observed to gain ~20% in body mass during the final 4–6 weeks of pregnancy. Factors thought to be important to eventual breeding success of resident captive animals included a reliable source of clean water, maintenance of water quality, enclosures of adequate size, separation of males from females and pups to minimize risks from dangerously aggressive behavior by the males, and provision of a variety of prey species, in particular including prey with ingestible hard shells thought to be vital to gastrointestinal health.

Behavioral ontogeny in dependent pups was characterized, based on observations of three resident captive sea otter pups in Washington and

15 free-ranging pups off California and Oregon (Payne and Jameson, 1984). Correlations were found between post-partum ages of dependent pups and the attainment of developmental milestones such as replacement of natal pelage (the woolly coat of pups; Fisher, 1940) by the adult coat and abilities to swim, groom, use tools, dive, and capture and process live prey.

Behavioral relationships were documented for an adult female sea otter and four successively born pups (Hanson et al., 1993). Initially, the mother's activity was dominated by grooming of pups and by periods of rest. The proportion of time spent grooming pups declined steadily with pup age, but was still substantial at 3 months post-partum. Pup activity was initially limited to nursing and resting, with sequential appearance of increases in self-grooming, swimming, diving, and feeding over the first 2 months. Pup activity patterns converged with the mother's pattern by age 3 months.

Data were reported for steroid metabolites in fecal material to evaluate patterns of reproductive endocrinology in a collaborative study of resident captive female sea otters in Washington, California, and Illinois, USA, and in Portugal (Larson et al., 2003). Significant differences were found in reproductive hormone levels among animals with differing reproductive status. In a related study done at the Lisbon Aquarium in Portugal, methods were developed for monitoring the progression and indicators of pregnancy phases in an adult female sea otter (Da Silva and Larson, 2005).

Reproductive studies in captive sea otters provided insights into monitoring of milestones in pregnancy and post-partum pup development, and in the relationship of dependent pups and their mothers, based on endocrine cues, patterns of gain in mass, and behavior. Conditions of captivity that facilitate reproductive success have been identified, and the timeline over which an adult female becomes reproductively receptive after loss of a dependent pup has been determined. Observations of pup longevity have provided insights into the importance of birth order in the survival potential of pups born in captivity.

Foraging, Dietary Preferences and Sensitivity to Natural Toxins in Prey

Dietary preferences of four adult resident captive sea otters were assessed in California (Antonelis et al., 1981). During a 17-h observation period the otters focused sequentially on particular prey taxa, switching to a different category only after numbers of the previous focal taxon were depleted. Subject sea otters primarily consumed crustaceans during the first ~4 h of the foraging period, followed in turn by sea urchins, bivalves, and gastropods. The otters did not consume sea stars or key-hole limpets. During the observation period sea otters were engaged in food processing ~48% of the time, grooming ~25%, resting ~23%, and searching for prey ~4% of the time. Dominance relationships among individuals also influenced foraging patterns.

The abilities of sea otters to detect and reject prey items contaminated by toxins known to cause paralytic shellfish poisoning (PSP) were assessed in Alaska (Kvitek et al., 1991). The butter clam (*Saxidomus giganteus* [Deshayes, 1839]) is frequently contaminated with PSP toxins. The species is abundant and widely distributed in Alaskan sea otter habitats and is frequently consumed as prey by sea otters. Kvitek and colleagues were interested in the hypothesis that sequestration of PSP toxins, particularly saxitoxin (STX), by butter clams might deter predation by sea otters. Sea otters were fed groups of butter clams known to contain STX and groups known to be uncontaminated. Rates of consumption of contaminated clams were lower than rates for uncontaminated clams. When presented with contaminated clams, otters significantly increased the rate at which siphons and gills with attached kidney and pericardial gland tissues were discarded, while continuing to consume other clam tissues. In contaminated clams the discarded tissues contained higher STX concentrations than tissues consumed by the otters. All captive otters survived feeding trials in good health. Kvitek and colleagues suggested that sea otters may be able to detect STX presence in prey by taste.

Mitigation of By-Catch Risks

There is ongoing concern regarding possible entrapment and drowning of sea otters in traps used for commercial harvests of crabs, spiny lobsters, and fish off the California mainland and at SNI. Nine captive sea otters at MBA were exposed to baited fish, lobster, and crab traps (Hatfield et al., 2011). During the trials otters attempted to enter the trap openings, with some becoming entrapped (all entrapped animals were released unharmed). The project included test deployments of traps with reduced opening sizes in crab fishing habitats off California, and found no differences in catch rates of harvestable sized crabs between standard and modified trap designs. The California Fish and Game Commission subsequently mandated new regulations requiring reduced diameters of openings of finfish and shellfish traps.

Experimental Studies of Effects of Fouling by Crude Oil

Effects of contamination of sea otter pelage with crude oil were evaluated in Prince William Sound, Alaska, during field work in 1978, 11 years prior to EVOS (Siniff et al., 1982). Two sea otters were placed in a 5 m diameter pool on shore, to which Prudhoe Bay crude oil was added, for 12 h. Both animals were quickly fouled with oil. After removal from the experimental pool one otter was cleaned of crude oil, dried, and placed in a floating holding pool free of oil in a nearby bay. The otter's pelage quickly became wetted to the skin, indicating that the cleaning procedure likely had removed natural oils essential to water repellency and thermoregulatory efficiency of the pelage. After the sea otter was held overnight in a dry pen, a transmitter was attached and the animal was released, but the otter was never relocated. The second sea otter was not cleaned of oil, and was also placed in the floating pen where it died 10 h later. *Post mortem* examination suggested that death was caused by hypothermia, with toxic effects of oil contamination recognized as a possible contributing factor.

Thermal conductance of sea otter pelts was assessed in the context of crude oil contamination by measuring heat flux through five sea otter pelts (four from adults, one from a pup; Kooyman et al., 1977). Heat flux increased in oiled pelts by 25% as compared to the pre-oiling data, and returned to near the pre-oiling rates after washing. In a comparable study of effects of crude oiling on sea otter pelt samples (Williams et al., 1988), oiling resulted in heat flux increases of two- to fourfold as compared with flux through unoiled control samples. After cleaning of oil, heat flux rates through treatment samples were similar to those for control samples. Data from the two studies suggested that oiling would increase heat loss rates in live animals and that cleaning of oiled animals would remove both the contaminating crude oil and a portion of the natural lipids present in sea otter fur. Results of related *in vivo* research (described below) were consistent with these *in vitro* studies.

Effects of external oiling and subsequent washing were examined with live sea otters captured off California (Costa and Kooyman, 1982). Deep core body temperatures were obtained from radio telemetric devices placed through the esophagus and into the stomachs of subject animals under anesthesia. Subcutaneous body temperatures were obtained from radio telemetric devices surgically implanted beneath the skin. Mean O_2 consumption rates were found to increase by 41% above control levels after 20% of the surface area of treatment animals was coated with crude oil. After fur washing the O_2 consumption rate increased by 106% compared to controls. Recovery to pre-oiling O_2 consumption rates required ≥8 days for experimental subjects. Core body temperatures after oiling were higher than those measured pre-oiling when oiled animals were active, but dropped below pre-oiling levels when subject animals were at rest. Subcutaneous body temperatures averaged 35.1°C pre-oiling, falling to 26.3°C beneath oiled portions of the pelage. Subcutaneous body temperatures returned to pre-oiling values soon after washing.

In a similar study (Davis et al., 1988, a companion project to the study of Williams et al., 1988) impacts of external crude oil exposure were examined using 12 live sea otters captured in Alaska and transported to California. Thermal conductivity of oiled animals was 1.8 times that of the pre-oiling rate. After oiling, the proportion of time spent grooming and swimming increased to 1.7 times that of the pre-oiled state. After cleaning, means of core body temperature, metabolic rate, and thermal conductivity returned to pre-oiling levels after 3–6 days. Concentrations of squalene, the primary component of natural lipid in the pelage, had not returned to normal levels 7 days after cleaning. The investigators also utilized surgically implanted intra-abdominal transmitters to provide thermal data in an experiment similar to previous research (Costa and Kooyman, 1982). Mean core body temperatures were 38.9°C prior to oiling and 39.0°C after oiling but before washing. When subject otters were returned to the holding pool after washing, mean core temperatures declined significantly by an average of 1°C, with recovery to pre-oiling temperatures in ~5 days. One in five of the treated animals had died after 11 days following exposure to oil and subsequent cleaning.

A study was conducted by the Office of Spill Prevention and Response (OSPR), part of the California Department of Fish and Game, to evaluate methods used for cleaning the pelts of live sea otters in the context of responses to contamination by spilled oil at sea (Jessup et al., 2012). Water type (sea water and fresh water) and temperature were varied in a series of protocols for washing the pelage of two resident captive males in California. Treatment responses were monitored with temperature-sensitive passive implantable transponder (PIT) tags and infrared thermographic imaging. Use of softened (i.e., low dissolved mineral content) fresh water instead of salt water reduced times of recovery to normal thermoregulatory function after washing. Subcutaneous temperatures typically returned to normal ranges sooner after washing than core body temperatures. Thermographic imaging methods used in the project facilitated identification of areas of the pelage retaining poor water repellency after washing, allowing consideration of follow-up steps to improve recovery of washed otters.

Collectively the studies of oil impacts were consistent with hypotheses that contact with crude oil reduces thermoregulatory efficiency of sea otter pelage, and that cleaning of the fur to remove oil from impacted animals can compromise the role of natural oils in the pelage in limiting heat loss. In the latter case several days may be required post-washing before the pelage fully recovers thermoregulatory efficiency. However, the *in vitro* studies did not anticipate all impacts of oiling to sea otter health resulting from the EVOS as described in a subsequent section.

PART TWO: TEMPORARY HOLDING OF LARGE NUMBERS OF WILD ANIMALS

Temporary captive aggregations of large numbers of sea otters have been assembled rarely. Information sufficient for evaluation and discussion of such events is available for only two cases. The first was the translocation of sea otters from the mainland coast of central California to SNI off southern California from 1987 through 1990, involving 150 sea otters. The second was the rescue and rehabilitation effort following the grounding and spillage of crude oil from the tank-ship *Exxon Valdez* in Alaska in 1989, involving

357 sea otters (see Chapter 4). Each event was a highly controversial and costly effort to contribute positively to the conservation of sea otters and the ecosystems of which they are part. Outcomes and lessons from both events were significant to the future of sea otter conservation.

The Translocation of Sea Otters to SNI, California: Benefits and Insights from the Captive Phase for Sea Otter Conservation

Background

The California sea otter translocation project of 1987–1990 incorporated several attributes unique in the history of translocation projects. First, sea otters destined for transport from the nearshore marine waters off the California mainland to SNI were accumulated in a temporal sequence of relatively large groups (up to 24 individuals at any one time) at an established, internationally recognized research, education, and public display institution (MBA). Facilities at MBA were "state of the art," maintained by a professional staff of facilities experts. Second, the project was supported by a number of temporary employees and volunteers, most of whom had prior experience with (a) sea otter husbandry, either at MBA or at facilities of the Monterey County Chapter of the Society for the Prevention of Cruelty to Animals, located near Salinas, California; or (b) capture of, handling of, and field research on free-ranging sea otters in natural habitats. Core project staff charged with managing the accumulating groups of sea otters at MBA also received consultation and support from permanent MBA veterinary, husbandry, and facilities staff. Third, the timing of transport of accumulating sea otter aggregations from MBA to SNI was such that opportunities emerged to evaluate feeding rates, water quality management issues, signs of stress related to captivity, and mortality patterns in a research context. Data and interpretations relating to the conservation implications of captive animals at MBA during the California translocation program have not been published previously.

Facilities and Protocols

Translocated sea otters were held primarily in the Quarantine Facility (QF) at MBA while en route to SNI. Each of two 6.1 m diameter pools received continuously running filtered fresh sea water taken from a nearby ocean intake located ~305 m offshore from MBA at approximately 15 m depth, with pool water turnover time estimated at ~100 min. Sea water provided to the QF pools ranged in temperature from 10°C to 15°C during the project, and in salinity from 33‰ to 34‰. Both pools contained haul-out platforms allowing otters to rest out of water.

Sea otters were transported to MBA in air-conditioned vans from landing sites in the vicinity of capture locations off Santa Cruz, Monterey, and San Luis Obispo Counties. On arrival at MBA captive sea otters were weighed and marked with external flipper tags (Ames et al., 1983; Rathbun et al., 1990) and an injected PIT tag (similar to the microchips now widely used to identify domestic animals and household pets; Thomas et al., 1987). A blood sample was drawn from either the popliteal or femoral vein (Geraci and Lounsbury, 1993).

Sea otters in QF pools at MBA were fed at 6-hour intervals. Primary foods were commercially available packaged frozen geoduck clam siphons and market squid, and live rock crabs. Amounts and types of food offered were adjusted based on observed consumption rates and apparent preferences of subject otters. Food was always provided in excess of demand by the otters.

Once in holding pools, sea otters were observed continuously by husbandry staff assigned to record timeline milestones, food consumption rates, and behavioral observations. Observational effort was focused on specific behaviors of greatest interest or concern, including apparent comfort level, willingness of otters to associate and interact with one another, fluidity of movement, attentiveness to grooming behavior, responsiveness to acoustic and visual stimuli and disturbances, and responses to food presentation and pool cleaning activity. Physical condition of behaviorally well-adjusted otters was assessed primarily by condition of the pelage as visually monitored (Davis and Hunter, 1995).

Observations of Feeding Rates

Feeding rates were estimated by weighing food prior to feeding, and by retrieving and weighing uneaten food after each feeding event once it was apparent that otters were sated. Data on food mass consumed, body mass of otters, and time at MBA allowed computation of a feeding rate (grams of food consumed per kilogram of otter body mass per hour) and an extrapolated estimate of the mean equivalent percentage of otter mass consumed as food mass per day by each group of otters (Table 8.1). The overall weighted mean consumption rate for 23 groups of otters, comprising 72 different individuals, was the equivalent of 38% (range 18–73%) of group body mass per day, substantially higher than expectation. Kenyon's (1969) data for a single resident captive otter indicated a range in daily rates of equivalents of 6–42% of body mass. The relatively high mean, the broad range among groups, and the seemingly implausible maximum estimate of 73% for one group suggest the following possible interpretations, not mutually exclusive: (a) sea otters may have been experiencing generalized food shortages in natural habitat off California at the time of the translocation project, such that over-consumption was a reaction to the abrupt availability of high-quality food on arrival at MBA; (b) of 23 groups assessed only two were present at MBA for >24 h. The brief residence times may not have been sufficient to allow the otters to become accommodated to MBA facilities and protocols, possibly resulting in over-consumption of food on average, reflecting stress

TABLE 8.1 Food Consumption Rates for Groups of Sea Otters Held in Captivity Prior to Translocation to San Nicolas Island, California, September 28, 1988 through September 13, 1989

Number of Sea Otters in Group ($N_{Total} = 72$)	Duration of Period Over Which Food Consumption was Monitored (Hours)	Summed Biomass of Sea Otters in Group (kg)	Calculated Food Consumption Rate	
			Food Consumed per Kilogram Sea Otter Biomass per Hour (g)	Equivalent Percentage of Summed Sea Otter Biomass Consumed per Day
1	19.25	13.6	14	34
4	24.0	54.5	19	45
9	17.0	122.7	19	46
6	21.5	81.8	14	34
7	16.5	89.1	15	36
2	18.5	24.5	18	42
2	21.0	28.2	15	36
1	21.0	12.3	24	57
1	15.5	12.3	13	31
2	17.5	29.5	30	73
1	39.8	11.8	9	22
2	7.5	23.6	7	18
2	13.3	27.3	21	50

(Continued)

TABLE 8.1 (Continued)

Number of Sea Otters in Group ($N_{Total} = 72$)	Duration of Period Over Which Food Consumption was Monitored (Hours)	Summed Biomass of Sea Otters in Group (kg)	Calculated Food Consumption Rate	
			Food Consumed per Kilogram Sea Otter Biomass per Hour (g)	Equivalent Percentage of Summed Sea Otter Biomass Consumed per Day
1	16.0	11.8	25	61
3	20.0	55.5	18	43
2	16.75	47.7	15	36
1	16.0	20.4	27	64
3	14.8	55.6	11	26
4	14.0	85.0	14	33
4	24.25	86.9	12	29
4	16.3	83.6	18	42
6	18.0	136.4	12	29
4	20.75	77.3	9	20
Weighted means	18.4	16.5	16	38

responses to capture, transport, and captivity at MBA. Data from early US studies of sea otters in captivity suggested that the animals required several days in captivity before resuming normal patterns of food consumption (Kenyon, 1969); and (c) our method for estimating food consumption rates was systematically biased high by unknown factors.

Impacts on Institutional Water Quality

Fecal coliform bacterial (FCB) concentrations were monitored from 24 August through September 16, 1987 as directed by the California State Water Quality Control Board (CSWQCB). Water samples were collected daily from the two QF pools in which sea otters were held, from the MBA seawater intake pipeline, and from the overflow reservoir from which MBA seawater effluent returned to the ocean. Sea water accessed for sampling had not been treated in any way for elimination of FCB contamination, except for effluent waters from the facility holding publicly displayed sea otters at MBA. The latter were subject to ultra-violet irradiation prior to mixing with unsterilized effluent waters from other MBA exhibits, and with unsterilized waters from QF pools holding translocation animals. Samples were analyzed at a commercial laboratory certified by CSWQCB. All FCB data were reported to CSWQCB as legally required, and as agreed prior to project initiation.

Results for each sample were reported as the most probable number (MPN) of FCB per 100 mL of sampled water (sensitivity range 2–2400). The CSWQCB standard for water released to the ocean was a maximum MPN value of 200. FCB counts in the MBA intake water were below the detectable minimum level on all but one of the 16 sampling dates. Numbers of translocation otters held at MBA ranged from 0 to 24 during the study period (maximum of 12 animals per pool at any one time). MPN levels in pools used for translocation otters were below the detectable minimum in samples taken when no sea otters were present, increasing substantially when sea otters were present. When 15 or more sea otters were present, MPN level at the MBA outfall often exceeded CSWQCB standards. Such concentrations raised serious concerns for the health of captive sea otters. The data also indicated that untreated outfalls from facilities housing large numbers of sea otters pose a risk of locally excessive FCB levels in coastal marine waters, possibly posing hazards to public health and local marine ecosystems as well as to the involved captive sea otters.

Stress Signs and "Captivity Stress Syndrome"

The health of sea otters held at MBA was monitored primarily by gathering and evaluating behavioral data. Some observed behavioral patterns indicated significant stress in some sea otters during stopovers at MBA, with 11 behavioral signs suggesting the existence of a "captivity stress syndrome" (CSS; Figure 8.4; Table 8.2). Selection of signs was based on prior

FIGURE 8.4 An adult female sea otter under treatment for "CSS" in the veterinary care facility of the Monterey Bay Aquarium, Monterey, California, in 1987 during the translocation to San Nicolas Island. About one-third of the otters involved in the translocation displayed one or more signs of CSS during the project. *(Photograph courtesy of the US Fish and Wildlife Service.)*

TABLE 8.2 Frequency of Occurrence (%) of Behavioral Signs of Captivity Stress Syndrome in Animals Held in Captivity Prior to Translocation to San Nicolas Island, California, 1987–1990

Sign	Year 1 (*N* = 74)	Year 2 (*N* = 60)	Years 3 and 4 (*N* = 16)	Totals (*N* = 150)
Refusal of food	31	5	0	17
Shivering	16	13	13	15
Frequent vocalization	5	15	19	11
Rigid posture	5	7	6	7
Inadequate grooming	5	5	13	7
Continuous swimming	3	7	19	7
Matted pelage	5	5	6	6
Disorientation	2	5	7	4
Lethargy	3	3	0	3
Coughing	3	2	0	3
Piercing vocalization	3	3	0	3
Any sign	45	20	19	33

experience of project staff with captive wild otters, signs displayed in fatal CSS cases (see below), accumulated experience and insight during the course of the project, and published behavioral literature (Packard and Ribic, 1982). Signs used to define CSS by project staff were undeniably subjective, resulting from the lack of applicable published data at the beginning of the project. One-third of the sea otters involved in the project displayed at least one behavioral sign associated with CSS (Table 8.2). Collective frequencies of signs were higher in year one of the project than in subsequent years. The most frequently occurring signs were shivering, refusal of food, and frequent vocalization.

Six otters died during the translocation project at MBA following observation of signs of CSS. Each carcass was given a thorough external and internal gross necropsy by experts not otherwise involved with the translocation program, and samples of tissue were taken for virological, bacteriological, and histopathological investigation. The six otters all exhibited some of the core behavioral signs of CSS at MBA prior to death. On *post mortem* examination one otter was found to have died of pneumonia likely contracted prior to capture. A second had received a surgically implanted radio transmitter during the project and apparently died of post-operative complications.

Four otters died while at MBA, apparently as a direct result of CSS or other causes, either pre-existing or associated with capture, ground transportation to MBA, or holding at MBA. Time of death ranged from 8.5 to 94 h after arrival at MBA. Necropsies done by staff of the National Wildlife Health Center, USFWS, indicated a number of internal abnormalities in one or more of the fatal cases, including signs of disease or injury in the gastrointestinal tract, central nervous system, kidneys, and adrenal glands, along with generalized inflammation and lack of fat stores. Three of the four deceased animals were found to have internal bruising, hematoma, or petechial hemorrhage.[7]

Rescue and Rehabilitation of Sea Otters Influenced by the EVOS

EVOS was an event of extraordinary magnitude, complexity, and controversy. Comprehensive coverage of all conservation implications of the event for sea otters extends well beyond the scope of this chapter. Major findings of event response activities involving sea otters in captivity are summarized below (based primarily on Williams and Davis, 1995).

Three hundred and thirty-nine animals were captured and transported to rehabilitation centers in Valdez or Seward and admitted alive for assessment and possible treatment (Bayha and Kormendy, 1990b). Captured animals

7. Petechial hemorrhages are "pinpoint" round spots (2–3 mm diameter) resulting from localized bleeding under the skin or on organ surfaces (www.mayoclinic.org).

recovering to apparently good health were eventually transported to pre-release outdoor holding facilities in a small embayment off southern Kachemak Bay. Eighteen sea otter pups were born in the rehabilitation centers to adult females pregnant at the time of capture. Thus, the rehabilitation effort assessed and in many cases treated 357 sea otters for impacts linked to the exposure of spilled oil in natural habitats. Of the total number of animals admitted to rehabilitation facilities, 123 (34.4%) died within the facilities, nearly all within 12 weeks of the initial spill event. Maximum mortality rates in facilities occurred 2–3 weeks after the spill (Williams et al., 1995a). Surviving animals (55.2%) were released back into natural habitats judged to be free of oil contamination, or were transferred to zoos or aquaria for permanent care and residence (10.4%; Bayha and Kormendy, 1990b).

Disorders common in animals admitted during the early phase of response to EVOS included hypothermia, hyperthermia (the latter often associated with transport from capture sites to treatment facilities), toxic effects of ingested petroleum hydrocarbons, injury to respiratory tracts, hypoglycemia, shock, and occurrence of seizure. Disorders frequently observed in both early and late phases included liver and kidney dysfunction, gastrointestinal disorders, anemia, and stress (Williams et al., 1995a). It was apparent that injuries and pathologies in sea otters exposed to oiling generally were in one or more of three categories:

1. Hypothermia in response to oiling of pelage and the consequent loss of thermoregulatory capability.
2. Damage to internal organs, primarily the liver, kidneys and lungs, caused by trans-dermal absorption or ingestion of crude oil through grooming of the pelage, in some cases possibly including grooming of oiled dependent pups prior to rescue.
3. Acute damage to lungs caused by inhalation of volatized components of spilled oil, primarily in the first 6 weeks following the grounding of the *Exxon Valdez* (Williams et al., 1995c).

DISCUSSION

There can be no doubt of the substantial value of sea otters in captivity for progress in marine conservation. The record of achievement in development of husbandry methods, restoration and rehabilitation strategies and techniques, and research findings over a range of topics supporting conservation objectives may be as strong for sea otters as for any marine mammal species that can be held under humane and healthy circumstances in captivity. Examples of such benefits include public engagement in conservation awareness, successful application of translocation as a powerful strategy for population restoration, and identification of biological impacts and response options for sea otters exposed to spilled crude oil in natural habitats.

Insights into successful husbandry of captive sea otters have been significant in facilitating successful studies with sea otters on topics relating to marine conservation. The pioneering research programs of the Soviet Union and the United States provided such insights with remarkable rapidity, and findings were consistent between the two programs. Although incentives for the two research programs were largely linked to interests in domestication or replenishment of wild populations of sea otters in support of a revitalized fur trade, the true legacy of both efforts has been in captive husbandry standards now widely and successfully applied, and in the effectiveness of translocation programs for population restoration in North America.

Data from the California translocation project corroborated the significant water quality challenges associated with fecal production of sea otters in a captive setting. Data from the project also indicated unusual patterns of food consumption in wild sea otters during the first hours following capture and transfer to holding facilities. The data confirmed prior observations of Kenyon (1969) that food consumption rates of newly captured sea otters may not stabilize for several days after introduction to captive settings. During the captive phase of the translocation project 11 behavioral signs were identified that suggested a "Captivity Stress Syndrome" in sea otters brought from natural habitats into captivity, sometimes with lethal consequences.

Assessments of degree of oiling in the field following EVOS were important in selection of otters for capture, transport, and admission to rehabilitation facilities by field capture teams following EVOS. As noted above, healthy wild animals subject to capture, transport, and temporary holding may develop significant, sometimes fatal cases of CSS. Thus, a healthy, uncontaminated animal, captured with the intention of treatment for oiling would, on average, fare better if left alone in the wild than if captured, transported, and admitted to a rehabilitation facility. However, personal experiences of one of us (GRVB) in post-EVOS capture activities south of the Kenai Peninsula highlighted difficulties in accurately selecting otters for capture and treatment. For example, sea otters were encountered in areas with surface oil slicks nearby that were lethargic and abnormally ambivalent to the present of boats and skiffs used in capture efforts, but did not have obvious oiling of the pelage. The decision to capture such animals was influenced by observations, elsewhere in the spill area, of sea otters secured with visible oiling of the pelage that in some cases were able to groom themselves free of externally visible oil during the period of transport to rehabilitation facilities. Thus, it was possible that an otter lacking external signs of contamination could have had undetected contact with spilled oil. There were also reports of oiled adult females captured with dependent pups lacking visible external oil, suggesting that the female had cleaned contaminating oil from the pup's pelage by grooming. In addition, it was possible that a sea otter observed near spilled oil had suffered injuries from inhalation of the volatile components of spilled oil (Williams et al., 1995a) without coming into physical contact with unevaporated oil. Thus, when sea otters with abnormal behavior were observed near floating oil, it was necessary to make arbitrary judgments regarding the decision to capture. Capture decisions were particularly difficult and subjective in cases where apparently healthy unoiled sea otters were observed in close proximity to floating spilled oil. Decisions relating to such uncertainties led to claims that some captures were unnecessary, with consequent published disagreements over capture of animals of ambiguous oiling status post-spill (Ames, 1990; VanBlaricom, 1990; Estes, 1991).

Design, construction, and operation of treatment and rehabilitation facilities for the benefit of sea otters known or perceived to be oiled by the EVOS event were quite costly (Davis and Davis, 1995; Davis et al., 1995; Chen-Valet and Camlin, 1995). Given the absence of facilities intended for treatment, rehabilitation, and long-term holding of oiled sea otters prior to EVOS, the early aftermath of EVOS involved caring for captive injured sea otters in the midst of hastily assembled facilities with impromptu construction activities often under way directly adjacent to critically ill animals. It was immediately apparent to all involved that treatment of oiled sea otters in a construction zone was not desirable or likely to be effective. Later in the spill response, efficient and functional facilities came into operation and sea otters were being admitted and processed at high rates, with the resultant risk of exceeding the capacity of facilities. The problem became obvious as sea otters that had survived the early phases of recovery from oiling progressed to a need for maintenance and monitoring rather than acute care. The prospect of overcrowding led quickly to recognition that facilities for treatment and rehabilitation should function as "flow-through" locations designated specifically for sea otters in need of immediate cleaning and critical veterinary care (Davis and Davis, 1995). Concurrently, sea otters past the need for critical care were to be moved to facilities designed for long-term maintenance and monitoring of largely healthy sea otters recovered from oiling effects.

The necessity of critical care, swiftly applied, was greatest in the first 1–2 weeks after EVOS (Williams et al., 1995b). It follows that an efficient facility intended to be effective in caring for acutely impaired oiled sea otters should anticipate an oil spill rather than undergo design and construction only after the spill occurs. The subject premise has been embraced most effectively by the OSPR of the California Department of Fish and Wildlife (Jessup et al., 1996). OSPR was established soon after EVOS, in part due to relentless encouragement by professional scientists and resource managers involved in the response to EVOS (see also Chapter 14). OSPR currently operates the Marine Wildlife Veterinary Care and Research Center in Santa Cruz, with a capacity for care of up to 150 sea otters. The Center is legally designated as the primary care facility for all oiled sea otters in California.

The facility has an ongoing role in research and management support, thereby justifying permanent operations, with infrastructural attributes allowing rapid conversion to rescue, treatment, and rehabilitation functions should an oil spill disaster occur offshore. Such a facility could not possibly be immediately effective in oil spill response on behalf of impaired marine wildlife if construction was initiated only after a spill disaster.

The record indicates that sea otter conservation has been well served by the various forms of captivity in which the otters have been involved since 1932. The continued survival of the species may well rely on continuation of the partnership among captive animals, the informed public, zoo and aquarium staffs, research scientists, managers, and conservation-oriented non-governmental advocacy groups. The value of captive sea otters has evolved over the decades as a legacy to Fomka, Bek, Buyan, Yashka, Susie, Dave, Gus, Tichuk, Homer, the translocation animals, and all the others. The partnership has provided myriad benefits to sea otter conservation, and as such should be sustained indefinitely.

ACKNOWLEDGMENTS

Financial support during preparation of this chapter was provided by the Ecosystems Branch of the US Geological Survey, the School of Aquatic and Fishery Sciences of the University of Washington, the Department of Life Sciences of The Seattle Aquarium, and the Aquatic Animal Department of the Point Defiance Zoo and Aquarium. J.L. Bodkin, L. Lahner, S.E Larson, S.G. Lio, C. Hempstead, and K.K. VanBlaricom provided comments on earlier versions of the chapter manuscript.

REFERENCES

Abegglen, C.E., 1977. Sea mammals: resources and population. In: Merritt, M.L., Fuller, R.G. (Eds.), The Environment of Amchitka Island, Alaska. Technical Information Center, Energy Research and Development Administration, Washington, DC, pp. 493–510.

Ames, J.A., 1990. Impetus for capturing, cleaning, and rehabilitating oiled or potentially oiled sea otters after the T/V *Exxon Valdez* oil spill. In: Bayha, K., Kormendy, J. (Eds.), Sea Otter Symposium: Proceedings of a Symposium to Evaluate the Response Effort on Behalf of Sea Otters after the T/V *Exxon Valdez* Oil Spill into Prince William Sound, Anchorage, Alaska, 17–19 April 1990. US Fish and Wildlife Service, Biological Report 90 (12). US Department of the Interior, Fish and Wildlife Service, Washington, DC, pp. 137–141.

Ames, J.A., Hardy, R.A., Wendell, F.E., 1983. Tagging materials and methods for sea otters, *Enhydra lutris*. Calif. Fish Game 69, 243–252.

Antonelis Jr., G.A., Leatherwood, S., Cornell, L.H., Antrim, J.G., 1981. Activity cycle and food selection in captive sea otters. Murrelet 62, 6–9.

Antrim, J.E., Cornell, L.H., 1980. Reproduction of the sea otter. Int. Zoo Yearb. 20, 76–80.

Barabash-Nikiforov, I.I., 1947. The sea otter (*Enhydra lutris* L.)—biology and economic problems of breeding. The Sea Otter (Kalan). Main Administration of Reserves, Council of Ministers of the Russian Soviet Federated Socialist Republic, Moscow, USSR, Jerusalem, Israel, pp. 3–202 (1962) (English Translation, Israel Program for Scientific Translations, Ltd.).

Bayha, K., Kormendy, J. (Eds.), 1990a. Sea Otter Symposium: Proceedings of a Symposium to Evaluate the Response Effort on Behalf of Sea Otters after the T/V *Exxon Valdez* Oil Spill into Prince William Sound, Anchorage, Alaska, 17–19 April 1990. US Fish and Wildlife Service, Biological Report 90 (12). US Department of the Interior, Fish and Wildlife Service, Washington, DC.

Bayha, K., Kormendy, J., 1990b. Summary of sea otter rescue inventory. In: Bayha, K., Kormendy, J. (Eds.), Sea Otter Symposium: Proceedings of a Symposium to Evaluate the Response Effort on Behalf of Sea Otters after the T/V *Exxon Valdez* Oil Spill into Prince William Sound, Anchorage, Alaska, 17–19 April 1990. US Fish and Wildlife Service, Biological Report 90 (12). US Department of the Interior, Fish and Wildlife Service, Washington, DC, pp. vi–x.

Berkh, V.N., 1823. A chronological history of the discovery of the Aleutian Islands, or, the exploits of Russian merchants with a supplement of historical data on the fur trade. Materials for the study of Alaska history, vol. 5. English translation, D. Krenov, subsequently edited by R.A. Pierce, 1974. The Limestone Press, Kingston, ON. 127 pages.

Brennan, E.J., Houck, J., 1996. Sea otters in captivity: the need for coordinated management as a conservation strategy. Endangered Species Update 13, 61–67.

Brosseau, C., Johnson, M.L., Kenyon, K.W., 1975. Breeding the sea otter (*Enhydra lutris*) at the Tacoma Aquarium. Int. Zoo Yearb. 15, 144–147.

Casson, C.J., 2013a. Marine Mammal Taxon Advisory Group Annual Report for Sea Otters. Association of Zoos and Aquariums, Silver Spring, MD.

Casson, C.J., 2013b. North American Regional Studbook for Sea Otters (*Enhydra lutris*). Association of Zoos and Aquariums, Silver Spring, MD.

Chance, P., 1999. Learning and Behavior, fourth ed. Brooks/Cole Publishing Company, Salt Lake City, UT.

Chen-Valet, P., Camlin, T., 1995. Occupational safety in the rehabilitation center. In: Williams, T.M., Davis, R.W. (Eds.), Emergency Care and Rehabilitation of Sea Otters: A Guide for Oil Spills Involving Fur-Bearing Marine Mammals. University of Alaska Press, Fairbanks, AK, pp. 187–193.

Clayton, S., Fraser, J., Saunders, C.D., 2009. Zoo experiences: conversations, connections and concern for animals. Zoo Biol. 28, 377–397.

Cobb, J.N., 1906. The commercial fisheries of Alaska in 1905. Report of the Commissioner of Fisheries for the Fiscal Year Ended June 30, 1905 and Special Papers. Government Printing Office, Washington, DC, Bureau of Fisheries Document 603 (G.M. Bowers, Commissioner), pp. 124–166.

Costa, D.P., 1982. Energy, nitrogen, and electrolyte flux and sea water drinking in the sea otter *Enhydra lutris*. Physiol. Zool. 55, 35–44.

Costa, D.P., Kooyman, G.L., 1982. Oxygen consumption, thermoregulation, and the effect of fur oiling and washing on the sea otter, *Enhydra lutris*. Can. J. Zool. 60, 2761–2767.

Costa, D.P., Kooyman, G.L., 1984. Contribution of specific dynamic action to heat balance and thermoregulation in the sea otter *Enhydra lutris*. Physiol. Zool. 57, 199–203.

Coxe, W., 1780. Account of the Russian Discoveries Between Asia and America, to Which Are Added, the Conquest of Siberia, and the History of the Transactions and Commerce Between Russia and China. T. Cadell, London.

Da Silva, I.M., Larson, S.E., 2005. Predicting reproduction in captive sea otters. Zoo Biol. 24, 73–81.

Davis, R.W., Davis, C.W., 1995. Facilities for oiled sea otters. In: Williams, T.M., Davis, R.W. (Eds.), Emergency Care and Rehabilitation of Sea Otters: A Guide for Oil Spills Involving Fur-Bearing Marine Mammals. University of Alaska Press, Fairbanks, AK, pp. 159–175.

Davis, R.W., Hunter, L., 1995. Cleaning and restoring the fur. In: Williams, T.M., Davis, R.W. (Eds.), Emergency Care and Rehabilitation of Sea Otters: A Guide for Oil Spills Involving Fur-Bearing Marine Mammals. University of Alaska Press, Fairbanks, AK, pp. 95–101.

Davis, R.W., Williams, T.M., Thomas, J.A., Kastelein, R.A., Cornell, L.H., 1988. The effects of oil contamination and cleaning on sea otters (*Enhydra lutris*). 2. Metabolism, thermoregulation, and behavior. Can. J. Zool. 66, 2782–2790.

Davis, R.W., Styers, J., Otten-Stanger, J., 1995. Facility management and personnel. In: Williams, T.M., Davis, R.W. (Eds.), Emergency Care and Rehabilitation of Sea Otters: A Guide for Oil Spills Involving Fur-Bearing Marine Mammals. University of Alaska Press, Fairbanks, AK, pp. 177–186.

Elliott, H.W., 1875. A Report Upon the Condition of Affairs in the Territory of Alaska. Government Printing Office, Washington, DC.

Estes, J.A., 1991. Catastrophes and conservation: lessons from sea otters and the *Exxon Valdez*. Science 254, 1596.

Estes, J.A., Palmisano, J.F., 1974. Sea otters: their role in structuring nearshore communities. Science 185, 1058–1060.

Eyerdam, W.J., 1933. Sea otters in the Aleutian Islands. J. Mammal. 14, 70–71.

Falk, J.H., Reinhard, E.M., Vernon, C.L., Bronnenkant, K., Deans, N.L., Heimlich, J.E., 2007. Why Zoos and Aquariums Matter: Assessing the Impact of a Visit to a Zoo or Aquarium. Association of Zoos and Aquariums, Silver Spring, MD.

Fisher, E.M., 1940. Early life of a sea otter pup. J. Mammal. 31, 132–137.

Foose, T.J., 1989. Species survival plans: the role of captive propagation in conservation strategies. In: Seal, U.S., Thorne, E.T., Bogan, M.A., Anderson, S.H. (Eds.), Conservation Biology and the Black-Footed Ferret. Yale University Press, New Haven, CT, pp. 210–222.

Gentry, R.L., Peterson, R.S., 1967. Underwater vision of the sea otter. Nature 216, 435–436.

Geraci, J.R., Lounsbury, V.J., 1993. Marine Mammals Ashore: A Field Guide for Strandings. Sea Grant College Program, Texas A & M University, Galveston, TX.

Ghoul, A., Reichmuth, C., 2011. Preliminary investigation of sound reception in southern sea otters (*Enhydra lutris nereis*). J. Acoust. Soc. Am. 129, 2433.

Ghoul, A., Reichmuth, C., 2012. Sound production and reception in southern sea otters (*Enhydra lutris nereis*). In: Popper, A.N., Hawkins, A. (Eds.), Advances in Experimental Medicine and Biology, vol. 730. Springer, New York, NY, pp. 157–159.

Gruber, J.A., Hogan, M.E., 1990. Transfer and placement of non-releasable sea otters in aquariums outside Alaska. In: Bayha, K., Kormendy, J. (Eds.), Sea Otter Symposium: Proceedings of a Symposium to Evaluate the Response Effort on Behalf of Sea Otters after the T/V *Exxon Valdez* Oil Spill into Prince William Sound, Anchorage, Alaska, 17–19 April 1990. US Fish and Wildlife Service, Biological Report 90 (12). US Department of the Interior, Fish and Wildlife Service, Washington, DC, pp. 428–431.

Hammock, J., 2005. Structure, Function and Context: The Impact of Morphometry and Ecology on Olfactory Sensitivity. Doctoral Dissertation, Massachusetts Institute of Technology, Boston, MA.

Hanson, M.B., Bledsoe, L.J., Kirkevold, B.C., Casson, C.J., Nightingale, J.W., 1993. Behavioral budgets of captive sea otter mother–pup pairs during pup development. Zoo Biol. 12, 459–477.

Hatfield, B., Ames, J.A., Estes, J.A., Tinker, M.T., Johnson, A.B., Staedler, M.M., Harris, M.D., 2011. Sea otter mortality in fish and shellfish traps: estimating potential impacts and exploring possible solutions. Endangered Species Res. 13, 219–229.

Iversen, J.A., Krog, J., 1973. Heat production and body surface area in seals and sea otters. Norw. J. Zool. 21, 51–54.

Jessup, D.A., Mazet, J.A.K., Ames, J.A., 1996. Oiled wildlife care for sea otters and other marine animals in California: a government, university, private sector, non-profit cooperative. Endangered Species Update 13, 53–56.

Jessup, D.A., Yeates, L.C., Toy-Choutka, S., Casper, D., Murray, M.J., Ziccardi, M.H., 2012. Washing oiled sea otters. Wildl. Soc. Bull. 36, 6–15.

Kenyon, K.W., 1969. The sea otter in the eastern Pacific Ocean. North Am. Fauna 68, 1–352.

Kenyon, K.W., Yunker, C.E., Newell, I.M., 1965. Nasal mites (Halarachnidae) in the sea otter. J. Parasitol. 51, 960.

Kirkpatrick, C.M., Stulken, D.E., Jones Jr., R.D., 1955. Notes on captive sea otters. J. Arctic Inst. North Am. 8, 46–59.

Kleiber, M., 1975. The Fire of Life: An Introduction to Animal Energetics (revised edition). Robert E. Krieger Publishing Company, Huntington, NY.

Kooyman, G.L., Davis, R.W., Castellini, M.A., 1977. Thermal conductance of immersed prinniped (*sic*) and sea otter pelts before and after oiling with Prudhoe Bay crude. In: Wolfe, D.A. (Ed.), Fate and Effects of Petroleum Hydrocarbons in Marine Ecosystems and Organisms. Pergamon Press, New York, NY, pp. 151–157.

Kvitek, R., DeGange, A., Beitler, M., 1991. Paralytic shellfish toxins mediate sea otter food preference. Limnol. Oceanogr. 36, 393–404.

Larson, S., Casson, C.J., Wasser, S., 2003. Noninvasive reproductive steroid hormone estimates from fecal samples of captive female sea otters (*Enhydra lutris*). Gen. Comp. Endocrinol. 134, 18–25.

Lenfant, C., Johansen, K., Torrance, J.D., 1970. Gas transport and oxygen storage capacity in some pinnipeds and the sea otter. Respir. Physiol. 9, 277–286.

Lensink, C.J., 1960. Status and distribution of sea otters in Alaska. J. Mammal. 41, 172–182.

McLean, J.H., 1962. Sublittoral ecology of kelp beds of the open coast area near Carmel, California. Biol. Bull. (Woods Hole, MA) 122, 95–114.

McShane, L.J., Estes, J.A., Riedman, M.L., Staedler, M.M., 1995. Repertoire, structure, and individual variation of vocalizations in the sea otter. J. Mammal. 76, 414–427.

Morrison, P., Rosenmann, M., Estes, J.A., 1974. Metabolism and thermoregulation in the sea otter. Physiol. Zool. 47, 218–229.

Packard, J.M., Ribic, C.A., 1982. Classification of the behavior of sea otters (*Enhydra lutris*). Can. J. Zool. 60, 1362–1373.

Payne, S.F., Jameson, R.J., 1984. Early behavioral development of the sea otter, *Enhydra lutris*. J. Mammal. 65, 527–531.

Rathbun, G.B., Jameson, R.J., VanBlaricom, G.R., Brownell Jr., R.L., 1990. Reintroduction of sea otters to San Nicolas Island, California: preliminary results for the first year. In: Bryant, P.J., Remington, J. (Eds.), Endangered Wildlife and Habitats in Southern California, vol. 3. Memoirs of the Natural History Foundation of Orange County, Newport Beach, CA, pp. 99–114.

Reshetkin, V.V., Shidlovskaya, N.K., 1947. Acclimatization of sea otters. The Sea Otter (Kalan). Main Administration of Reserves, Council of Ministers of the Russian Soviet Federated Socialist Republic, Moscow, USSR, Jerusalem, Israel, pp. 175–224 (1962) (English Translation, Israel Program for Scientific Translations, Ltd.).

Shidlovskaya, N.K., 1947. Directions for the feeding and care of the male sea otter. The Sea Otter (Kalan). Main Administration of Reserves, Council of Ministers of the Russian Soviet Federated Socialist Republic, Moscow, USSR, Jerusalem, Israel, pp. 225–227 (1962) (English Translation, Israel Program for Scientific Translations, Ltd.).

Siniff, D.B., Williams, T.D., Johnson, A.M., Garshelis, D.L., 1982. Experiments on the response of sea otters *Enhydra lutris* to oil contamination. Biol. Conserv. 23, 261–272.

Stulken, D.E., Kirkpatrick, C.M., 1955. Physiological investigation of captive mortality in the sea otter (*Enhydra lutris*). In: Trefethen, J.B. (Ed.), Transactions of the Twentieth North American Wildlife Conference. Wildlife Management Institute, Washington, DC, pp. 476–494.

Tarasoff, F.J., 1974. Anatomical adaptations in the river otter, sea otter and harp seal with reference to thermal regulation. In: Harrison, R.J. (Ed.), Functional Anatomy of Marine Mammals, vol. 2. Academic Press, London, pp. 111–141.

Thomas, J.A., Cornell, L.H., Joseph, B.E., Williams, T.D., Dreischman, S., 1987. An implanted transponder chip used as a tag for sea otters (*Enhydra lutris*). Mar. Mammal Sci. 3, 271–274.

Townsend, R., Heneman, B., 1989. The *Exxon Valdez* Oil Spill: A Management Analysis. Center for Marine Conservation, Washington, DC.

VanBlaricom, G.R., 1990. Capture of lightly oiled sea otters for rehabilitation: a review of decisions and issues. In: Bayha, K., Kormendy, J. (Eds.), Sea Otter Symposium: Proceedings of a Symposium to Evaluate the Response Effort on Behalf of Sea Otters after the T/V *Exxon Valdez* Oil Spill into Prince William Sound, Anchorage, Alaska, 17–19 April 1990. US Fish and Wildlife Service, Biological Report 90 (12). US Department of the Interior, Fish and Wildlife Service, Washington, DC, pp. 130–136.

Vincenzi, F., 1962. The sea otter (*Enhydra lutris*) at Woodland Park Zoo. Int. Zoo Yearb. 3, 27–29.

Visscher, N.C., R. Snider, R., Stoep, G.V., 2009. Comparative analysis of knowledge gain between interpretive and fact-only presentations at an animal training session: an exploratory study. Zoo Biol. 28, 488–495.

Wagner, K., Chessler, M., York, P., Raynor, J., 2009. Development and implementation of an evaluation strategy for measuring conservation outcomes. Zoo Biol. 28, 473–487.

Williams, T.D., Williams, T.M., 1996. The role of rehabilitation in sea otter conservation efforts. Endangered Species Update 13, 50–52.

Williams, T.M., 1989. Swimming by sea otters: adaptations for low energetic cost of locomotion. J. Comp. Physiol. A 164, 815–824.

Williams, T.M., Davis, R.W. (Eds.), 1990. Sea Otter Rehabilitation Program. 1989 *Exxon Valdez* Oil Spill. International Wildlife Research, Galveston, TX.

Williams, T.M., Davis, R.W. (Eds.), 1995. Emergency Care and Rehabilitation of Oiled Sea Otters: A Guide for Oil Spills Involving Fur-Bearing Marine Mammals. University of Alaska Press, Fairbanks, AK, 279 pages.

Williams, T.M., Kastelein, R.A., Davis, R.W., Thomas, J.A., 1988. The effects of oil contamination and cleaning on sea otters (*Enhydra lutris*). 1. Thermoregulatory implications based on pelt studies. Can. J. Zool. 66, 2776–2781.

Williams, T.M., Davis, R.W., McBain, J.F., Tuomi, P.A., Wilson, R.K., McCormick, C.R., Donoghue, S., 1995a. Diagnosing and treating common clinical disorders of oiled sea otters. In: Williams, T.M., Davis, R.W. (Eds.), Emergency Care and Rehabilitation of Sea Otters: A Guide for Oil Spills Involving Fur-Bearing Marine Mammals. University of Alaska Press, Fairbanks, AK, pp. 59–94.

Williams, T.M., McBain, J.F., Tuomi, P.A., Wilson, R.K., 1995b. Initial clinical evaluation, emergency treatments, and assessment of oil exposure. In: Williams, T.M., Davis, R.W. (Eds.), Emergency Care and Rehabilitation of Sea Otters: A Guide for Oil Spills Involving Fur-Bearing Marine Mammals. University of Alaska Press, Fairbanks, AK, pp. 45–57.

Williams, T.M., O'Conner, D.J., Nielsen, S.W., 1995c. The effects of oil on sea otters: histopathology, toxicology, and clinical history. In: Williams, T.M., Davis, R.W. (Eds.), Emergency Care and Rehabilitation of Sea Otters: A Guide for Oil Spills Involving Fur-Bearing Marine Mammals. University of Alaska Press, Fairbanks, AK, pp. 3–22.

Yeates, L.C., Williams, T.M., Fink, T.L., 2007. Diving and foraging energetics of the smallest marine mammal, the sea otter (*Enhydra lutris*). J. Exp. Biol. 210, 1960–1970.

Chapter 9

The Value of Rescuing, Treating, and Releasing Live-Stranded Sea Otters

Andrew Johnson and Karl Mayer
Monterey Bay Aquarium, Monterey, CA, USA

INTRODUCTION

When a sea otter strands alive along the shore, what obligation do humans have to care for it? Should we treat ill or injured sea otters and release them or remove them from the population? What can we learn from live-stranded sea otters that will support the conservation of the species? Should we use live-stranded sea otters as research subjects before releasing them to the

Sea Otter Conservation. DOI: http://dx.doi.org/10.1016/B978-0-12-801402-8.00009-3

wild? How should we balance the cost of caring for individual sea otters and the value of other conservation outcomes (e.g., preserving nearshore marine habitat)? When should we euthanize live-stranded sea otters? These questions have guided the development of the Monterey Bay Aquarium's (MBA) response to live-stranded sea otters over the past decade, and they represent the kind of inquiry that wildlife rescue programs should undertake to generate programmatic and conservation value.

Many organizations and facilities seek to aid live-stranded marine mammals and effect feasible outcomes for them (e.g., release to the wild, placement in zoos and aquariums, use in research programs). The National Wildlife Rehabilitators Association and the International Wildlife Rehabilitation Council define "wildlife rehabilitation" as "[t]he treatment and temporary care of injured, diseased, and displaced indigenous animals, and the subsequent release of healthy animals to appropriate habitats in the wild" (Miller, 2012); however, different organizations harbor divergent perspectives on what "rehabilitation" means, on what constitutes a rehabilitated animal, and on what ethical obligations humans incur by rescuing an ill, injured, or orphaned marine mammal. Moore et al. (2007) asserted that "[t]he vagueness of even this most basic concept [i.e., rehabilitation] adds to the conflicting values and options within which the stranding organizations work." According to these authors, the enterprise of rehabilitating marine mammals "lacks a coherent central set of core values, ethics, or goals." For live-stranded southern sea otters (i.e., sea otters in California, *Enhydra lutris nereis*), classified as a threatened species under the US Endangered Species Act (ESA), caregivers and managers at the MBA must make hard ethical and practical choices each time a sea otter strands alive along the shore.

In the early days of the MBA's rescue and care program for sea otters, animal care personnel aided orphaned, injured, or ill sea otters from all age classes and targeted release for most cases. The program included lofty ambitions for undertaking scientific research and developing better care techniques; however, the labor intensity of managing an increasing load of live-stranded neonates and debilitated adult otters hindered progress in this area. In addition, caregivers had to take sea otter pups for time-consuming swims in Monterey Bay as a means of stimulating their fitness, helping them acquire swimming and diving skills, and developing their foraging abilities. Although this swim program accomplished the intended objectives, the pups bonded with, and remained acclimated to, the swimmers and never established a natural wariness of people.

Program managers have long believed that treating stranded, prime-age (i.e., reproductive) female sea otters and releasing them provided some value to the California population, but that treating stranded otters with geriatric problems or irresolvable health issues did not make good sense, since, in most cases, those otters could no longer reproduce and tended to fall ill within a short time after release. Of the more than 600 southern sea otters received by the MBA from 1984 through 2013, about 30% died, veterinary personnel euthanized about 40% for various reasons, animal care personnel released about one third, and program managers transferred a small percentage of healthy but non-releasable animals to accredited aquariums and zoos in the United States (MBA, unpublished data). Many subadult and adult otters did not receive radio tags upon release, so the program generated little data on the long-term survival and reproduction of these animals. Almost all otters that stranded as pups did receive radio tags; however, until 2001, most of these animals did not survive long in the wild—many died or vanished within a short time, and caregivers had to recapture many others because they failed to thrive or engaged in undesirable behavior (e.g., interacting with humans). Finally, few of them entered the reproductive population.

Williams (1990), Styers and McCormick (1990), Williams and Hymer (1992), Williams et al. (1995), and Tuomi (2001) have published general protocols for providing care to captive-held and live-stranded sea otters, including pups. Although pup-rearing strategies at the MBA evolved in increments through the years to enhance the survival and behavioral competence of these animals following release, failures in achieving these objectives risked imposing adverse effects on the wild population—for instance, by releasing otters that might never integrate into the reproductive population—and wasting program resources. To achieve better results, caregivers wanted to establish a fostering program for pups that would resolve methodological shortcomings—the undesirable behavioral effects of proximity to human caregivers, the long duration of captive holding, and the inability of released juvenile otters to adapt to the wild.

A major concern of managers and the MBA involved the low conservation value of returning live-stranded southern sea otters to the parent population. Releasing a small number of otters, regardless of their age or reproductive status, into established areas along the California coast each year had little or no impact on population numbers. This reality mirrored a frequent criticism of many wildlife treatment-and-release programs. As Quackenbush et al. (2009) have asserted, "...there are no marine mammal populations in Alaska where the release of small numbers of rehabilitated individuals will benefit a population." Managers at the MBA wanted to move beyond the feel-good aspect of releasing a few treated sea otters and find ways to derive meaning and value from handling these animals within an ethics-based, research-based program.

SEA OTTER STRANDINGS AND STRANDING RESPONSE

Congress codified a stranding response for marine mammals in US waters in the 1992 amendments to the Marine Mammal Protection Act (MMPA). Since then, authorized organizations have retrieved thousands of pinnipeds, cetaceans, manatees, and sea otters, attempted to return them to health, and released many of them to the wild. In general, these facilities can offer safe havens for wildlife in distress; can provide a service to communities concerned about injured wildlife; can help ensure an ethical response to animals injured by anthropogenic causes; can effect the humane care of injured and orphaned animals until the point of release, euthanasia, or other disposition; can foster a greater sense of responsibility for animals that need help; can prompt an increasing environmental awareness and concern for wildlife among people; and can encourage reductions in intentional or unintentional harassment of wildlife by people or domestic animals. But evaluating when and how these responding facilities make legitimate contributions to the welfare and conservation of wildlife populations represents a major challenge.

The MMPA defines the term "stranding" as "an event in the wild in which. . .a marine mammal is alive and is on a beach or shore of the United States and unable to return to the water." In their book, *Marine Mammals Ashore*, Geraci and Lounsbury (2005) describe a stranded animal as "any creature left in a helpless position, such as a marine mammal that falters ashore ill, weak, or simply lost." These definitions conjure scenarios in which an individual marine mammal faces imminent death. When people find marine mammals in distress along the shore, they expect that some group or agency will rescue and treat the animals; however, each stranding activates a convoluted and imprecise sequence of events in which responders must locate an animal, evaluate its condition, determine whether it requires assistance, transport it to a qualified facility, and embark upon a humane and appropriate course of treatment, which may or may not involve returning the animal to the wild.

Over the past two decades, the overall population of sea otters in California has experienced periods of sluggish growth and periods of stagnation or decline, increasing about 1.5% per year during that time. The influx of live-stranded southern sea otters has progressed at a higher rate, close to 8% per year, an increase based on more than the slow population growth can explain. Public awareness about marine mammals and about what actions to take in the event of live strandings account for some level of increase in the reporting numbers; however, despite the 25% increase in the human population residing in Santa Cruz, Monterey, and San Luis Obispo Counties over the past two decades, along with the substantial draw of tourists to Central California each year, the volume of people who live in and visit these coastal communities does not offer an obvious explanation for the fourfold increase in the annual number of live-stranded sea otters during that same period. In short, whether the sea otter population has increased or declined, the number of live-stranding cases has climbed (Figure 9.1).

Sea otters give birth to helpless pups. These newborns cannot swim, groom their fur, or feed themselves. They float upon the surface of the ocean, reliant on their mothers for food and care. When a pup comes ashore without its mother, something catastrophic has occurred in that pup's life. The mother might have died; illness or injury might have caused her to abandon her pup; a strong wave might have carried the pup beyond the hearing range of its mother.

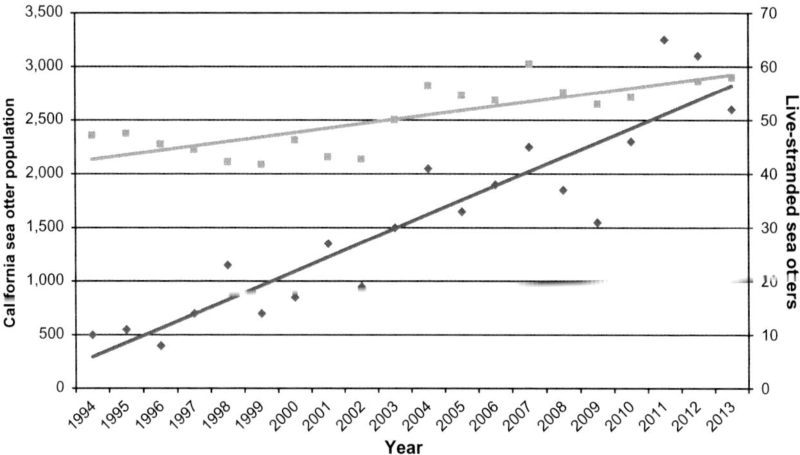

FIGURE 9.1 The annual counts of southern sea otters along the coast of California compared with the number of live-stranded sea otters received by the MBA, showing the strong upward trend in live strandings over the past 20 years.

Regardless of the cause of separation, a newborn sea otter cannot survive much longer than 24 h without care and feeding—it will die unless humans intervene. Juvenile otters face different challenges. Once weaned from their mothers, "recently weaned juvenile sea otters just learning to feed independently" may seek less-desirable, easier-to-find prey that carry harmful parasites (Mayer et al., 2003). Adult otters succumb to a range of natural and human-caused diseases and injuries, and they come ashore when they can no longer forage or maintain their body temperature in the cold ocean. Regardless of the age class or condition of these animals, the MBA has a long-standing commitment to respond when these otters come ashore in distress.

Sea otters contribute immeasurable benefits to the health and productivity of nearshore marine systems, including kelp forests (Estes et al., 2010) and estuarine habitats (Hughes et al., 2013). Along with their value to these ecosystems, sea otters generate significant economic benefits for coastal communities in the way of tourism-related employment (Loomis, 2006). Unfortunately, because of their presence in these nearshore areas, sea otters suffer significant morbidity and mortality from human-generated impacts along California's coastline and other range areas—a factor that contributes to the increasing number of live strandings.

For more than two decades, pathologists at the National Wildlife Health Center (NWHC) (Madison, WI) and the Marine Wildlife Veterinary Care and Research Center (MWVCRC) (Santa Cruz, CA) have documented numerous associations between sea otter mortality and the exposure of sea otters or their prey items to toxins, parasites, bacteria, fungi, and anthropogenic chemicals. Jessup and Miller (2012) stated that some of these pathogens and pollutants "appear to be particularly noxious to sea otters due to unique aspects of their metabolism and biology," such as their high consumption of filter-feeding prey items that tend to bioaccumulate contaminants that wash into nearshore waters. The annual rate of infectious disease as the primary cause of death in the southern sea otter population has fluctuated between 40% and 60%, and from 1998 to 2001, "parasitic disease alone caused death in 38.1 percent of otters examined" (Kreuder et al., 2003). Several studies have documented the land-to-sea transfer of material harmful to sea otters, such as "enteric bacterial pathogens" (Oates et al., 2012); "bacteria and protozoal parasites from sewage, agricultural, and street runoff; contaminants; nutrients that cause algal blooms and intoxications" (Jessup et al., 2007); and "concentrations of organochlorine pesticides" (Kannan et al., 2008). Many other investigators have chronicled the degradation of harbors and nearshore habitat through inputs of agricultural chemicals, toxic contaminants, and disease-causing pathogens (e.g., Dubey et al., 2001; Jessup et al., 2010; Kannan et al., 2004, 2006; Miller et al., 2002, 2010a,b). In a paper looking at the overlap of epidemiology and conservation theory, Lafferty and Gerber (2002) concluded that "disease may be an important factor limiting the growth of otter populations." Other recent studies (Tinker et al., 2008; Johnson et al., 2009) have highlighted the potential impact that food limitation and nutritional stress could have on sea otters residing in the central portion of the California range. "High levels of infection with protozoal pathogens may be an adverse consequence of dietary specialization in this threatened species, with both depleted resources and disease working synergistically to limit recovery" (Johnson et al., 2009).

Given the high level of morbidity and mortality in the southern sea otter population stemming from land-based sources, scientists have called for policy changes and better water-management practices to reduce or control the problems caused by contaminated runoff. Results from Shapiro et al. (2012) "highlight the complex mechanisms that influence transmission of terrestrially derived pathogens to marine wildlife" and demonstrate that the health of sea otters and humans remains "intimately linked to adequate water quality of coastal habitats." Of course, sea otters die from other natural and anthropogenic causes: intraspecific aggression and mating wounds; oceanographic and environmental perturbations; boat strikes; cardiac disease; fisheries-related conflicts; shark bites; gunshots; and starvation (Foott, 1970; Staedler and Riedman, 1993; Ames et al., 1996; Pattison et al., 1997; Kreuder et al., 2003; Kreuder et al., 2005; Dau et al., 2009; Hatfield et al., 2011). In general, sea otters in California face inordinate threats to their survival and recovery, and the MBA has had to deal with increasing numbers of near-dead animals coming ashore—a circumstance that has challenged program managers to evaluate the relative merits of every case and to measure the relative importance of individual otters against the viability of the wild population.

The US Fish and Wildlife Service (FWS), which has management authority for sea otters in US waters, issues permits and letters of authorization for the MBA and others to conduct work with these animals. Decisions to intervene in live-stranding cases occur within a complex conceptual framework. In general, sea otters that emerge onto land in public areas require some form of response to ascertain their problems and confirm the need for human intervention. At the MBA, whenever responding personnel believe that a stranded otter will not survive another 24 h, or when conditions represent a specific danger to the health of the animal (e.g., the presence of dogs), they will capture the otter and transport it to the Aquarium for evaluation.

Responders to sea otter live-stranding events must consider several logistical factors—condition of the animal, environmental conditions (tides, swell, weather), location of the stranding, beach/shoreline access, time of day, proximity to humans and human activities, etc.—before deciding to pick up a sea otter. In all cases, though, responders must consider a broad array of decision-making criteria and guidelines to achieve program objectives.

CASELOAD MANAGEMENT AND ETHICAL CHALLENGES

In 2002, the Aquarium convened a blue ribbon panel of marine biologists, marine ecologists, marine mammal rehabilitation experts, and wildlife managers to evaluate its programmatic approach to the care of live-stranded southern sea otters. Three recommendations for modifying the live-stranding program emerged from the panel's discussions: consider rehabilitation as a research component of a comprehensive recovery effort; reposition and transform the focus of the rehabilitation program to support research, conservation, and education initiatives; adopt stricter acceptance and rehabilitation criteria to reduce the numbers of live-stranded sea otters that require long-term or costly care.

In 2004, following a 2-year review of methods, MBA managers shifted the focus of program activities from the health of individual sea otters to the health and recovery of the wild population and the nearshore ecosystem. To accomplish this philosophical change, staff members aligned animal care efforts with the activities outlined in the *Final Revised Recovery Plan for the Southern Sea Otter* (US Fish and Wildlife Service, 2003), reemphasized the need to undertake scientific research, and created a system for managing the increasing caseload of live-stranded sea otters. The strategy for making the change viable and successful included an open release to the media of the intent to euthanize more sea otters, including pups, as part of a comprehensive triage and caseload-management strategy.

The plan for managing an increasing caseload involved treating and releasing those animals that care and veterinary personnel believed have the best odds of survival on release. Caregivers and medical staff remained committed to releasing animals that would pose no direct threat to the wild population (e.g., by introducing disease) and that would provide additional conservation value through their involvement in research projects. Because handling live-stranded sea otters can afford opportunities for increasing knowledge about the health of specific sea otter populations and nearshore ecosystems, program activities targeted those opportunities, while minimizing the suffering and discomfort of ill or injured animals. To achieve program goals, personnel continued to respond to all reports of live-stranded sea otters—in association with The Marine Mammal Center (TMMC) in Sausalito, California, and other partners—and determined subsequent actions using a rigorous system of evaluation and triage.

Responding personnel have learned to evaluate all stranding situations and to determine whether stranded otters require some form of intervention. When responders decide to bring a live-stranded sea otter to the Aquarium, it enters one of four caseload-management streams: care for release to the wild; care for long- or short-term research and possible release to the wild; permanent transfer to a zoo or aquarium for long-term management; or euthanasia.

When the condition of an animal reaches a point of "insurmountable suffering" (AVMA, 2013), the veterinarian will decide, in consultation with other program staff members, to euthanize it. When a live-stranded sea otter does not qualify for release to the wild and no facility can offer it a permanent home within the foreseeable future, euthanasia remains the sole option. To facilitate research projects and to archive tissues for future testing, veterinary personnel employ chemical restraint techniques to collect a variety of biologic samples, such as blood, urine, and other tissues associated with authorized requests, before euthanasia.

To improve the efficacy of rescuing and treating live-stranded sea otters, veterinary, managerial, and animal care staff members evaluate each new case to determine the appropriate action for each animal. The increasing caseload has forced program personnel to make tougher decisions about geriatric males, diseased juveniles, prime-age females in poor condition, animals with neurological impairment, and animals with shark- or human-caused trauma. Although the attending veterinarian may opt to euthanize a suffering animal, euthanasia of any sea otter remains the last resort. Program managers have developed interrelated criteria for making decisions about the treatability and potential releasability of live-stranded sea otters. When a live-stranded sea otter arrives at the Aquarium, the medical team and care staff evaluate its condition, obtain basic morphometrics, acquire samples for diagnostic testing, perform an extensive physical examination, and triage the animal based on the caseload criteria. The staff veterinarian determines whether the condition of an animal requires euthanasia for health and humane reasons. The program manager, the animal care coordinator, and other personnel use the triage criteria to determine when to euthanize an otherwise treatable animal. This philosophical approach for managing the live-stranding program ensures a humane response for each live-stranded sea otter, supports conservation research initiatives, advances zoological care programs, and protects the wild population.

Of course, critics often question whether the well-intentioned activities of wildlife rehabilitation centers should proceed when program personnel have no mechanisms for evaluating release success and the effects of releasing treated animals in wild populations. In many cases, the application of ethical principles conflicts with the reality of operating wildlife treatment facilities, and program managers have to adjust their objectives to meet the practical and economic challenges of responding to ill or injured wildlife. A recent survey of mammal-rehabilitation programs by Guy et al. (2013) found that only one-third of respondents had criteria for assessing the success of a release and that 38% of respondents did not undertake post-release monitoring of any kind. This information suggests that without methods for evaluating care protocols and release success, many programs lack a framework for weighing the ethical options for managing borderline cases, and they don't have the resources to obtain sufficient information to render consistent, values-based judgments on future cases.

In a review of wildlife rehabilitation research undertaken through the Royal Society for the Prevention of Cruelty to Animals in the United Kingdom, Grogan and Kelly (2013) inquired about "which individual animals can, or should, be treated, which animals will survive treatment and which animals will survive after they have been released back to the wild." In their view, "the welfare of wildlife casualties can be improved by investigating which injuries or illnesses are most likely to result in a successful release for each species, and by collecting data on postrelease survivorship" (p. 211). This sort of evaluation of methods and outcomes has become foundational to the approach of the sea otter live-stranding program at the MBA. Given the ESA-listed status of the southern sea otter and the myriad challenges to population health and recovery, decisions to treat these animals and return them to the parent population must receive careful thought. Will borderline cases consume limited prey resources but fail to integrate within or contribute to the wild population? Will aged animals suffer and succumb to illness soon after release? Should program managers consider euthanasia for non-releasable cases? Scientists, population managers, and stranding coordinators have started to advocate for a stricter approach. "Rehabilitation efforts should be evaluated on whether the likely benefits to science, nature, or knowledge outweigh the potential harm to individuals or populations" (Moore et al., 2007); however, many wildlife rehabilitation programs have trouble weighing the benefits and value of their activities.

Kirkwood and Sainsbury (1996) have suggested that "the decision to treat sick or injured free-living wild animals should not be based on welfare grounds alone." In a similar vein, the Veterinary Association for Wildlife Management in the United Kingdom believes that a wildlife casualty "...must be released with a chance of survival at least equivalent to that of other free-living members of its species" (Kirkwood and Best, 1998). The International Union for Conservation of Nature (IUCN) has warned of the risks—introduction of disease, competition for food—of reintroducing confiscated or captive-held animals to the wild because the "conservation of the species as a whole, and of other animals already living free, must take precedent [sic] over the welfare of individual animals that are already in captivity" (IUCN, 2002). Albrecht (2003) discussed methods to evaluate the success of rehabilitation and release programs, proposing that such programs should focus on biodiversity and ecosystem preservation and must ensure that adequate habitat exists to support released animals. The Wildlife Rehabilitators Code of Ethics (National Wildlife Rehabilitators Association) includes stronger sentiments, stating that "non-releasable animals, which are inappropriate for education, foster parenting, or captive breeding have a right to euthanasia" and that "wildlife rehabilitator[s] should work on the basis of sound ecological principles, incorporating appropriate conservation ethics and an attitude of stewardship" (Miller, 2012). As rehabilitation programs have grown in their sophistication, the familiarity of program personnel with ethical issues has expanded and integrated into core procedures. Still, practical elements must factor into all decisions relating to release.

Few people doubt the need for a response when a sea otter or other marine mammal comes ashore in distress; however, personnel working for responsible organizations must set their decisions and actions within a context that considers more than the individual animals and use science results and common sense to create ethical and values-based programmatic criteria for managing ill and injured wildlife (Johnson, 2000). Oversimplifying or disregarding the realities that surround stranding events and the release of captive-held animals risk bigger ethical challenges, such as introducing disease or aberrant behavior into wildlife populations (Beck et al., 1993). According to the IUCN Guidelines for the Placement of Confiscated Animals (2002), "...a growing body of scientific study of re-introduction of captive animals, the nature and dynamics of wildlife diseases, and the nature and extent of the problems associated with invasive species suggests that such actions may be among the least appropriate options" (p. 10).

In an article in *Marine Mammal Science*, Moore et al. (2007) summarized the state of marine mammal rehabilitation programs in the United States and outlined criteria for limiting the number of rehabilitated marine mammals released to the wild. The authors explicated some of the MBA's dilemmas for managing live-stranded sea otters in the following assessment:

> *Recently [2004], the Monterey Bay Aquarium adopted a policy to euthanize stranded sea otter (*Enhydra lutris*) pups that could not be placed into a surrogate hand-rearing program. On the basis of their experience of 20 yr rehabilitating sea otter pups, this organization determined that the rehabilitation of preweaned pups without a surrogate female was unlikely to be successful and*

> *decided to use limited funds on other aspects of sea otter conservation rather than rehabilitation. This decision has been very controversial and illustrates the conflict between an agenda that encompasses the sum of economic, practical, and survival expectancies vs. one that is driven more by the overarching value of the welfare of an individual animal without looking at the broader contexts. It also highlights the public outreach dilemma for an organization encouraging coastal conservation while "killing" a charismatic keystone species from the same ecosystem. (p. 742)*

All facilities that offer care to ill, injured, or orphaned wildlife should make sure that they address the full range of ethical and practical considerations to manage their caseload and avoid actions that could interfere with the effective care of treatable animals or harm wild populations.

THE SURROGACY PROJECT

From 1986 to 2001, staff at the MBA hand raised sea otter pups and attempted to return each of them to the wild. Several techniques proved successful in rearing pups to a releasable age; however, none of the methodologies resulted in consistent, positive release outcomes. Given the small number of animals involved, the release of individual sea otters back to the core population didn't offer much, if any, conservation value. In addition, in the late 1990s, the number of older sea otters stranding alive with severe injuries, infectious diseases, and toxin-induced medical problems started to increase. Program managers recognized the need to develop a more sophisticated approach toward live strandings and to realize additional value within a broader research and conservation framework.

Evidence exists that adoption, alloparenting, and allosuckling occur in numerous phocid and otariid species (Flatz and Gerber, 2010; Gemmell, 2003; Maniscalco et al., 2007; Perry et al., 1998; Riedman and Le Boeuf, 1982; Stirling, 1975) and, on occasion, in sea otters (Kenyon, 1975; Staedler and Riedman, 1989). According to Riedman and Estes (1990), the natural history of sea otters does not select for alloparenting, and the advantages that may accrue to adult otters through fostering—gaining parental experience, exploiting fostered young—do not materialize with any regularity. "Given their social organization and breeding behavior, fostering opportunities are rare for sea otters, unlike other mammals that give birth synchronously and raise their young in crowded breeding colonies, such as several species of bats, pinnipeds, and ungulates" (Riedman and Estes, 1990). Because sea otters in California give birth throughout the year and rear their pups in low-density groups, "[t]he chances of an orphaned sea otter pup encountering a potential foster mother—particularly one that has recently lost her own pup—are probably remote" (Riedman and Estes, 1990). Staedler and Riedman (1989) witnessed one adoption along the Monterey Peninsula in which a tagged adult female nursed an abandoned pup, shared food with it, and rested with it for a few weeks before it died, probably from starvation.

Kenyon (1975) documented one case of allomothering in a captive setting, in which a female sea otter showed maternal behavior toward a juvenile for several days. Kenyon concluded that "under certain conditions a tolerant adult otter might contribute to the survival of an orphaned juvenile" (p. 102). In a zoological setting, caregivers can reduce the metabolic cost to female sea otters of investing in alloparental behavior toward live-stranded pups by offering a safe living environment, protection from the elements, and abundant food. These circumstances reinforce the apparent inclination of female otters to foster pups.

In developing a surrogate-rearing program for live-stranded sea otter pups, program managers at the MBA had to gain some sense about the factors that would lead to release success and find ways to overcome the effects of a captive environment, such as acclimation to people and delayed development of survival skills. In a comprehensive review, Jule et al. (2008) "found evidence to support that reintroduction projects using wild-caught animals are significantly more likely to succeed than projects using captive-born animals." Other studies (McPhee, 2004) have shown that the length of captive holding, especially over multiple generations, leads to a reduction in predator avoidance behaviors and other survival skills in reintroduced species. Given these conclusions, caregivers at the MBA felt confident that techniques for managing live-stranded sea otter pups and for mitigating the detrimental effects of captive experience on rearing procedures could occur under strict guidelines.

Through the years, approximately 90% of live-stranded sea otter pups arrived at the MBA without consequential or long-lasting health problems. Program personnel had long predicted that creating a mother–pup environment would help override the detrimental effects of captivity on live-stranded pups (e.g., growing up in a small pool, having close contact with humans), and they postulated that these otters, after release as juveniles, would maintain their wariness of people and avoid interactions with humans, exhibit greater success adapting to life in the wild (compared to non-surrogate-reared pups), and demonstrate survival in the wild at a rate comparable to free-ranging juveniles. In a review of techniques used for rearing young sea otters for release to the wild over a 20-year period, Nicholson et al. (2007) compared the effectiveness of a trial surrogate-rearing program with methods relying on human care. Before embarking on a large-scale surrogacy study, investigators wanted to confirm whether surrogates would provide "a social environment that stimulates natural behavior and facilitates learning among young sea otters," and whether pups reared in this manner would develop foraging skills at a younger age than pups reared by traditional captive methods (i.e., using human caregivers). Between 2001 and 2003, a trial group of seven male sea otter pups (1–10 weeks old), reared by surrogate sea otter mothers and released to the wild, foraged well, integrated with the wild population, avoided interactions with humans, and survived to a year after release at a comparable rate (71% versus 75%) to wild-reared pups (Hanni et al., 2003) and at a better rate (71% versus 31%) than pups reared without surrogates (Nicholson et al., 2007). This preliminary data suggested that surrogacy could serve as a less labor-intensive and more effective method for treating, rearing, and releasing sea otter pups when compared to rearing methods that rely on human care.

Nicholson et al. (2007) envisioned that a surrogate-rearing program could benefit the study and conservation of southern sea otters by helping investigators address specific research objectives, such as increasing understanding of pup behavioral and physiological development, measuring the energetic costs of rearing pups, and determining why the survival rate of prime-age females has declined, along with improving techniques to reintroduce sea otters in the event of a catastrophic population decline and educating the public regarding threats to sea otters and nearshore marine habitat.

In practice, work on the project was broken into three phases: a treatment phase, in which caregivers provided care, rearing, and management corresponding to the dependency and early post-weaning phases in wild sea otter pups; a transition phase, in which caregivers attempted a conditional release to the wild and tracked the otters during an intensive monitoring period; and an evaluation phase, in which caregivers compared post-release survival and reproductive success over several years with the best available data from wild southern sea otter population studies. Pups received care from humans until they reached about 8 weeks old. During that time, caregivers fed the pups formula and supplemented them with pieces of solid food equivalent to 25–35% of body weight daily. By necessity, the protocols for introducing pups and surrogates remained flexible—some surrogates accepted pups during the first introduction; others required multiple introductions over several days to bond with pups. Starting to house pups with surrogates by 8 weeks of age helped reduce the potential for pups to habituate to humans and promoted normal social and physical development.

To document the successes and failures of the program, personnel had to monitor activity and behavior following release to the wild. To accomplish this, the MBA veterinarian implanted a VHF radio transmitter in the abdomen of each surrogate-reared otter (Ralls et al., 1989) when it reached 6 months of age to facilitate tracking of these released animals.

Project protocols called for releasing surrogate-reared animals into Elkhorn Slough—a large, shallow tidal embayment in the center of the Monterey Bay coastline that connects the open ocean through Moss Landing Harbor. Program personnel used this estuarine environment as a release site for surrogate-reared animals because it offered a protected area with abundant, easy-to-capture prey items, such as shore crabs, cancer crabs, fat innkeeper worms, and mussels, and helped discourage immediate emigration to the open ocean. Trackers detected radio signals from instrumented otters with a VHF receiver and a directional three-element Yagi antenna (Ralls et al., 1989). Daily resights from shore, boat, and/or plane took place for the first 2 weeks after release in order to determine whether released juveniles obtained enough food to survive their transition to the wild, whether juveniles traveled too far—along the shore or offshore—to maintain their body condition, and whether juveniles interacted with humans.

Trackers and caregivers considered recapturing a released sea otter whenever it exhibited an inability to forage effectively or seemed stressed beyond its capacity to function, as evinced by constant swimming or traveling, movement offshore, failure to forage, obvious weight loss, or the presence of aberrant or unproductive behavior. If a released otter demonstrated marked weight loss over a short period (2–4 days), repeatedly beached itself and seemed unable or unwilling to return to the water, displayed ineffective grooming, or exhibited improper behavior toward humans, program personnel would attempt to recapture the otter. Decisions to recapture juvenile otters did not preclude future releases; in fact, the release success of surrogate-reared animals tends to increase with successive releases, with almost all otters succeeding on a third release (MBA, unpublished data).

Goals for the immediate post-release phase included minimizing mortality; maintaining daily observations of foraging and movement patterns; making recapture attempts in the event of inadequate foraging, dispersal offshore, or interactions with humans; and attempting re-release of recaptured otters. Investigators considered surrogate-reared otters independent if they located food, maintained their body condition, displayed normal behavior, and avoided humans for 2 weeks following release (or re-release). Once a juvenile otter achieved independent status and entered the post-release phase of the study, resight attempts and foraging observations (via shore or boat) occurred at least once a week, and aerial surveys for missing animals took place based on budget and pilot availability. Program personnel recovered otters that died as soon as possible after receiving mortality signals and submitted the carcasses for necropsy to determine causes of death.

To document how well surrogate-reared individuals integrated into the wild population, field personnel attempted to track and monitor surrogate-reared and released sea otters for a minimum of 2 years and, when possible, up to a decade or more. With the initial sample group of 20 surrogate-reared otters, survival to 1 year post-release occurred at the same rate as post-weaning survival of wild-reared otters—about 75% for surrogate-reared versus 71–80% for wild-reared (Monnett et al., 2000; Bodkin et al., 2007; Tinker et al., 2008). By comparison, less than a third of the pups reared by human caregivers using original program methods survived to 1 year post-release (Nicholson et al., 2007).

Care and release methods developed by the MBA over the course of the surrogacy study yielded effective and reproducible means of returning stranded sea otter pups to the wild as competent juveniles, and these otters entered the reproductive population. By the end of 2013, seven surrogate-reared and released females had delivered at least 22 pups, and about 65% of the pups have survived to weaning (MBA, unpublished data), compared to 50–64% of wild-reared pups in prior studies in California (Siniff and Ralls, 1991; Jameson and Johnson, 1993; Riedman et al., 1994).

CREATING VALUE

An unpublished February 2007 document titled "Research Plan—California sea otter recovery," created by an ad hoc committee of scientists, made a compelling case for expanding captive research with southern sea otters:

> *The value of captive animal research to sea otter conservation and management is in the understanding of processes and the development of techniques that cannot be achieved by working on wild animals. Although there are many potential directions for captive research, two stand out as being of particular importance: obtaining a clearer and more comprehensive understanding of various aspects of the functional biology of sea otters, and developing improved methods for tagging and instrumentation.*

To realize the research potential of working with captive-held sea otters, the MBA obtained a Federal Fish and Wildlife Permit under the MMPA from the US Fish and Wildlife Service's Division of Management Authority; acquired a license as a registered research facility under the Animal Welfare Act; and established an Institutional Animal Care and Use Committee (IACUC) to evaluate and approve a range of projects using releasable and non-releasable live-stranded southern sea otters. The permit allows specific research activities under the proviso that those activities must not compromise the potential releasability of an otter. Most studies looking at responses of marine mammals to research procedures have concluded that the activities produced no effects that might contravene the release of animals to the wild. For example, Baker and Johanos (2002) determined that restraint, blood sampling, and flipper tagging using careful handling techniques had no deleterious effects on Hawaiian monk seals. Other investigators have espoused the use of free-ranging marine mammals during periods of temporary holding in captive settings, because they can serve as "an alternative to traditional field methods for some types of focused physiological studies" (Mellish et al., 2006). Whenever possible, wildlife rehabilitation facilities should investigate whether and how they can integrate conservation-related research into their core programs.

The surrogacy project represented a groundbreaking effort—a more efficient and effective means of managing live-stranded sea otter pups and a potential avenue for expanding research and conservation opportunities with this species using young otters. Investigators have proposed a number of goals using surrogate-reared sea otters in lieu of wild-reared counterparts. One project would use these otters to sample the health of the environment in Elkhorn Slough as a way to understand factors affecting survival of juvenile and subadult sea otters in that area. Another would investigate the feasibility and efficacy of releasing surrogate-reared animals in areas near the periphery of the southern sea otter range—areas with better habitat and abundant food—as a way to augment the population.

Overall, managing a dedicated treatment-and-release program for live-stranded sea otters has allowed the MBA to participate at a higher level in population research and recovery. Operating the program has advanced the overall knowledge base for providing quality care to sea otters in captive settings, live-stranding settings, and research settings. Veterinary medical personnel have helped achieve additional advances using live-stranded and non-releasable sea otters (Chapter 7), including developing and evaluating gene transcription as a diagnostic tool (Chapter 6, Bowen et al., 2012); determining the pharmacokinetics of antibiotic drugs in sea otters (Brownstein et al., 2011); evaluating the safety of using a preventive vaccine in southern sea otters (Jessup et al., 2009); measuring serum vitamin A concentrations in sea otters (Righton et al., 2011); and assessing sea otter hearing thresholds (Ghoul and Reichmuth, 2012).

Some researchers (Estes, 1991, 1998; Paine et al., 1996) have described the exorbitant cost and low conservation return of trying to rehabilitate and release sea otters affected by oil spills, and they have wondered about the relative merits of attending to oiled otters. But the appeal of sea otters and their intrinsic value to humans can supersede a scientific rationale. As Estes (1998) summarized, "[T]he answer to the question 'why rehabilitate oiled wildlife?' is that we have to, not to enhance populations but to meet a public demand." Over time, though, investigators have used non-releasable sea otters to improve techniques for managing oiled sea otters. From 2004 to 2008, collaborating personnel conducted partial washings on healthy, non-releasable sea otters; on the same otters dipped in canola oil; and on a single live-stranded otter exposed to crude oil in Monterey Bay. They determined that "[p]roviding soft freshwater in recovery pools reduced recovery time substantially. Warming the freshwater appeared to offer additional benefits in some cases" (Jessup et al., 2012). While the value of treating and releasing oiled sea otters remains an elusive target, finding ways to reduce treatment times and improve care protocols for oiled sea otters has demonstrated the potential value that research using non-releasable animals can have for the wild population.

CONCLUSION

For many years, the MBA managed live-stranded sea otters in ways that did not help the recovery of the southern sea otter population. Williams and Williams

(1996) warned against limiting rehabilitation programs "to serving the emotional needs of the public rather than the species affected" (p. 52); thus, along with providing humane care for individual animals, facilities that engage in the treatment and release of ill, injured, or orphaned wildlife must grasp and act upon ideas that support meaningful research and conservation outcomes.

Sleeman (2008) summarized the potential role that a professional live-stranding program can provide:

> *Wildlife rehabilitation centers may contribute to a wide variety of important activities that may not only enhance animal welfare, but also advance veterinary science and veterinary education, biodiversity conservation and ecosystem health, wildlife health monitoring, public health, and biosecurity, as well as public education and conservation-related public policy. (p. 102)*

Although the primary goal of most wildlife rehabilitation programs involves treating ill, injured, or orphaned animals and returning them to the wild, managers need to appreciate a wider-ranging set of objectives and deliver value within each facet of a concerted program. From the care and humane treatment of ill or injured animals to the use of live-stranded animals in addressing important research questions to developing ways to release to the wild and monitor fit, capable animals, the quest to find programmatic and conservation value has driven the MBA's work with live-stranded southern sea otters. Using non-releasable adult female sea otters to rear live-stranded pups has spawned a colossal advancement in the social development of stranded pups and their potential for survival and reproductive success following release. Using live-stranded sea otters to address research questions has provided information not obtainable from free-ranging otters. In the end, pursuing value within the live-stranding program has ensured the best care of individual sea otters and established a much higher capacity within the program for achieving research and population-conservation goals—circumstances that other wildlife rehabilitation programs should try to emulate.

REFERENCES

Albrecht, G.A., 2003. Thinking like an ecosystem: the ethics of the relocation, rehabilitation and release of wildlife. In: Armstrong, S.J., Botzler, R.G. (Eds.), The Animal Ethics Reader. Routledge, London, pp. 422–425.

American Veterinary Medical Association, 2013. AVMA Guidelines for the Euthanasia of Animals: 2013 Edition. American Veterinary Medical Association, Schaumburg, IL (102 pp.).

Ames, J.A., Geibel, J.J., Wendell, F.E., Pattison, C.A., 1996. White shark-inflicted wounds of sea otters in California, 1968–1992. In: Klimley, A.P., Ainley, D.G. (Eds.), Great White Sharks: The Biology of *Carcharodon carcharias*. Academic Press, San Diego, CA.

Baker, J.D., Johanos, T.C., 2002. Effects of research handling on the endangered Hawaiian monk seal. Mar. Mammal Sci. 18 (2), 500–512.

Beck, B., Cooper, M., Griffith, B., 1993. Infectious disease considerations in reintroduction programs for captive wildlife. J. Zoo Wildl. Med. 24 (3), 394–397.

Bodkin, J.L., Monson, D.H., Esslinger, G.G., 2007. Activity budgets derived from time-depth recorders in a diving mammal. J. Wildl. Manage. 71 (6), 2034–2044.

Bowen, L., Miles, A.K., Murray, M., Haulena, M., Tuttle, J., Van Bonn, W., et al., 2012. Gene transcription in sea otters (*Enhydra lutris*); development of a diagnostic tool for sea otter and ecosystem health. Mol. Ecol. Resour. 12 (1), 67–74.

Brownstein, D., Miller, M.A., Oates, S.C., Byrne, B.A., Jang, S., Murray, M.J., et al., 2011. Antimicrobial susceptibility of bacterial isolates from sea otters (*Enhydra lutris*). J. Wildl. Dis. 47 (2), 278–292.

Dau, B.K., Gilardi, K.V.K., Gulland, F.M., Higgins, A., Holcomb, J.B., Leger, J.S., et al., 2009. Fishing gear-related injury in California marine wildlife. J. Wildl. Dis. 45 (2), 355–362.

Dubey, J.P., Rosypal, A.C., Rosenthal, B.M., Thomas, N.J., Lindsay, D.S., Stanek, J.F., et al., 2001. *Sarcocystis neurona* infections in sea otter (*Enhydra lutris*): evidence for natural infections with sarcocysts and transmission of infection to opossums (*Didelphis virginiana*). J. Parasitol. 87 (6), 1387–1393.

Estes, J.A., 1991. Catastrophes and conservation: lessons from sea otters and the *Exxon Valdez*. Science 254 (5038), 1596.

Estes, J.A., 1998. Concerns about rehabilitation of oiled wildlife. Conserv. Biol. 12 (5), 1156–1157.

Estes, J.A., Tinker, M.T., Bodkin, J.L., 2010. Using ecological function to develop recovery criteria for depleted species: sea otters and kelp forests in the Aleutian Archipelago. Conserv. Biol. 24 (3), 852–860.

Flatz, R., Gerber, L.R., 2010. First evidence for adoption in California sea lions. PLoS One 5 (11), e13873.

Foott, J.O., 1970. Nose scars in female sea otters. J. Mammal 51 (3), 621–622.

Gemmell, N.J., 2003. Kin selection may influence fostering behaviour in Antarctic fur seals (*Arctocephalus gazella*). Proc. R. Soc. Lond., B., Biol. Sci. 270 (1528), 2033–2037.

Geraci, J.R., Lounsbury, V.J., 2005. Marine Mammals Ashore: A Field Guide for Strandings, second ed. National Aquarium in Baltimore, Baltimore, MD.

Ghoul, A., Reichmuth, C., 2012. Sound production and reception in southern sea otters (*Enhydra lutris nereis*). In: Popper, A.N., Hawkins, A. (Eds.), The Effects of Noise on Aquatic Life, vol. 730. New York: Springer, US, pp. 157–159.

Grogan, A., Kelly, A., 2013. A review of RSPCA research into wildlife rehabilitation. Vet. Rec. 172 (8), 211–244.

Guy, A., Curnoe, D., Banks, P., 2013. A survey of current mammal rehabilitation and release practices. Biodivers. Conserv. 22 (4), 825–837.

Hanni, K.D., Mazet, J.A.K., Gulland, F.M.D., Estes, J., Staedler, M., Murray, M.J., et al., 2003. Clinical pathology and assessment of pathogen exposure in southern and Alaskan sea otters. J. Wildl. Dis. 39 (4), 837–850.

Hatfield, B.B., Ames, J.A., Estes, J.A., Tinker, M.T., Johnson, A.B., Staedler, M.M., et al., 2011. Sea otter mortality in fish and shellfish traps: estimating potential impacts and exploring possible solutions. Endangered Species Res. 13 (3), 219–229.

Hughes, B.B., Eby, R., Van Dyke, E., Tinker, M.T., Marks, C.I., Johnson, K.S., et al., 2013. Recovery of a top predator mediates negative eutrophic effects on seagrass. Proc. Natl. Acad. Sci. 110 (38), 15313–15318.

International Union for Conservation of Nature, 2002. IUCN guidelines for the placement of confiscated animals. Prepared by the IUCN/SSC Re-introduction Specialist Group. Gland, Switzerland, 27 pp.

Jameson, R.J., Johnson, A.M., 1993. Reproductive characteristics of female sea otters. Mar. Mammal Sci. 9 (2), 156–167.

Jessup, D.A., Miller, M.A., 2012. Southern sea otters as sentinels for land-sea pathogens and pollutants. In: Aguirre, A., Ostfeld, R.S., Daszak, P. (Eds.), Conservation Medicine: Applied Cases of Ecological Health. Oxford University Press, New York, pp. 328–342.

Jessup, D.A., Miller, M.A., Kreuder-Johnson, C., Conrad, P.A., Tinker, M.T., Estes, J., et al., 2007. Sea otters in a dirty ocean. J. Am. Vet. Med. Assoc. 231 (11), 1648–1652.

Jessup, D.A., Murray, M.J., Casper, D.R., Brownstein, D., Kreuder-Johnson, C., 2009. Canine distemper vaccination is a safe and useful preventive procedure for southern sea otters (*Enhydra lutra nereis*). J. Zoo Wildl. Med. 40 (4), 705–710.

Jessup, D.A., Johnson, C.K., Estes, J., Carlson-Bremer, D., Jarman, W.M., Reese, S., et al., 2010. Persistent organic pollutants in the blood of free-ranging sea otters (*Enhydra lutris* ssp.) in Alaska and California. J. Wildl. Dis. 46 (4), 1214–1233.

Jessup, D.A., Yeates, L.C., Toy-Choutka, S., Casper, D., Murray, M.J., Ziccardi, M.H., 2012. Washing oiled sea otters. Wildl. Soc. Bull. 36 (1), 6–15.

Johnson, A.B., 2000. Rehabilitation in a Proper Context. *AZA Communiqué*. Association of Zoos and Aquariums, Wheeling, WV.

Johnson, C.K., Tinker, M.T., Estes, J.A., Conrad, P.A., Staedler, M., Miller, M.A., et al., 2009. Prey choice and habitat use drive sea otter pathogen exposure in a resource-limited coastal system. Proc. Natl. Acad. Sci. USA 106 (7), 2242–2247.

Jule, K.R., Leaver, L.A., Lea, S.E.G., 2008. The effects of captive experience on reintroduction survival in carnivores: a review and analysis. Biol. Conserv. 141 (2), 355–363.

Kannan, K., Kajiwara, N., Watanabe, M., Nakata, H., Thomas, N.J., Stephenson, M., et al., 2004. Profiles of polychlorinated biphenyl congeners, organochlorine pesticides, and butyltins in southern sea otters and their prey. Environ. Toxicol. Chem. 23 (1), 49–56.

Kannan, K., Moon, H.B., Yun, S.H., Agusa, T., Thomas, N.J., Tanabe, S., 2008. Chlorinated, brominated, and perfluorinated compounds, polycyclic aromatic hydrocarbons and trace elements in livers of sea otters from California, Washington, and Alaska (USA), and Kamchatka (Russia). J. Environ. Monitor. 10 (4), 552–558.

Kannan, K., Perrotta, E., Thomas, N.J., 2006. Association between perfluorinated compounds and pathological conditions in southern sea otters. Environ. Sci. Technol. 40 (16), 4943–4948.

Kenyon, K.W., 1975. The Sea Otter in the Eastern Pacific Ocean. Dover Publications, Inc., New York, NY.

Kirkwood, J., Best, R., 1998. Treatment and rehabilitation of wildlife casualties: legal and ethical aspects. In Pract. 20 (4), 214–216.

Kirkwood, J.K., Sainsbury, A.W., 1996. Ethics of interventions for the welfare of free-living wild animals. Anim. Welf. 5 (3), 235–243.

Kreuder, C., Miller, M., Jessup, D.A., Lowenstine, L.J., Harris, M.D., Ames, J.A., et al., 2003. Patterns of mortality in the southern sea otter (*Enhydra lutris nereis*) from 1998–2001. J. Wildl. Dis. 39 (3), 495–509.

Kreuder, C., Miller, M.A., Lowenstine, L.J., Conrad, P.A., Carpenter, T.E., Jessup, D.A., et al., 2005. Evaluation of cardiac lesions and risk factors associated with myocarditis and dilated cardiomyopathy in southern sea otters (*Enhydra lutris nereis*). Am. J. Vet. Res. 66 (2), 289–299.

Lafferty, K.D., Gerber, L.R., 2002. Good medicine for conservation biology: the intersection of epidemiology and conservation theory. Conserv. Biol. 16 (3), 593–604.

Loomis, J., 2006. Estimating recreation and existence values of sea otter expansion in California using benefit transfer. Coast. Manage. 34 (4), 387–404.

Maniscalco, J.M., Harris, K.R., Atkinson, S., Parker, P., 2007. Alloparenting in Steller sea lions (*Eumetopias jubatus*): correlations with misdirected care and other observations. J. Ethol. 25 (2), 125–131.

Mayer, K.A., Dailey, M.D., Miller, M.A., 2003. Helminth parasites of the southern sea otter *Enhydra lutris nereis* in central California: abundance, distribution and pathology. Dis. Aquat. Organ. 53 (1), 77–88.

McPhee, E.M., 2004. Generations in captivity increases behavioral variance: considerations for captive breeding and reintroduction programs. Biol. Conserv. 115 (1), 71–77.

Mellish, J.-A. E., Calkins, D.G., Christen, D.R., Horning, M., Rea, L.D., Atkinson, S.K., 2006. Temporary captivity as a research tool: comprehensive study of wild pinnipeds under controlled conditions. Aquat. Mammals 32 (1), 58–65.

Miller, E.A. (Ed.), 2012. Minimum Standards for Wildlife Rehabilitation. fourth ed. National Wildlife Rehabilitators Association, St. Cloud, MN.

Miller, M.A., Gardner, I.A., Kreuder, C., Paradies, D.M., Worcester, K.R., Jessup, D.A., et al., 2002. Coastal freshwater runoff is a risk factor for *Toxoplasma gondii* infection of southern sea otters (*Enhydra lutris nereis*). Int. J. Parasitol. 32 (8), 997–1006.

Miller, M.A., Conrad, P.A., Harris, M., Hatfield, B., Langlois, G., Jessup, D.A., et al., 2010a. A protozoal-associated epizootic impacting marine wildlife: mass-mortality of southern sea otters (*Enhydra lutris nereis*) due to Sarcocystis neurona infection. Vet. Parasitol. 172 (3–4), 183–194.

Miller, M.A., Kudela, R.M., Mekebri, A., Crane, D., Oates, S.C., Tinker, M.T., et al., 2010b. Evidence for a novel marine harmful algal bloom: cyanotoxin (microcystin) transfer from land to sea otters. PLoS One 5 (9), e12576.

Monnett, C., Rotterman, L.M., Monson, D.H., Estes, J.A., Bodkin, J.L., Siniff, D.B., 2000. Survival rates of sea otter pups in Alaska and California. Mar. Mammal Sci. 16 (4), 794–810.

Moore, M., Early, G., Touhey, K., Barco, S., Gulland, F., Wells, R., 2007. Rehabilitation and release of marine mammals in the United States: risks and benefits. Mar. Mammal Sci. 23 (4), 731–750.

Nicholson, T.E., Mayer, K.A., Staedler, M.M., Johnson, A.B., 2007. Effects of rearing methods on survival of released free-ranging juvenile southern sea otters. Biol. Conserv. 138 (3–4), 313–320.

Oates, S.C., Miller, M.A., Byrne, B.A., Chouicha, N., Hardin, D., Jessup, D., et al., 2012. Epidemiology and potential land–sea transfer of enteric bacteria from terrestrial to marine species in the Monterey Bay region of California. J. Wildl. Dis. 48 (3), 654–668.

Paine, R.T., Ruesink, J.L., Sun, A., Soulanille, E.L., Wonham, M.J., Harley, C.D.G., et al., 1996. Trouble on oiled waters: lessons from the *Exxon Valdez* oil spill. Annu. Rev. Ecol. Syst. 27 (1), 197–235.

Pattison, C.A., Harris, M.D., Wendell, F.E., 1997. Sea otter, *Enhydra lutris*, mortalities in California, 1968 through 1993. Marine Resources Division Administrative Report. California Department of Fish and Game Marine Resources Division, Morro Park, CA (48 pp.).

Perry, E.A., Boness, D.J., Fleischer, R.C., 1998. DNA fingerprinting evidence of nonfilial nursing in grey seals. Mol. Ecol. 7 (1), 81–85.

Quakenbush, L., Beckmen, K., Brower, C.D.N., 2009. Rehabilitation and release of marine mammals in the United States: concerns from Alaska. Mar. Mammal Sci. 25 (4), 994–999.

Ralls, K., Siniff, D.B., Williams, T.D., Keuchle, V.B., 1989. An intraperitoneal radio transmitter for sea otters. Mar. Mammal Sci. 5 (4), 376–381.

Riedman, M.L., Estes, J.A., 1990. The sea otter: behavior, ecology, and natural history, US Fish and Wildlife Service. Biol. Rep. 90 (14), 1–26.

Riedman, M.L., Le Boeuf, B.J., 1982. Mother–pup separation and adoption in northern elephant seals. Behav. Ecol. Sociobiol. 11, 203–215.

Riedman, M.L., Estes, J.A., Staedler, M.M., Giles, A.A., Carlson., D.R., 1994. Breeding patterns and reproductive success of California sea otter. J. Wildl. Manage. 58 (3), 391–399.

Righton, A.L., Leger, J.A.S., Schmitt, T., Murray, M.J., Adams, L., Fascetti, A.J., 2011. Serum vitamin A concentrations in captive sea otters (*Enhydra lutris*). J. Zoo Wildl. Med. 42 (1), 124–127.

Shapiro, K., Miller, M., Mazet, J., 2012. Temporal association between land-based runoff events and California sea otter (*Enhydra lutris nereis*) protozoal mortalities. J. Wildl. Dis. 48 (2), 394–404.

Siniff, D.B., Ralls, K., 1991. Reproduction, survival and tag loss in California sea otters. Mar. Mammal Sci. 7 (3), 211–229.

Sleeman, J.M., 2008. Use of wildlife rehabilitation centers as monitors of ecosystem health. In: Fowler, M.E., Miller, R.E. (Eds.), Zoo and Wild Animal Medicine: Current Therapy. Saunders/Elsevier, St. Louis, MO, pp. 97–104.

Staedler, M., Riedman, M., 1993. Fatal mating injuries in female sea otters (*Enhydra lutris nereis*). Mammalia 57 (1), 135–139.

Staedler, M.M., Riedman, M.L., 1989. A case of adoption in the California sea otter. Mar. Mammal Sci. 5 (4), 391–394.

Stirling, I., 1975. Adoptive suckling in pinnipeds. J. Aust. Mammal Soc. 1, 389–391.

Styers, J., McCormick, C., 1990. Pup nursery at the Seward otter rehabilitation center. In: Bayha, K., Kormendy, J. (Eds.), Sea Otter Symposium: Proceedings of a Symposium to Evaluate the Response Effort on Behalf of Sea Otters after the T/V *Exxon Valdez* Oil Spill into Prince William Sound, Anchorage, AK, 17–19 April 1990. US Fish and Wildlife Service, Biological Report 90(12). 485 pp.

Tinker, M.T., Bentall, G., Estes, J.A., 2008. Food limitation leads to behavioral diversification and dietary specialization in sea otters. Proc. Natl. Acad. Sci. USA 105 (2), 560–565.

Tuomi, P., 2001. Sea otters. In: Dierauf, L.A., Gulland, F.M.D. (Eds.), CRC Handbook of Marine Mammal Medicine, second ed. CRC Press, Boca Raton, FL, pp. 961–987.

US Fish and Wildlife Service, 2003. Final Revised Recovery Plan for the Southern Sea Otter (*Enhydra lutris nereis*). Portland, OR. xi + 165 pp.

Williams, T.D., 1990. Sea otter biology and medicine. In: Dierauf, Leslie A. (Ed.), CRC Handbook of Marine Mammal Medicine: Health, Disease, and Rehabilitation. CRC Press, Boca Raton, FL, pp. 625–648.

Williams, T.D., Hymer, J., 1992. Raising orphaned sea otter pups. J. Am. Vet. Med. Assoc. 201 (5), 688–691.

Williams, T.D., Williams, T.M., 1996. The role of rehabilitation in sea otter conservation efforts. Endangered Species Update 13 (12), 50–52.

Williams, T.D., Styers, D., Hymer, J., Rainville, S., McCormick, C.R., 1995. Care of sea otter pups. Emergency Care and Rehabilitation of Oiled Sea Otters: A Guide for Oil Spills Involving Fur-Bearing Marine Mammals. University of Alaska Press, Fairbanks, AK, pp. 133–140.

Chapter 10

The Use of Quantitative Models in Sea Otter Conservation

M. Tim Tinker
US Geological Survey, Western Ecological Research Center, Long Marine Laboratory, Santa Cruz, CA, USA

INTRODUCTION

The use of quantitative or mathematical models as tools in wildlife conservation has a rich history, dating back to the birth of conservation biology as a distinct branch of science (Gilpin and Soulé, 1986; Soule, 1985). Researchers working within sub-disciplines of conservation science—population biology, genetics, community ecology, landscape ecology—make extensive use of

Sea Otter Conservation. DOI: http://dx.doi.org/10.1016/B978-0-12-801402-8.00010-X

mathematical models for bridging ecological theory and empirical data sets (Alvarez-Buylla et al., 1996; Brook et al., 2000; Lamberson et al., 1992; Mac Nally, 2000; Morris and Doak, 2002; Wennergren et al., 1995). This has certainly been the case for research focused on the conservation of sea otters where mathematical models have played a central role in many advances of our understanding of sea otters and their ecosystems. However, before delving into examples of the use of models in sea otter conservation, it may be helpful to take a step back to explain what I mean by the term "model," since this term may convey different things to different people, and indeed there are many different kinds of models that are used in many different ways.

Within the scientific disciplines of ecology and conservation biology, a model is first and foremost an analytical tool used to achieve a specific aim (Hilborn and Mangel, 1997). According to the Merriam-Webster's dictionary, a model is defined as "a description or analogy used to help visualize something (such as an atom) that cannot be directly observed" or "a system of postulates, data, and inferences presented as a mathematical description of an entity or state of affairs." Together these two definitions encompass the key features of most mathematical models used by conservation biologists. Fundamentally, models are conceptual or mathematical abstractions, stylized representations, or simplifications of reality that are meant to capture the most important elements or dynamics of a phenomenon of interest, while stripping away extraneous details. The distinction between models and "hypotheses" is worth noting here, as they are closely related but not synonymous (Levins, 1966). A scientific hypothesis is a proposition or unproved theory about some phenomenon of interest, tentatively accepted to explain certain facts or observations, or to provide a basis for further investigation. A good hypothesis can be verified or falsified by experiments, while a model cannot really be verified because *all* models are false by definition (since models are intentional simplifications or characterizations of reality). However, a hypothesis may be represented or described by one or more mathematical models, which may be used as part of the hypothesis testing process, or for generating new hypotheses (Hilborn and Mangel, 1997).

But that still doesn't answer the question of why models are necessary at all. In practice, scientists use models as tools to achieve a number of different objectives, including (1) to formulate a simplified statistical or quantitative description of a phenomenon; (2) to gain or improve a mechanistic understanding of a phenomenon; (3) to act as conceptual tools for generating new hypotheses; (4) to evaluate the feasibility of alternate explanations or hypotheses about the cause of an observed event (e.g., a population decline); (5) to make predictions about the future, by synthesizing the data in hand about a situation, or (6) to provide a systematic framework for the decision-making process (Maynard-Smith, 1978). In this last case, a model can provide a means for evaluating the potential effects or implications of various kinds of decisions, as well as evaluating which data sets are adequate and which need to be improved for better decision making. So long as their limitations are clearly recognized, models can provide a scientist with powerful mathematical tools to help guide intuition about how various processes interact, evaluate testable hypotheses, generate key predictions, suggest appropriate experiments, and provide novel insights or new ways of thinking about a problem (Caswell, 1988).

The effective conservation of sea otters, and most other species, requires that scientists solve problems across a wide range of topics, from behavior and physiology to complex population dynamics, genetics, disease ecology, food-web interactions and physical/biological oceanography. These are complicated subjects, and the data sets collected by researchers in each subject area are vast and complex in their own right. Models provide one of the key tools for simplifying and integrating all these data, allowing scientists to test hypotheses, elucidate the underlying mechanisms of a particular problem, and make sound decisions. As computing power increases and new analytical techniques such as Bayesian approaches become more readily available, it is inevitable that models will become even more valuable for solving conservation challenges (Clark and Gelfand, 2006). In the following sections, I will discuss several branches of conservation science in which sea otter researchers have used mathematical models to aid in their research. For each subject area, I will explain some of the key research questions, provide a few examples of how models have been used to help answer those questions, and then highlight promising areas for future and ongoing work in sea otter conservation. I will end by exploring some of the general lessons that can be drawn from these examples, focusing on points that may be useful for wildlife conservation more broadly.

MODELS OF BEHAVIOR

Animal behavior may seem at first blush to be a rather esoteric subject in the context of wildlife conservation, as the term "behavioral science" often conjures up images of lab-coated scientists watching rats running through mazes. However, applied studies of animal behavior have actually played a vital role in animal conservation, mainly because knowledge of how animals behave in different environments is often necessary to inform management decisions. To provide just a few examples relevant to sea otters: (1) studies of foraging behavior and diet choices can provide important insights as to the role of food resources in regulating population abundance at a particular location; (2) information on individual movement behavior is vital for determining the habitat needs for a threatened population, and may also provide clues about exposure to human-caused threats such as harvests, oil spills, or pollution; and (3) an understanding of social behavior and reproductive strategies can be informative for interpreting why and how populations grow and expand the way they do, and can shed light on the factors contributing to declining or threatened

FIGURE 10.1 Sea otters are unique among marine mammals in that they bring all captured prey to the surface to handle and consume, and they do this while lying on their backs. This behavior, in combination with their nearshore distribution, makes them ideal subjects for studies of feeding behavior. Equipped with powerful telescopes, biologists are able to record each item of prey that sea otters capture and consume (such as this kelp crab, captured by a female sea otter near Monterey, CA), as well as the time it takes to locate and process those prey, and then use these data to estimate diet composition and energy consumption rates. *(Used by Permission of Nicole LaRoche.)*

populations. Quantitative models have been—and will continue to be—useful tools for exploring and understanding all these themes.

Sea otters are uniquely amenable to studies of foraging behavior and diet, due to a number of aspects of their behavior and biology. Because they feed exclusively on benthic invertebrates in shallow nearshore coastal waters (most feeding dives are made to depths of less than 40 m) and because they bring all their food to the surface to consume while lying on their backs (Figure 10.1), they are unique among all marine mammals (and perhaps among all large carnivores) in that it is possible for an observer equipped with a high-powered telescope to directly observe, identify, and record all prey items as they are consumed. There is a rich tradition of biologists capitalizing on this unusual behavior, with the result that there is now more known about the foraging behavior and diet of sea otters than almost any other carnivore (see, e.g., Estes et al., 1982; Garshelis et al., 1986; Laidre and Jameson, 2006; Ostfeld, 1982; Watt et al., 2000). This wealth of information has been used to test hypotheses associated with some of the central theories of ecology, such as optimal diet choice, and the effect of competition for food on the range of different prey types consumed by predators (often referred to by ecologists as "niche diversity"). The intersection of theoretical models with empirical data on sea otter diets and feeding behavior has revealed some unexpected but fascinating patterns, and it has become clear that sea otter foraging behavior holds clues to solving many challenges of sea otter conservation.

Models of optimal diet choice (see Box 10.1) have been applied to sea otter populations, resulting in a number of specific predictions, as described by Estes et al. (1981). When sea otters first begin to colonize a new habitat,

Box 10.1 Primer on Models of Foraging Behavior and Diet Choice

Some of the most fundamental questions of ecology concern predator–prey interactions, and in particular (1) why predators choose to prey on some species but not others, (2) how predators affect the abundance of their prey populations, (3) whether there are predictable ways in which predators adjust their behavior and diet diversity in response to reductions in abundance of their preferred prey (which may be caused in part by their own predation), and (4) whether the dynamics described in questions 1–3 may set into motion a broader suite of changes in the larger food web. Questions 1–3 have been explored by ecologists within a set of mathematical models together referred to as "foraging theory." In brief, these models adopt a cost-benefit approach to predicting the foraging decisions of predators, similar to the models of human consumer behavior used by economists (and indeed, fundamental economic models and foraging theory models utilize the same mathematical and conceptual framework). Unlike economic models, where consumers are predicted to make decisions so as to minimize monetary costs while maximizing their material benefits, the currency of many foraging models is food energy, or calories. This makes sense because energy represents a scarce but critically important resource for wild animals: they require energy to grow, keep warm, move through their environment, hunt for food, and reproduce and raise their young. Energy-based models are premised on the idea that, all else being equal, predators will select prey species and utilize prey "patches" in the environment in such a way so as to minimize the energetic "expenditures" of feeding while maximizing their energy intake (Schoener, 1971).

One of the building blocks of foraging theory is the diet choice model (Charnov, 1976; Emlen, 1966). This model, while fairly simplistic, provides the basis for more elaborate models and illustrates many of the basic principles of foraging theory, so it is worth taking the time to examine and understand the model mathematically. In Eq. (10.1), E represents total consumed food energy and T represents time, so that E/T is the rate of energy intake (e.g., the number of kcal the predator consumes per minute), calculated as

$$E/T = \frac{\lambda_1 e_1 P_1 + \lambda_2 e_2 P_2 + \ldots \lambda_n e_n P_n}{1 +_1 \lambda_1 h_1 P_1 + \lambda_2 h_2 P_2 + \ldots \lambda_n h_n P_n} \qquad (10.1)$$

Other symbols in Eq. (10.1) include λ_i, representing the relative abundance of a particular type of prey and thus the rate with which it is encountered by the predator (note that the subscript numbers in Eq. (10.1) are used to keep track of

(*Continued*)

Box 10.1 (Continued)

the parameter values for different types of prey, $i = 1, 2, \ldots, n$, where n is the number of potential prey types); e_i, representing the net energy content of each prey type; h_i, representing the time it takes to acquire, handle, and consume each prey type; and P_i, the probability that a predator will attack and consume items of each type of prey when encountered. If we assume that prey 1 is the most profitable and thus highest ranked prey (its ratio of net energy content to handling time, e_i/h_i, is higher than for all other prey types), it makes intuitive sense that a predator should always attack it when encountered ($P_1 = 1$). But what about the other prey types: should the predator include them in its diet, or ignore them when encountered and continue searching? We assume that predators will tend to make decisions so as to maximize their overall rate of energy gain, all else being equal (since this will allow them to maximize their lifetime reproductive output). Mathematically, this is achieved by adjusting the attack probability values (P_i) for each prey type until we find the maximum value of E/T. It turns out that the overall rate of energy intake (E/T) is maximized if the predator makes its decision to attack or ignore each prey type based on the relative abundance of all higher-ranked prey types. Specifically, if we assume that prey 2 is ranked just below prey 1 in terms of profitability, the predator should include prey 2 in its diet ($P_2 = 1$) if the following inequality holds:

$$e_2/h_2 \geq \frac{\lambda_1 e_1}{1 +_1 \lambda_1 h_1} \tag{10.2}$$

Otherwise, the predator should ignore items of prey 2 when encountered ($P_2 = 0$). A similar equation can be used to predict whether or not prey 3 will be included, and so on. Some algebra reveals that Eq. (10.2) will be satisfied—and prey 2 included in the diet—if the abundance of prey 1 (λ_1) is sufficiently low. Likewise Eq. (10.2) will not hold—and prey 2 will be excluded from the diet—if the abundance of prey 1 (λ_1) is sufficiently high. However, you might notice that the abundance of prey 2 (λ_2) does not appear in Eq. (10.2): this means that the abundance of prey 2 is itself predicted to be irrelevant to the predator's decision to include it in the diet! This prediction may seem surprising, and indeed it was quite controversial when first proposed; however, years of data from experiments and wild predator populations have largely supported these predictions. When highly preferred prey is abundant in the environment, predators tend to rely primarily on these profitable prey types and ignore less profitable prey, resulting in low diet diversity. As the relative abundance of the preferred prey decreases, predators add additional lower-ranked prey types to their diet, resulting in more diverse diets. What would cause the relative abundance of preferred prey to decline? This can occur when the predator population is capable of depleting its own prey populations, thereby reducing prey density. But even if the absolute prey density remains the same, a decrease in relative *per capita* abundance of prey (i.e., number of items per individual predator) can occur if there is an increase in the predator population density: this is because the same number of prey items per unit area are now competed for by a larger number of predators (there are more diners at the table). And of course prey populations can vary over time in response to factors independent of predation.

they are expected to limit their diet primarily to highly profitable prey types (e.g., abalone and large red sea urchins in kelp habitats, or large Dungeness crabs and butter clams in soft sediment habitats) that allow a high rate of energy intake. As the sea otter population grows, and per capita availability of these preferred prey types goes down, it is expected that the rate of energy intake will decline and diet diversity will go up (as less profitable prey are sequentially added to the diet). Ostfield (1982) tested these predictions by studying the diets of sea otters as they first recolonized pristine habitat near Santa Cruz, CA. He found that, consistent with model predictions, sea otters initially preyed mostly on large red sea urchins, the prey species with the highest energy content to handling time ratio. These highly profitable prey were quickly depleted, after which sea otters increased their usage of the next most profitable prey (abalone and cancer crabs) and eventually began to add lower-ranked prey (kelp crabs and smaller clams) to their diet. The resulting pattern was an increase in diet diversity and a decrease in energy intake rates over time. A similar pattern was found by Estes et al. (1981) near Monterey, CA. Moreover, going back to the same area of California three decades later, Tinker et al. (2012) found that diet diversity is even greater, with a high frequency of low profitability prey types in the average sea otter diet. In soft sediment habitats, similar but more subtle patterns have been reported: in southeast Alaska, Kvitek and Oliver (1992) and Weitzman (2013), found that sea otters preyed mostly on large butter clams, mussels, and urchins when those energy-rich prey were sufficiently abundant; however, as otters became more abundant and depleted these profitable prey, their energy intake rates declined and they began to add less profitable prey to their diets (diet switching was less dramatic than in kelp habitats, possibly because soft sediment clam communities, in some cases, provide a more sustainable prey base over the long term). Thus the predictions of the basic diet model appear to hold up well for sea otters, at least when examined at the population level—but more on this below.

At this point we should pause and note that in the above discussion of changes in sea otter diets and energy intake rates over time, we have assumed that it is possible to obtain unbiased estimates of sea otter diet composition and intake rates. For most free-ranging carnivores, such a requirement represents a very tall order; in fact, there are few systems for which this is even remotely possible. Sea otters are, fortunately, an exception, although there are still some substantial challenges involved in obtaining unbiased estimates of diet composition, and even greater challenges to overcome in estimating energy intake rates. The reason that estimating these statistics is so challenging is not due to paucity of data—indeed, with a little practice and effort it is possible to amass vast quantities of observational data on sea otter feeding behavior and prey capture rates. Rather, it is because the data records collected are quite often incomplete, and there tends to be a non-random pattern to the bits of data that are missing—for example, prey that are very small, or consumed quickly, are more frequently unidentified by the observer, as are prey that are captured and consumed far from shore—and this non-random nature of missing data introduces the potential for biased estimates of diet. Overcoming this potential bias has been made possible by a very different type of model, something called (intriguingly) a "Monte Carlo simulation model."

The term "Monte Carlo simulation" is derived, as the name suggests, from the Monte Carlo Casino, and like the games of chance played at the casino, Monte Carlo simulations take advantage of the aggregate statistical properties of many random events. Monte Carlo models are especially useful for simulating complex systems, where there is substantial uncertainty and interactions between model inputs, making more traditional "deterministic" models inappropriate (a deterministic model is illustrated by Eq. (10.1) in Box 10.1). The general approach of the Monte Carlo method involves defining the range and "probability distribution" of possible values for a series of input statistics, generating large numbers of random samples from these probability distributions, performing a series of computations on each set of random input values (the computations correspond to the presumed dynamics of the system being modeled), and then aggregating and describing the results of these computations. Now, unless you are very comfortable with probability and statistics, the above explanation may not make that much sense to you, so let's consider a simplified model of sea otter foraging to better illustrate the Monte Carlo method. Imagine we are observing sea otters feeding at an island in the Aleutian Archipelago where sea urchins make up almost all the prey. Let's say that we decide to record the number of urchins captured by sea otters during each of 1000 feeding dives. We also record the size of the urchins captured on each dive (it is possible to estimate the size of sea otter prey by comparing each item to the width of the feeding sea otter's paw, which is approximately 45 mm). However, we notice that the size of the urchins depends on the number of items captured: sea otters tend to either capture a few large urchins or many small urchins on each dive (this is because prey patches on the bottom tend to be made up of many urchins of a similar size). Luckily, because we kept track of both number of urchins AND size of urchins on each dive, we can account for the interaction between prey number and prey size in our model, as we'll see shortly. Finally, we record the duration (in seconds) of each dive cycle, which includes the time underwater finding prey plus the subsequent time at the surface handling and consuming the prey. Tallying up the results of our recorded feeding dives, we can plot histograms of the three types of data (Figure 10.2). If we have recorded enough data, these histograms can be easily translated into "probability distributions," which are useful mathematical formulas that tell us the likelihood of observing a particular value of a given statistic in the future. For example, if we find that the size of the urchins captured on feeding dives with three prey items is well

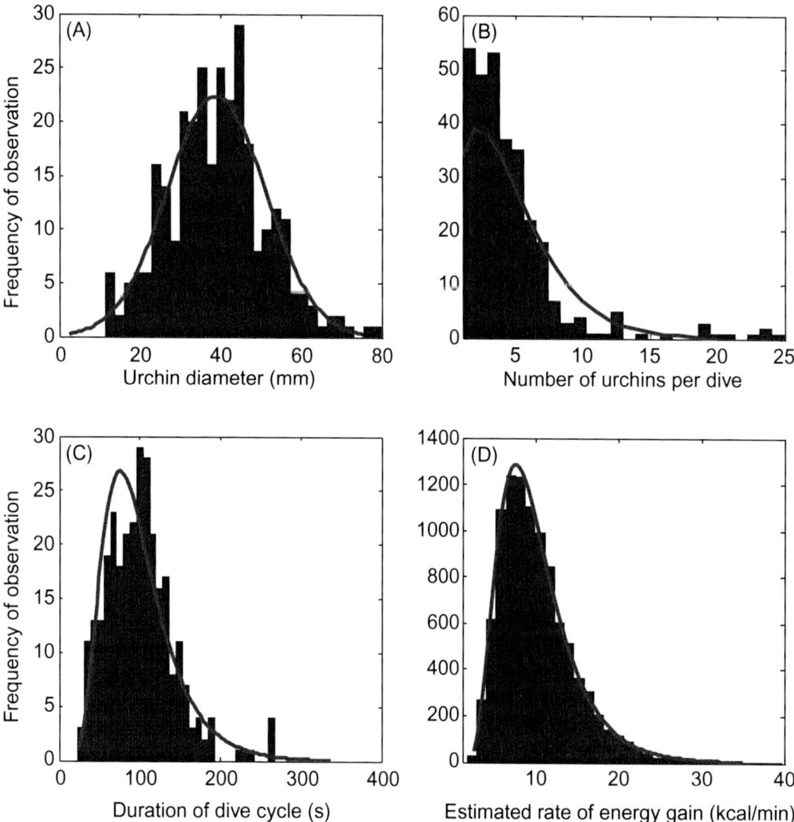

FIGURE 10.2 Frequency distributions and fitted probability density functions for several variables (and one estimated parameter) recorded during observations of sea otter foraging in the Aleutian Islands, AK. (A) Blue bars show a frequency histogram of the size (test diameter) of sea urchins captured on feeding dives, and the red line shows a normal probability density function that was fit to these data using maximum likelihood methods. (B) Blue bars show a frequency histogram of the number of sea urchins captured per feeding dive, and the red line shows a negative-binomial probability density function that was fit to these data using maximum likelihood methods. (C) Blue bars show a frequency histogram of dive cycle durations (i.e., the time spent under water finding prey plus the subsequent time at surface spent handling and consuming that prey), and the red line shows a log-normal probability density function that was fit to these data using maximum likelihood methods. (D) Blue bars show a frequency histogram of estimates of the rate of energy gain, calculated from the variables shown in panels A–C using Eq. (10.3) in the text. The red line shows a log-normal probability density function that was fit to these estimated values using maximum likelihood methods.

described by a normal probability distribution having mean of 38 mm and standard deviation of 13 mm, then we know that the probability that the urchins captured on the next three-urchin dive will have a size of 50–55 mm is 8.2% (Figure 10.2A). The normal distribution is just one type of probability distribution (although the one that is familiar to most people), but there are other distributions as well, and it turns out that the number of prey items per dive is better described by the "negative binomial distribution" (Figure 10.2B) while the duration of the dive cycle is better described by the "log-normal" distribution (Figure 10.2C). Next we use a random number generator to create a sample of new values drawn at random from each of the above-described probability distributions, and we repeat this until we have 10,000 sets of random values for our three statistics (number of urchins/dive, size of urchins per dive *given* the number of items captured, and the duration of the dive cycle). For each set of randomly generated values, we perform the following calculations to calculate the rate of energy intake (*E*/*T*), or the number of kcal consumed per minute in this "simulated future dive":

$$E/T = \frac{n \times 0.396 \times 0.0005(s|n)^{2.9035}}{d} \tag{10.3}$$

In Eq. (10.3), the symbols used are n for the number of items captured, s for urchin size (the expression $s|n$ simply indicates that the value of s is dependent on n because there is a different probability distribution of s for each value of n) and d for the duration of the dive cycle (in minutes). Note that the calculations performed in Eq. (10.3) allow urchin energy content to be estimated from urchin diameter: this is possible because we have previously measured the relationship between urchin diameter and urchin biomass (in grams), as well as the number of kilocalories per gram of urchin biomass. Equation (10.3) is solved for each of the 10,000 sets of random values, and the results are tabulated in a histogram (Figure 10.2D). Thus by using this basic Monte Carlo model, we are able to estimate that sea otters consume, on average, approximately 10.4 kcal/min of time spent feeding at our Aleutian Island. Just as importantly, we also have an idea of how *precise* our estimate is (or how much uncertainty there is associated with this mean value, as represented by how "spread out" the Monte Carlo estimates are in Figure 10.2D).

The above example is of course highly simplified, but illustrates the key concepts of how the Monte Carlo method has been applied to sea otter foraging data sets. In reality there are many more characteristics of feeding behaviors and prey that are recorded, and many more interactions among these statistics that have to be accounted for in the calculations. The power of the Monte Carlo method in these cases is that (a) all the interactions between recorded variables can be accounted for in the calculations, no matter how complex, and (b) the algorithm can properly incorporate the additional uncertainty and biases caused by missing data points. For example, prey species is less likely to be recorded for small items that are handled very fast: we can account for this by incorporating the *context-specific* probability that prey species is unidentified within our Monte Carlo simulation. On a given iteration of the simulation, if prey species is drawn as "unidentified," we then randomly assign a value by drawing from the list all recorded prey species for that area (or individual otter) having the *appropriate dive attributes* (e.g., if our randomly generated dive has a short surface interval and small prey size, then we draw randomly from all recorded prey that were captured on dives with short surface intervals and small prey size). In this way, the more data points that are missing the more uncertainty there is in our resulting estimates of diet composition and energy gain, and we can remove the biases that are inherent in the raw data sets due to non-randomness of missing data.

Dean et al. (2002) used a Monte Carlo model to estimate diet and prey consumption rates at two sites within Prince William Sound: Knight Island, which had been heavily impacted by the *Exxon Valdez* oil spill (EVOS), and Montague Island, which had escaped oiling (Chapter 4). They found that prey consumption rates were significantly higher at Knight Island, and thus were able to reject the hypothesis that food limitation was preventing further population recovery of otters at Knight Island. A similar comparison was made by Tinker et al. (2008a, 2012) in California: comparing long-established, high-density sea otter populations in central California to a recently established, low-density population at San Nicolas Island, they found that diet diversity was high and rate of energy intake was very low in the high-density central California sites, in contrast to the low-density San Nicolas site where diet diversity was low (and diets were dominated by energy-rich red urchins) and the rate of energy intake was high. The examples mentioned above demonstrate that using Monte Carlo models to analyze foraging data can provide valuable insights into the role of food resources in limiting growth of sea otter populations. In fact, if we apply this Monte Carlo model to the combined database of foraging data collected by many researchers at sites across the sea otter's range, from California to Alaska to the Commander Islands in Russia, we find a consistent pattern in terms of how foraging success (energy intake per minute) varies as a function of sea otter population density and population growth rate (Figure 10.3). This is important for two reasons: first, it tells us that in most cases, the relative per-capita abundance of energy-rich prey is ultimately the factor that determines sea otter population growth and equilibrium abundance. Second, it means that it is important to understand the status of a population with respect to foraging success because it can highlight situations where something other than food abundance is limiting a population. In Figure 10.3, for example, we notice that two populations fall outside of the expected relationship between foraging success and population density and growth rate: sea otters in the central Aleutian Islands and Alaska Peninsula are currently at low density (having declined by more than 80% in the last two decades), and yet have a high rate of energy intake. If food abundance were the only important factor then we would expect them to have a very high rate of growth, but instead we find populations that are stable or declining. We can therefore surmise that some factor other than

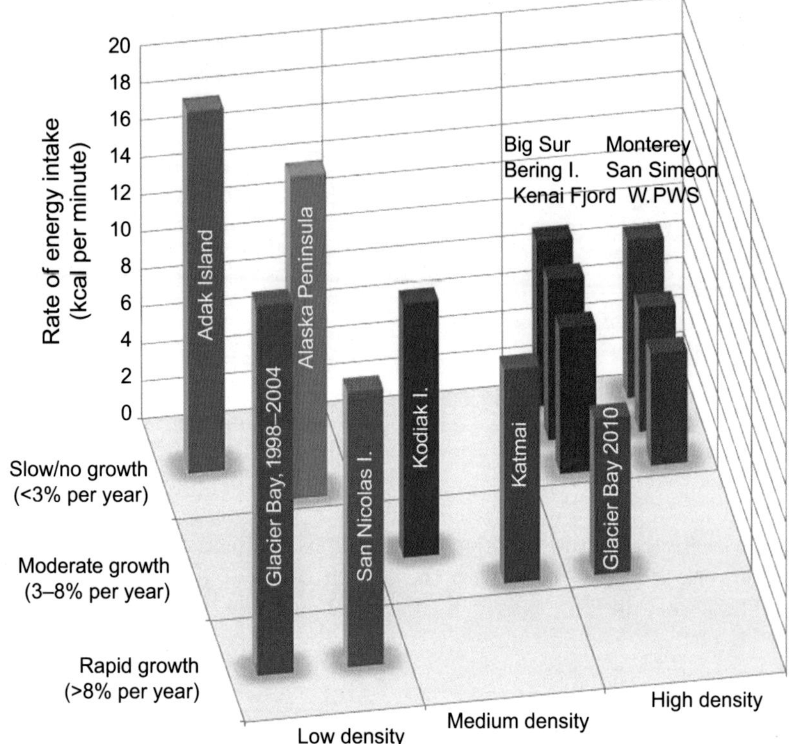

FIGURE 10.3 Estimated rate of energy intake while feeding for 13 sea otter populations, ranging from the Russian Commander Islands to southern California (see Figure 10.2 and text for description of Monte Carlo model used to estimate energy gain from observational feeding data). Each population has been classified into one of three categories organized along two independent axes: (1) rate of population growth (slow, moderate, rapid) and (2) population density (low, medium, high). In general, rate of energy gain tends to decrease as populations vary from low-density, rapidly growing populations to high-density, slowly growing (or stable) populations. Two sites that deviate from this axis are Adak Island and the Alaska Peninsula during the mid-2000s, where populations had been reduced to low density by factors other than resource abundance.

food abundance is limiting abundance in the Aleutians and Alaska Peninsula. At this time, the preponderance of evidence points toward elevated mortality due to predation by killer whales (Estes et al., 1998; Chapters 4 and 6). Similar logic can be applied to other situations as well, for example to determine places in California where land-based pollution may be affecting sea otter health and reducing population growth even though food resources may be relatively abundant (Jessup et al., 2007).

The models and patterns discussed above are useful for understanding the factors that affect sea otters at the population level—that is, they focus on an "average" individual within the population. However, a study by Estes et al.

(2003) demonstrated that the population average diet was not necessarily reflective of individual diets: to the contrary, individual sea otters tended to specialize on just a few prey types, and it was the differences between these specialized individual diets that gave rise to diverse diets at the population level. While this finding at first appeared to contradict the predictions of optimal diet theory, further investigations have shown that this is not the case. Specifically, theoretical predictions that individual prey choices should tend to converge upon an "optimal diet" are based on the assumption that individuals within a population are more or less similar with respect to foraging ability, or are "phenotypically similar." From a distance, one sea otter may look much like another, but in fact if one carefully measures the behavior and prey-specific capture and handling efficiency of tagged sea otters, it turns out that individuals differ considerably. A sea otter that specializes on feeding on turban snails, for example, can handle 42% more snails per hour than can a sea otter that specializes on some other prey type (Tinker et al., 2012). This variability in foraging skills appears to represent the effect of learned traits in sea otters, rather than genetic variation (Estes et al., 2003). Tinker et al. (2009) developed a stochastic dynamic model (a type of behavioral model that can incorporate learning: Mangel and Clark, 1988) that showed that the ability to learn foraging skills through extensive practice, or to lose those skills through lack of practice, would lead to the maintenance of diet specialization within a population, and moreover that the ability of females to pass on some of these skills to their offspring (a form of cultural transmission, also observed in sea otters) would further increase the degree of diet specialization observed. Another analysis of these data used network theory (the developing science of analytical approaches for studying networks, or graphical representations of the pattern of relationships between discreet objects) to examine how resource abundance affected the degree of specialization (Tinker et al., 2012). In this analysis, individual resource use was modeled as a bipartite network in which two sets of nodes, one representing individual consumers and the other representing their prey resources, were connected by links representing utilization of each prey by each individual. It was found that network modularity (the diversification of individual sea otters into specialized "dietary modules") increased greatly as preferred food resources became less abundant. This finding has important implications for sea otter conservation, because many of the pollutants and disease-causing parasites that affect sea otter health and survival are acquired by sea otters through their diet. Indeed, in an analysis of the factors that potentially increased the risk of infection in sea otters from two protozoan parasites that can cause lethal brain infections, it was found that a sea otter's diet specialization was one of the most important predictive factors (Johnson et al., 2009). Thus understanding and measuring individual diet specialization in sea otters provided a useful tool for elucidating the mechanisms and pathways of disease exposure.

Models of foraging behavior are perhaps the best examples of behavioral models for sea otters in the published literature; however, other types of behavior can also be investigated using models, such as time-activity budgets, habitat selection patterns, and movement behavior. Describing and quantifying the movement of animals in marine environments can be challenging, but technological advances in bio-logging instruments have greatly improved our ability to regularly and reliably geo-locate individuals and investigate how they move within their environment. As these types of data have become more common, a number of different models have been brought to bear to understand the factors that underlie the observed patterns and answer questions such as "what habitats are critical for this population?" or "how far will a typical individual move within a specified period of time?" The former question has been investigated with models of individual home range and habitat use (an individual's home range is the spatial extent or outside boundary comprising all or most of that animal's movements over some standardized period of time, such as a year). Home range models frequently make use of probabilistic utilization distributions, calculated from repeated observations of an animal's locations over time (see Borger et al., 2008). The utilization distribution is a probability density surface representing the likelihood of finding the individual at any particular point in space. Non-parametric kernel density models have been used to calculate home range utilization distributions and home range boundaries (corresponding to the area encompassing 95% of the kernel probability density function) for sea otters in California (Tinker et al., 2008a) and in Washington (Laidre et al., 2009). In the latter example, individual home ranges were estimated as kernel probability distributions along the one-dimensional axis represented by the coast. Future work will need to focus on estimating two-dimensional utilization distributions that account for the particular habitat restrictions of this coastal marine species.

Related to the concept of home range and habitat use are questions about individual movements and dispersal behavior: that is, given an animal's location right now, how far away is it likely to have moved by some future time (e.g., by next year)? The answer to this question is critical for developing models of population growth and range expansions (see "Population Dynamics," below), and can be represented by probability distributions of "net linear displacement," also called "dispersal kernels." Empirically derived dispersal kernels have been calculated by fitting exponential probability distributions to data on the movements of radio-tagged sea otters in California (Ralls et al., 1996; Tinker et al., 2008b). Expected dispersal distances can also be predicted using "random walk" models, which are a family of Lagrangian, individual-based models of animal movements (Turchin, 1998). These models simulate animal movements as a series of steps whose direction and distance are either random (random walk model), random but correlated (correlated random walk model), or random and correlated and biased in some way (biased correlated random walk model), such as toward the center of a home range (Borger et al., 2008). A correlated random walk model fit to data on the movements of radio-tagged sea otters in California demonstrated a strong bias in the observed versus expected net linear displacement over time, suggesting that individuals exhibited a strong site fidelity and were less likely than expected to move very far from the center of their home range (Kage, 2004; Tinker et al., 2006b). Future models will build on this approach to develop biased correlated random walk models that incorporate multiple centers of use (i.e., more than one home range center) and the unique bathymetric habitat restrictions of sea otters.

MODELS OF POPULATION DYNAMICS

There is perhaps no other subject in wildlife conservation to which the use of models is more central than the analysis of wildlife populations, or population biology. The reasons for this are fairly straightforward: although the concept of a population is simple enough (the assemblage of a specific type of organism living within a defined area at a given time), it is very difficult to say useful things about how that population is currently behaving, or will behave in the future. Actually measuring the characteristics of a population—its abundance, its spatial extent, and how individuals of the population are distributed over the landscape—is challenging enough for wide-ranging animals such as sea otters, but actually being able to interpret how those characteristics are changing over time, and why, is far more challenging. To accomplish the latter, we need to understand all the processes that cause a population to change in abundance and distribution: taken to the extreme, this would entail being able to predict the movements, reproduction, and survival (and time of death) for each member of the population! Obviously that is not practical or even possible for any wildlife population; however, it is possible to do the next best thing: to use mathematical equations that correspond to simplified cartoons of how typical individuals within the population will move, produce offspring, live, and die. In other words, we can describe aggregate properties of the population using a model. Of course we obviously have to sacrifice an awful lot of detail in so doing, but it turns out that even simple models can be quite useful both for understanding how a population came to have the abundance and distribution that we currently observe, and (more importantly for conservation) for forecasting how those characteristics might change in the future, under different management scenarios.

Relatively simple, unstructured population models can provide a useful way of interpreting the patterns of growth seen in many recovering sea otter populations. Exponential models (see Box 10.2) form the basic tool for interpreting short-term rates of change—for example, estimates of r derived from trend data collected by wildlife surveys can be used to assess current

Box 10.2 Primer on Models of Population Dynamics

The simplest type of model for a population can be summarized by a single number, N, the number of individuals in the population. Or to be more precise, N_t, where the subscript "t" indicates that we are interested in the number of organisms in the population at a particular point in time. The simplest type of model that is used to describe the dynamics of N_t is something called a differential equation, which describes how N_t changes over time (as represented by the expression "dN/dt"). Specifically, the change in population size that can be expected over a small increment of time is calculated as

$$\frac{\mathrm{d}N}{\mathrm{d}t} = rN_t \tag{10.4}$$

Equation (10.4) is pretty easy to interpret: the rate of change in population size is equal to the product of the current population size and r, which is known variously as the instantaneous growth rate or the *intrinsic rate of population growth.* Biologically speaking, r represents per-capita contributions to population growth, which is simply the difference between births and deaths for an average individual in the population, $r = b - d$. If the number of births per individual per unit time (b) exceeds the number of deaths per individual per unit time (d), r will be positive and the population will grow; if deaths exceed births, r will be negative and the population will shrink. We can also re-express Eq. (10.4) in a different form to predict the population size at any time in the future:

$$N_t = N_0 \times \mathrm{e}^{rt} \tag{10.5}$$

In Eq. (10.5), N_0 represents the initial population size, and e^{rt} means to take the exponential function of the product of r and t (where t is the number of time steps into the future). Note that the exponential function of a number means to raise the natural logarithm base e, which is approximately 2.718, to the power of that number. Equation (10.5) seems simple enough, but there is a slight problem: this simple model would suggest that any population with a positive value of r, however small, will eventually grow to massive size. For example, try plugging in an initial population size of 100 and an instantaneous growth rate of $r = 0.2$ (a pretty typical maximum growth rate for a large mammal). Assuming our time step is a year, you will find that after just 100 years our population would number about 48,517,000,000! Thus the exponential population growth described by Eqs. (10.4)–(10.5) may be useful for predicting dynamics over short time intervals, but clearly we need to add a little bit more realism to this model before it is useful for longer periods. Biologically speaking, we have to account for the fact that individuals in a population can have negative impacts on other individuals, frequently referred to as "negative density dependence"; for example, they may compete for limited resources, or spread diseases to each other. We can represent this mathematically by adding a new term, α, representing the incremental amount by which each new individual added to the population reduces the rate of births (or increases the chance of death) of every other individual in the population. As more and more individuals are added these tiny

(*Continued*)

Box 10.2 (Continued)

incremental effects add up and reduce the per capita birth rate (or increase the death rate), so that the realized rate of population growth will slow over time:

$$\frac{dN}{dt} = rN_t(1 - \alpha N_t) \quad (10.6)$$

Equation (10.6), the logistic growth equation, is one of the simplest but most powerful ways of modeling the long-term dynamics of population growth. With a bit of algebra one can solve Eq. (10.6) to find the value of N_t at which the rate of change = 0, that is, the abundance at which growth will stop and the population will stabilize. It turns out this happens when $N_t = 1/\alpha$, so we can define the value $1/\alpha$ as the equilibrium abundance of N_t, also known as the carrying capacity or "K" (thus $K = 1/\alpha$ = the point at which births and deaths cancel each other out). Another way to express Eq. (10.6) is as a "discrete equation," meaning that time is represented as discrete steps (e.g., 1-year steps) rather than as a continuous process:

$$N_{t+1} = N_t e^{r(1-(N_t/K))} \quad (10.7)$$

Equation (10.7) tells us that the abundance of the population next year can be calculated from three pieces of information: (a) the abundance this year, (b) the intrinsic rate of growth, r, and (c) the carrying capacity, K. When population abundance is very low, the fraction N/K is approximately 0 and so drops out of the model, with the result that Eq. (10.7) becomes effectively identical to Eq. (10.5), the exponential growth model. This implies that the per capita growth rate of the population will be at its maximum, r, when N is very small, and indeed in this context it is common to refer to the parameter r as "r_{max}," the maximum possible rate of growth.

population status and/or population impacts of various perturbations. Gerrodette (1987) developed a statistical model for gauging the effectiveness of survey data for estimating r, and demonstrated that a long-term time series of regular counts can be an extremely powerful means of detecting even subtle changes in growth. Estes (1990) used this approach to analyze time series from multiple sea otter populations across the North Pacific, including Amchitka Island (where the population had apparently reached carrying capacity by the early 1970s), Attu Island (which sea otters had only recently recolonized), California, and three populations established by translocation, Washington, British Columbia, and southeast Alaska. The results were fascinating: populations at Attu, Washington, British Columbia, and southeast Alaska were all growing at remarkably high rates, with estimated values r of 0.17–0.25, in contrast to a slower growth rate of 0.05 in California. Estes (1990) also employed a life table model derived by Cole (1954) to estimate the theoretical maximum rate of growth (r_{max}) for an age-structured population, given a specified age at first reproduction (a = 2–4 years for sea otters), age of last reproduction ($w = 12-18$ years for sea otters), female birth rate ($b = 0.43-0.49$ female pups per year), and assuming near-perfect survival of females up until the age of last reproduction:

$$1 = e^{-r_{max}} + be^{-r_{max}a} - be^{-r_{max}(w+1)} \tag{10.8}$$

By solving Eq. (10.8) with the parameter values listed above, it can be shown that the theoretical value of r_{max} for sea otters is somewhere between 0.20 and 0.25. Thus the recovering population at Attu and the translocated populations at Washington, British Columbia, and southeast Alaska were found to be increasing at (or near to) the theoretical r_{max}, while the recovering population in California was growing at only one quarter of that rate, at best. This result has been taken to suggest that the California population has experienced a chronically higher mortality rate than northern populations, although an alternate explanation for the difference may be related to the narrow, almost one-dimensional configuration of habitat in California, which, when combined with the high degree of spatial structure of the population (Tinker et al., 2008b), results in more constrained range expansion and thus slower population growth (this is the case because a larger proportion of the population more quickly becomes resource limited than northern populations where there is a two-dimensional matrix of available habitat for range expansion).

The other key parameter of the logistic model shown in Eq. (10.7) is of course *K*, the carrying capacity. Estes (1990) used a qualitative assessment of the time series at Amchitka to conclude that the population at that island had reached a carrying capacity of between 5245 and 6597 otters by the early 1970s. A similar qualitative assessment was made by Laidre et al. (2001) to identify areas in central California assumed to be at or near carrying capacity. Laidre and colleagues then used a spatial habitat model, developed in a geographic information system (GIS), to extrapolate these equilibrium densities to areas of similar habitat type throughout the rest of coastal California and thereby estimate a potential *total* carrying capacity for southern sea otters (assuming that sea otters were to eventually recolonize all nearshore habitat between the Oregon border and Mexico). Gregr and colleagues (2008) used a similar spatial habitat model to estimate potential carrying capacity for sea otters in British Columbia. Coletti (2006) developed an even more elaborate model for Prince William Sound, using stepwise logistic regression to describe relationships among behavior, diving activity, and habitat attributes, including both nearshore and offshore habitat, and then applied those relationships to predict equilibrium densities throughout the Sound, and in particular to those areas affected by the EVOS in order to determine the potential for further post-spill recovery.

All the population models discussed to this point have focused on predicting changes in population abundance overall, while effectively ignoring any details about the animals that make up the population. Often such simplified models are sufficient and appropriate for addressing the questions of interest. But what if we are interested in learning about a specific segment of the population, or we want to know about the impacts of a factor that only affects certain animals (i.e., a disease that primarily affects juveniles or females). Accounting for this complexity requires a different kind of population model, one which explicitly incorporates "population structure," or differences within the population (see Box 10.3).

Box 10.3 Primer on Matrix Models of Structured Populations

The most common type of model used to investigate structured populations is the projection matrix model (a thorough treatment of this subject is provided by Caswell, 2001). A projection matrix is a mathematical tool for describing "transitions between states," and is basically a series of linear equations organized to simplify algebraic operations such as multiplication. If we were to classify all the individuals from a population into one of a limited number of "states" (where states could be defined by sex, age, size, or any other attribute deemed to be important), then the matrix describes all the ways that individuals of a given state can transition to (or contribute to) other possible states over the course of a single time-step (the time-step is usually assumed to be 1 year, but could also be a month, or a week, or any other appropriate interval). For example, if we decide that it is important to classify individuals by age, where the possible state values are juvenile, sub-adult, adult, and old adult, then our matrix will reflect the fact that it is possible for a juvenile to make a transition to sub-adult simply by surviving and growing a year older. Conversely an adult cannot grow into a sub-adult (so this transition will not be allowed in our matrix); however, an adult *can* contribute to the juvenile age class by reproducing and successfully rearing an offspring to become a juvenile. These examples illustrate the three main types of transition that are described by a projection matrix: growth, survival, and reproduction (Figure 10.4A).

The values that are actually stored in the matrix are "per capita transition rates": in the case of survival, this corresponds to the probability that a typical individual of a given state (e.g., age class) will survive from 1 year to the next, while for reproduction this corresponds to the average number of offspring produced by a typical individual over the course of 1 year. The value of the matrix in column j and row i (which we will refer to as a_{ij}) represents the annual transition rate from state j to state i (a population projection matrix is always square, so both i and j take on values from 1 to the number of different possible states). If we know the number of individuals in the population assigned to each state value at the start of this year, we can use *matrix multiplication* (or matrix projection) to calculate the expected number of individuals in each state at the start of next year, and similarly for all future years: this operation is illustrated in Figure 10.4B. Additional matrix algebraic operations can be used to obtain useful properties of the population such as the asymptotic annual population growth rate, usually represented by the symbol λ (not to be mistaken with the

(*Continued*)

Box 10.3 (Continued)

instantaneous growth rate r discussed previously, but which is related to r by the equation $\lambda = e^r$). In Figure 10.4B, $\lambda = 1.018$, meaning that a population with these vital rates will grow at an annual rate of 1.8%. Other advantages include the ability to determine which state transitions are most responsible for causing λ to go up or down. This latter property of population matrices, referred to as *sensitivity analysis,* is clearly a useful property, because it allows us to identify the demographic segments of the population that are most important to focus conservation efforts on if we wish to ensure population recovery.

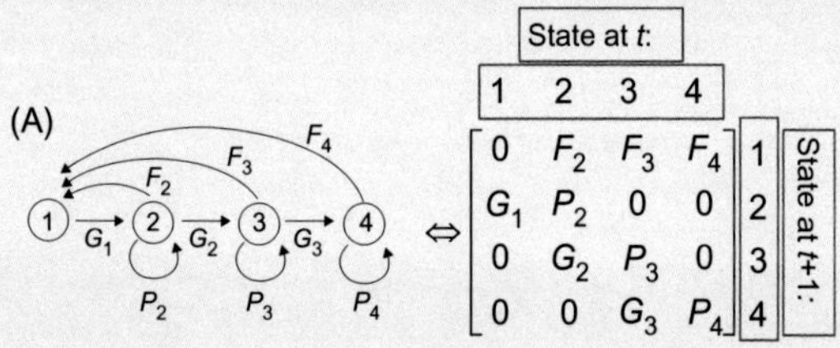

(A)

State at t: 1 2 3 4

$$\begin{bmatrix} 0 & F_2 & F_3 & F_4 \\ G_1 & P_2 & 0 & 0 \\ 0 & G_2 & P_3 & 0 \\ 0 & 0 & G_3 & P_4 \end{bmatrix}$$

State at $t+1$: 1 2 3 4

(B)

$$\begin{pmatrix} n_1(t+1) \\ n_2(t+1) \\ n_3(t+1) \\ n_4(t+1) \end{pmatrix} = \begin{pmatrix} 0 & F_2 & F_3 & F_4 \\ G_1 & P_2 & 0 & 0 \\ 0 & G_2 & P_3 & 0 \\ 0 & 0 & G_3 & P_4 \end{pmatrix} \times \begin{pmatrix} n_1(t) \\ n_2(t) \\ n_3(t) \\ n_4(t) \end{pmatrix}$$

$$\begin{pmatrix} 480 \\ 676 \\ 1238 \\ 389 \end{pmatrix} = \begin{pmatrix} 0 & 0.15 & 0.25 & 0.20 \\ 0.80 & 0.45 & 0 & 0 \\ 0 & 0.40 & 0.80 & 0 \\ 0 & 0 & 0.10 & 0.70 \end{pmatrix} \times \begin{pmatrix} 471 \\ 664 \\ 1216 \\ 382 \end{pmatrix}$$

FIGURE 10.4 Illustration of population projection matrix methods. (A) Loop diagram showing the possible life history transitions between each of four demographic "states," in this case age-classes (1 = juveniles, 2 = sub-adults, 3 = adults, 4 = aged adults), and a matrix formulation of this same life history. The three types of transition at each time-step include survival and growth to next age-class (G), survival without growth (P), and contribution to the first age-class via reproduction (F). In the matrix formulation, a transition from state j at time t to state i at time $t + 1$ is entered into row i and column j. (B) The operation of matrix multiplication (also called matrix projection) is demonstrated, whereby the number of animals in each age-class at time $t + 1$ is calculated based on the number of animals in each stage at time t, given the particular combination of vital rate estimates used to parameterize the matrix model.

Demographic models of sea otter populations have commonly used age as the state variable (generally modeling only the females in the population), although a few models have used both sex and age. Eberhardt (1995) was one of the first to apply an age-structured demographic model to sea otters, using a Lotka-Leslie age-structured matrix (see Box 10.3) within a re-sampling simulation to develop a method for estimating reproductive rates (and associated variance estimates) that accounted for the fact that sea otters give birth year round. Eberhardt concluded that the matrix model approach could be used to evaluate population recovery from events such as the EVOS, and his model integrated previously published information on sea otter vital rates (e.g., Jameson and Johnson, 1993; Siniff and Ralls, 1991) and laid the groundwork for a number of later matrix models used to analyze sea otter population dynamics. For example, Estes and colleagues (1998) used a similar projection matrix model to estimate the number of sea otters that would have to have been lost to killer whale predation over a 6-year period (1990–1996) to account for the dramatic decline in sea otter populations that occurred throughout southwest Alaska. In this case, the model projections were used as part of a simulation analysis to determine whether it was feasible for killer whale predation to account for the 70–90% decline in sea otters that occurred over this period across the Aleutian Archipelago and the Alaskan Peninsula. Amazingly, the model results indicated that just 3–4 killer whales consuming 3–5 sea otters per day (the number of otters required to meet energetic requirements of an adult killer whale) would have been sufficient to drive the observed decline. It is worth noting that the estimates of sea otter vital rates (age-specific survival and reproduction) used for this feasibility analysis were taken from the literature, but matrices can also be useful for estimating vital rates from empirical data sets using standard data fitting techniques.

The use of maximum likelihood or Bayesian methods to fit age-specific survival rates to empirical data on the ages of either living animals or those that have died (in many areas it is possible to retrieve dead sea otters when they wash up on beaches) is one of the most powerful applications of projection matrices. Although this may sound intimidating if you are unfamiliar with data fitting techniques, the concept is actually quite simple, and can be summed up as follows: a key property of projection matrices is the ability to calculate the expected proportion of the population that can be found in each age class, as well as the proportional age distribution of the animals that die (the "death assemblage"). The precise nature of this output depends on the survival rate estimates in the matrix itself: tweaking these values will have the effect of changing the expected proportion of animals in each age class in both the living population and the death assemblage. If we were to systematically tweak all the survival rate values (and birth rate values) in the matrix model, running through the almost infinite set of possible permutations, we would eventually stumble upon the exact combination of values that result in the projected age distributions we find in nature. This may sound like a lot of work for us, but that is essentially what maximum likelihood and Bayesian analyses do. Udevitz and Ballachey (1998) developed a novel method for conducting this type of data fitting, using the ages of both living animals and beach cast carcasses from a population where the growth rate was varying over time (the assumption of stable growth rates was a limitation of earlier techniques). Monson and colleagues (2000) used this type of age-structured model to assess whether there were lingering population effects from the 1989 EVOS. By using maximum likelihood methods to fit time-varying age-specific survival rates to age distributions of sea otters found dead on beaches of western Prince William Sound, Alaska, they found strong support for continued reductions in survival in the areas impacted by the oil spill. This model was rerun and elaborated upon 10 years later (Monson et al., 2011), to determine whether the oil spill effects had dissipated. This updated model indicated a gradual reduction of the survival effects over time; however, it was also shown that the oil-affected areas represented a "sink" population, where there was a net reduction in abundance, but that this was counterbalanced by the "source population" of unaffected areas, with the net result of slow recovery of the population of western Prince William Sound. A similar methodological approach was used by Tinker and colleagues (2006a) to examine the spatial and temporal shifts in demographic rates that were responsible for the period of decline seen in the California population in the late 1990s. The California model built on the approach of Udevitz and Ballachey (1998) and Monson et al. (2000, 2011), fitting survival rates to a time series of annual census counts as well as age at death data, and incorporating two sexes and spatial structure into the projection matrix. The results indicated that the period of decline that occurred between 1995 and 2000 was the result of decreased survival of sub-adult and prime-age adult females, particularly in the north and center of the sea otter's range (Figure 10.5).

Projection matrix models have also been used to answer an even wider range of applied conservation questions about sea otter populations. For example, Gerber et al. (2004) used a matrix model to assess the relative contribution of various causes of death to variation in the growth rate of the California sea otter population. Their sensitivity analysis revealed that infectious disease and emaciation had the greatest impact on population growth during the period of decline in the late 1990s, although shark bite wounds and human-caused trauma also had substantial impacts. Another common application of projection matrices includes the broad category of models known as "population viability analyses," or PVA. PVA models are used to assess the future viability of a given population under specific management scenarios, or to assess the potential impacts of a new threat or perturbation on the viability of a population. A PVA was developed by Tinker and colleagues for the southwest Alaska sea otter population, and is included as part of the recovery plan for this population (USFWS, 2010, 2013). As part of that analysis, trend data from skiff surveys of index sites across the Aleutian Archipelago were combined with data on vital rates collected during telemetry studies at Amchitka, Adak, and Kodiak Islands, and these were used to estimate mean and variance in per-capita mortality rates in southwest Alaska over the period of 1992–2007. Simulations were then run to investigate the

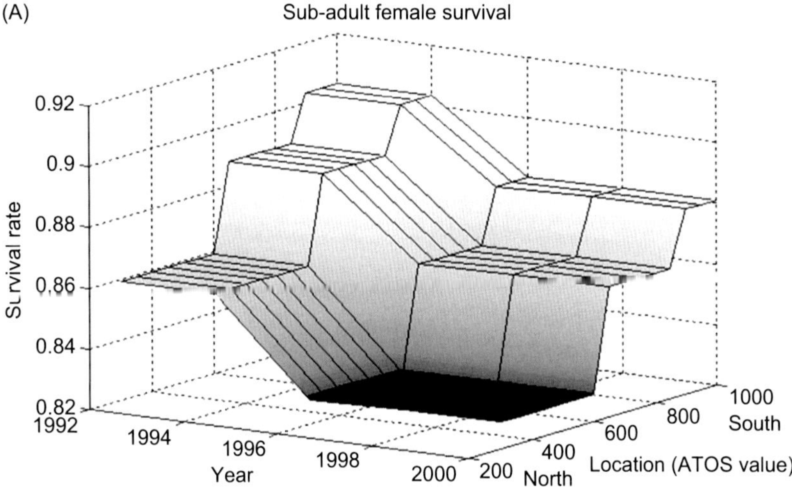

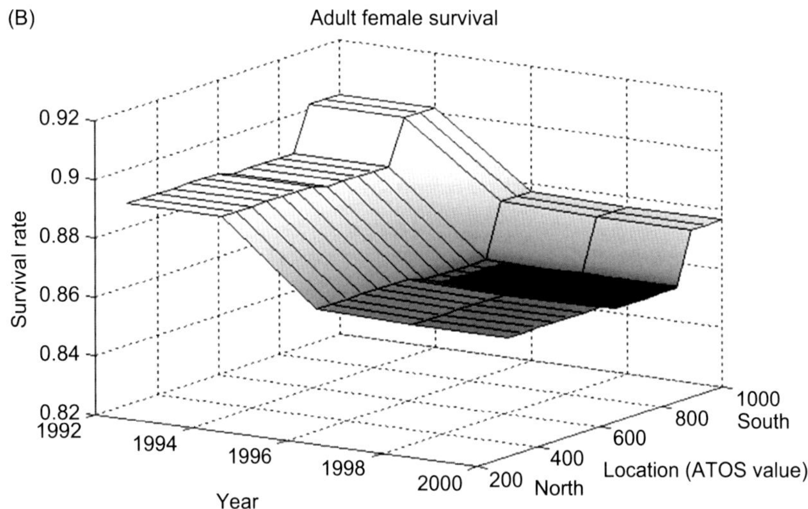

FIGURE 10.5 Spatial and temporal variation in sea otter survival rates in California during the 1990s plotted for sub-adult females (A) and adult females (B). The two horizontal axes are time (in yearly increments, 1992–2000) and geographic location within the range (units are 500 m increments along the coastline, ranging from Santa Cruz in the north to Pt. Conception in the south). The vertical axis is the model-estimated annual rate of survival for the indicated life-stage. *(Figure from Tinker et al. (2006a). Used by permission of The Ecological Society of America.)*

probability of persistence under various scenarios about initial population size and the nature of predator-induced mortality. The results of these simulations provided guidance on appropriate abundance thresholds for the up-listing and de-listing criteria (i.e., the abundance at which up-listing from

"Threatened" to "Endangered" status should be considered, and the abundance at which de-listing from "Threatened" status should be considered). Another PVA for Alaskan sea otters was developed by Bodkin and Ballachey (2010), designed to investigate the potential impacts of human harvests on local sea otter populations in southeast Alaska (Chapter 4). By parameterizing a matrix-based model with vital rates derived from stable, slowly growing and rapidly growing populations, and then simulating harvests of varying magnitudes and sex ratios, Bodkin and Ballachey determined what levels of take would be sustainable (i.e., would not drive the population to decline). They found that the level of additional mortality from harvest that was sustainable varied considerably, from just 0.5% increased mortality in stable populations with 1:1 harvest sex ratio, up to 27% increased mortality in rapidly growing populations with a 3:1 ratio of males to females in the harvest. Interestingly, although they found that male-only mortality maximized annual harvest in stable populations, they also noted that highly male-biased mortality in all simulations eventually led to low proportions of males in localized areas, leading to instability in projected populations over time. This result highlighted the need to account for localized sources of mortality in sea otter PVA models, due to the fact that sea otter populations tend to be structured at relatively small spatial scales.

The above examples demonstrate that it can be important to include spatial differences in population density and vital rates in conservation models. Another aspect of spatial dynamics that has been explored using models is changes in the distribution of sea otter populations. This was the focus of an analysis by Lubina and Levin (1988), who used a diffusion-based model to describe the rate with which the sea otter population in California was spreading northwards and southwards along the coast. This relatively simple model combines behavior (movement) and population growth, treating population spread as a diffusion process, more or less equivalent to the outward diffusion of a gas cloud released into a vacuum. The diffusion model estimates "invasion speed," V, or the rate at which the "population front" moves into unoccupied habitat:

$$V = 2(r_{max}D)^{1/2} \tag{10.9}$$

Equation (10.9) predicts the rate of invasion into new habitat with only two parameters: r_{max} and D, the diffusion parameter, which has units of km^2 per unit time, and is a property of the relative mobility of individuals within the population. Specifically, in the case of population diffusion along a one-dimensional axis (i.e., the California coastline), D can be estimated as one half of the mean square net displacement per unit time of a typical individual (the net displacement of an individual is calculated as the straight-line distance between its location at time $t = 0$ and its location at time $t = 1$, and this is then squared to estimate mean square displacement). The diffusion model of Lubina and Levin provides a good first approximation of how sea otter populations spread into unoccupied habitat, generally matching observed patterns at large scales although omitting many details. Some of the important details that the diffusion model does not incorporate include differences in how sea otters utilize different types of habitat, and the fact that male and female sea otters, and otters of different ages, tend to have different patterns of habitat use, mobility, and survival rates. Many of these sources of variation were later included in more elaborate integrodifference equation models of population spread in California (Krkosek et al., 2007; Smith et al., 2009; Tinker et al., 2008b). As with the simple diffusion models, integrodifference equation models are used to estimate the asymptotic wave speed, the rate at which a population front will advance into unoccupied habitat, based on the mobility and demographic rates of individuals within the population. Unlike the diffusion models, however, integrodifference equations can accommodate differences in the mobility, survival, and/or reproductive rates between individuals of different ages (or sexes), and also can handle frequency distributions of individual movement distances that are not normally distributed (e.g., skewed distributions where a few individuals move very great distances: Kot et al., 1996; Neubert and Caswell, 2000). By applying integrodifference equation models to the California sea otter population it was found that a skewed distribution of net displacement distances could explain the accelerating nature of range expansion in the earlier part of the twentieth century (Krkosek et al., 2007), and that differences between northern and southern rates of range expansion were potentially explained by differences in survival rates (Smith et al., 2009). A specific management application of an integrodifference equation model will be discussed below in "Applied Conservation Questions."

MODELS OF COMMUNITY DYNAMICS

Sea otters have long been recognized as keystone predators in nearshore ecosystems (Estes and Palmisano, 1974), which means that their presence and abundance in the ecosystem has inordinately strong impacts on the structure and dynamics of sub-tidal and inter-tidal community assemblages (see Chapter 2). The substantial ecosystem impacts of sea otters can be attributed to both direct and indirect effects of their predator–prey interactions with other species. Note that in ecology, the terms "direct effect" and "indirect effect" have very specific meanings: a species has a *direct effect* on another species if it actually eats (or is eaten by) that species, or otherwise directly affects the abundance of that species through a non-feeding interaction (e.g., parasitism and pollination). A species is said to have an *indirect effect* on another species if it affects the abundance of that species without actually eating (or being eaten by) that species, but the effect is mediated through one or more other species interactions (e.g., if species A eats species B, and species

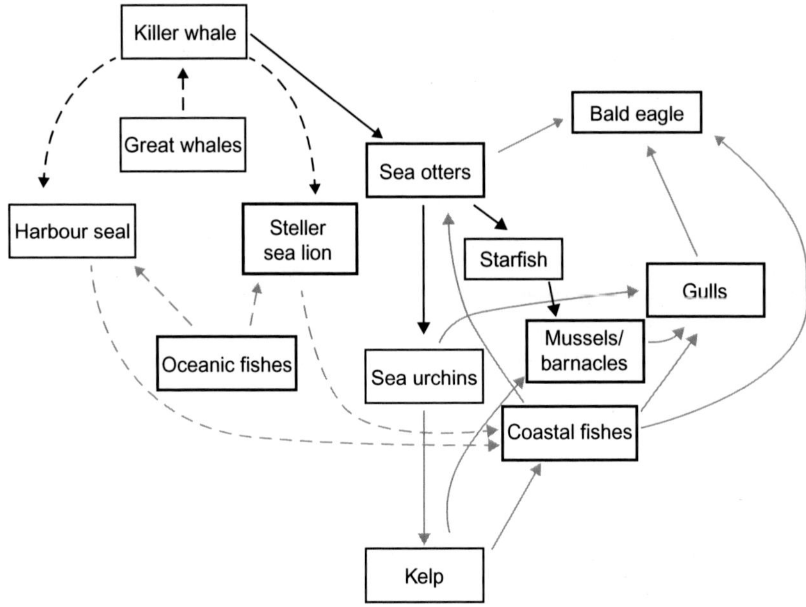

FIGURE 10.6 Food-web relationships among selected species in the North Pacific Ocean and southern Bering Sea. The arrows represent feeding linkages for which there are known (solid lines) or suspected (dashed lines) dynamic interactions. See text for explanations. *(Figure from Estes et al. (2009). Used by permission of the authors.)*

C also eats species B, then species A can affect species C indirectly via its impacts on species B). Direct and indirect effects can best be pictured using a food web (Figure 10.6), a graphical representation of the network of feeding interactions occurring within a specified community of plants and animals. The direct effects in a food web are represented by the arrows that indicate a feeding link between two species. Indirect effects are represented as the connections between two species that can be made by following a pathway through more than one arrow: for example, in Figure 10.6 an indirect effect between sea otters and coastal fish can be traced through links from sea otters to sea urchins to kelp to coastal fishes. If one begins to describe all possible indirect effects in a food web, it quickly becomes apparent that this kind of interaction network contains a great deal of potential complexity. How, then, do we begin to predict which interactions will be most important for maintaining stability of the community (or conversely, which interactions are most likely to de-stabilize the community)? This is exactly the type of question posed by a community ecologist, and multi-species interaction models have been used for many years as tools to help answer such questions.

One of the simplest and most well-known multi-species models to be applied to sea otters is a "trophic cascade" model, consisting of a vertical cascade of feeding links across two or more "trophic levels" (e.g., a primary consumer feeds on a herbivore that feeds on a plant). In most rocky-substrate habitats across the North Pacific, sea otters tend to prey heavily on sea urchins, a relatively abundant and energy-rich invertebrate, and sea urchins in turn are one of the most voracious herbivores to consume the various algal species that make up kelp forests. The quantitative predictions of the three species trophic cascade model are straightforward and easy to test: when the primary consumer (sea otters) increases in abundance, we expect to see a decrease in the herbivore (urchins) and an increase in the primary producer (kelp), because sea otters have a negative *direct effect* on urchins and thus a positive *indirect effect* on kelp (Estes and Palmisano, 1974). These predictions have been tested repeatedly through "opportunistic experiments" (specifically, the reintroduction or natural recolonization of sea otters into unoccupied habitat) in many regions, from Alaska's Aleutian Islands to Vancouver Island in British Columbia, with results generally supporting the model predictions (Estes and Duggins, 1995; Watson and Estes, 2011). The trophic cascade model can be extended by adding additional levels: for example, adding a fourth trophic level (a secondary consumer) is predicted to reduce the abundance of the primary consumer, increase herbivore abundance, and thus decrease plant biomass. An opportunity to test the predictions of this model occurred in the early 1990s, when killer whales began to prey upon sea otters in southwest Alaska (Estes et al., 1998). Consistent with the model predictions, sea otters declined, sea urchins proliferated, and kelp forests across the Aleutian Archipelago were dramatically reduced, all within a period of 5–10 years (Estes et al., 2004).

While the trophic cascade model has been very useful for predicting changes in kelp forest habitat caused by the addition or loss of sea otters, it is obviously somewhat limited in scope because it accounts for just a handful of species. More elaborate food web models, such as the one depicted in Figure 10.6, are more difficult to interpret in terms of their predictions because of the vast number of possible indirect effects. In some cases qualitative interpretations of model dynamics are possible, such as the review by Estes et al. (2009) of the cascade of indirect effects that occurred after killer whale predation reduced sea otter populations across the Aleutian Islands: this strong direct effect apparently resulted from a dietary shift of certain mammal-eating killer whales, presumably in response to earlier declines in their preferred prey of Steller sea lions and harbor seals (Estes et al., 1998, 2009). Indirect effects of the killer whale–sea otter interaction included a two- to threefold decrease in the growth rates of barnacles and mussels (Duggins et al., 1989), a 10-fold decrease in abundance of some kelp forest fish such as greenling (Reisewitz et al., 2006), and a shift in the diet of bald eagles from largely fish and mammals to largely gulls and other marine birds (Anthony et al., 2008). These particular indirect effects were interpretable because the mechanisms were well understood (Estes et al., 2009); however, because the number of potential indirect effects increases geometrically with the number of species in a food web, qualitative interpretations of more species-rich food web models can be problematic. There are mathematical tools for analyzing more rigorously the expected dynamics of multi-species assemblages (Allesina et al., 2008; Dunne, 2006; McCann and Hastings, 1997; McCann et al., 1998; Pascual and Dunne, 2006; Wootton, 2001), although these tools can also become overwhelmed by complexity and the uncertainty associated with quantifying each direct species interaction (Novak et al., 2011). One such multi-species model is "Ecopath," which traces the flow of energy through a food web using a series of mass-balance equations that describe the production, consumption, and loss of biomass for each species in the system at equilibrium (Pauly et al., 2000). An Ecopath model that included sea otters was assembled for the Prince William Sound ecosystem (Okey and Pauly, 1998); however, that model was focused primarily on the open-water food-web interactions in order to predict fisheries dynamics, and so the treatment of sea otters and their invertebrate prey was cursory at best.

SOME EXAMPLES: APPLIED CONSERVATION MODELS

In the history of sea otter conservation there have been a number of particular challenges for which very specialized models have been designed and brought to bear. These applied models often incorporate more than one element of the subjects discussed above (e.g., both behavior and population dynamics) as well as other subject areas, and were generally designed to address a very specific question. I will briefly discuss two examples—understanding oil spills and predicting population spread—to provide a flavor of some of the ways in which specialized models can be designed to explore and help answer challenging conservation questions.

One of the most devastating anthropogenic impacts to sea otters, and to coastal ecosystems more generally, has been the occurrence of large oil spills. Perhaps the most well known of these to affect sea otters was the 1989 EVOS. Sea otters are particularly susceptible to coastal oil spills because of their nearshore distribution, but also because of idiosyncratic aspects of their biology that make them particularly vulnerable to oil in two ways: (1) they are dependent upon the integrity of their fur for thermoregulation, and oiling of their fur compromises its integrity as an insulator (Davis et al., 1988) and can lead to hypothermia and death (thermal challenges can be exacerbated by acute poisoning when they groom their fur and incidentally ingest and inhale oil); and (2) they feed on bottom-dwelling invertebrates in the subtidal and inter-tidal zones, which often require extraction from the substrate by digging, and this makes them susceptible to direct exposure to and ingestion of oil while foraging, potentially leading to chronic health effects. The first of these impacts tends to manifest itself immediately after an oil spill and can result in extensive mortality over large areas (Bodkin et al., 2002; Bodkin and Weltz, 1990; Degange and Lensink, 1990; Garshelis, 1997;

Williams et al., 1988). The second of these impacts, toxicological effects from oil ingestion, can continue to affect sea otters for months or years after the oil spill because oil can remain in the environment in inter-tidal sediments and sea otters may be repeatedly exposed during foraging (Ballachey et al., 2002). Using archival time depth recorder data and models to assign dive types, Bodkin et al. (2012) estimated sea otters foraging in oiled areas of Prince William Sound were exposed to lingering oil from 2 to 24 times per year, more than 15 years after the spill. Effects of such exposure can be difficult to measure and the health effects may be sub-acute and effectively "invisible," yet they can still have population-level impacts by reducing survival. To address this conservation concern, Harwell and colleagues (2010) developed a risk assessment model to assess the potential for toxicological harm to sea otters from buried EVOS subsurface oil residues (SSOR). Their model was an "individually based model," meaning that they used computer-based simulations of individual behavior, iterated many times. They incorporated multiple pathways of potential exposure to SSOR into their model, including oil that becomes suspended in the water column and is taken up by filter feeding invertebrates (and thus otters are exposed through diet); otters are directly exposed to SSOR as they dig pits in the inter-tidal zone to extract clams and other infaunal prey. To parameterize their model they used data on sea otter behavior (specifically activity budgets, diet composition, and feeding dive locations) as well as environmental data on shoreline substrate composition and oil residue levels in different substrate types. They ran 500,000 simulations of individual foraging otters, estimated exposure levels to SSOR for each simulated otter, and compared these to chronic toxicity reference values approved by the US Environmental Protection Agency. They concluded that although sea otters continued to be exposed to oil, the maximum exposure rate would not result in sufficient exposure to cause any health effects (Harwell et al., 2010). These results suggest that direct exposure to residual oil from EVOS should not be a factor in limiting sea otter populations in western Prince William Sound. This conclusion was contradicted by the results of Monson et al. (2000, 2011), summarized above in the discussion of population models, which found evidence for continued demographic impacts from EVOS. Further complicating the picture are analyses conducted by Ballachey et al. (2003), which examined the relation of various blood parameters to survival rates of juveniles in eastern and western Prince William Sound, and by Miles et al. (2012), which examined relations between oil exposure and the expression of genes. These authors found that differences in blood parameters and gene transcription in the area affected by EVOS were consistent with continued exposure to oil, and also possibly contributed to the lower survival rates in this group. So how should we proceed when different models produce contradictory findings? This is perhaps a good time to bear in mind the famous quote by the late George Edward Pelham Box: "Essentially, all models are wrong, but some are useful." None of these models should be confused with reality, though all of them seek to provide useful explorations of some specific aspect of the real world. In this case there are empirical data which suggest some continued biological exposure to oil (Ballachey et al., 2003), contradicting at least some of the predictions of the risk assessment model developed by Harwell et al., but the question of actual population-level effects is more complex. Both the Harwell et al. (2010) model and the Monson et al. (2000, 2011) models make a number of simplifying assumptions, and these must be examined closely in order to evaluate the degree of reliability in the results of each paper. Ultimately, when two models that appear to be founded on reasonable assumptions produce contradictory results, the safest conclusion is that additional investigation into the question is warranted.

Despite continued debate over the long-term impacts of EVOS on sea otters in Prince William Sound, the EVOS has resulted in an enormous amount of baseline information about the Prince William Sound ecosystem, as well as the short-term and long-term effects of oil spills on ecosystems in general (Peterson et al., 2003). Some of this information has been used to inform conservation efforts aimed at sea otter populations elsewhere, particularly in California. Brody et al. (1996) used data on how EVOS had affected sea otter survival in Prince William Sound, specifically how those impacts varied with time and distance from the point of origin of the spill, to parameterize a simulation-based risk assessment model for California sea otters. Their model sought to estimate the population-level impacts of oil spills of similar magnitude if they should occur at different locations along the California coast, predicting morality of sea otters as a function of how an oil slick would disperse along the coast, how many otters it would encounter, duration of oil-otter encounters, and time since the spill. They found that an oil spill occurring at the Monterey peninsula could expose some 90% of the California sea otter population at that time to oil and kill at least 50% of individual otters (Brody et al., 1996). These model results supported an earlier, somewhat unusual model by VanBlaricom and Jameson (1982), which took advantage of an accidental lumber spill off central California during the winter of 1978, and spread to cover most of the sea otter's distribution within 4 weeks. These authors noted that the movement rates of lumber were very similar to those of oil slicks observed elsewhere, and their observations indicated that a major oil spill could expose most of the population to oil contamination within a month of a spill. These sobering findings were considered by the Endangered Species Recovery Team in setting de-listing criteria for southern sea otters, and in advising the US Fish and Wildlife Service on Management options (USFWS, 2003).

Management and conservation of wildlife species nearly always demands that we try to forecast the future based on past events, to gauge how a population will respond to some anticipated conservation challenge. The PVA described above represent one type of forecasting model, one that seeks to assess the future viability of a population given a range of known or anticipated threats. A more specific management question in California required a more specialized forecasting model, one that combined information on population vital rates and individual behavior in order to forecast the future range expansion of southern sea otters. In particular, as part of the Environmental Impact Statement on the translocation of southern sea otters to San Nicolas Island (USFWS, 2012), the US Fish and Wildlife Service requested a model to forecast the future spatial extent of the sea otter population in southern California, and the number of animals projected to be south of Pt. Conception at various future times. This information was required as part of the Services' re-evaluation of the "No Otter Zone" policy that, in theory, would restrict sea otter range expansion south of Pt. Conception. Tinker and colleagues (2008b) developed a spatially structured integrodifference equation model with which to produce such a forecast. As discussed above (see "Population Models"), integrodifference equations can be used to estimate the rate at which a population will "spread" by making use of data on stage-specific vital rates and mobility. This matrix-based model utilized previously published estimates of survival and reproductive rates from various parts of the sea otter distribution (Tinker et al., 2006a) and combined these with data on individual movement behavior collected from radio telemetry studies, to predict both the speed with which the population would spread into habitat south of Pt. Conception, as well as the growth in numbers in this region. The model was initiated using data from the 2004 census, and predicted that by 2014 there would be 112 sea otters southeast of Pt. Conception (95% confidence limits = 69 − 163) and the southern range limit would be between Goleta and Carpentaria, California. These forecast estimates were used by Fish and Wildlife to estimate biological and socioeconomic impacts of allowing sea otters to naturally recolonize habitat south of Pt. Conception (USFWS, 2012), and factored into the decision to end the "No Otter" zone (Chapter 12).

CONCLUSION

The examples described above (and summarized in Table 10.1) are by no means an exhaustive review of all models developed by researchers investigating sea otters and their ecosystems; however, they do provide an overview of the diversity of topics and sea otter studies in which models have played an important role. At this point it may be useful to take a figurative step back and evaluate some of the more general lessons that may be gleaned from these examples and applied to other species and other systems. Models are an important tool for wildlife conservation, and when one considers the sea otter models described above, some little pearls of wisdom emerge that explain why they were possible and often successful, and what we learned from them.

TABLE 10.1 A Summary of Select Modeling Studies in Sea Otter Conservation

Published Paper (see References for Full Citations)	Subject Area Addressed									
	Behavior		Population				Community		Applied	
	Foraging	Movement Habitat Use	Growth/K	Demography Matrices	PVA	Spatial	Trophic Cascade	Food Web	Oil Spill	Risk Assessment
Ballachey et al., 2003, Can. J. Zool. 81	X			X					X	
Bodkin and Ballachey, 2010, USGS Report 2010				X	X					
Bodkin et al., 2012, Marine Ecology-Progress Series 447	X	X								X
Brody et al., 1996, Marine Mammal Science 12				X					X	
Coletti, 2006 Masters Thesis, U. New Hampshire		X								
Dean et al., 2002, Marine Ecology-Progress Series 241	X									

Eberhardt,1995, Journal of Wildlife Management 59				X				
Estes et al., 1981, Worldwide Furbearer Conference	X							
Estes et al., 1998, Science (Washington DC) 282						X	X	
Estes et al., 2003, Journal of Animal Ecology 72	X							
Estes et al., 2004, Bulletin of Marine Science 74						X	X	
Estes et al., 2009, Phil. Trans. Royal Soc. 364						X	X	
Estes, 1990, Journal of Animal Ecology 59			X					
Garshelis, 1997, Conservation Biology 11				X				
Gerrodette, 1987, Ecology 68			X					

(Continued)

TABLE 10.1 (Continued)

Published Paper (see References for Full Citations)	Subject Area Addressed									
	Behavior		Population				Community		Applied	
	Foraging	Movement Habitat Use	Growth/K	Demography Matrices	PVA	Spatial	Trophic Cascade	Food Web	Oil Spill	Risk Assessment
Gregr et al., 2008, JWM 72		X								
Harwell et al., 2010, Hum. Ecol. Risk Assess. 16									X	X
Johnson et al., 2009, PNAS 106		X								X
Kage, 2004, Masters Dissertation, UC Santa Cruz		X								
Krkosek et al., 2007, Theor. Popul. Biol. 71						X				
Kvitek and Oliver, 1992, MEPS 82	X									
Laidre et al., 2009, J. Mammal. 90		X								
Laidre and Jameson, 2006, J. Mammal. 87	X									

Lubira and Levin, 1988, American Naturalist 131				X	
Monson et al., 2000, PNAS 97			X		X
Monson et al., 2011, Ecol. Appl. 21			X	X	X
Ostfeld, 1982, Oecologia 53	X				
Siniff and Ralls, 1991, Marine Mammal Science 7			X		
Smith et al., 2009, Ecology 90				X	
Tinker et al., 2006, Ecol. Appl. 16			X		
Tinker et al., 2008b, Ecol. Appl. 18		X	X	X	
Tinker et al., 2008a, PNAS 105	X	X			
Tinker et al., 2009, Evol. Ecology Research 11	X				

(*Continued*)

TABLE 10.1 (Continued)

Published Paper (see References for Full Citations)	Subject Area Addressed									
	Behavior		Population				Community		Applied	
	Foraging	Movement Habitat Use	Growth/K	Demography Matrices	PVA	Spatial	Trophic Cascade	Food Web	Oil Spill	Risk Assessment
Tinker et al., 2012, Ecology Letters 15	X									
Udevitz and Ballachey, 1998, JWM 62				X						
USFWS, 2010, SWAK sea otter Draft Recovery Plan				X	X			X		
VanBlaricom and Jameson, 1982, Science 215									X	

1. Long-term data sets are powerful. One of the most rewarding aspects of working in the world of sea otter conservation is that our predecessors in this field had the foresight to initiate long-term monitoring programs, and as a result there is an embarrassment of riches when it comes to long-term data sets. Models are most useful in conservation if they are based on a solid understanding of the system in question AND can be parameterized with values that reflect something close to the full range of variation and uncertainty that exists in that system. For example, having 5 years of data on which to base a population growth model is certainly better than nothing; however, it is unlikely that those 5 years will provide a full picture of the range of variation in growth rates (i.e., really good years and really bad years). In contrast, many of the examples discussed above were based on >30 years of sequential population estimates, and in many cases time series of similar duration for other components of the ecosystem (e.g., kelp and urchins). Instinctively we tend to place more confidence in the projections of models that are built on such long-term data sets, and for good reason. What if that type of data set is simply not available for a species of concern? For species in which long-term data sets simply don't exist, it is often appropriate to look at closely related species where data may be more plentiful and use these data sets to parameterize models. As long as one allows for suitable amounts of uncertainty (either explicitly in the model, or implicitly in terms of interpretation of results), this approach can be useful for designing conservation models for data-poor species.
2. Combining multiple data types can be a good thing. Particularly in the modern age of information theory and Bayesian estimation techniques, it is very possible and indeed desirable to bring together and integrate multiple data types, as doing so can make model results more robust and often provide more insight. For example, estimation of vital rate parameters used in the matrix projection models developed for Alaska and California (Monson et al., 2011; Tinker et al., 2006a) was based on multiple independent data sets (in the case of the Alaska model, age distributions of both living and dead animals and survey estimates of population size; in the case of the California model, age distributions of dead animals, survey estimates of population size, and mark-recapture data from tagged animals). Maximum likelihood methods were used to extract far more information from these combined data sets than could be gained from any one by itself. The synthesis of diverse data types is indeed one of the main advantages provided by mechanistic models, in contrast to more traditional statistical analyses of single data sets, and in many cases the results that emerge can be surprising and non-intuitive.
3. In complex systems, models can be invaluable for elucidating mechanisms. Community models such as those discussed above provide an excellent illustration of how models can provide insights into complex

and otherwise intractable systems. In the Aleutian Island example, many apparently unrelated phenomena had been observed independently: declines in Steller sea lions and other pinnipeds in the pelagic realm, sea otter declines in the nearshore zone, reductions in kelp abundance, and declines in certain fish populations. Because none of these species have direct interactions with each other, it was a stretch to imagine that they were all connected; however, a food web model provided the conceptual and quantitative means of describing the network of indirect effects that linked them all together. In this sense models provided a feasibility test, answering the question of how likely or unlikely it was that killer whales could cause the sea otter decline, or that reductions in sea otters could lead to the observed changes in kelp and fish abundance. Just as importantly, models can be critical in understanding and interpreting "emergent dynamics" of complex systems. Emergent dynamics are, as the name suggests, phenomena that emerge from the complex suite of interactions of the system as a whole and are not predictable from examining any individual part of the system. It is generally not possible to examine a single pair of interacting species and predict whether impacts of that species' interaction on the community as a whole will be large or small. However, a food web model that incorporates multiple species (and thus includes both direct and indirect effects) can be used to assess emergent properties such as the relative importance of each pairwise interaction, and how the loss of any one species will affect the stability and resilience of the entire community. Although such model predictions obviously need to be empirically tested, they are nonetheless an important tool in conservation because they provide insight into when and how a threat to a single species will propagate throughout the community and have ecosystem-level impacts.

4. Models designed to answer questions about basic biology can provide unexpected insights into conservation questions. Some of the sea otter models described above would not initially seem to be tools for conservation. For example, models used to understand foraging ecology might seem interesting from a purely academic standpoint, but are perhaps less so when it comes to species conservation. And indeed, many models are initially designed to answer basic questions about organismal biology or ecology: how does an individual sea otter select prey, how does diet change as competition for food increases, how far away will a sea otter move over the course of a year? However, it is well known that effective conservation strategies are built on a solid understanding of natural history, and in addition to answering the ecological questions for which they were designed, such models may also provide unexpected but important insights into conservation challenges. For example, models designed to investigate individual foraging decisions (Tinker et al., 2009) and the network structure of predator–prey interactions at the individual level (Tinker et al., 2008a,

2012) showed that sea otters in high-density populations exhibited pronounced diet specialization, with individuals falling into one of a number of distinct diet modules. Making use of this information, Johnson et al. (2009) developed an epidemiological model to investigate whether diet specialization was a risk factor for exposure to land-based disease pathogens. The results of this model showed that certain prey specializations, including a diet high in marine turban snails, greatly increased risk of infection with protozoan parasites. Thus models initially developed to understand foraging behavior unexpectedly led to information that is helping conservation biologists determine how disease-causing parasites on land are finding their way into coastal marine food webs.

As computing capabilities become ever greater, along with technological innovations that increase our ability to collect and access large amounts of data, it is inevitable that quantitative models will become more and more critical for synthesizing and making sense of this information in order to solve conservation problems. The examples discussed here from the field of sea otter conservation provide a synopsis of how models have been used in the past, and many of the basic principles and lessons learned will apply to future models as well. However, more computationally intensive techniques such as individually based or agent-based models will likely become more widespread. Future directions in quantitative models also promise to bridge multiple conceptual levels, linking cellular-level processes to physiological and behavioral dynamics at the organismal level, and linking these to population-, community-, and ultimately landscape-level phenomena. Such models will play a crucial part in solving some of the most challenging conservation problems we now face, such as climate change, loss of bio-diversity, invasive species, and other large-scale environmental drivers. And given their important role as an apex predator in nearshore ecosystems, and their tractability for scientific study, sea otters will likely continue to provide an excellent model system for developing new conservation models.

REFERENCES

Allesina, S., Alonso, D., Pascual, M., 2008. A general model for food web structure. Science 320, 658–661.

Alvarez-Buylla, E.R., Garcia-Barrios, R., Lara-Moreno, C., Martinez-Ramos, M., 1996. Demographic and genetic models in conservation biology: applications and perspectives for tropical rain forest tree species. Annu. Rev. Ecol. Syst. 27, 387–421.

Anthony, R.G., Estes, J.A., Ricca, M.A., Miles, A.K., Forsman, E.D., 2008. Bald eagles and sea otters in the Aleutian Archipelago: indirect effects of trophic cascades. Ecology 89, 2725–2735.

Ballachey, B., Bodkin, J., Howlin, S., Kloecker, K., Monson, D., Rebar, A., et al., 2002. Hematology and serum chemistry of sea otters in oiled and unoiled areas of Prince William Sound, Alaska, 1996–98. In: Holland-Bartels, L.E. (Ed.), Exxon Valdez Oil Spill

Restoration Project 99025 Final Report: Mechanisms of Impact and Potential Recovery of Nearshore Vertebrate Predators Following the 1989 Exxon Valdez Oil Spill, Volume 2-Appendices. US Geological Survey, Alaska Biological Science Center, Anchorage, AK, pp. Appendix BIO-01.

Ballachey, B.E., Bodkin, J.L., Howlin, S., Doroff, A.M., Rebar, A.H., 2003. Correlates to survival of juvenile sea otters in Prince William Sound, Alaska, 1992–1993. Can. J. Zool. 81, 1494–1510.

Bodkin, J.L., Ballachey, B.E., 2010. Modeling the Effects of Mortality on Sea Otter Populations. US Geological Survey Scientific Investigations Report 2010, pp. 1–12.

Bodkin, J.L., Ballachey, B.E., Coletti, H.A., Esslinger, G.G., Kloecker, K.A., Rice, S.D., et al., 2012. Long-term effects of the "Exxon Valdez" oil spill: sea otter foraging in the intertidal as a pathway of exposure to lingering oil. Mar. Ecol. Prog. Ser. 447, 273–287.

Bodkin, J.L., Ballachey, B.E., Dean, T.A., Fukuyama, A.K., Jewett, S.C., McDonald, L., et al., 2002. Sea otter population status and the process of recovery from the 1989 "Exxon Valdez" oil spill. Mar. Ecol. Prog. Ser. 241, 237–253.

Bodkin, J.L., Weltz, F., 1990. Evaluation of sea otter capture after the T-V Exxon Valdez oil spill Prince William Sound Alaska USA. US Fish and Wildlife Service Biological Report 90, pp. 61–69.

Borger, L., Dalziel, B.D., Fryxell, J.M., 2008. Are there general mechanisms of animal home range behaviour? A review and prospects for future research. Ecol. Lett. 11, 637–650.

Brody, A.J., Ralls, K., Siniff, D.B., 1996. Potential impact of oil spills on California sea otters: implications of the Exxon Valdez spill in Alaska. Mar. Mammal Sci. 12, 38–53.

Brook, B.W., O'Grady, J.J., Chapman, A.P., Burgman, M.A., Akçakaya, H.R., Frankham, R., 2000. Predictive accuracy of population viability analysis in conservation biology. Nature 404, 385–387.

Caswell, H., 1988. Theory and models in ecology: a different perspective. Ecol. Modell. 43, 33–44.

Caswell, H., 2001. Matrix Population Models: Construction, Analysis, and Interpretation, second ed. Sinauer Associates, Sunderland, MA.

Charnov, E.L., 1976. Optimal foraging: the marginal value theorem. Theor. Popul. Biol. 9, 129–136.

Clark, J.S., Gelfand, A.E., 2006. A future for models and data in environmental science. Trends Ecol. Evol. 21, 375–380.

Cole, L., 1954. The population consequences of life history phenomena. Quart. Rev. Biol. 29, 103–137.

Coletti, H., 2006. Correlating sea otter density and behavior to habitat attributes in Prince William Sound, Alaska: a model for prediction. Natural Resources Policy and Management. University of New Hampshire, Durham, NH, p. 86.

Davis, R.W., Williams, T.M., Thomas, J.A., Kastelein, R.A., Cornell, L.H., 1988. The effects of oil contamination and cleaning on sea otters (*Enhydra lutris*). Metabolism, thermoregulation, and behavior. Can. J. Zool. 66, 2782–2790.

Dean, T.A., Bodkin, J.L., Fukuyama, A.K., Jewett, S.C., Monson, D.H., O'Clair, C.E., et al., 2002. Food limitation and the recovery of sea otters following the "Exxon Valdez" oil spill. Mar. Ecol. Prog. Ser. 241, 255–270.

Degange, A.R., Lensink, C.J., 1990. Distribution age and sex composition of sea otter carcasses recovered during the response to the T-V Exxon Valdez oil spill. US Fish and Wildlife Service Biological Report 90, pp. 124–129.

Duggins, D.O., Simenstad, C.A., Estes, J.A., 1989. Magnification of secondary production by kelp detritus in coastal marine ecosystems. Science 245, 170–173.

Dunne, J.D., 2006. The network structure of food webs. In: Pascual, M., Dunne, J.D. (Eds.), Ecological Networks: Linking Structure to Dynamics in Food Webs. Oxford University Press, Oxford, UK, pp. 27–86.

Eberhardt, L.L., 1995. Using the Lotka-Leslie model for sea otters. J. Wildl. Manage. 59, 222–227.

Emlen, J.M., 1966. The role of time and energy in food preference. Am. Nat. 100, 611–617.

Estes, J.A., 1990. Growth and equilibrium in sea otter populations. J. Anim. Ecol. 59, 385–402.

Estes, J.A., Danner, E.M., Doak, D.F., Konar, B., Springer, A.M., Steinberg, P.D., et al., 2004. Complex trophic interactions in kelp forest ecosystems. Bull. Mar. Sci. 74, 621–638.

Estes, J.A., Doak, D.F., Springer, A.M., Williams, T.M., 2009. Causes and consequences of marine mammal population declines in southwest Alaska: a food-web perspective. Philos. Trans. R Soc. B Biol. Sci. 364, 1647–1658.

Estes, J.A., Duggins, D.O., 1995. Sea otters and kelp forests in Alaska: generality and variation in a community ecological paradigm. Ecol. Monogr. 65, 75–100.

Estes, J.A., Jameson, R.J., Johnson, A.M., 1981. Food selection and some foraging tactics of sea otters. In: Chapman, J.A., Pursley, D. (Eds.), Worldwide Furbearer Conference Proceedings, vol. 1. University of Maryland Press, Baltimore, MD, pp. 606–641.

Estes, J.A., Jameson, R.J., Rhode, E.B., 1982. Activity and prey selection in the sea otter: influence of population status on community structure. Am. Nat. 120, 242–258.

Estes, J.A., Palmisano, J.F., 1974. Sea otters: their role in structuring nearshore communities. Science 185, 1058–1060.

Estes, J.A., Riedman, M.L., Staedler, M.M., Tinker, M.T., Lyon, B.E., 2003. Individual variation in prey selection by sea otters: patterns, causes and implications. J. Anim. Ecol. 72, 144–155.

Estes, J.A., Tinker, M.T., Williams, T.M., Doak, D.F., 1998. Killer whale predation on sea otters linking oceanic and nearshore ecosystems. Science (Washington, DC) 282, 473–476.

Garshelis, D.L., 1997. Sea otter mortality estimated from carcasses collected after the Exxon Valdez oil spill. Conserv. Biol. 11, 905–916.

Garshelis, D.L., Garshelis, J.A., Kimker, A.T., 1986. Sea otter time budgets and prey relationships in Alaska. J. Wildl. Manage. 50, 637–647.

Gerber, L.R., Tinker, M.T., Doak, D.F., Estes, J.A., Jessup, D.A., 2004. Mortality sensitivity in life-stage simulation analysis: a case study of southern sea otters. Ecol. Appl. 14, 1554–1565.

Gerrodette, T., 1987. A power analysis for detecting trends. Ecology 68, 1364–1372.

Gilpin, M.E., Soulé, M.E., 1986. Minimum viable populations: processes of species extinction. In: Soulé, M.E. (Ed.), Conservation Biology: The Science of Scarcity and Diversity. Sinauer Associates, Sunderland, MA, pp. 19–34.

Gregr, E.J., Nichol, L.M., Watson, J.C., Ford, J.K.B., Ellis, G.M., 2008. Estimating carrying capacity for sea otters in British Columbia. J. Wildl. Manage. 72, 382–388.

Harwell, M.A., Gentile, J.H., Johnson, C.B., Garshelis, D.L., Parker, K.R., 2010. A quantitative ecological risk assessment of the toxicological risks from Exxon Valdez subsurface oil residues to sea otters at Northern Knight Island, Prince William Sound, Alaska. Hum. Ecol. Risk. Assess. 16, 727–761.

Hilborn, R., Mangel, M., 1997. The Ecological Detective: Confronting Models with Data. Princeton University Press, Princeton, NJ.

Jameson, R.J., Johnson, A.M., 1993. Reproductive characteristics of female sea otters. Mar. Mammal Sci. 9, 156–167.

Jessup, D.A., Miller, M.A., Kreuder-Johnson, C., Conrad, P.A., Tinker, M.T., Estes, J., et al., 2007. Sea otters in a dirty ocean. J. Am. Vet. Med. Assoc. 231, 1648–1652.

Johnson, C.K., Tinker, M.T., Estes, J.A., Conrad, P.A., Staedler, M., Miller, M.A., et al., 2009. Prey choice and habitat use drive sea otter pathogen exposure in a resource-limited coastal system. Proc. Natl. Acad. Sci. USA 106, 2242–2247.

Kage, A.H., 2004. Temporal and Spatial Variation in Movement Patterns of the California Sea Otter, *Enhydra lutris nereis.* Department of Ecology and Evolutionary Biology, University of California, Santa Cruz, CA, p. 80.

Kot, M., Lewis, M.A., Van Den Driessche, P., 1996. Dispersal data and the spread of invading organisms. Ecology (Washington, DC) 77, 2027–2042.

Krkosek, M., Lauzon-Guay, J.S., Lewis, M.A., 2007. Relating dispersal and range expansion of California sea otters. Theor. Popul. Biol. 71, 401–407.

Kvitek, R.G., Oliver, J.S., 1992. Influence of sea otters on soft-bottom prey communities in Southeast Alaska. Mar. Ecol. Prog. Ser. 82, 103–113.

Laidre, K.L., Jameson, R.J., 2006. Foraging patterns and prey selection in an increasing and expanding sea otter population. J. Mammal. 87, 799–807.

Laidre, K.L., Jameson, R.J., DeMaster, D.P., 2001. An estimation of carrying capacity for sea otters along the California coast. Mar. Mammal Sci. 17, 294–309.

Laidre, K.L., Jameson, R.J., Gurarie, E., Jeffries, S.J., Allen, H., 2009. Spatial habitat use patterns of sea otters in Coastal Washington. J. Mammal. 90, 906–917.

Lamberson, R.H., McKelvey, R., Noon, B.R., Voss, C., 1992. A dynamic analysis of northern spotted owl viability in a fragmented forest landscape. Conserv. Biol. 6, 505–512.

Levins, R., 1966. The strategy of model building in population biology. Am. Sci. 54, 421–431.

Lubina, J.A., Levin, S.A., 1988. The spread of a reinvading species: range expansion in the California sea otter. Am. Nat. 131, 526–543.

Mac Nally, R., 2000. Regression and model-building in conservation biology, biogeography and ecology: the distinction between—and reconciliation of—"predictive" and "explanatory" models. Biodivers. Conserv. 9, 655–671.

Mangel, M., Clark, C., 1988. Dynamic Modeling in Behavioral Ecology. Princeton University Press, Princeton, NJ.

Maynard-Smith, J., 1978. Models in Ecology. Cambridge University Press, New York, NY.

McCann, K., Hastings, A., 1997. Re-evaluating the omnivory-stability relationship in food webs. Proc. R. Soc. Lond. B Biol. Sci. 264, 1249–1254.

McCann, K., Hastings, A., Huxel, G.R., 1998. Weak trophic interactions and the balance of nature. Nature 395, 794–798.

Miles, A., Bowen, L., Ballachey, B., Bodkin, J., Murray, M., Estes, J., et al., 2012. Variations of transcript profiles between sea otters *Enhydra lutris* from Prince William Sound, Alaska, and clinically normal reference otters. Mar. Ecol. Prog. Ser. 451, 201–212.

Monson, D.H., Doak, D.F., Ballachey, B.E., Bodkin, J.L., 2011. Could residual oil from the Exxon Valdez spill create a long-term population "sink" for sea otters in Alaska? Ecol. Appl. 21, 2917–2932.

Monson, D.H., Doak, D.F., Ballachey, B.E., Johnson, A., Bodkin, J.L., 2000. Long-term impacts of the Exxon Valdez oil spill on sea otters, assessed through age-dependent mortality patterns. Proc. Natl. Acad. Sci. USA 97, 65626567.

Morris, W.F., Doak, D.F., 2002. Quantitative Conservation Biology: Theory and Practice of Population Viability Analysis. Sinauer Associates, Sunderland, MA.

Neubert, M.G., Caswell, H., 2000. Demography and dispersal: calculation and sensitivity analysis of invasion speed for structured populations. Ecology 81, 1613–1628.

Novak, M., Wootton, J.T., Doak, D.F., Emmerson, M., Estes, J.A., Tinker, M.T., 2011. Predicting community responses to perturbations in the face of imperfect knowledge and network complexity. Ecology 92, 836–846.

Okey, T.A., Pauly, D., 1998. Trophic Mass-Balance Model of Alaska's Prince William Sound Ecosystem, for the Post-spill Period 1994–1996. Fisheries Centre, University of British Columbia, Vancouver, BC.

Ostfeld, R.S., 1982. Foraging strategies and prey switching in the California sea otter. Oecologia 53, 170–178.

Pascual, M., Dunne, J.A., 2006. Ecological Networks: Linking Structure to Dynamics in Food Webs. Oxford University Press, Oxford, New York.

Pauly, D., Christensen, V., Walters, C., 2000. Ecopath, ecosim, and ecospace as tools for evaluating ecosystem impact of fisheries. ICES J. Mar. Sci.: Journal du Conseil 57, 697–706.

Peterson, C.H., Rice, S.D., Short, J.W., Esler, D., Bodkin, J.L., Ballachey, B.E., et al., 2003. Long-term ecosystem response to the Exxon Valdez oil spill. Science 302, 2082–2086.

Ralls, K., Eagle, T.C., Siniff, D.B., 1996. Movement and spatial use patterns of California sea otters. Can. J. Zool. 74, 1841–1849.

Reisewitz, S.E., Estes, J.A., Simenstad, C.A., 2006. Indirect food web interactions: sea otters and kelp forest fishes in the Aleutian Archipelago. Oceologia 146, 623–631.

Schoener, T.W., 1971. Theory of feeding strategies. Annu. Rev. Ecol. Syst. 2, 369–404.

Siniff, D.B., Ralls, K., 1991. Reproduction, survival and tag loss in California sea otters. Mar. Mammal Sci. 7, 211–229.

Smith, C.A., Giladi, I., Lee, Y.S., 2009. A reanalysis of competing hypotheses for the spread of the California sea otter. Ecology 90, 2503–2512.

Soule, M.E., 1985. What is conservation biology? BioScience 35, 727–734.

Tinker, M.T., Bentall, G., Estes, J.A., 2008a. Food limitation leads to behavioral diversification and dietary specialization in sea otters. Proc. Natl. Acad. Sci. USA 105, 560–565.

Tinker, M.T., Doak, D.F., Estes, J.A., 2008b. Using demography and movement behavior to predict range expansion of the southern sea otter. Ecol. Appl. 18, 1781–1794.

Tinker, M.T., Doak, D.F., Estes, J.A., Hatfield, B.B., Staedler, M.M., Bodkin James, L., 2006a. Incorporating diverse data and realistic complexity into demographic estimation procedures for sea otters. Ecol. Appl. 16, 2293–2312.

Tinker, M.T., Estes, J.A., Ralls, K., Williams, T.M., Jessup, D., Costa, D.P., 2006b. Population Dynamics and Biology of the California Sea Otter (*Enhydra lutris nereis*) at the Southern End of its Range. MMS OCS Study 2006-007. Coastal Research Center, Marine Science Institute, University of California, Santa Barbara, CA. MMS Cooperative Agreement Number 14-35-0001-31063, p. 253.

Tinker, M.T., Guimarães, P.R., Novak, M., Marquitti, F.M.D., Bodkin, J.L., Staedler, M., et al., 2012. Structure and mechanism of diet specialisation: testing models of individual variation in resource use with sea otters. Ecol. Lett. 15, 475–483.

Tinker, M.T., Mangel, M., Estes, J.A., 2009. Learning to be different: acquired skills, social learning, frequency dependence, and environmental variation can cause behaviourally mediated foraging specializations. Evol. Ecol. Res. 11, 841–869.

Turchin, P., 1998. Quantitative Analysis of Movement: Measuring and Modeling Population Redistribution in Animals and Plants. Sinauer Associates, Inc., Sunderland, MA.

Udevitz, M.S., Ballachey, B.E., 1998. Estimating survival rates with age-structure data. J. Wildl. Manage. 62, 779–792.

USFWS, 2003. Final revised recovery plan for the southern sea otter (*Enhydra lutris nereis*). US Fish and Wildlife Service, Portland, OR, p. xi + 165 pp.

USFWS, 2013. Southwest Alaska Distinct Population Segment of the Northern Sea Otter (*Enhydra lutris kenyonii*)—Recovery Plan US Fish and Wildlife Service, Region 7 Alaska, 171 pp.

U.S. Fish and Wildlife Service, 2012. Final Supplemental Environmental Impact Statement on the Translocation of Southern Sea Otters. Ventura Fish and Wildlife Office, Ventura, California, 348 pp. + front matter and appendices.

VanBlaricom, G.R., Jameson, R.J., 1982. Lumber spill in central california waters: implications for oil spills and sea otters. Science 215, 1503–1505.

Watson, J., Estes, J.A., 2011. Stability, resilience, and phase shifts in rocky subtidal communities along the west coast of Vancouver Island, Canada. Ecol. Monogr. 81, 215239.

Watt, J., Siniff, D.B., Estes, J.A., 2000. Inter-decadal patterns of population and dietary change in sea otters at Amchitka Island, Alaska. Oecologia (Berlin) 124, 289–298.

Weitzman, B.P., 2013. Effects of Sea Otter Colonization on Soft-Sediment Intertidal Prey Assemblages in Glacier Bay, Alaska, Ecology and Evolutionary Biology. University of California, Santa Cruz, CA.

Wennergren, U., Ruckelshaus, M., Kareiva, P., 1995. The promise and limitations of spatial models in conservation biology. Oikos 74, 349–356.

Williams, T.M., Kastelein, R.A., Davis, R.W., Thomas, J.A., 1988. The effects of oil contamination and cleaning on sea otters (*Enhydra lutris*): I. Thermoregulatory implications based on pelt studies. Can. J. Zool. 66, 2776–2781.

Wootton, J.T., 2001. Prediction in complex communities: analysis of empirically derived Markov models. Ecology 82, 580–598.

Chapter 11

First Nations Perspectives on Sea Otter Conservation in British Columbia and Alaska: Insights into Coupled Human–Ocean Systems

Anne K. Salomon[1], Kii'iljuus Barb J. Wilson[2], Xanius Elroy White[3], Nick Tanape Sr.[4] and Tom Mexsis Happynook[5]

[1]School of Resource and Environmental Management, Simon Fraser University, Burnaby, BC, Canada, [2]Skidegate, Haida Gwaii, BC, Canada, [3]Bella Bella, BC, Canada, [4]Nanwalek, AK, USA, [5]Uu-a-thluk Council of Ha'wiih, Huu-ay-aht, BC, Canada

Sea Otter Conservation. DOI: http://dx.doi.org/10.1016/B978-0-12-801402-8.00011-1

INTRODUCTION: REGIME SHIFTS AND TRANSFORMATIONS ALONG NORTH AMERICA'S NORTHWEST COAST

> *One of our legends explains that the sea otter was originally a man. While collecting chitons he was trapped by an incoming tide. To save himself, he wished to become an otter. His transformation created all otters.*
>
> Alutiiq Museum and Archaeological Repository (2005)

Human interactions with sea otters and kelp forest ecosystems have spanned millennia (Figure 11.1; Rick et al., 2011). In fact, archeological evidence suggests that the highly productive kelp forests of the Pacific Rim may have sustained the original coastal ocean migration route of maritime people to the Americas near the end of the Pleistocene (Erlandson et al., 2007). Similarly, many coastal First Nations stories speak of ancestors who came from the sea (Boas, 1932; Brown and Brown, 2009; Guujaaw, 2005; Swanton, 1909). Yet this vast and aqueous "kelp highway," providing food, tools, trade goods, and safe anchorage for sophisticated watercraft, would have been highly susceptible to overgrazing by sea urchins had it not been

FIGURE 11.1 Sea otter pictographs from Kachemak Bay, Alaska. The origin and age of these rock paintings are not known with certainty, but ancestors of the Sugpiat are thought to have painted these images as many as 1500–3000 years ago. According to archeologists and local knowledge holders, it is thought that sea otters are among the animals depicted including this image of a sea otter struck by a harpoon. *(Originally reproduced and drawn by de Laguna, 1934; in Klein, 1996.)*

for the existence of an extremely effective urchin predator: the sea otter (*Enhydra lutris*). Highly valued, hunted, controlled, and traded by indigenous people for at least some 12,000 years (Braje and Rick, 2011a; Erlandson and Rick, 2010; Fedje and Mathewes, 2005; Fedje et al., 2001; Szpak et al., 2012), this fur-bearing keystone predator, which may have indirectly facilitated the peopling of North America, later drew Europeans to the northeast Pacific, forever transforming the coast ecologically, socially, and culturally.

The Pacific maritime fur trade of the eighteenth and nineteenth centuries had profound effects on the ecosystems, social systems, and management systems of the northwest coast, leaving a legacy we continue to observe, experience, and grapple with today. Following Bering's 1741 contact with the Sugpiat in south central Alaska, Perez's 1774 exchange with the Haida off Langara Island, British Columbia, and Cook's interactions with the Nuu-chah-nulth of Yuquot (Friendly Cove) on Vancouver Island 4 years later, came a steady stream of European and American trading vessels. The commercial trade in sea otter pelts began in earnest by the 1750s in Alaska and by the 1780s in British Columbia (Cook and Norris, 1998; Gibson, 1988; Gibson, 1992), prompting the introduction of the western economic system and opening the door to colonial settlement and laws. This transformation led to the erosion of First Nations economies and governance structures that had been in place for over 2000 years (Trosper, 2009). These structures included well-established trade networks, spatially explicit marine tenures, and complex traditional resource management protocols (Brown and Brown, 2009; Happynook, 2000; K̲ii'iljuus Wilson and Luu Gaahlandaay Borserio, 2011; Trosper, 2009). By the middle of the nineteenth century, sea otters were extirpated from many regions of the northwest coast as a result of the maritime fur trade (Gibson, 1988), with populations eventually declining in excess of 99% of their pre-contact numbers (Kenyon, 1969; Chapter 3). The commercial trade and subsequent extirpation of sea otters from the northwest coast irrevocably changed coastal indigenous societies and triggered a cascade of indirect ecological effects that propagated throughout coastal food webs (Chapter 2).

> *Below high water mark in some places the large urchins are very thickly strewn over the bottom. (George M. Dawson, 1878. Traveling along the coast-line of Ḱunghit Island, in southern Haida Gwaii, British Columbia. In Cole and Lockner 1993.)*

With the elimination of sea otters from much of their former range came a release in the predation pressure they once exerted. Consequently, their herbivorous macroinvertebrate prey, such as abalone, clams, crabs, and sea urchins, expanded in numbers, size, and depth range (Tegner and Dayton, 2000). Because sea urchins are universally the most significant temperate reef herbivore, capable of controlling the distribution, abundance, and diversity of benthic macroalgae (Dayton, 1985), their marked increase caused many coastal rocky reefs in Alaska and British Columbia to shift from

FIGURE 11.2 On Haida Gwaii, British Columbia, where sea otters have been ecologically absent since the 1850s, (A) red and (B) purple sea urchin barrens dominate subtidal rocky reefs at depths below 3–7 m. Above active feeding fronts of sea urchins, (C) early successional, fringing kelp forests, composed primarily of the short-lived annual bull kelp, exist in shallow waters. In contrast, on the central coast of British Columbia, where sea otters have been recovering for over 25 years, (D) "old growth," structurally complex kelp forests, composed of a high diversity of both short and long-lived kelp species, expand to greater depths in the sustained absence of grazing herbivores, sea urchins in particular. Along this same mainland coast further south, where sea otters have yet to fully recover, kelp forests resemble those surrounding the archipelago of Haida Gwaii. North America's kelp forests are influenced by multiple and interacting drivers of change, both predator-driven from the top-down (Estes and Palmisano, 1974; Estes et al., 1998; Lafferty, 2004; Tegner and Dayton, 2000), nutrient-driven from the bottom-up (Parnell et al., 2010), and physically driven by waves and storm events (Dayton, 1985; Dayton et al., 1984; Dayton and Tegner, 1984). Furthermore, the relative strength of these drivers varies with depth, wave exposure and oceanographic context, which itself changes dramatically with latitude (Graham et al., 2010; Steneck et al., 2002). Although the cascading effects of sea otter extirpation from Alaska, through British Columbia, down to Baja, California, may vary (Carter et al., 2007; Foster, 1990), at northern latitudes in British Columbia, these indirect effects are pronounced. *(Photos by Anne Salomon (A), Lynn Lee (B), Mark Wunsch (C) and Brittany Keeling (D).)*

predominantly kelp forest to invertebrate-dominated (Figure 11.2; Estes and Palmisano, 1974; Watson and Estes, 2011). Switches from one ecosystem state to another likely occurred at different time periods along the northwest coast, varying with local extirpation date and pre-fur trade sea otter densities. Historical records suggest these fur trade-induced ecosystem flips may have occurred as early as the 1870s, if not sooner, along the coast of British Columbia (see quote above in Cole and Lockner, 1993) and possibly several decades earlier in Alaska following the onset of the Russian fur trade (Cook and Norris, 1998). Accompanying these remarkable switches in high-latitude temperate reefs (Figure 11.2) in the mid-1800s came several key transformations to coastal traditional and commercial fisheries.

> *We survived by the ocean and beach. That's what sustained us.*
>
> Walter Meganack Jr., Sugpiaq Elder, 2004 (Salomon et al., 2011)

With the increase in shellfish following the extirpation of sea otters came the expansion of shellfish fisheries. Although invertebrates were always an integral part of harvest and trade among coastal First Nations, made clear by the sheer number and depth of prehistoric shell middens up and down the coast, the ecological extirpation of sea otters by the 1850s would have enhanced shellfish harvest opportunities amongst coastal indigenous people (Sloan, 2003, 2004). Yet the subsequent commercialization of both shellfish and finfish fisheries in the early 1900s eventually restricted the ability of many indigenous people to access these resources.

In British Columbia, for example, as new commercial finfish fisheries emerged and developed along the coast, government agencies became increasingly challenged to manage this fishing effort. By the late 1960s, commercial fishing fleets were restricted through the use of vessel buy-back programs specifically designed to reduce fishing capacity. This was later followed by individual vessel quotas and area licensing policies in the 1990s. These fleet rationalization policies ended up excluding small boat operators, a high proportion of which were indigenous. With the loss of commercial salmon, halibut, and herring licenses once held by indigenous people, these policies indirectly excluded indigenous participation in what had become commercialized finfish fisheries. While salmon abundance declined through the late 1980s and early 1990s, the market demand for shellfish from Asia increased as did its market value. Individual fishing quotas for shellfish and market conditions radically raised the value of commercial shellfish licenses, making them financially inaccessible for many First Nations. Consequently, complex socio-economic forces from the late 1960s onwards posed a risk to First Nations food security, food sovereignty, and economic livelihoods, as did a looming biological force.

> *They came back in the late 1950s, early '60s. The population exploded in the late '70s early '80s.*
>
> John Moonin, Sugpiaq Elder, 2004 (Salomon et al., 2011)

Sea otter populations began to recover along some stretches of the northwest coast throughout the mid- to late twentieth century both naturally, owing to various government protection measures, and via intentional translocation campaigns (Jameson et al., 1982; Chapter 3). For example, recovery began along the south central coast of Alaska, traditional territory of the Sugpiat, as early as the 1950s via natural range expansion (Salomon et al., 2007). Accompanying this recovery came the sequential decline of highly valued subsistence and commercially important shellfish, a phenomenon that has been attributed to the synergistic effects of both predation by sea otters and shellfish harvest by humans (Salomon et al., 2011). Following this came a reported increase in the spatial extent of kelp forests (Salomon et al., 2011). In British Columbia, however, between 1969 and 1972, 89 Aleutian Island sea otters were intentionally translocated, with no First Nation consultation (Osborne, 2007), to Checleset Bay on the northwest coast of Vancouver Island, traditional territory of the Ka:'yu:'k't'h and Che:k'tleset'h' First Nations, two of the 14 Nuu-chah-nulth Nations (Bigg and Macaskie, 1978). Between 1977 and 1995, population growth was initially rapid at 19% per year and has since slowed substantially to 8.4% per year (Nichol et al., 2009; Chapter 13). Along British Columbia's central coast, traditional territory of the Heiltsuk Nation, sea otters were first reported in 1989 and have been growing at a rate of 11.4% per year (Nichol et al., 2009). Recovery of this keystone predator in parts of Alaska and British Columbia's shoreline has caused some high-latitude temperate reef ecosystems to flip back from macroinvertebrate-dominated to kelp-dominated (Figure 11.2; Breen et al., 1982; Estes and Palmisano, 1974; Markel, 2011; Watson and Estes, 2011).

> *The urchins were the first to go, then crab and clams. [Chitons], they're the most recent change, now they're declining.*
>
> Richard Moonin, Port Graham, 2004 (Salomon et al., 2011)

> *In Checleset where sea otters were first transplanted in the 1960s, we have noticed declines in most shellfish. We have not been able to harvest urchins, geoduck, clams, scallops, or abalone for the last 10 years and the children have not had an opportunity to have these foods.*
>
> Peter Hansen, Kyuquot/Checlesaht Nation Treaty Manager, 2003 (Dovetail, 2003)

> *Since sea otters were introduced in Kyuquot Sound, there have been changes in kelp beds, maybe influencing rockfish, baitfish and herring spawn.*
>
> Anthony Oscar, Kyuquot Fisheries, 2003 (Dovetail, 2003)

Because their effects are often rapid, pronounced, direct and indirect, the recovery of sea otters and the trade-offs they induce have elicited complex social-ecological conflicts among coastal communities (Salomon et al., 2011). With the reintroduction of this apex predator, some parts of the ecosystem stand to gain, while others stand to lose. For example, while the return of sea otters has been shown to indirectly increase catch rates of kelp-associated fish in the Aleutian Archipelago (Reisewitz et al., 2006), enhance rockfish recruitment in

British Columbia (Markel, 2011), magnify secondary production in the Aleutians (Duggins et al., 1989), and facilitate atmospheric carbon storage along high-latitude coastlines (Wilmers et al., 2012), these indirect effects typically come at the expense of economically and culturally valuable shellfish (Salomon et al., 2007; Singh et al., 2013). Thus, sea otter recovery can pose a threat to coastal communities due to the immediate cultural and economic loss associated with reduced shellfish harvesting opportunities, as well as perceived and real threats to food security. On the other hand, the recovery of this charismatic species can often elicit tourist dollars, yet profits are rarely distributed to community members who are affected by sea otter recovery. Finally, some of these positive indirect ecological effects described above have been shown to be context dependent, occurring in some places but not others (Singh, 2010). And yet, while the recovery of sea otters has ignited controversy and will continue to do so, it also presents an opportunity for scientists and coastal communities to expand our understanding of this predator's role in kelp forest food webs and First Nations cultures, forcing us to confront our views on the role of humans in ecosystems and our notion of what is natural (Dayton et al., 1998).

> *...now we find that sea otters are once again playing a large role in a shifting Nuu-chah-nulth society, as we see the impacts that their increased presence is having in the nearshore marine environment. Nuu-chah-nulth are challenged to ask themselves difficult questions about the economic, social, and spiritual impacts sea otters are having on their communities.*
>
> Nuu-chah-nulth Tribal Council Fisheries, 2011 (Uu-a-thluk, 2011)

Crossing tipping points are well known to induce turmoil across coupled social-ecological systems (Gunderson and Holling, 2002). Consequently, conservation and management practices are increasingly seeking to maintain social-ecological system resilience, that is, a system's capacity to absorb shocks, learn, re-organize, and adapt (Folke, 2006; Folke et al., 2004). In the case of sea otter recovery, coastal communities are increasingly asking if they can withstand the return of sea otters and simultaneously prevent the decline of other marine resources upon which they depend. How should the limits be decided and who will be the decision makers? At the heart of these questions lie fundamental issues of food security, food sovereignty, and indigenous rights, title, and self-determination. Fortunately, archeological evidence suggests that sea otter-induced tipping points are not new to coastal communities on the northwest coast (Corbett et al., 2008; Erlandson et al., 2005; Simenstad et al., 1978) and ancient marine management strategies likely evolved to cope with some of these trade-offs (Rick and Erlandson, 2009; Trosper, 2009). Consequently, in this case as in many others, establishing well-informed reference points based on archeological and historical records (Braje and Rick, 2011b; Dayton et al., 1998; Tegner and Dayton, 2000) and learning from the vast archive of expertise accumulated in traditional knowledge

(Davis, 2009; Huntington, 2000) may offer innovative solutions to the contemporary and future conservation challenges associated with sea otter recovery.

By integrating and synthesizing evidence from archeological faunal records, historical records, oral histories, contemporary ecological data, and traditional knowledge from British Columbia and Alaska, we ask the following questions. First, to what extent, and how, were sea otters used by coastal indigenous people of the northwest coast prior to European contact? Second, how were sea otter hunting practices managed in the past and how did these ancient management practices fit within the broader governance structures and management protocols of coastal marine resource use during prehistoric times? Finally, we ask, What are First Nations' perspectives on sea otter conservation and management today and can traditional management practices be applied in an effort to balance the needs of people and nature?

SEA OTTER USE IN ANCIENT TIMES

> *People have to understand how valuable the sea otter is to our people. We have great histories. We have been with them for years and years, thousands of years. Big chiefs use sea otters to recognize a great chief amongst our people... The sea otter can bring back all the histories of people before.*
>
> Tsah-seets (Stanley Sam), Ahousaht Elder (Nuu-chah-nulth Tribal Council, 2012)

Although the magnitude of sea otter hunting by indigenous people in Alaska and British Columbia prior to European contact is not known with certainty, zooarcheological data and historical records offer some insights and alternative hypotheses. Traditionally, sea otters were highly valued and traded among First Nations for their uniquely soft and warm fur. Sea otter pelts were crafted into ceremonial robes and adornments, worn by chiefs and other high-ranking people (Figure 11.2; Drucker, 1951; Uu-a-thluk, 2011). Sea otter pelts were also used in day-to-day life as bedding and insulation by chiefs and high-ranking people who had the canoes, technical skills, and customary right to hunt sea otters (K̲ii'iljuus Wilson, 2012). Sea otter furs were an important trade item among coastal First Nations well before contact with European trading vessels and the emergence of the global market for sea otter fur (Murdock, 1934). Historic accounts of the maritime fur trade document that early traders were accessing substantial fur supplies already in use among well-established indigenous trade networks (Bartlett, 1925; Beresford, 1968; Hoskins, 1969). Furthermore, the occurrence of sea otter bones in archeological faunal records from Alaska and British Columbia suggest that although the magnitude of use varied spatially among coastal indigenous people (McKechnie and Wigen, 2011), sea otter hunting was a significant and widespread practice of aboriginal people in this area for at least the past 12,000 years (Corbett et al., 2008; Fedje et al., 2001; Szpak et al., 2012).

Evidence of Prehistoric Sea Otter Population Reduction

Several lines of evidence from ancient sea otter bones excavated from 10 late-Holocene (ca. 5200 years BP to 1900 AD) archeological sites in central and northern British Columbia suggest that sea otter populations were hunted and may have been reduced in numbers below carrying capacity in this area (Szpak et al., 2012). First, the chemical signature of these bones suggests that these sea otter diets may have been primarily composed of benthic invertebrates, with a very low contribution of benthic fish. Evidence from contemporary sea otter populations in Alaska suggests that fish become an important prey item only when otter populations reach very high densities (Estes, 1990; Estes et al., 1981) in response to reduced macroinvertebrate prey and increased kelp habitat which can support greater fish abundances (Bodkin, 1988; Reisewitz et al., 2006). Second, low variability in the chemical signature among these late-Holocene sea otter bones may indicate low dietary diversity, a feature common among contemporary sea otters existing at lower population densities (Tinker et al., 2008). Consequently, a lack of piscivory, in addition to evidence of low dietary variability, implies that these late-Holocene sea otters may have existed at relatively low population densities (Szpak et al., 2012). Sea otter bones represented the most common marine mammal species harvested at several of these sites and were present throughout all examined time periods. These data reveal both the importance and temporal depth of sea otter hunting among First Nations on the central and north coast of British Columbia for millennia prior to the maritime fur trade. Furthermore, sea otter reduction by hunters might have occurred down the entire west coast of North America. For example, Erlandson and colleagues (2005) postulate that the abundance and large size of red abalone shells and sea urchin tests in several early middens on the Californian Channel Islands suggest that sea otter populations were likely limited by Native American Chumash hunters beginning as early as ~8000 years before present (Erlandson et al., 2005).

It has also been hypothesized that throughout the Holocene, 12,000 years ago until early contact, sea otters in coastal Alaska and British Columbia likely occupied a patchy spatial distribution, with reduced numbers of individuals in the vicinity of village sites and greater numbers along coastlines far from centers of human occupation (Corbett et al., 2008; Simenstad et al., 1978; Szpak et al., 2012). In addition to direct mortality due to hunting, sea otters may have been suppressed in numbers near human settlements due to human avoidance behavior and competition with humans for shellfish. Evidence suggests that at least 15,000 people (Acheson, 1998; Boyd, 1990, 1999), and perhaps as many as 30,000 (K̲ii'iljuus Wilson, 2012), once occupied the islands of Haida Gwaii. Furthermore, it has been told by Haida Elders that no land around the islands was unknown; rather, the entire coastline was owned and managed by different family clans (K̲ii'iljuus Wilson and Luu Gaahlandaay Borserio, 2011). Given the high density of village sites and indigenous populations spread throughout the coast, this would have meant kilometers of coastline that may have been otter free. Reduced sea otter abundance surrounding ancient human settlements has also been suggested for California's Northern Channel Islands (Rick et al., 2008). Consequently, the exclusion of sea otters near human habitations may have had localized cascading effects on nearshore subtidal ecosystems as early as the Holocene (Erlandson et al., 2005).

Evidence of Prehistoric Trophic Cascades

Mounting evidence suggests that harvesting by indigenous people during prehistoric times profoundly altered coastal ecosystems long before European contact (Erlandson and Rick, 2010; Jackson et al., 2001; Steneck et al., 2004). Faunal evidence from a prehistoric midden excavated on the Aleutian Island of Amchitka indicates a dramatic shift in harvested resources during 2500 years of occupation (Simenstad et al., 1978). In this archeological time series, the abundance of sea otters is positively related to the abundance of marine fish and seals, and inversely related to the abundance of sea urchins and limpets. This alternating pattern of abundance suggests that Aleuts were technically capable of locally reducing sea otters, thereby increasing the availability of harvestable macroinvertebrates and inducing a shift in nearshore community states, from one dominated by marine mammals (sea otters, harbor seals) and reef-associated fish to another dominated by herbivorous invertebrates (sea urchins, limpets, chitons, and snails). Furthermore, reconstructed size-frequency distributions of sea urchins through time were constant and encompassed large urchins, implying a nearshore species assemblage that excluded sea otters and persisted throughout much of the Aleut occupancy. Additional archeological evidence from other Aleutian Island sites suggest that urchin barrens may have surrounded human settlements while kelp forests may have flourished away from human occupations, creating a mosaic of kelp forests and urchin barrens along the coastline (Corbett et al., 2008). This mosaic hypothesis has also been invoked for British Columbia's northern and central rocky reefs (Szpak et al., 2013) and the Californian Channel Islands (Erlandson and Rick, 2010; Erlandson et al., 2005). Compared with contemporary urchin barrens, evidence from the Californian Channel Islands suggests that ancient urchin barrens were probably more localized and possibly shorter lived, with harvest of shellfish by humans in the intertidal and shallow subtidal (Figure 11.3) replacing the predatory control once conferred by sea otters (Rick and Erlandson 2009; Erlandson and Rick 2010). However, prehistoric sea otter hunting and its cascading effects during the Holocene were likely much less pronounced and more spatially constrained in comparison with the widespread cascading effects triggered by the historic extirpation of sea otters by the maritime fur trade.

FIGURE 11.3 Nuu-chah-nulth Chief Maquinna (muk^{w}ina) and his brother Callicum of the Mowachat First Nation, cloaked in robes made from the furs of sea otters (from Mears, 1790).

Trophic Cascades of the Nineteenth Century

Although ancient sea otter hunting could trigger localized trophic cascades, it was the industrial extirpation of sea otters by the maritime fur trade that had sweeping, long-term, and profound cascading effects on the kelp forest ecosystems of North America's high-latitude rocky reefs (Estes and Palmisano, 1974). Evidence from the chemical signature of rockfish (*Sebastes* spp.) bones from late-Holocene archeological sites in southern Haida Gwaii that span the maritime fur trade (ca. 1500 BP to 1880 AD) reveal that kelp-derived carbon in rockfish diets decreased in post-contact rockfish compared with pre-contact rockfish, likely due to the reduction of kelp forests associated with the extirpation of sea otters (Szpak et al., 2013). This implies an unprecedented shift in the local ecosystem following the maritime fur trade due to the cascading effects of sea otter hunting that did not occur to the same magnitude or spatial extent prior to European contact. The temporal consistency of sea otter bones recovered in British Columbian and Alaskan middens indicates a degree of continuity and capacity for sustained harvest by indigenous people during the Holocene (Corbett et al., 2008; McKechnie and Wigen, 2011; Simenstad et al., 1978; Szpak et al., 2012).

ANCIENT GOVERNANCE AND MANAGEMENT PROTOCOLS OF COASTAL MARINE RESOURCES AND SEA OTTERS

Evidence of Coastal Conservation and Management in Deep Time

[An] example of a stewardship practice was the design of halibut fish hooks, which were made to catch only specific sizes – not the small ones that still needed to grow or the large ones that were needed to reproduce.

(Brown and Brown, 2009)

Emerging evidence from North America's west coast suggests that its First People developed diverse technologies to conserve and manage coastal marine resources, including selective harvesting, seasonal restrictions on use or consumption, and proprietorship that was contingent on sustained productivity (Berkes and Turner, 2006; Trosper, 2009). In many cases, technologies were used to maintain or enhance coastal resource productivity. For example, traditional ecological knowledge from Heiltsuk and Haida knowledge holders suggests that the construction of stone fish traps at the mouths of rivers (Brown and Brown, 2009; Xanius White, 2006) and wooden fish weirs within streams (K̲ii'iljuus Wilson, 2012; K̲ii'iljuus Wilson and Luu G̲aahlandaay Borserio, 2011) were used to selectively harvest all five species and specific sizes of salmon. It is told that knowledgeable fishers inspected the catch inside these traps during high tides and harvested fish of smaller sizes, leaving larger, more robust fish to continue up the rivers to become part of the breeding stock (Hilistis Waterfall, 2009). Ancient clam gardens, human-made intertidal rock-walled terraces recorded from Alaska through British Columbia to Washington State, were designed to increase clam yields (Caldwellet al., 2012; Williams, 2006) and contemporary experimental evidence has confirmed this (Groesbeck, 2013). Herring spawn on hemlock bows and kelp were transplanted by the Haida, Heiltsuk, and Tlingit for population restoration purposes (Boas, 1932; Brown and Brown, 2009). Estuarine root gardens at the mouths of rivers were tended to increase crop yields of northern rice root, springbank clover, and Pacific silverweed (Deur, 2005). Higher up on the shores, root gardens were cultivated by routine clearing and burning to promote the productivity of blue camas lily, one of the most widely traded food resources in the Pacific Northwest whose bulbs were also transplanted outside of their range to make them more widely accessible (Turner and Turner, 2007). Salmon and eulachon were translocated among streams to enhance production (Brown and Brown, 2009), while abalone were transplanted close to villages to increase access. The intentional reduction and exclusion of sea otters by indigenous people of the northwest coast would have greatly increased the productivity of nearshore shellfish (Dayton, 1985; Erlandson et al., 2005). Consequently, sea otters may have been merely one of numerous managed coastal species, considered the property and responsibility of particular families.

Ancient Marine Tenure System

"There was no land lying vacant." (Swanton, 1905) Translated and interpreted as: "Gam tluu tllgaay aa k'iixa Gang ga; There is no land strange."

(K̲ii'iljuus Wilson and Luu G̲aahlandaay Borserio, 2011)

...There was clear ownership of streams and hunting areas, so that only a limited number of people had the right to fish in certain rivers or to hunt in specific areas. A song that my father sang to us was about our river, which made it clear that we belonged there and no one else.

Namgis Matriarch Gloria Cranmer—Webster 'Wika lalisame 'ga'
(Brown and Brown, 2009)

In the old days, Ha-houlthee was very strict in terms of boundaries. You had to formally ask for permission. If there was a shortage of resources in one's territory, you made an arrangement with the chief to secure access elsewhere.

Wickaninnish Cliff Atleo, President Nuu-chah-nulth Tribal Council, 2012
(Uu-a-thluk, 2012)

Indigenous peoples of the northwest coast had a territorial governance system and complex protocols that delineated access rights to the land and sea (Trosper, 2009). Unlike the contemporary notion of "property," titleholders could not sell their territories and were obligated to share its resources. Proprietorship of these territories was organized through a system of "houses" or "clans," which consisted of a number of related families who lived under the direction of a head titleholder. Each house (or clan) has a territory containing seasonal fishing and hunting sites, gardens, and berrybush picking and tending sites—in sum, areas of land and sea that produced food, trade goods, medicines, and other important resources. Among the Nisg̲a'a and Gitxsan, the name for these houses is "wilp." Oral histories indicate that marine tenures were owned and managed by chiefs on Haida Gwaii with specific rules of ownership and responsibilities (K̲ii'iljuus Wilson, 2012). Similarly, on the west coast of Vancouver Island, the 'Ha' houlthee' are chiefly territories of the Nuu-chah-nulth First Nations, owned and managed by hereditary Chiefs (Ha'wiih). Every place in the entire territory of a society belonged to the chief and his clan/house (K̲ii'iljuus Wilson, 2012; K̲ii'iljuus Wilson and Luu G̲aahlandaay Borserio, 2011; Trosper, 2009) and clan-based seasonal camps were distributed throughout the territory for specific species, such as abalone, salmon, halibut, and seaweed (K̲ii'iljuus Wilson, 2012).

Contingent Proprietorship, Public Accountability, and Reciprocity

Territorial access rights, in combination with rules about the behavior of chiefs, created a system of governance over common pool fisheries resources that conferred resilience to societies on the northwest coast for over 2000 years (Trosper, 2009). First, proprietorship over territories was contingent on proper management, specifically defined in terms of maintaining the productivity of a territory's resources for future generations (Trosper, 2009). Second, the chief holding the territory needed to publicly demonstrate the continued productivity of resources and was accountable to the house, otherwise they could be overthrown. This public accountability created an important feedback loop that provided a strong incentive for learning how to manage for continued use. Ethnographic and historic evidence suggests that both raiding among houses and slavery occurred on the northwest coast of North America (Donald, 1997), the latter potentially providing a means for commoners to discipline titleholders by cooperating in allowing their leaders to be captured (Trosper, 2009).

> *.... It is necessary to give in order to receive.... Consequently generosity can be viewed as a natural law of reciprocity. The ancient Nuu-chah-nulth felt so strongly about the importance of the relationship between generosity and the quality of life that the opposite of generosity was equated to death.*
>
> Nuu-chah-nulth Chief UmeeK, 2004 (Trosper, 2009)

> *Each tribe owned specific resource sites, which could be shared if, for example, salmon were abundant in one area, but not in another. As inter-marriage meant extended kinship groups among several villages, those from the village experiencing a poor fishing year, would ask for permission to share in the bounty of another, with which they had some family connection.*
>
> Gloria Cranmer-Webster 'Wika lalisame 'ga', Namgis Matriarch (Brown and Brown, 2009)

> *When people from other villages ran out of smoked and dried salmon, they would come to Mauwash and ask my great-great-great grandfather to fish there.... When we had everything we needed – seaweed, fish, eggs etc. – we would send some to our relatives at Bella Coola, Rivers Inlet, Kitimaat, and the Nass River. When it was their time, when the eulachons and herrings started running, they would send up smoked fish and grease. This trading back and forth was the way of our people, and the First Generation. This is the reason we never knew hunger.*
>
> Angus Campbell, Heiltsuk, 1968 (Brown and Brown, 2009)

Finally, the requirement of reciprocity, clearly held by the Nuu-chah-nulth, Namgis, Haida, Heiltsuk, and most indigenous people of the northwest coast, provided two additional incentives for sustainable management. First, it provides social insurance against misfortune. Neighbors would be asked for support when resources were low (i.e., when a salmon run failed, or when herring spawned in another's territory but not one's own), knowing that such support would be returned later if and when needed. Second, the sharing and exchange of the net returns of a territory reduces competition and any incentive to overharvest amongst resource users, thereby providing a solution to the tragedy of the commons. The enforcement of reciprocity was made legal via the potlatch system, a public governance system among coastal First Nations whose ubiquity suggests that its advantages became widely recognized. In sum, ancient marine tenure systems and governance protocols founded in reciprocity were used to conserve and manage most marine species, including sea otters.

Ancient Sea Otter Hunting Practices and Evidence for Spatial Management

> *In our history, certain people had rights to certain areas of Hesquiaht Harbour. They had licences to take sea otters for harvesting. Sea otters were hunted at certain times of the year, not every month. Certain people had rights to the pelts. The pelts were distributed to other First Nations for their regalia; these were people with hierarchy.*
>
> Hesquiaht Chief Dominic Andrews 'Matlaha,' 2003 (Dovetail, 2003)

According to oral historical accounts, sea otter hunting was a respected skill and an honored tradition among coastal aboriginal communities and only certain people had the privilege to hunt. Among the Nuu-chah-nulth, sea otter hunters were either chiefs who had access rights to the water where the hunt was to occur or a noted sea otter hunter delegated by the territory's chief (Drucker, 1951). Hunting took place with a bow made from yew wood, arrows whose shafts were carved of cedar, a spear or harpoon, a club, and specially crafted canoes whose polished hulls allowed hunters to approach soundlessly within range (Drucker, 1951). Hunters and their steersman typically set out before daybreak to kelp beds where an otter might be found asleep (Drucker, 1951). Sea otter hunting required stealth and precision, skills that were refined with much training. Adult Haida and Sugpait hunters often engaged in target practice and children played games to hone their hunting skills for precision and accuracy from an early age (K̲ii'iljuus Wilson, 2012; Tanape, 2012).

Like most coastal resources, it is possible that hereditary leaders who owned territories also managed sea otters spatially. It has been hypothesized that male sea otters may have been targeted to control the distribution of sea otter populations in space (K̲ii'iljuus Wilson, 2012). This proposed traditional management practice would have made use of an important natural history characteristic of sea otters. Rafts of sea otters are typically segregated by sex (Riedman and Estes, 1990). While both male and female rafts persist within existing core habitats, male rafts tend to occupy the periphery of the population range. Range expansion typically occurs in growing populations when rafts of males move into previously unoccupied habitat (Garshelis and Garshelis, 1984; Garshelis et al., 1984). Female rafts then follow and establish in newly occupied areas once male rafts have left (Garshelis et al., 1984; Loughlin, 1980; Wendell et al., 1986). Thus, the targeting of roving male sea otters by indigenous hunters would have allowed chiefs to spatially manage sea otter occupation in their territories, reducing or excluding sea otters from some areas while maintaining population numbers through the persistence of male and female rafts elsewhere within their territories. Sea otter fur, and the pelts of other mammals such as ermine, were highly valued and used for chiefly regalia, bedding, and insulation and to demonstrate societal status and good standing among First Nations (K̲ii'iljuus Wilson, 2012). Consequently, we surmise that the incentive to maintain a viable population of sea otters in part of a chief's territory existed.

Traditional Principles of Stewardship and Sustainability

> *In the old stories, the Haidas, as with other coastal people, were taught to use all parts of whatever was hunted or gathered. If one did not treat the ocean, sky and land with respect, they would leave us without their presence in our world. Whether it is on the land or oceans, never taking more than you needed, sharing and using cyclical harvesting methods were major aspects of cultivation and conservation. For example, we cultivated and managed crab-apple orchards, berry patches, clam gardens, octopus houses, and fish traps, whether out of rock or fibre.*
>
> (K̲ii'iljuus Wilson, 2012)

Traditional principles of stewardship and sustainability were common amongst coastal First Nations. For example, the principles of "Gvi'ilas," laws of the ancestors, guided the marine harvest activities of British Columbia's Heiltsuk First Nation over the past 10,000 years (Brown and Brown, 2009). This was a stewardship model based on a social responsibility to the Nation's members. Coastal communities up and down the northwest coast were guided by governance protocols designating access rights to resources that were contingent on intergenerational accountability and reciprocity (Trosper, 2009). On the west coast of Vancouver Island, the Nuu-chah-nulth principle of "Hishukish tsa'walk" (everything is one, everything is interconnected) underscores that the Nuu-chah-nulth, like most coastal indigenous people, view themselves and all humans as an integrally linked component of the ecosystem (Happynook, 2000). Thus, this stewardship principle underscores the notion that sustaining species, and the ecosystems in which they are embedded, is a requirement for sustaining humanity and human well-being. Consequently, setting sea otter recovery goals based on the notion of an ideal state of nature without humans fits poorly with those whose cultural foundation is inseparable from the landscapes and seascapes within their traditional territories (Sloan and Dick, 2012).

BALANCING THE NEEDS OF PEOPLE AND NATURE: FIRST NATIONS PERSPECTIVES

Perspectives on sea otter conservation and management vary among First Nations, as they do among non-native people and even government agencies. For example, in 1987, on Haida Gwaii, where sea otters have yet to recover since their extirpation in the 1850s, the Council of Haida Nation passed a resolution supporting the reintroduction of Alaskan sea otters to the islands specifically for ecosystem restoration (Sloan and Dick, 2012). That same year, the Haida Nation and the BC Ministry of the Environment jointly applied to the provincial BC Wildlife Branch for a sea otter translocation permit. The provincial government formally proposed the notion to the federal Department of Fisheries and Oceans (Sloan and Dick, 2012). At a public meeting held in 1988 on Haida Gwaii, both support and opposition was expressed by First Nations and non-native island residents. That year, the Department of Fisheries and Oceans wrote a memo to the federal parks service, Parks Canada, which had tabled the idea of active sea otter restoration. The letter stated the department's lack of support and recommended natural reintroduction over active restoration (Sloan and Dick, 2012). Of course, First Nations perspectives on sea otter conservation and management also vary through time as sea otter population status changes and as new legal frameworks are modified to reflect this.

In Canada, as of 2009, sea otters were down-listed from "Threatened" status to "Special Concern," removing previous prohibitions and allowing some level of hunting by First Nations with an Aboriginal Communal Fishing License issued under the Fisheries Act for food, social, and ceremonial purposes. Sea otter range expansion, reintroduction, and active management is discussed today on Haida Gwaii among some Haida leaders and marine planning groups as a way to restore both kelp forest ecosystems and the relationship that the Haida had with this species (K̲ii'iljuus Wilson, 2012; Sloan and Dick, 2012). Conservation and active management is being addressed directly by Nuu-chah-nulth Nations, who have been coping with an expanding sea otter population since their intentional reintroduction in the early 1970s. Although sea otter recovery is a controversial topic within Nuu-chah-nulth communities (Nuu-chah-nulth Tribal Council, 2012), it has also created the opportunity to reinvigorate traditional laws and customs and engage in collaborative research.

> *The return of sea otters have opened the doors to discussions throughout Nuu-chah-nulth Ha'houlthee (chiefly territories) about the role of Nuu-chah-nulth Ha'wiih (traditional chiefs) to maintain ecosystem balance through the principles of Hishukish tsa'walk (everything is one) and Iisaak (respect with caring). In the spirit of these important principles, Nuu-chah-nulth have supported and initiated sea otter recovery efforts. Together Nuu-chah-nulth*

now turn to management of sea otter populations for ceremonial use opportunities for Nuu-chah-nulth First Nation communities, to re-establish the sacred relationship that once existed for the benefit of all Nuu-chah-nulth people.

(Nuu-chah-nulth Tribal Council, 2012)

In 2012, the Nuu-chah-nulth Tribal Council Fisheries Department, Uu-a-thluk, drafted a comprehensive management plan for $K^w ak^w atl$ (sea otters) in the Nuu-chah-nulth Ha' houlthee, with the overarching goal of "maintain[ing] healthy sea resources, including a healthy and sustainable sea otter population, while providing ceremonial use[of] sea otter[s] ... for Nuu-chah-nulth First Nation communities, in Nuu-chah-nulth Territories" (Nuu-chah-nulth Tribal Council, 2012). Specific objectives of this plan include the following:

- Ensuring that the management plan does not conflict with the principles of Hishukish Tsa'walk (everything is one) and Iisaak (respect with caring), the principles vested in Uu-a-thluk.
- Fully involving Nuu-chah-nulth First Nations in all aspects of sea otter management and related initiatives including Nuu-chah-nulth direct participation in any recovery and management planning processes.
- Maintaining a viable and sustainable sea otter population in the Nuu-chah-nulth area.
- Ensuring the availability of sea otters to Ha'wiih and their representatives for ceremonial use.

This detailed plan outlines the current status, range, threats, and regulatory protection of sea otters. It then presents the details of a Nuu-chah-nulth harvest including quantitative estimates for an annual allowable harvest rate, based in turn on estimates of human-caused mortality and its uncertainty from all sources that can be sustained by a population while still allowing it to grow or remain at a target level (Wade, 1998). Harvest regulations include spatial boundaries, allowable hunter designations, harvest protocols, bio-sampling, compliance monitoring, and enforcement. The management plan specifically includes annual population monitoring and a biological sampling program, as well as a standard tagging certificate form and additional measures for a communal harvest plan. Last, this plan commits to working collaboratively with the 14 Nuu-chah-nulth First Nations, the federal Department of Fisheries and Oceans (DFO), and other relevant agencies on meeting these objectives as well as research and education programs. This plan is not designed to simultaneously manage shellfish resources and traditional shellfish harvest despite their localized declines.

Sea otters are part of the problem. They eat everything we eat. But bidarkis [chitons can adjust to nature. It's us they can't adjust to.

Walter Meganack Jr., Port Graham, Alaska 2004 (Salomon et al., 2011)

I wouldn't blame the sea otters, it's us. Our exhaust, gas, and oil. We are the ones damaging all that. The problem now is human impact, it's a heavy impact.

Nick Tanape Sr., Elder, Nanwalek, 2004 (Salomon et al., 2011)

In south central Alaska, on the tip of the Kenai Peninsula, the Sugpiat of Port Graham and Nanwalek, who have lived with sea otter recovery and range expansion since the 1950s, recognize that sea otters are but one of many factors driving the decline in shellfish resources. In these villages, it is widely acknowledged that both local subsistence and regional commercial harvest by humans, and changing ocean conditions have a role (Salomon et al., 2011). Although Alaskan Natives are legally exempt from the marine mammal protection act and are legally sanctioned to take sea otters, no intentional culls have occurred over the past 20 years in these two villages. The Sugpiat have, however, been involved in boat-based sea otter counts in parts of their traditional territory since 1999 as active members of the Alaskan Sea Otter and Sea Lion Commission (Tanape, 2012). This tribal consortium was established in 1988 to promote Alaska Native involvement in policy decisions pertaining to sea otters and, 10 years later, Steller sea lions. Working directly with coastal Alaska Natives, the goal of this organization is to further the conservation and local management of marine mammals, as well as local research. However, like many programs, these desired outcomes are perpetually constrained by sporadic and limited funding.

It's time to call the Russians back again!

Comment at Port Graham Elder's meeting, 2004 (Salomon et al., 2011)

And yet, because people and sea otters compete for similar foods, sea otter recovery is often a source of frustration among First Nations despite the clear interest in rekindling traditional sea otter and human relationships. Management and conservation decisions pertaining to sea otters, or any resource or ecosystem for that matter, are ethical decisions, informed by scientific information but driven by citizens and their values. Consequently, people will need to draw upon a wide range of knowledge (ecological, archeological, economic, cultural, experiential) and their own ethical beliefs and worldviews when weighing the costs and benefits of co-existing with sea otters (Sloan and Dick, 2012).

Reconciling Worldviews

How can sea otters be protected from trappers and aquariums, but at the same time we do not protect the shellfish? What is the balance?... What's the cost of recovering sea otters and where's the balance to it? Man is part of the ecosystem too. The aboriginals of Hesquiaht are part of the ecosystem and also have rights.

Paul Lucas, Hesquiaht Fisheries Technician, 2003 (Dovetail, 2003)

Differences in the worldviews typically held by "Western" and indigenous people mean that native and non-native people and government agencies do not always share a common perspective on sea otter recovery (Osborne, 2007). One of the most fundamental distinctions separating these worldviews is the perspective on the role humans play in ecosystems. Indigenous societies tend to view themselves as a component of the ecosystem. This runs counter to the view of people as external disrupters of an otherwise pristine ecosystem, a view once held by many scientists and government managers. Today, non-native societies and government policies tend largely to ignore or undervalue the cultural, economic, and ecological relationship indigenous people had with sea otters and wish to revitalize. Thus, excluding or minimizing the cultural and ecological roles that indigenous people play in sea otter ecology and recovery runs counter to an indigenous worldview. As we describe above, archeological evidence and oral histories tell us that indigenous people likely played a significant role in driving the spatial distribution and population dynamics of sea otters in Alaska, British Columbia, and California, at least at small spatial scales. Sea otters were an integral part of indigenous culture and economies for millennia, well before contact. Recognizing and acknowledging the different perspectives society has on the role humans play in ecosystems and the notion of "what is natural" is an important step towards reaching a mutual understanding of different recovery objectives for sea otters held by western and indigenous communities. In both Canada and the United States, First Nations have constitutional rights and have hunted sea otters for millennia. Consequently, restoring sea otter populations should justly be followed by restoring the relationships between First Nations and sea otters.

Who Decides How Much?

> *Many of the Chiefs want to exercise their rights and authorities in areas in regard to aquatic resources and sea otters.*
>
> Ron Frank, Nuu-chah-nulth Tribal Council, 2003 (Dovetail, 2003)

The imbalance of power between native and non-native governments along the northwest coast has a deep history and legal issues of rights and title remain unresolved. In Canada, the government's outlawing of the potlatch in 1885 and removal of the authority of chiefs to manage the fisheries within their territories struck down a system of fisheries management that had persisted for more than 2000 years (Trosper, 2009). Less than 100 years later, the Nuu-chah-nulth people were not included in the Canadian government's decision to reintroduce sea otters to Checleset Bay from 1969 to 1972 (Osborne, 2007). In spite of this, recent efforts seeking reconciliation have started to take shape. After a long history of social injustice and inequity in resource access rights to coastal First Nations, reconciliation protocol agreements were signed in 2009 between provincial and coastal First Nation governments in an unprecedented move to support First Nations' rights to co-manage coastal resources (British Columbia, 2009a,b).

In the ocean, this is manifested in a collaborative, ecosystem-based, marine use planning process that is currently under way along British Columbia's northern and central coasts. This represents a remarkable opportunity to transform coastal management and implement aboriginal constitutional rights; however, how this will affect sea otter recovery targets and management plans has yet to be determined.

Restoring to What?

People's notions and perceptions of "what is natural" often suffer from the "sliding baselines" syndrome, in which a lack of information about the past can lead to misinterpretations of a species' or an ecosystem's overall status (Dayton et al., 1998; Pauly, 1995). This syndrome plagues scientists, policy makers, and the public alike. Above, we provide evidence that humans have been exploiting, modifying, and managing coastal species and ecosystems, including sea otters, for millennia. Given this, what sea otter population size is "natural"? What should our recovery targets be?

In British Columbia, the current status of sea otters, which are currently estimated to occupy 25–33% of their "historic" range, is based on a habitat-suitability model and records of pelts purchased during the maritime fur trade (Nichol, 2007). Given the archeological and oral history evidence on the extent of prehistoric sea otter population reductions, and the stockpiling of and trade in sea otter pelts throughout the Holocene, a high degree of uncertainty surrounding "pre-contact" baseline estimates of sea otters remains. However, recovery targets and baseline estimates for most species at risk appear to be based on population estimates in the absence of humans or without the explicit recognition or knowledge of the history and prehistory of human occupation of the northwest coast.

Deep into human prehistory, human–environment interactions in coastal ecosystems have spanned a continuum, from degradation to active management and enhancement (Rick and Erlandson, 2009). This blurs the separation between the natural and anthropogenic worlds. Increasing evidence on the antiquity of human alteration of marine ecosystems requires us to reassess our baselines for the long-term management, restoration, and sustainability of coastal marine species and ecosystems in which humans are included. Consequently, scientists, managers, and policy makers involved in the recovery of all species and ecosystems at risk must confront their own assumptions and worldviews in order to help define what "recovery" means to all members of society.

Can Traditional Governance and Management Practices Be Applied Today?

The concept of designating access privileges to participants in a fishery was not unique to the indigenous people of North America's northwest coast, nor is it new to contemporary fisheries management. In fact, there is a long

FIGURE 11.4 Coastal First Nations used spears and staffs to collect intertidal and shallow subtidal fish and shellfish. (A) A Nuu-chah-nulth fisherman with spear in Clayoquot Sound, BC. The notch in the top of the canoe prow was often used to rest the shaft of a spear or harpoon (Curtis, 1916). (B) A Kwakwa̲ka̲'wakw First Nation gathering northern abalone (*Halitois kamtschatkana*) from the rocky intertidal with a staff in hand (Curtis, 1915). Northwestern University Library, *Edward S. Curtis's 'The North American Indian': the Photographic Images*, 2001. (see http://memory.loc.gov/ammem/award98/ienhtml/curthome.html)

history of territorial use rights for fishing (TURFs) across the Pacific Islands of Polynesia and Micronesia. Today, rights-based marine tenures are being used along the coasts of Chile, Baja, California, and Kenya as part of contemporary ecosystem-based marine management plans (Gelcich et al., 2010). Multiple examples of traditional, community-based marine resource management techniques continue to grow across the Pacific Islands, from limited entry zones, closed areas, and seasonal closures, to restrictions on damaging or overly efficient fishing methods (Johannes, 2002; Figure 11.4). Recent evidence suggests that the implementation of rights-based catch shares, specifically individual-transferable fishing quotas, can slow and even stop the global trend towards commercial fisheries collapses (Costello et al., 2008). Contemporary transformational changes in the governance of marine resources, including the establishment of community quotas and spatial allocation of user rights and responsibilities to community collectives, have been shown to

FIGURE 11.5 While sea otters have been functionally absent from the shores of Haida Gwaii since the 1830s, the occasional sighting of a lone individual, like this one floating behind a pack of Stellar sea lions off Garcon Rocks, reminds us that sea otter range expansion is just a matter of time. Unassisted recovery of sea otters to Haida Gwaii is likely within this century (Sloan and Dick, 2012). *(Photo by Nadine Schoderer, June 17, 2010.)*

prevent fishery-induced population collapses and maintain the resilience of coupled human–ocean ecosystems (Gelcich et al., 2010). If this is the case for a variety of fisheries worldwide, how can we transition back to traditional harvest methods and principles like Gvi'ilas and Hishukish tsa'walk along North America's northwest coast and what might this look like when applied to sea otter conservation and management?

The Future: Preparing for and Adapting to Change

> *Nature changes. Man changes. Is it natural? I feel that changes are more pronounced now. Change is happening at a faster pace now than before.*
>
> Walter Meganack Jr., Sugpiaq Elder, 2004 (Salomon et al., 2011)

> *Prior to all the sea-otter being extirpated, kelp was never an issue. Now with warmer water, an overabundance of sea-urchins, preferable kelp isn't always available.*
>
> K̲ii'iljus Barb Wilson, Haida Matriarch, 2009 (Brown and Brown, 2009)

Change was and will continue to be inevitable. Importantly, sea otter recovery and range expansion (Figure 11.5) is one change that is occurring within the context of other ecological and social changes: the decline of commercial fisheries for species such as herring, eulachon, and salmon, the rise of introduced species, global climate change, and increased public and legal acknowledgment of indigenous rights. Because the recovery of sea otters is occurring amidst emerging reconciliation agreements between native and non-native governments in Canada and the United States, a window of opportunity exists for policy innovation and change in the governance of marine resources. We suggest that coastal communities can begin to prepare for these transformations by revisiting their old management systems and ways of thinking, while learning about and participating in the latest science examining the ecological, economic, and socio-cultural ripple effects triggered by sea otter range expansion. We cannot manage out of ignorance. Only when we know what a species does, what it eats, and what role it has within coupled human–ocean systems, can we begin to make some intelligent decisions. Dialogues among resource users, scientists, policy makers, and the public will help identify and introduce old and new pathways forward to prepare for these transformations.

NAVIGATING TOWARDS ECOLOGICAL AND SOCIAL RESILIENCE ON THE NORTHWEST COAST

> *[I]ndigenous science has developed over millennia providing principles which reflect an acute awareness of the necessity of including social, cultural, spiritual and economic considerations within our understanding of the ecological world.*
>
> Tom Mexsis Happynook Huu-ay-aht First Nations Ha'wiih and Chairman of the Nuu-chah-nulth Council of Ha'wiih 2013 (Happynook, 2000)

As the magnitude of our impacts on marine ecosystems intensifies and becomes more apparent, there is an increasing appreciation of the strong links between the social and ecological processes that support human well-being. This has initiated a shift in approaches to marine conservation and management, from one that is single species based and focused on optimizing yields to one that focuses on maintaining social and ecological resilience by recognizing the reciprocal relationships between interlinked systems of people and nature (Folke, 2006; Folke et al., 2004). These complex systems are often characterized by variability, cross-scale dynamics, and thresholds. This new approach aligns well with traditional ingenious worldviews and principles of stewardship and sustainability, as well as the sea otter-induced tipping points observed on the high-latitude temperate reefs of the northwest coast (Figure 11.2). The challenge is in finding compromise between appropriate levels of human use while sustaining restored coastal ecosystems which include sea otters (Sloan and Dick, 2012).

> *Our ancestors lived side by side with all the creatures on land, sea and air via complex and even simple guidelines and rules. Today, collaboration is a preferred method of managing our resources.*
>
> Xanius Elroy White, Heiltsuk Cultural Historian and Archaeologist, 2013

When it comes to designing coastal conservation and management policies, sea otters being one element, each coupled human–ocean system will have its unique ecological, cultural, and socio-economic features to address. However, several broad principles can be applied. First, engaging and collaborating with coastal indigenous communities to identify relevant research questions and to participate in the design, implementation, monitoring, and evaluation of alternative policy options will vastly improve their likelihood of success. Reconstructing prehistoric and historic kelp forest baselines and documenting the evolution of socio-cultural values associated with sea otter recovery will help diagnose and treat the symptoms of sliding baselines. This will allow researchers and communities to co-establish appropriate and regionally specific reference points and recovery targets based on both empirical data and human values. Invoking ethics and justice in marine conservation and management parallels indigenous beliefs of respect and responsibility. If we are to restore ecosystems to an earlier state of biodiversity, productivity, and ecosystem completeness with humans, then the ethical beliefs of First Nations are integral to this restoration (Happynook, 2000; Sloan and Dick, 2012).

To design scientifically sound conservation and management policies that are tailored to the ecological, social, and cultural nuances of each region, we recommend synthesizing data on the regional variation in the ecological and social effects triggered by sea otter recovery. Furthermore, integrating western science and local knowledge will improve our ability to determine the ecological and socio-economic drivers of coastal ecosystem change (Salomon et al., 2007) and will lend legitimacy to both parties' data and their inferences based on them. For example, predictive ecosystems models of kelp forest food web interactions (Salomon et al., 2002), based on western and traditional knowledge, will allow scientists and communities to make predictions and evaluate the trade-offs associated with alternative management policies. Integrating design features of western and traditional ecosystem-based management could be used to develop alternative experimental management strategies that address direct and indirect effects of sea otter predation on benthic fisheries. Moreover, creating responsive governance structures that support flexible and adaptive management approaches would allow these policies to be trialed as experiments through pre-existing marine planning process. These policies could then be monitored, evaluated, modified, and reassessed. Finally, equitable governance means sharing power through joint decision making and co-management. This means democratizing conservation science and management. Finally, because no coupled human–ocean system can ever be fully understood, all marine conservation and management decisions should be approached with humility (Sloan and Dick, 2012). Abiding by these principles will help coastal communities transition back to a path of sustainability and towards rebuilding resilient ecosystems and communities.

ACKNOWLEDGMENTS

With deep appreciation, we acknowledge the indigenous people whose voices contributed to and illuminated this challenging topic. Valuable guidance, knowledge, and inspiration were shared by several coastal First Nations organizations and community members, particularly the Heiltsulk Integrated Resource Management Department, Nuu-chah-nulth Council of Ha'wiih, Coastal Guardian Watchmen, Gwaii Haanas, Port Graham and Nanwalek Village Councils, Mike Reid, Ross Wilson, Julie Carpenter, Jennifer Carpenter, Robert Russ, and Stu Humchitt. Don Hall and Roger Dunlop kindly shared the 2012 draft "Management plan for $k^{w}ak^{w}atl$ (sea otter) in the Nuu-chah-nulth Ha' houlthee, West Coast Vancouver Island." Norm Sloan, Shawn Larson, Glenn VanBlaricom, Jim Bodkin, and Lynn Lee carefully reviewed this manuscript and provided important feedback that improved this contribution. We would like to thank Leah Honka, Josh Silberg, and Christine Gruman for their assistance in locating and compiling key references. Thanks to Britt Keeling, Lynn Lee, and Mark Wunsch for sharing their photographs. This research was supported by funding from the Natural Sciences and Engineering Research Council of Canada and Pew Fellows Program in Marine Conservation to AKS.

REFERENCES

Acheson, S., 1998. In the wake of the Ya'aats'xaatgaay [Iron People]: a study of changing settlement strategies among the Kunghit Haida. British Archaeological Reports, International Series, Oxford, UK.

Alutiiq Museum and Archaeological Repository, 2005. Land otter—Aaquyaq/Sea Otter—Arhnaq. In: Repository, A.M.A.A. (Ed.), Kodiak, AK.

Bartlett, J., 1925. A narrative of events in the life of John Bartlett of Boston, Massachusetts, in the years 1790–93, during voyages to Canton and the Northwest Coast of North America. In: Snow, E. (Ed.), The Sea, the Ship and the Sailor. Marine Research Society, Salem, MA.

Beresford, W., 1968. A Voyage Round the World: But More Particularly to the Northwest Coast of America Performed in 1785, 1786, 1787, and 1788, in the King George and Queen Charlotte, Captains Portlock and Dixon. Da Capo Press, New York, NY.

Berkes, F., Turner, N.J., 2006. Knowledge, learning and the evolution of conservation practice for social-ecological system resilience. Hum. Ecol. 34, 479–494.

Bigg, M.A., Macaskie, I.B., 1978. Sea otters re-established in British Columbia. J. Mammal. 59, 874–876.

Boas, F., 1932. Bella Bella Tales. G.E. Stechert & Co., New York, NY.

Bodkin, J.L., 1988. Effects of kelp forest removal on associated fish assemblages in central California. J. Exp. Mar. Biol. Ecol. 117, 227–238.

Boyd, R., 1990. Demographic history, 1774–1874. In: Suttles, W. (Ed.), Handbook of North American Indians: Northwest Coast, vol. 7. Smithsonian Institution, Washington, DC, pp. 135–148.

Boyd, R., 1999. The Coming of the Spirit of Pestilence—Introduced Infectious Diseases and Population Decline Among Northwest Indians, 1774–1874. UBC Press, Vancouver, BC.

Braje, T.J., Rick, T.C., 2011a. Human Impacts on Seals, Sea lions, and Sea Otters: Integrating Archaeology and Ecology in the Northeast Pacific. University of California Press, Berkeley, pp. 297–308.

Braje, T.J., Rick, T.C., 2011b. Perspectives from the Past Archaeology, historical ecology, and northeastern Pacific pinnipeds and sea otters. In: Braje, T.J., Rick, T.C. (Eds.), Human

Impacts on Seals, Sea Lions, and Sea Otters: Integrating Archaeology and Ecology in the Northeast Pacific. University of California Press, Berkeley, pp. 297–308.

Breen, P.A., Carson, T.A., Foster, J.B., Stewart, E.A., 1982. Changes in subtidal community structure associated with British Columbia sea otter transplants. Mar. Ecol. Prog. Ser. 7, 13–20.

British Columbia, 2009a. Coastal First Nation/British Columbia Reconciliation Protocol, in: Her Majesty the Queen in right of the province of British Columbia, W.N., Metlakatla First Nation, Kitasoo Indian Band, Heiltsuk Nation, Haisla Nation, Gitga'at First Nation (Ed.), p. 28.

British Columbia, 2009b. Kunst'aa guu—Kunst'aayah Reconciliation Protocol. In: Her Majesty the Queen in right of the province of British Columbia, H.N. (Ed.), p. 18

Brown, F., Brown, Y.K., 2009. Staying the course, staying alive—Coastal First Nations fundamental truths: biodiversity, stewardship and sustainability. Biodiversity BC, Victoria, BC, p. 82.

Caldwell, M., Lepofsky, D., Combes, G., Harper, J., Welch, J., Washington, M., 2012. A bird's eye view of Northern Coast Salish Intertidal resource management features. J. Island Coastal Archaeol. 7, 219–233.

Carter, S.K., VanBlaricom, G.R., Allen, B.L., 2007. Testing the generality of the trophic cascade paradigm for sea otters: a case study with kelp forests in northern Washington, USA. Hydrobiologia 579, 233–249.

Cole, D., Lockner, B., 1993. To the Charlottes; George Dawson's 1878 Survey of the Queen Charlotte Islands. UBC Press, Vancouver, BC.

Cook, L., Norris, F., 1998. A Stern and Rock-Bound Coast; Kenai Fjords National Park Historic Resource Study. National Parks Service, Anchorage, AK.

Corbett, D.G., Causey, D., Clementz, M., Koch, P.L., Doroff, A., Lefevre, C., et al., 2008. Aleut hunters, sea otters, and sea cows: Three thousand years of interactions in the western Aleutian islands, Alaska. In: Rick, T.C., Erlandson, J.M. (Eds.), Human Impacts on Ancient Marine Ecosystems: A Global Perspective. University of Utah Press, Salt Lake City, pp. 43–75.

Costello, C., Gaines, S.D., Lynham, J., 2008. Can catch shares prevent fisheries collapse? Science 321, 1678–1681.

Davis, W., 2009. The Wayfinders. House of Anansi Press, Toronto, ON.

Dayton, P.K., 1985. Ecology of kelp communities. Annu. Rev. Ecol. Syst. 16, 215–245.

Dayton, P.K., Tegner, M.J., 1984. Catastrophic storms, El Nino, and patch stability in a southern California kelp community. Science 224, 283–285.

Dayton, P.K., Currie, V., Gerrodette, T., Keller, B.D., Rosenthal, R., Ventresca, D., 1984. Patch dynamics and stability of some Californian kelp communities. Ecol. Monogr. 54, 253–289.

Dayton, P.K., Tegner, M.J., Edwards, P.B., Riser, K.L., 1998. Sliding baselines, ghosts, and reduced expectations in kelp forest communities. Ecol. Appl. 8, 309–322.

de Laguna, F., 1934. The Archaeology of Cook Inlet, Alaska. University of Pennsylvania Museum, Philadelphia, PA.

Deur, D., 2005. Tending the garden, making the soil: Northwest coast estuarine gardens as engineered enrivonments. In: Deur, D., Turner, N. (Eds.), "Keeping it Living": Traditions of Plant Use and Cultivation on the Northwest Coast of North America. University of Washington Press, UBC Press, Seattle, Vancouver, BC.

Donald, L., 1997. Aboriginal Slavery on the Northwest Coast of North America. University of California Press, Berkeley and Los Angeles, CA.

Dovetail, C.I., 2003. Proceedings of the public workshop on the draft sea otter recovery strategy. Prepared for Fisheries and Oceans Canada, Port Alberni, BC.

Drucker, P., 1951. The Northern and Central Nootkan Tribes. Smithsonian Institution. Bureau of American Ethnology, Washington, DC.

Duggins, D.O., Simenstad, C.A., Estes, J.A., 1989. Magnification of secondary production by kelp detritus in coastal marine ecosystems. Science 245, 170–173.

Erlandson, J.M., Rick, T.C., 2010. Archaeology meets marine ecology: the antiquity of maritime cultures and human impacts on marine fisheries and ecosystems. Annu. Rev. Mar. Sci. 2, 231–251.

Erlandson, J.M., Rick, T.C., Graham, M., Estes, J., Braje, T., Vellanoweth, R., 2005. Sea otters, shellfish, and humans: 10,000 years of ecological interaction on San Miguel Island, California. In: Garcelon, D.K., Schwemm, C.A. (Eds.), Proceedings of the Sixth California Islands Symposium. Institute for Wildlife Studies and National Park Service, Arcata, CA.

Erlandson, J.M., Graham, M.H., Bourque, B.J., Corbett, D., Estes, J.A., Steneck, R.S., 2007. The kelp highway hypothesis: marine ecology, the coastal migration theory, and the peopling of the Americas. J. Island Coastal Archaeol. 2, 161–174.

Estes, J.A., 1990. Growth and equilibrium in sea otter populations. J. Anim. Ecol. 59, 385–401.

Estes, J.A., Palmisano, J.F., 1974. Sea otters: their role in structuring nearshore communities. Science 185, 1058–1060.

Estes, J.A., Jameson, R.J., Johnson, A.M., 1981. Food selection and some foraging tactics of sea otters. Proceedings of the Worldwide Furbearer Conference 1, 606–641.

Estes, J.A., Tinker, M.T., Williams, T.M., Doak, D.F., 1998. Killer whale predation on sea otters linking oceanic and nearshore ecosystems. Science 282, 473–476.

Fedje, D.W., Mathewes, R.M., 2005. Haida Gwaii: Human History and Environment from the Time of Loon to the Time of the Iron People. University of British Columbia Press, Vancouver, BC, p. 426.

Fedje, D.W., Wigen, R.J., Mackie, Q., Lake, C.R., Sumpter, I.D., 2001. Preliminary results from investigations at Kilgii Gwaay: an early holocene archaeological site on Ellen Island, Haida Gwaii, British Columbia. Can. J. Archaeol./J. Canadien d'Archéologie 25, 98–120.

Folke, C., 2006. Resilience: the emergence of a perspective for social–ecological systems analyses. Global Environ. Change 16, 253–267.

Folke, C., Carpenter, S., Walker, B., Scheffer, M., Elmqvist, T., Gunderson, L., et al., 2004. Regime shifts, resilience, and biodiversity in ecosystem management. Annu. Rev. Ecol. Evol. Syst. 35, 557–581.

Foster, M.S., 1990. Organisation of macroalgal assemblages in the Northeast Pacific: the assumption of homogeneity and the illusion of generality. Hydrobiologia 192, 21–33.

Garshelis, D.L., Garshelis, J.A., 1984. Movements and management of sea otters in Alaska. J. Wildl. Manage. 48, 665–678.

Garshelis, D.L., Johnson, A.M., Garshelis, J.A., 1984. Social organization of sea otters in Prince William Sound. Can. J. Zool. 62, 2648–2658.

Gelcich, S., Hughes, T.P., Olsson, P., Folke, C., Defeo, O., Fernandez, M., et al., 2010. Navigating transformations in governance of Chilean marine coastal resources. Proc. Natl. Acad. Sci. USA 107, 16794–16799.

Gibson, J., 1988. The maritime trade of the North Pacific coast. In: Washburn, W. (Ed.), History of Indian-White Relations. Smithsonian Institution, Washington, DC, pp. 375–390.

Gibson, J.R., 1992. Otter Skins, Boston Ships, and China Goods. The Maritime Fur Trade of the Northwest Coast 1785–1841. McGill–Queen's University Press, Montreal, QC, Kingston, ON.

Graham, M., Halpern, B., Carr, M., 2010. Diversity and dynamics of Californian subtidal kelp forests. In: McClanahan, T.R., Branch, G.M. (Eds.), Food Webs and the Dynamics of Marine Reefs. Oxford University Press, Oxford.

Groesbeck, A., 2013. Ancient Clam Gardens Increased Production: Adaptive Strategies from the Past Can Inform Food Security Today, School of Resource and Environmental Management. Simon Fraser University, Burnaby, BC, p. 47.

Gunderson, L.H., Holling, C.S., 2002. Panarchy: Understanding Transformations in Human and Natural Systems. Island Press, Washington, DC.

Guujaaw, 2005. Foreward. In: Fedje, D.W., Mathewes, R.W. (Eds.), Haida Gwaii Human History and Environment from the Time of Loon to the Time of the Iron People. UBC Press, Vancouver, BC.

Happynook, T.M., 2000. Securing Food, Health and Traditional Values through the Sustainable Use of Marine Resources, Presentation to Oregon State University. World Council of Whalers, Brentwood Bay, BC.

Hilistis Waterfall, P., 2009. Fish traps. In: Brown, F., Brown, K. (Eds.), Staying the Course Staying Alive—Coastal First Nations Fundamental Truths: Biodiversity, Stewardship and Sustainability. Biodiversity, BC, Victoria, BC, p. 47.

Hoskins, J., 1969. The narrative of a voyage etc: John Hoskins' narrative of the second voyage of the "Columbia". In: Howay, F.W. (Ed.), Voyages of the "Columbia" to the Northwest Coast 1787–1790 and 1790–1793. De Capo Press, New York, NY.

Huntington, H.P., 2000. Using traditional ecological knowledge in science: methods and applications. Ecol. Appl. 10, 1270–1274.

Jackson, J.B.C., Kirby, M.X., Berger, W.H., Bjorndal, K.A., Botsford, L.W., Bourque, B.J., et al., 2001. Historical overfishing and the recent collapse of coastal ecosystems. Science 293, 629–638.

Jameson, R.J., Kenyon, K.W., Johnson, A.M., Wight, H.M., 1982. History and status of translocated sea otter populations in North America. Wildl. Soc. Bull. 10, 100–107.

Johannes, R.E., 2002. The renaissance of community-based marine resource management in Oceania. Annu. Rev. Ecol. Syst. 33, 317–340.

Kenyon, K., 1969. The sea otter in the Eastern Pacific ocean. In: Wildlife, U.S.B.o.S.F.a. (Ed.). U.S. Government Printing Office, Washington, DC, p. 366.

K̲ii'iljuus Wilson, B., 2012. Personal Communication.

K̲ii'iljuus Wilson, B., Luu Gaahlandaay Borserio, K.J., 2011. Gam tluu tllgaay aa k'iixa Gang ga: there is no land strange. In: Steedman, S., Jisgang, Collison, N. (Eds.), That which makes us Haida—The Haida Language Book. Haida Gwaii Museum, Skidegate.

Klein, J.R., 1996. Archaeology of Kachemak Bay, Alaska. Kachemak Country Publications, Homer.

Lafferty, K.D., 2004. Fishing for lobster indirectly increases epidemics in sea urchins. Ecol. Appl. 14, 1566–1573.

Loughlin, T.R., 1980. Home range and territoriality of sea otters near Monterey, California. J. Wildl. Manag. 44, 576–582.

Markel, R., 2011. Rockfish Recruitment and Trophic Dynamics on the West Coast of Vancouver Island: Fishing, Ocean Climate, and Sea Otters. University of British Columbia, Vancouver, BC.

McKechnie, I., Wigen, R.J., 2011. Toward a historical ecology of pinniped and sea otter hunting traditions on the coast of southern british columbia. In: Braje, T.J., Rick, T.C. (Eds.), Human Impacts on Seals, Sea Lions, and Sea Otters: Integrating Archaeology and Ecology in the Northeast Pacific. University of California Press, Berkeley, pp. 129–166.

Mears, J., 1790. Chiefs of nootka sound. In: Combe, W. (Ed.), Voyages made in the years 1788 and 1789, from China to the North West Coast of America. Logographic Press, London.

Murdock, G., 1934. The Haidas of British Columbia. In: Murdock, G. (Ed.), In Our Primitive Contemporaries. The MacMillan Company, New York, NY, pp. 221–263.

Nichol, L.M., 2007. Recovery potential assessment for sea otters (Enhydra lutris) in Canada. DFO Can. Sci. Advis. Sec. Res. Doc. 2007/034.

Nichol, L.M., Boogaards, M.D., Abernethy, R., 2009. Recent trends in the abundance and distribution of sea otters (Enhydra lutris) in British Columbia. DFO Can. Sci. Advis. Sec. Res. Doc. 2009/016. iv + p.16

Nuu-chah-nulth Tribal Council, 2012. DRAFT—A management plan for k^{w}ak^{w}atl (sea otter) in the Nuu-chah-nulth Ha' houlthee, West Coast Vancouver Island. Nuu-chah-nulth Fisheries Program, Port Alberni, BC, p. 28.

Osborne, J., 2007. Restoring to what? Using Nuu-chah-nulth knowledge in the recovery of K^{w}ak^{w}atl (sea otter) populations on the west coast of Vancouver Island, Summer Institute 2007: Advances in Ecological Restoration. Saving the pieces—restoring species at risk, University of Victoria, BC.

Parnell, P.E., Miller, E.F., Lennert-Cody, C.E., Dayton, P.K., Carter, M.L., Stebbins, T.D., 2010. The response of giant kelp (*Macrocystis pyrifera*) in southern California to low-frequency climate forcing. Limnol. Oceanogr. 55, 2686–2702.

Pauly, D., 1995. Anecdotes and the shifting baseline syndrome of fisheries. Trends Ecol. Evol. 10, 430.

Reisewitz, S.E., Estes, J.A., Simenstad, C.A., 2006. Indirect food web interactions: sea otters and kelp forest fishes in the Aleutian Archipelago. Oecologia 146, 623–631.

Rick, T.C., Erlandson, J.M., 2009. Coastal exploitation. Science 325, 952–953.

Rick, T.C., Erlandson, J.M., Braje, T.J., Estes, J.A., Graham, M.H., Vellanoweth, R.L., 2008. Historical ecology and human impacts on coastal ecosystems of the santa barbara channel region, california. In: Rick, T.C., Erlandson, J.M. (Eds.), Human Impacts on Ancient Marine Ecosystems: A Global Perspective. University of California Press, Berkeley, pp. 77–101.

Rick, T.C., Braje, T.J., DeLong, R.L., 2011. People, Pinnipeds, and Sea Otters of the Northeast Pacific. In: Braje, T.J., Rick, T.C. (Eds.), Human Impacts on Seals, Sea Lions, and Sea Otters: Integrating Archaeology and Ecology in the Northeast Pacific. University of California Press, Berkeley, pp. 1–17.

Riedman, M.L., Estes, J.A., 1990. The Sea Otter (*Enhydra lutris*): Behavior, Ecology, and Natural History, Biological Report. US Department of the Interior. Fish and Wildlife Service.

Salomon, A.K., Waller, N., McIlhagga, C., Yung, R., Walters, C., 2002. Modeling the trophic effects of marine protected area zoning policies: a case study. Aquat. Ecol. 36, 85–95.

Salomon, A.K., Tanape, N., Huntington, H., 2007. Serial depletion of marine invertebrates leads to the decline of a strongly interacting grazer. Ecol. Appl. 17, 1752–1770.

Salomon, A.K., Huntington, H., Tanape Sr., N., 2011. Imam Cimiucia; Our Changing Sea. Alaska Sea Grant, University of Alaska Press, Fairbanks, AK.

Simenstad, C.A., Estes, J.A., Kenyon, K.W., 1978. Aleuts, sea otters, and alternate stable-state communities. Science 200, 403–411.

Singh, G.G., 2010. Effects of Sea Otters on Nearshore Ecosystem Functions with Implications for Ecosystem Services, Institute for Resource and Environmental Sustainability. University of British Columbia, Vancouver, BC.

Singh, G.G., Markel, R.W., Martone, R.G., Salomon, A.K., Harley, C.D.G., Chan, K.M.A., 2013. Sea otters homogenize mussel beds and reduce habitat provisioning in a rocky intertidal ecosystem. Plos One 8, e65435.

Sloan, N.A., 2003. Evidence of California—Area Abalone shell in Haida trade and culture. Can. J. Archaeol./J. Canadien d'Archéologie 27, 273–286.

Sloan, N.A., 2004. Northern abalone: using an invertebrate to focus marine conservation ideas and values. Coast. Manage. 32, 129–143.

Sloan, N.A., Dick, L., 2012. Sea Otters of Haida Gwaii: Icons in Human–Oceans Relations. Archipelago Management Board, Haida Gwaii Museam, Skidegate, BC.

Steneck, R.S., Graham, M.H., Bourque, B.J., Corbett, D., Erlandson, J.M., Estes, J.A., et al., 2002. Kelp forest ecosystems: biodiversity, stability, resilience and future. Environ. Conserv. 29, 436–459.

Steneck, R.S., Vavrinec, J., Leland, A.V., 2004. Accelerating trophic-level dysfunction in kelp forest ecosystems of the western North Atlantic. Ecosystems 7, 323–332.

Swanton, J.R., 1905. In: Brill, E.J., Stechert, G.E. (Eds.), Contributions to the Ethnology of the Haida. Memoirs of the American Museum of Natural History, New York, NY.

Swanton, J.R., 1909. In: Brill, E.J., Stechert, G.E. (Eds.), Contributions to the Ethnology of the Haida. Memoirs of the American Museum of Natural History, New York, NY, pp. 1–300.

Szpak, P., Orchard, T.J., McKechnie, I., Grocke, D.R., 2012. Historical ecology of late Holocene sea otters (*Enhydra lutris*) from northern British Columbia: isotopic and zooarchaeological perspectives. J. Archaeol. Sci. 39, 1553–1571.

Szpak, P., Orchard, T.J., Salomon, A.K., Gröcke, D.R., 2013. Regional ecological variability and impact of the maritime fur trade on nearshore ecosystems in Southern Haida Gwaii (British Columbia, Canada): evidence from stable isotope analysis of rockfish (*Sebastes* spp.) bone collagen. Archaeol. Anthropol. Sci.

Tanape, N.S., 2012. Personal Communication.

Tegner, M.J., Dayton, P.K., 2000. Ecosystem effects of fishing in kelp forest communities. ICES J. Mar. Sci. 57, 579–589.

Tinker, M.T., Bentall, G., Estes, J.A., 2008. Food limitation leads to behavioral diversification and dietary specialization in sea otters. Proc. Natl. Acad. Sci. USA 105, 560–565.

Trosper, R., 2009. Resilience, Reciprocity and Ecological Economics. Routledge, New York, NY.

Turner, N.J., Turner, K.L., 2007. Traditional food systems, erosion and renewal in Northwestern North America. Indian J. Tradit. Knowl. 6, 57–68.

Uu-a-thluk, 2011. Nuu-chah-nulth's historical relationship with sea otters. Council of Ha'wiih Forum on Fisheries. Nuu-chah-nulth Tribal Council Fisheries, Port Alberni, BC.

Uu-a-thluk, 2012. Nations make alliances for better access to fish. Council of Ha'wiih Forum on Fisheries. Nuu-chah-nulth Tribal Council Fisheries, Port Alberni, BC.

Wade, P.R., 1998. Calculating limits to the allowable human-caused mortality of cetaceans and pinnipeds. Mar. Mammal Sci. 14, 1–37.

Watson, J., Estes, J.A., 2011. Stability, resilience, and phase shifts in rocky subtidal communities along the west coast of Vancouver Island, Canada. Ecol. Monogr. 81, 215–239.

Wendell, F.E., Hardy, R.A., Ames, J.A., Burge, R.T., 1986. Temporal and spatial patterns in sea otter, *Enhydra lutris*, range expansion and the loss of Pismo clam fisheries. Calif. Fish Game 72, 197–212.

Williams, J., 2006. Clam gardens: aboriginal mariculture on Canada's West Coast. Transmontanus.

Wilmers, C.C., Estes, J.A., Edwards, M., Laidre, K.L., Konar, B., 2012. Do trophic cascades affect the storage and flux of atmospheric carbon? An analysis of sea otters and kelp forests. Front. Ecol. Environ. 10, 409–415.

Xanius White, E.A.F., 2006. Heiltsuk Stone Fish Traps: Products of My Ancestors' Labour. Department of Archaeology. Simon Fraser University, Burnaby, BC.

Chapter 12

Shellfish Fishery Conflicts and Perceptions of Sea Otters in California and Alaska

Lilian P. Carswell[1], Suzann G. Speckman[2] and Verena A. Gill[3]
[1]US Fish and Wildlife Service, Ventura Fish and Wildlife Office, Ventura, CA, USA, [2]PO Box 244145, Anchorage, AK, USA, [3]US Fish and Wildlife Service, Marine Mammals Management, Anchorage, AK, USA

Public perceptions of sea otters have changed dramatically since the species' near extirpation worldwide and enactment of the first legislative safeguards on its behalf (see Chapters 13 and 14). Prized initially for the economic value of their pelts, sea otters are now largely protected from hunting internationally and embraced by many as beloved icons of a healthy nearshore environment. Like other top predators such as wolves, however, sea otters inspire extremes

Sea Otter Conservation. DOI: http://dx.doi.org/10.1016/B978-0-12-801402-8.00012-3

of emotion, and sentiment toward them tends to coalesce into camps of hatred or adoration. In this chapter, we trace the history of conflicts between sea otters and shellfish fisheries in California and Alaska and accompanying changes in management concepts and public attitudes toward sea otters. Although conflicts occur in other parts of the sea otter's range, we focus on debates in California and Southeast Alaska because they provide a compelling contrast between two very different prevailing views of sea otters and, by extension, the natural world.

Federal laws enacted in the early 1970s, namely the Marine Mammal Protection Act (MMPA) of 1972 and the Endangered Species Act (ESA) of 1973, reflected a major shift in American attitudes toward nature. Departing from the "wise-use" philosophy that originated with Gifford Pinchot in the early 1900s—whereby the natural world is seen as a storehouse of resources that should be scientifically managed to sustain their continued exploitation by human beings—the MMPA and ESA acknowledge and codify values in non-human organisms that go well beyond economic value. The MMPA states that "marine mammals have proven themselves to be resources of great international significance, esthetic and recreational as well as economic." The ESA extends the list of non-economic values to include not just "esthetic" and "recreational" value but also the "ecological, educational, historical, [...] and scientific value to the Nation and its people" of species of fish, wildlife, and plants. Even more explicitly than the MMPA, the ESA articulates the potential opposition between economic and conservation values, recognizing that "various species [...] have been rendered extinct as a consequence of economic growth and development untempered by adequate concern and conservation." Both acts also reflect a new appreciation of the intrinsic importance of biological species and the roles they play in their respective ecosystems. The MMPA affirms that "species and population stocks should not be permitted to diminish beyond the point at which they cease to be a significant functioning element in the ecosystem of which they are a part," and the ESA states as one of its purposes "to provide a means whereby the ecosystems upon which endangered species and threatened species depend may be conserved."

It is the interplay between economic forces and these alternate means of valuation of the natural world—esthetic, ecological, educational, historical, recreational, and scientific, but also ethical and empathetic—that has shaped modern debates surrounding sea otter conservation and management. Although the International Fur Seal Treaty of 1911 and the US Fur Seal Act of 1912 (which protected only those sea otters in waters beyond 3 miles offshore) did less to protect sea otters than did their post-exploitation scarcity, which made it economically infeasible to find and kill them for their skins (Baur et al., 1996), local protective laws and, later, the MMPA's broad moratorium on the taking of marine mammals in US waters (with exceptions for Alaska Natives, discussed further below) removed the primary driver that had influenced the fate of the species since the mid-eighteenth century: the international market value of sea otter pelts. The maritime fur trade reduced the original sea otter population from perhaps a few hundred thousand animals to less than a few thousand, but with protection, sea otters have repopulated large portions of their former range throughout the North Pacific rim (Chapter 3). However, as predators that consume approximately one-quarter of their body weight daily (Costa and Kooyman, 1984), they have encountered new threats as they have come into competition with shellfish fisheries seeking to exploit the same prey species. In the eyes of some people, primarily those who have an economic stake in these fisheries, sea otters fall into the category of destructive pests that should be restricted to, if not eliminated from, portions of their range. For others, each stepwise reclamation by sea otters of their historical range is an event to be celebrated, not just for the hope it represents for the future of the species and for biodiversity in general, but also for the release from culpability it offers humanity for nearly exterminating the species worldwide. Because of the intensity of feelings they inspire, sea otters have played a significant role in mobilizing cultural shifts in perspective on the relationship between human beings and the rest of the natural world.

CALIFORNIA

During the time of landmark environmental legislating at the federal level in the early 1970s, a number of forces were at play in California that would later test the limits of federal protections and shape the direction of sea otter conservation in the state. The southern sea otter population had grown in size and expanded in range from a tiny remnant colony off Big Sur that numbered about 50 animals (Bryant, 1915) at the time that the International Fur Seal Treaty and state protections (under section 4700 of the California Fish and Game Code) were enacted. By 1973, approximately 1600–1800 sea otters ranged from Santa Cruz to Point Buchon in San Luis Obispo County. The California Sea Otter Game Refuge, within which the possession of firearms was prohibited, had been expanded in 1959 from its original 1941 boundaries to the Carmel River mouth in the north and the Santa Rosa Creek mouth (near Cambria) in the south. However, by the early 1970s, sea otters had again recolonized habitat outside refuge boundaries to both the north and south. Expansion to the south in particular had brought the animals into competition with lucrative commercial and sport fisheries for abalone from Cape San Martin to Cayucos (Wild and Ames, 1974). The conflict that ensued was the first major clash in North America between a recovering sea otter population and fisheries that had developed to exploit the unnaturally high abundance of shellfish resulting from the sea otter's protracted absence.

The Sea Otter–Abalone Fishery Conflict of the 1960s–1970s

As a result of the scarcity of sea otters during the early twentieth century, several generations of Californians had grown up with a marine ecosystem that did not include these marine mammals. People perceived what was familiar—in this case the lack of a top predator and an abundance of exposed (non-cryptic) herbivorous marine invertebrates—as the natural environmental condition. As early as the late 1940s, some central California fishermen had begun to view sea otters as the cause of depletion of the abalone stocks they were seeking to exploit. By the 1960s, the conflict between sea otters and the commercial abalone fishery had gained the attention of politicians. On November 19, 1963, the California Senate Fact Finding Committee on Natural Resources, Subcommittee on Sea Otters, Chaired by Senator Fred S. Farr of Monterey (State Senator 1955–1967), held a hearing in San Luis Obispo on the "Affect [sic] of the Sea Otter on the Abalone Resource" (California Legislature, 1963). Its purpose was to receive comments on the biology and ecology of sea otters and the "relative economic impact, alleged and actual of the sea otter on the abalone." In attendance were representatives of environmental and fishing groups, commercial and sport abalone divers, academic biologists, employees of the California Department of Fish and Game (CDFG), students, concerned citizens, and others. Senator Farr had also invited Big Sur resident and conservationist Margaret Owings. The ideas expressed at the 1963 hearing would influence the debate about sea otters for the next half century.

Abalone fishermen testified to the immediate effects of sea otters on commercially exploited abalone beds, conjuring images of plenitude before the arrival of sea otters and of destruction and waste in their wake. Buzz Owen, a commercial abalone diver of four years from Morro Bay, said that after sea otters moved into an area, fishermen would find, instead of abalone, their "broken empty shells." Thomas Reviea, a longtime abalone fisherman also from Morro Bay, testified that in 1947, before the arrival of sea otters in Gorda (between Monterey and Morro Bay), he had taken, during the course of one overnight trip, "one hundred seventy-six dozen abalone." Ernest Porter, an abalone diver of five years from Atascadero, had regularly taken "forty dozen" abalone every few months for the previous "several years," but after a group of sea otters appeared off the San Simeon Lighthouse, the area was ruined for fishing and "littered with abalone, sea urchin shells" (California Legislature, 1963).

Despite the vast numbers of abalone taken by the fishery, the fishermen maintained an unshakable confidence in the infallibility of the legal size limit to protect abalone so long as cheaters, namely sea otters (they downplayed the effects of human poaching), were excluded. John Gilchrist, General Manager of the California Seafood Institute, noted that seasonal, area, and size restrictions, combined with weather, made it "virtually [. . .] impossible for the commercial fisherman alone to completely destroy" the commercial abalone fishery. Kent Williams, a commercial abalone operator, characterized fishermen as conservationists preserving "California's abalone resource" and sea otters as outlaws violating the principles of wise use. "The problem of the otter and the abalone must be thought of in terms of conservation of natural resources," stated Williams, or "the intelligent economic and recreational utilization of natural resources." Owen demanded to know "why one valuable resource should be permitted to destroy another when, through proper methods, both can be preserved" (California Legislature, 1963).

Resource managers and environmentalists were careful to correct the impression left by the fishermen that sea otters might exterminate abalone entirely. Distinguishing between commercial and ecological viability, CDFG Deputy Director Harry Anderson pointed out that "although abalones are taken, the beds are not wiped out." For centuries, he noted, "sea otters have been feeding along the California coast on [. . .] marine life, without eradicating any species." Tom Meyers, a photographer commissioned by National Geographic and the Associated Press to do an extensive story on sea otters, also emphasized the importance of a historical perspective: "[T]he sea otter and the abalone live[d] together for who knows how many millions of years before us and the sea otter never wiped out the abalone." Owings observed that sea otters may in fact be better stewards of abalone than fishermen are, pointing out that for the previous two decades in the Point Lobos Marine Reserve (which had formerly suffered heavy exploitation of abalone but was subsequently closed to human take), "the sea otter [had] been the prime consumer of abalone yet the abalone [were] building up their beds" (California Legislature, 1963).

Some speakers suggested that the sea otter was being made a scapegoat for problems that were the result of excessive human competition. The more than 400-fold increase in the number of licensed commercial abalone fishermen in the previous three and a half decades (from 11 in 1928 to 505 in 1963) would necessarily be expected to result in a reduced average catch per fisherman, noted Deputy Director Anderson, and "it is very natural for fishermen to be unhappy with a lower average catch [. . .] [b]ut I do not believe we should blame the sea otter for this." Unfairly or not, it was clear that sea otters were bearing the brunt of antagonism and anger over the depletion of abalone stocks. According to Captain Howard Shebley, CDFG Wildlife Protection for the central coastal counties of California, "the sea otter has a large—let's say a non-buyer group. There are several groups of people dead set against the sea otter and it is quite an emotional thing. [. . .] [T]here are groups of people who do definitely shoot this animal" (California Legislature, 1963).

The 1963 hearing was intended to settle questions of fact regarding the effect of sea otters on the abalone fishery, but deeper and more intractable issues were rooted in the unsettled consensus on questions of value. Early twentieth century prohibitions on hunting sea otters may have been conceived as an eleventh-hour bid to preserve a valuable fur-bearing species for future exploitation, but during their decades of protection in California, the animals had ceased to be regarded as natural resources in the traditional sense of furnishing physical goods for human consumption. Deputy Director

Anderson recognized the sea otter's return from near-extinction in California as "one of the great stories of conservation," in which strong legal protection and public cooperation "has helped bring back to its old haunts a prized [furbearer] at one time thought to be extinct," but his allusion to the sea otter's esteemed fur was not meant to imply any intention or legal possibility of resuming the exploitation of California sea otters for their pelts (California Legislature, 1963). The wise-use scheme of valuation no longer mapped easily onto the situation in California. No matter that the language of "conservation" was used loosely in reference to the sea otter; the conservationist ethic of Pinchot had come into conflict with the preservationist ethic of John Muir.

The fishermen viewed it as self-evident that the economic value of the abalone fishery should outweigh any other, less tangible considerations, and they attempted to represent this economic value as the greater societal good. Williams stressed the importance of abalone, not as the source of private gain for a small group of fishermen and processors but as "a resource of economic and recreational value to tens of thousands of Californians." Gilchrist summed up the conflict as one between two user-groups: "those who enjoy the aesthetic values of our animals and creatures and those who enjoy and eat sea food or meat or whatever it may be." He acknowledged both types of value, but the appreciation of sea otters was amorphous and insubstantial, whereas the value of the abalone industry was "real" and "definite." Recognizing the obvious sway of economic arguments, Myers countered Gilchrist's devaluation of esthetic appreciation with figures demonstrating that the "cash value of the abalone industry in California [...] is peanuts" compared to the tourist industry of the central California coast. Other speakers rejected the unquestioned primacy of economic value. Dr. Richard Boolootian, of the Department of Zoology at the University of California, Los Angeles, asserted an intrinsic biological value in the unique evolutionary status of the sea otter, and Owings defended the importance of esthetic values to the long-term sustainability of society (California Legislature, 1963).

Despite their disagreements over the relative value of sea otters and abalone fisheries to society, everyone present seemed to agree that any true resolution of the conflict would require the separation of the realms in which sea otters and the human exploitation of abalone occurred. The biologists and environmentalists proposed that abalone be raised in hatcheries or seeded in suitable areas well outside the range of sea otters. The fishermen, though supportive of the hatchery idea, insisted on the importance of immediately removing sea otters from areas where they were competing with the abalone fishery. Williams presented the fishermen's proposed solution: federal wildlife officials should "determine methods of relocating the sea otter to an area in which the taking of abalone by man is inconsequential" and "carry out such a program." To placate sea otter user-groups, he suggested what amounted to a sort of wild-animal park: "a small group of otters" should be kept "in the 17-Mile Drive area, so that the many friends of the sea otter may observe them in their natural habitat." Gilchrist emphasized the benign nature of the non-lethal predator control they were suggesting, noting that "no one engaged in the fishing industry [...] has in any way suggested that they should destroy any of the existing sea otters." Likening it to the "situation with the California bear," Gilchrist represented as simply unavoidable the need to direct the distribution of sea otters, just as for "lions and every other type of predator with that nature where it becomes necessary to control" (California Legislature, 1963).

The fishermen took advantage of the idea, deeply entrenched in American thinking, that predator control was a necessary accompaniment to human enterprise. After all, the federal government had used public funds to eradicate wolves from the lower 48 states during the first half of the twentieth century at the behest of the livestock lobby (Robinson, 2005). A proposal that federal officials non-lethally remove sea otters from abalone fishing grounds seemed mild by comparison. The biologists and environmentalists opposed the proposal but avoided challenging the philosophy of predator control outright. Instead, they opposed its application to an entity in such a precarious situation as the recovering California sea otter. Myers took a practical stance, questioning the viability of such a plan. Having reviewed the accounts of federal efforts to relocate Alaskan sea otters, he rejected the notion that moving them was non-lethal and pointed out that sea otters, even if they were conceived as objects that could be moved around to suit human convenience, might not turn out to be so compliant. He asked a prescient question: "[I]f, by chance, you are successful in moving the sea otter, how would you keep him from coming back down? They are fast swimmers and they seem to want to go wherever they wish to go" (California Legislature, 1963).

Although the fishermen had presented what they hoped would be seen as a reasonable proposal, the views of the biologists and environmentalists gained temporary ascendancy. The Senate Committee concluded, in light of the evidence presented, that "a program of trapping and transplanting would be most unwise at this time" (Senate Permanent Factfinding Committee on Natural Resources, 1965). The Committee's conclusions did nothing to quell the conflict, however. Owings recalled that period as "a time when a single fishing boat could come in with 500 pounds of abalone," but "one little otter floating along on its back" eating an abalone "was all [the fishermen] needed to see absolute red" (Owings, 1991). Continuing demands for action ensured that a steady stream of political proposals would continue to buffet CDFG. In 1967, Donald Grunsky, California State Senator from 1953 to 1976, introduced Senate Concurrent Resolution No. 74, which directed CDFG to take up the very proposal that the Senate Committee had deemed in 1965 to be "most unwise." The department was to "determine the feasibility and possible means of confining sea otters within the protection of the existing refuge or other means that will [...] lessen the possibilities of resource conflicts" (Senate Concurrent Resolution No. 74). In areas distant from the conflict, the improbable image of state wildlife officials herding sea otters generated mild levity. One Pennsylvania newspaper ran the Associated Press story on the subject under the title "Wanted: Suggestions for Otter Herding" (Associated Press, 1967).

The political efforts of the fishermen, though clearly advancing their agenda, also had the effect of galvanizing resistance. In March 1968, Owings wrote a letter to the editor of the *Monterey Peninsula Herald*. What abalone fishermen viewed as the "vicious little otter" that was preventing them from making a living was declining in population size, and carcasses were washing up with evidence of gunshots and knife wounds (Owings, 1968). Owings questioned the assumption implicit in the fishermen's calls for predator control: that the abalone belonged to them. "As has been true with practically every wild animal in the vicinity of man," she wrote, "the otter is thought to compete with an economic value that man claims as his own" (Owings, 1968). The newspaper printed Owing's editorial on the front page under the title "Do Sea Otters Have Any Friends?" That headline, Owings later recounted, inspired her to form an organization to fight for sea otter protection (one that continues to spearhead campaigns on behalf of sea otters today): "When I read that, and I read the title of it, [. . .] I said, 'I'm going to have to form Friends of the Sea Otter' " (Owings, 1991).

In compliance with Senator Grunsky's resolution, CDFG produced a report establishing the framework of a "Sea Otter Research Project" (CDFG, 1968). The project would consist of an initial phase to gather information and "to provide a measure of relief to the commercial abalone fishery" and a subsequent implementation phase. Up to 20 sea otters would be removed from the site of greatest conflict (the Cambria-Pt. Estero area) and either placed in captivity or relocated to the northern part of the sea otter refuge. If these efforts were deemed safe and successful, additional sea otters could be relocated. Despite the report's required focus on the means of removing sea otters from areas of conflict with fisheries, it reflected CDFG's continuing concern for the preservation of the California sea otter, which it recognized as a unique subspecies, *Enhydra lutris nereis*, and acknowledged the sea otter's non-consumptive and ecological value. The report concluded that "relocation should have as its major objective not only the relief of conflict with the abalone industry" but also the "ultimate stabilization of the California sea otter population in a variety of habitats in distinctly separate areas so the population will be more secure from harassment or disasters." It cautioned that the potential consequences to the subspecies "should remain the foremost consideration" when contemplating any management programs (CDFG, 1968). In terms of practical implementation, the report was not particularly optimistic. Of six potential translocation sites in California, each turned out to have the potential for conflict with sport or commercial shellfish fisheries (CDFG, 1968). It seemed that any place one might put sea otters in California was already claimed by a human contingent asserting its rights to the benthic invertebrates that lived there.

Whereas the Friends of the Sea Otter had fervently opposed the translocation of sea otters for the purpose of preserving the commercial abalone fishery, they came to view the experimental translocations about to be undertaken by CDFG as potentially useful for the preservation of sea otters themselves. The January 1969 Santa Barbara oil spill, a blowout of Union Oil's Platform A, released 100,000 barrels of oil into the Santa Barbara Channel, killing at least 37,000 birds, and was the worst incident involving an offshore oil platform that had occurred in the United States to date (NOAA, 1992). Although the sea otter's range did not yet extend that far south, the potential implications of such a spill were self-evident. The first issue of *The Otter Raft*, the newsletter of the Friends of the Sea Otter, acknowledged that in light of the disastrous spill, a "program which originally was intended to solve a political problem" might have an unanticipated "positive value for sea otters." "Several populations, scattered widely along the coast," the author explained, "would better [e]nsure survival of the species" (Judson, 1969).

The Santa Barbara oil spill initiated a national conversation about the environment that brought non-economic values to the fore. President Richard Nixon summed up its profound impact upon the minds of Americans. The "Santa Barbara incident," he remarked, "has frankly touched the conscience of the American people" (Nixon, 1969). A Californian who had spent his youth on the southern coast, the President was personally moved by witnessing the spill's effects on the beaches he remembered so well. He pledged that the federal government would do more to prevent environmental destruction and to provide environmental leadership to the states. He had set up a group for the environment within the Cabinet to ensure that "we have all the material progress that we need, but [. . .] not at the cost of the destruction of all those things of beauty without which all the material progress is meaningless" (Nixon, 1969). While the federal government did provide the promised environmental leadership (Nixon's administration signed into law the National Environmental Policy Act of 1969, the MMPA, and the ESA within 5 years of the spill), the political winds were blowing in the opposite direction in the State of California. Republican Ronald Reagan had defeated the incumbent democratic Governor Edmund Gerald "Pat" Brown and would subsequently serve as Governor of California for two terms (1967–1975). In 1969, Reagan appointed Raymond Arnett as Director of CDFG, replacing Walter Shannon, who had served in that role since 1960. Whereas Shannon had been sympathetic toward sea otters, Arnett was not. According to Owings, "Ray Arnett saw red, white, and blue when he heard of otters. He also saw other colors when he heard of mountain lions. In other words, everything we were working towards, he was on the opposite side" (Owings, 1991).

CDFG's "Sea Otter Research Project" was under way under the new leadership of Arnett when on February 22, 1970, Senator Grunsky introduced into the state legislature Senate Bill (SB) 442. The bill allowed sea otters found outside of the California Sea Otter Game Refuge to be "taken," either by permit or by CDFG. Soon on the defensive for his use of the ambiguous term, Grunsky amended the bill to clarify that "sea otters may be pursued, caught, or captured but not killed or destroyed" (Pentony, 1970). Not appeased, the Friends of the Sea Otter presented a petition at the hearing with approximately 15,000 signatures that declared their opposition to the "taking of the sea otter outside the sea otter refuge" and their belief "in a balanced ecology which the sea otter and abalone have long shared." Its signatories, it stated, "place higher value on the presence of the sea otter along our shores than on the $ sign of the commercial abalone industry" (Pentony, 1970). The Sierra Club, also opposed to the bill, recommended that measures be developed "to protect a balanced ecology" by "an institution of recognized competence and impartiality to resist pressures originating from narrow viewpoints" (Condit, 1970).

No action was taken on SB 442, but the importance of the responses by environmental groups went beyond mere opposition to the bill. They called for a broader notion of the public than just those with a direct interest in the consumptive exploitation of marine organisms. These other stakeholders had a very different view of the role of public agencies as trustees of the natural world. CDFG "statements repeatedly emphasize that management programs are 'for the best interest of the public of the State,'" Owings noted, but "an informed public is becoming increasingly convinced that natural balances have permitted diverse forms of coastal life to arise and flourish without benefit of man's management" (Owings, 1970). That the natural world required human intervention to operate optimally was a presumption that reflected the wise-use goal of ensuring maximum extractive value to human users. However, it could no longer be assumed that achieving maximum extractive value, at whatever cost to other members of the ecosystem, was a goal shared by all elements of the public. Owings cited the "strong-willed battle" over SB 442 as evidence that the "public outlook is changing." SB 442 was "pressured by commercial abalone interests," Owings noted, but "authoritative scientific minds and a broad expanse of public sentiment" both "placed a higher value on the otter" (Owings, 1970). This new scheme of valuation demanded a re-envisioning of the role of management agencies and new conclusions regarding which human activities should be encouraged or even allowed. CDFG's "prejudicial statement that 'the presence of large numbers of sea otters is not compatible with abalone fishing' might well be reversed," Owings suggested, "for a more enlightened view would hold that 'commercial abalone fishing is not compatible with the welfare of otters'" (Owings, 1970).[1]

The 1974 report documenting CDFG's "Sea Otter Research Project" and outlining its new management recommendations did not embody the enlightened view that Owings had hoped to see. Rather, it revealed a stark shift in the

1. The State's Abalone Recovery and Management Plan now recognizes that abalone populations subject to sea otter predation will not support fisheries and excludes any area within sea otter range from consideration for a potential reopening of a sport or commercial abalone fishery (CDFG, 2005).

relative value CDFG placed on the protection of sea otters and shellfish fisheries. Relying on Roest (1973), CDFG no longer regarded *E.l. nereis* as a distinct subspecies, in need of protection due to its small population size and distribution. Instead, the report declared it "essential that the sea otter population be restricted within geographical limits" because "the recreational and commercial uses of our resources are important and beneficial and we cannot allow them to be destroyed by sea otters" (Wild and Ames, 1974). This management concept, subsequently known as "zonal management," envisioned the establishment of geographically separate zones for sea otters and human users of shellfish. However, the report was singularly vague on how this separation might be achieved. Based on its experimental translocations, during which about one-third of the sea otters returned to their location of capture, CDFG no longer viewed translocation as an effective means of controlling range expansion. The report recommended that a "program be developed" to restrict the sea otter's range, but it left to inference the means by which sea otters, if not removed to some other location, would be eliminated from the areas where they were not wanted (Wild and Ames, 1974). The portent of CDFG's recommendations was not lost on the sea otter's defenders. "This may appear to be a reasonable proposal to some," Owings wrote in *The Otter Raft*, "but like the word 'take' (meaning 'kill') which the Department used 5 years ago when they advised on a Legislative Bill [SB 442] to control otters outside their Refuge, the word 'restrict' now carries a depth of drastic manipulations and potentially high mortality" (Owings, 1974).

Such ominous implications were not, however, to be realized. Passage of the MMPA placed sea otters under the jurisdiction of the US Department of Interior and compelled termination of the state's trapping and tagging activities as of the law's effective date (Wild and Ames, 1974). Additional activities could proceed only under a federal permit in accordance with the protective provisions of the MMPA. Passage of the MMPA thus marked the end of one phase of sea otter management and conservation in California, but the management concepts and rhetoric that had developed in the course of the sea otter–abalone fishery conflict would prove to be remarkably resilient. The dominant theme—one that both parties sought to use to their advantage—was "balance." The fishermen and CDFG (under Arnett) referred to a "balance" of uses, whereby one area of the coast would be used to support enough sea otters to forestall their extinction, and other areas would be used to sustain commercial and sport fisheries. The environmentalists advocated the restoration of a "balanced ecology," which had been, and continued to be, disrupted by the actions of human beings.

A Plan to Move Sea Otters: The Southern Sea Otter Translocation Program, 1987–2012

The transfer of management authority for sea otters to the federal government included a mandate under the MMPA to increase sea otter numbers to a level at which they would resume their role as a "significant functioning element of the ecosystem of which they are a part" and remain at or above their "optimum sustainable population" size. In 1974, soon after this transfer of authority, CDFG submitted a request under the MMPA for the return of management to the state and a waiver of the MMPA's moratorium on the taking of sea otters (USFWS, 1982). The state's plan was to restrict the sea otter population to the coastline between Seaside and Cayucos. USFWS deemed the plan inadequate, suggesting instead that the state might remove sea otters from areas where it wished to establish shellfish reserves and release them in other areas to help the population restore itself more quickly (USFWS, 1982). CDFG submitted a revised proposal in 1976 that reflected USFWS's suggestion to request a scientific research permit as an interim measure (CDFG, 1976). The state's proposed "experimental management procedures" were intended "to contain a non-threatened and healthy sea otter population within a segment of the California coastline until the secondary effects of sea otter foraging [are] clearly documented." CDFG would translocate any sea otters moving south of Avila to the northern periphery of the range and decide upon future management when the "entire proposed range from Miramontes Point to Avila" was occupied. CDFG justified its bid to restrict the sea otter's range despite its apparent contradiction with the MMPA's primary objective—to increase the "health and stability of the ecosystem"—by arguing that the objective was "an elusive and poorly defined concept" that was "presently expressed in value judgments" (CDFG, 1976).

Shortly after submitting its revised proposal, the state determined, after consulting with USFWS and the Marine Mammal Commission (MMC) (an independent agency of the Executive Branch, created by the MMPA to review federal actions regarding marine mammals), that a scientific research permit could authorize its proposed "experimental" translocations and that a waiver of the MMPA's moratorium on the taking of marine mammals was unnecessary. The state withdrew its waiver application and applied for a research permit instead (USFWS, 1982). USFWS issued a permit for the state's "experimental management plan" authorizing the translocation of up to 40 sea otters from the southern end of the range to its northern end (42 FR 44,314, September 2, 1977).

In the meantime, USFWS proposed to list the California sea otter, which it identified by the common name "southern sea otter" to distinguish it from the northern sea otter in Alaska, as an endangered species under the ESA (40 FR 44329, September 26, 1975). The interested public viewed the proposal favorably, but the state opposed listing the southern sea otter as either endangered or threatened, arguing that it did not meet any of the listing criteria, that the population was at an optimum level, that continued range expansion was likely, and that no major natural or human-caused threats to the population as a whole existed. The Friends of the Sea Otter argued extensively in support of the "endangered" listing. In accordance with the MMC's recommendation, USFWS ultimately listed the southern sea otter as threatened on the basis of its small population size, its severely reduced range, and the potential risk posed by oil spills (42 FR 2965, January 14, 1977). USFWS convened a recovery team for the species in 1980 and approved a recovery plan in 1982. In light of the pace of offshore oil development, the difficulty of containing an oil spill on the open ocean, and the difficulty of preventing sea otters from entering an area in the path of a spill, the recovery team concluded that translocation was the "most effective and reasonable management action" (USFWS, 1982).

The management concept that fishermen had proposed during the 1963 hearing and that CDFG had subsequently embraced gained additional legitimacy, this time at the federal level, through the advocacy of the MMC. In a 1980 letter, the MMC urged USFWS not only to "establish at least one additional group of sea otters at a site that is secure from oil spills" as soon as possible but to "recognize the ultimate need for 'zonal management' of the California sea otter." This management scheme, it noted, "would be based upon a determination that the [MMPA's] goal of optimum sustainable population can and should be achieved with reference to the 'health and stability of the marine ecosystem' and that historic levels and distribution are not necessary to satisfy that goal." The MMC envisioned the designation of numerous zones where sea otters would either be protected or excluded, the reasons for which might include that "conflicts between otters and fisheries would be substantial" (Twiss, 1980). How these zones might practically be maintained remained, as always, an outstanding question. The MMC pointed out that sites could be chosen that were "naturally self-limiting" because of adjacent unsuitable habitat but noted that zonal management would "almost certainly require the development and utilization of effective techniques" to keep sea otters within or out of zones (Twiss, 1980). Despite the MMC's urgings, the USFWS recovery plan, although recognizing the potential for conflict between an expanding sea otter population and shellfish fisheries, mentioned the possibility of "some level of zonal management" only once, as a potential ultimate strategy in the long-term context of managing the southern sea otter to obtain and maintain its "optimum sustainable population" size; it did not suggest the containment of any translocated population (USFWS, 1982).

The MMC, however, continued to press for zonal management, seeing it as a way to address the incompatibility of certain human activities in the marine environment with sea otters. One such activity was commercial gill and trammel net fishing, which was rumored for several years to be drowning sea otters before the first direct observation occurred in 1982. An average estimated 80 sea otters per year were killed in nets from 1982 to 1984, prompting CDFG to institute successive depth and area closures within the sea otter's range (Wendell et al., 1986; MMC, 1988). In 1984, amidst calls for action to address the sea otter drownings, USFWS announced its intent to prepare an environmental impact statement (EIS) on the translocation of southern sea otters. The notice identified a proposed action—translocation to San Nicolas Island—and several possible alternatives. Translocation in conjunction with restriction of the existing population had "substantial" support from fishing groups concerned about the impacts of range expansion, the notice acknowledged, but "would be very controversial because of strong opposition to it by other interested members of the public" and would require "significant amendments" to the ESA and MMPA (49 FR 26313, June 27, 1984).

The scoping sessions that followed demonstrated the resolve of the fishermen not to allow a translocation to go forward without significant concessions (such as restriction of the parent population in conjunction with any translocation) and the willingness of the environmental groups, unnerved by the potential of a catastrophic oil spill to decimate the sea otter population, to yield to concessions (such as containment of the translocated population) (USFWS, 1984a). Whereas the environmental groups and their supporters continued to foreground the value of the sea otter as a "keystone species" for its positive effect on kelp abundance and associated biodiversity, the fishermen reframed their earlier rhetorical emphasis on a balance of uses. They now represented themselves as defending the MMPA's goal of promoting the "health and stability of the marine ecosystem" and characterized the proposed translocation as emotionally driven, unbalanced, single-species management (USFWS, 1984a). The Western Oil and Gas Association, intimately involved in the planning process for the translocation program and powerful enough to ignore rhetorical niceties, was determined to ensure that no additional restrictions on oil and gas activities would be put in place because of sea otters (USFWS, 1984b); unlike the fishermen, it had the means to secure legal counsel to remedy the impediments the ESA and MMPA presented to zonal management.

Years of additional meetings between USFWS, CDFG, and the various stakeholders resulted in what Owings termed "an uncomfortable tradeoff" (Owings, 1986). USFWS initiated the Southern Sea Otter Translocation Program in 1987. On one hand, the program was intended to serve as a "primary recovery action" for the species (52 FR 29754, August 11, 1987). It put into effect the main recommendation of the 1982 recovery plan for the southern sea otter, the creation of a second population that could buffer the fate of the subspecies in the event of a catastrophe (USFWS, 1982). On the other hand, it included a "management zone," which was to be kept otter free as long as the program remained in effect (52 FR 29754, August 11, 1987). This latter provision became a mandatory feature of the plan with the passage of Public Law 99–625, signed into law by President Ronald Reagan on November 7, 1986. The law authorized USFWS to undertake a sea otter translocation program but stipulated that any such plan must have a translocation zone, to which sea otters would be relocated, and a surrounding management zone, from which any straying sea otters would be removed by "all feasible non-lethal means" (P.L. 99-625). It also stipulated that reduced federal protections for sea otters would apply to defense-related activities in the translocation zone (San Nicolas Island was owned by the US Navy) and to all otherwise legal activities (such as fishing and oil and gas development) in the management zone. The final rule establishing the program thus included a management zone, which spanned all Southern California waters from Point Conception to the Mexican border (except the translocation zone at San Nicolas Island) and nearly abutted the then-current southern edge of the sea otter's range. The final rule also included criteria that would be used to determine whether the program had failed (52 FR 29754, August 11, 1987). While the program appeared on paper to be an effective compromise between competing interests, the problems inherent in the zonal management scheme became clear during its implementation.

The program's unworkability stemmed from the fact that sea otters were biological organisms with a strong homing impulse and interests of their own. As Myers had warned at the 1963 hearing, they could not necessarily be moved and expected to stay where they were put. Despite the belief of USFWS and others that San Nicolas Island would be a site that was largely "self-limiting" due to the deep ocean channels that surrounded it, most of the 140 sea otters that were brought to San Nicolas Island from 1987 to 1991 left immediately or simply disappeared. Many entered the management zone, from which they were required to be removed, however impractical or contrary to sea otter recovery such removals turned out to be. In 1993, USFWS suspended sea otter captures in the management zone, citing the need to reevaluate its methods in light of sea otter deaths that had apparently resulted from capture and removal.

Whereas the emigration of translocated sea otters from San Nicolas Island had finally ceased, resulting in a population of approximately a dozen animals, sea otters began entering the management zone again by the late 1990s, this time from the mainland. The seasonal appearance of large groups of over 100 animals generated new calls by fishermen for sea otter removals from the zone (USFWS, 2012, Appendix C). In 2001, USFWS issued a policy statement explaining that it had determined the containment of southern sea otters to be inconsistent with the ESA requirement to avoid jeopardy to the species and therefore would not remove any sea otters until it had completed a full evaluation of the program, and alternatives to it, under the National Environmental Policy Act (66 FR 6649, January 22, 2001). In 2003, USFWS released a revised recovery plan for the southern sea otter that recommended a fundamentally different recovery strategy, one that relied on natural range expansion instead of additional translocations. The *Exxon Valdez* oil spill had demonstrated that San Nicolas Island was not sufficiently far from the mainland range to ensure the survival of the subspecies in the case of a catastrophic oil spill. The recovery team advised that, given the population decline that had been noted in the mainland population and its range expansion into the management zone, it would be "in the best interest of recovery of the southern sea otter" to declare the program a failure and to discontinue zonal management (USFWS, 2003). After more than a decade of formal review, during which environmental groups sued to speed a final decision and CDFG, MMC, and the California Coastal Commission expressed support for termination, USFWS ended the program in 2012 (77 FR 75266, December 19, 2012).

Whereas fishermen had initially opposed the translocation program, they came to see it as preferable to having no program at all, despite the fact that USFWS had suspended containment operations in 1993, because it maintained the concept of zonal management, afforded hope that capture and removal operations would resume, and provided continuing incidental take exemptions under the MMPA and ESA in the management zone. Additionally, the small population at San Nicolas Island had had very limited impacts on fisheries compared to those projected in the original EIS under the hopeful scenario of a thriving population of hundreds of animals. The environmental groups had initially advocated for the translocation program but became opposed to it once it became clear that sea otters were not staying at San Nicolas Island, that some were dying as a result of being removed from the management zone, that the island colony would not provide the hoped-for security for the subspecies in the event of a massive oil spill, and that natural range expansion to the south was in danger of being curtailed because of the requirements of the management zone. They and USFWS also recognized that the high rate of dispersal of translocated sea otters from San Nicolas Island undermined the premise of the program: that recovery and persistence of the southern sea otter could be secured, in the face of a catastrophic event that decimated the mainland range, by means of translocations of small numbers of sea otters from a single reserve colony (77 FR 75266, December 19, 2012).

Despite these reversals of position, the arguments for and against range restriction remained strikingly consistent with those of previous years. This was the case, in part, because many of the same people had been involved in the debate for decades. However, the number and proportion of people expressing interest in marine policy and support for sea otters had increased dramatically in concert with Internet-age campaigns conducted by Defenders of Wildlife, The Otter Project, the Sierra Club, Friends of the Sea Otter, and other organizations. During the comment periods on the 2005 draft and 2011 revised draft supplemental EIS on the future of the translocation program, commenters submitted approximately 27,000 comments and petitions with more than 13,000 signatures.

An overwhelming majority of the comments expressed support for ending the translocation program, specifically the management zone component, for reasons including that range expansion is important for sea otter recovery, that sea otters are a native, keystone species whose presence enhances biodiversity, that sea otters benefit tourism and other industries that depend on ocean health, and that sea otters have an intrinsic right to recolonize their historical habitat. Some commenters expressed disbelief that a taxpayer-funded "no-otter" management zone had ever been put in place. Despite years of discussions with USFWS to try to identify a mutually agreeable resolution, the US Navy remained opposed to termination of the program because it wanted to retain the ESA and MMPA exemptions connected with the zones. Standing to benefit from the efforts of this powerful, incidental ally, shellfish fishermen and Southern California gillnet fishermen (fearing additional depth or area restrictions) also called for the perpetuation of the program in some form. They argued that unrestricted sea otter range expansion could cause the extinction of the now federally endangered white abalone and black abalone,[2] harm precariously low stocks of other abalone species in California, and destroy shellfish fisheries throughout Southern California. They advocated a "balanced" solution that would "allow for the co-existence of fishermen and sea otters" through a continuation of zonal management, but they offered no explicit suggestions about how non-lethal containment could realistically be achieved (USFWS, 2011, Appendix G; USFWS, 2012, Appendix G).

The quarter century during which the translocation program remained in effect witnessed a marked increase in the engagement of California's public in matters of marine policy, in part due to fisheries collapses. The sport and commercial abalone fisheries were closed in 1997, not just in the sea otter's range, but throughout California (except for a limited sport fishery north of San Francisco). Citizens' initiatives asserted the right of the broader public to engage in issues that had formerly been dominated by what many now regarded as special-interest groups. The Marine Life Management Act of 1998 mandated a new approach to fisheries management that emphasized the conservation of entire ecosystems, recognized the importance of esthetic, recreational, scientific, and educational values, and foregrounded long-term sustainability over short-term gain. The Marine Life Protection Act of 1999 directed the state to redesign the system of marine protected areas in California to function as a scientifically designed network in accordance with goals that included protecting marine ecosystems, improving opportunities for non-consumptive uses, and protecting representative and unique marine life habitats for their intrinsic value. The expansion of public engagement in marine policy, increased understanding of cascading ecological effects, and formal recognition of the intrinsic value of intact marine ecosystems created an environment in which the narrow goals of maximum short-term resource extraction could no longer rule the day. Termination of the translocation program brought federal sea otter policy in California in line with the goals of the California public and the ecosystem vision of the MMPA and ESA.

2. White abalone and black abalone were listed as endangered under the ESA in 2001 and 2009, respectively, due primarily to reductions caused by overharvesting for human consumption and, in the case of black abalone, also disease (66 FR 29046, May 29, 2001; 74 FR 1937, January 14, 2009).

Alaska

If the prevailing public attitude toward sea otters in California is now one of interest and good will, Alaskan attitudes are more culturally diverse and less amenable to broad characterization. The maritime fur trade not only nearly extirpated sea otters but also profoundly affected coastal indigenous peoples, leaving in its wake a complex mixture of cultural attitudes and emotions among Alaska Natives toward sea otter conservation and hunting (see Chapter 11). Coastal Alaska Natives traditionally used sea otter furs for clothing and also hunted them for food (Gross and Khera, 1980; Jones, 1985; Lech et al., 2011). Although hunting may have led to localized reductions in some areas (Simenstad et al., 1978), sea otters remained abundant throughout their range until their discovery by Russian seafarers in the eighteenth century. In their relentless pursuit of sea otter furs, the Russians, who pressed Aleut and Alutiq men into service as hunters, and later the British and Americans, who traded for furs (Gibson, 1976), fundamentally transformed Alaska's indigenous cultures. Many traditional Native practices were lost during this time, and interest in using sea otters for handicrafts, clothing, and trade under an MMPA exemption may be seen as part of a broader effort at cultural reclamation.

For most of Alaska's history following European contact, the non-Native population was transient and almost exclusively concerned with resource extraction in the fur, fishing, mining, timber, and oil industries; Alaska was a place to go, get rich, and leave (Morehouse et al., 1984). In more recent decades, Alaska's non-Native population has become less mobile, and there are now multiple generations of non-Native people, many of whom have a different perspective regarding the natural world and environmental stewardship. In parts of South Central Alaska, such as Homer, sea otters are embraced as the center of a thriving tourist industry based on wildlife viewing (see, e.g., Klouda, 2012). Although wildlife viewing is also an important part of the economy in Southeast Alaska, the conflict between sea otters and shellfish fisheries there has produced an attitude of hostility that has come to dominate other views of sea otters on the Alaskan political stage.

The Sea Otter–Shellfish Fishery Conflict in Southeast Alaska

In the aftermath of the fur trade, sea otters persisted in Alaska in only a few isolated locations across their northern range and were absent south of Prince William Sound. When Alaska attained statehood in 1959, it gained jurisdiction over sea otters, and in the mid-1960s, in concert with translocations elsewhere, the Alaska Department of Fish and Game (ADF&G) reintroduced sea otters to Southeast Alaska (see Chapter 3). Although neither the State of Alaska nor USFWS documented clear objectives for the reintroductions, both the translocations and the "experimental harvests" that occurred in the

1960s (which yielded 1000 pelts for public auction in 1968, the first that had occurred since 1911) were apparently viewed in the practical context of game management, with incidental benefits for scientific study (Kenyon, 1969). Enactment of the MMPA brought the State of Alaska's short-lived management authority for sea otters to an end and returned it to the federal government.

Sea otter abundance in Southeast Alaska has subsequently been estimated many times by a variety of methods, including aerial and boat-based surveys. Unlike California, Alaska has a complex coastline and little human infrastructure to support surveys or other forms of monitoring. In general, surveys show that population growth was slow during the first decade after translocation but gradually increased over time (Esslinger and Bodkin, 2009). Pitcher (1989) estimated a population size of 4462 sea otters, and Agler et al. (1995) estimated a population size of 8180 sea otters. An aerial survey by the US Geological Survey (USGS) of all known sea otter habitats in Southeast Alaska in 2002 and 2003, using a correction factor to account for sea otters under water, resulted in an estimated population size of 8949 (Esslinger and Bodkin, 2009). In 2010–2012, USFWS and USGS replicated the 2002–2003 survey and found that the population had increased in range and size but not yet recolonized all of Southeast Alaska. An estimated 25,712 sea otters inhabited Southeast Alaska as of 2012 (USFWS, 2013), a number roughly nine times the size of the sea otter population in California distributed over a larger and much more complex habitat.

What might have been viewed as a remarkable conservation success story in Southeast Alaska has been met with anger and hostility by a vocal segment of the population. As in California, the absence of sea otters from much of their habitat in the aftermath of the fur trade initiated profound changes in the nearshore marine environment. Released from the foraging pressure of a top predator, macroinvertebrate populations increased, community structure shifted, and lucrative fisheries developed to exploit the accumulated abundance of Dungeness crabs, California sea cucumbers, geoduck clams, and red sea urchins (hereafter referred to collectively as "shellfish"). In the early 1980s, Ancel Johnson of USFWS observed that "the combination of perceiving sea otters as a competitor and not being able to take them for their pelts is causing increasing feelings of resentment and animosity toward sea otters and resource management agencies" (Johnson, 1982). The conflict between sea otters and shellfish fisheries in Southeast Alaska intensified in the 1990s, as reflected in a *Los Angeles Times* article that described the complaints of fishermen and subsistence hunters who saw the growing sea otter population as a threat to their livelihood (Mader, 1993). As sea otters in Southeast Alaska continued to increase in number and to expand their range into new areas during the 2000s, fishermen became increasingly agitated about competition with sea otters for shellfish.

In late 2005, the Southeast Alaska Regional Dive Fisheries Association (SARDFA), a private non-profit economic development organization, released a report it had commissioned to estimate economic losses to Southeast Alaska commercial shellfish fisheries caused by sea otter predation (McDowell, 2005). Considering every California sea cucumber, red sea urchin, geoduck, and Dungeness crab consumed by a sea otter to be a direct loss to the income of fishermen, the McDowell report estimated an impact of $11.2 million in total economic activity from 1996 through 2005 (McDowell, 2005). The report's findings ignited the fury of Southeast Alaska fishermen, who denounced what they viewed as poor management of this "voracious" predator by USFWS. Fishermen began to organize and to call collectively for USFWS to "manage" sea otters to reduce their impacts on the prey species for which fishermen compete. They demanded reduced numbers of sea otters overall and even the extirpation of sea otters in areas of known shellfish concentrations.

Governed by the protective provisions of the MMPA, USFWS could not legally reduce sea otter numbers to protect a commercial fishery even if it deemed it desirable to do so, nor could fishermen. However, when the MMPA was enacted, Ted Stevens of Alaska (United States Senator, 1968–2009) successfully advocated for an amendment that created an exemption for Alaska Natives in recognition of their historical reliance upon marine mammals. The exemption allows coastal Alaska Natives to "take" marine mammals for subsistence or for the purpose of creating or selling authentic native articles of handicrafts and clothing. Describing Alaska Native handicrafts as finished goods exhibiting a high degree of workmanship and skill, Senator Stevens explained, "We have sought a solution that would protect the mammals, yet not wipe out the Eskimo culture and several important Native handicraft activities in the process" [118 Cong. Rec. 8,400 (1972)]. Under the exemption, coastal Alaska Natives may legally hunt sea otters at any time and place in Alaska for the specified purposes but are required to have the skull and hide tagged within 30 days to facilitate monitoring of hunting-related mortality. Without a tag from USFWS, hunters cannot have the hide commercially tanned, although some do their own tanning. Removing the hide from a sea otter is sufficient to avoid "wasteful take" under the MMPA; unlike other marine mammals hunted under this provision, no meat is required to be salvaged.

The Alaska Native exemption affords USFWS few management options to control or limit the hunting of sea otters unless a marine mammal stock is declared "depleted"; only then can regulations be established to institute seasonal closures, bag limits, or other methods that have traditionally been used to manage the human exploitation of wildlife populations. A depleted designation under the MMPA means that a stock is listed as threatened or endangered under the ESA or that a stock is below its optimum sustainable population size. The Alaska Native exemption is limited, however, in that it applies only to hunting for the specified purposes; it does not allow Natives to kill sea otters for other reasons, such as to reduce population size.

Searching for ways to reduce sea otter numbers, non-Native fishermen in Southeast Alaska approached USFWS to determine the legality of numerous scenarios intended to encourage increased Native hunting of sea otters. None of these scenarios were deemed legal because the MMPA contains no provisions to allow for non-Native participation in the hunting of marine mammals. The fishermen's advances generated uneasiness among Alaska Native hunters and handicrafters, who feared they could be arrested by USFWS for making articles from sea otter hide that were not considered to be "significantly altered." If Alaska Natives could not sell the products they made from sea otters, there was no point in hunting more of them. These outcomes increased the frustration of the fishermen and escalated their perception of USFWS as the agency that, instead of supporting their efforts to protect fisheries by reducing sea otter numbers, was actually preventing sea otters from being "managed."

Increased focus on the intent of the MMPA's exemption for Alaska Natives reignited long-standing frustrations within the Alaska Native community over the federal government's enforcement of the term "significantly altered," which USFWS had introduced in its implementing regulations for the MMPA in 1974. The MMPA defines "[a]uthentic native articles of handicrafts and clothing" as "items composed wholly or in some significant respect of natural materials, and which are produced, decorated, or fashioned in the exercise of traditional native handicrafts without the use of pantographs, multiple carvers, or other mass copying devices." According to the MMPA, "[t]raditional native handicrafts include, but are not limited to weaving, carving, stitching, sewing, lacing, beading, drawing and painting." In its implementing regulations, USFWS defined "authentic native articles of handicrafts and clothing" as items made by Alaska Natives that are "composed wholly or in some significant respect of natural materials *and are significantly altered from their natural form* [italics added] and are produced, decorated, or fashioned in the exercise of traditional native handicrafts" (50 CFR Part 18). The term "significantly altered" applies only to the sale of handicrafts made from sea otters when sold by an Alaska Native to a non-Native; Alaska Natives may sell and trade raw pelts and handicrafts to other Alaska Natives without restriction. The interjection of the term "significantly altered" into enforceable regulations created a legacy of confusion, hostility, and fear within the Alaska Native community. Because it is a poorly defined and subjective term, it has been applied unequally by the USFWS Office of Law Enforcement to a variety of sea otter handicrafts, resulting in a series of confiscations and court decisions over the past decades.

The terms "authentic native articles of handicrafts and clothing" and "traditional" have also caused confusion, and both have been the subject of court decisions. In 1985 and 1986, USFWS law enforcement agents confiscated items of handicrafts and clothing made from sea otter pelts and fur from Marina Rena Katelnikoff Beck, an Aleut, and Boyd Didrickson, a Tlingit.

Katelnikoff Beck's items, which included a fur teddy bear, were confiscated "on the basis that they did not fall within the native handicraft exception because they were not of a nature commonly made or produced prior to the passage of the MMPA in 1972," and Didrickson's items were confiscated by USFWS law enforcement agents on the basis that the sea otter fur parka and hat contained metal snaps and zippers, "which did not qualify as traditional" (*Didrickson v. U.S. Dep't. of Interior*, 982 F.2d 1332 (9th Cir. 1992)). The two cases were merged and decided together. The Friends of the Sea Otter, Greenpeace, Alaska Wildlife Alliance, and the Humane Society of the United States (collectively referred to as "Friends of the Sea Otter" for this court case) were permitted by the district court to intervene on behalf of the government. The case was complicated and protracted, with several motions and appeals over a period of years.

In April 1990, after extensive hearings in which Alaska Natives and Friends of the Sea Otter testified and submitted evidence, USFWS amended its implementing regulations, determining that items created in whole or in part from the sea otter did not meet the requirements of the MMPA exemption (55 FR 14,973, April 20, 1990). This determination essentially made it illegal for Alaska Natives to create any handicrafts or clothing from sea otters. USFWS based its conclusion on the finding that "no handicraft trade using sea otters by Alaska Natives was in existence prior to passage of the Act that would allow the utilization of sea otters under the handicraft exemption" (55 FR 14,973, April 20, 1990). Didrickson and Katelnikoff Beck filed complaints challenging the regulation, and the court determined in September 1991 that the part of the USFWS regulation that excluded any handicrafts made from sea otters was contrary to Congress' express definition of "authentic native articles of handicraft or clothing" in the MMPA (*Didrickson v. U.S. Dept. of the Interior*, 796 F. Supp. 1281 (D. Alaska 1991)). The requirement that items "were commonly produced on or before December 21, 1972" (the date the MMPA was enacted) was also struck down. The US Court of Appeals upheld the district court's judgment. The courts agreed with the Alaska Natives' contention that there was "extensive evidence of centuries of use by Alaska Natives of the sea otter for many purposes, including clothing, handicrafts and items of barter and trade," and that continued subsistence practices were interrupted and precluded first by the Russians who, while "using the Alaska Natives to harvest the sea otter, banned the sea otters' use by Alaskan Natives," then by American traders, who imposed similar restrictions, and finally by the US government, which imposed restrictions on killing sea otters once they were nearly extinct (*Didrickson v. U.S. Dep't. of Interior*, 982 F.2d 1332 (9th Cir. 1992)).

Although these decisions provided some clarity regarding what Alaska Natives could legally do with sea otter pelts, fear of enforcement action for selling handicrafts that do not rise to the standard of these ambiguous terms has been blamed by many Alaska Natives for quelling their right to practice subsistence from sea otters. The distrust engendered by USFWS law enforcement actions remains acute within the Alaska Native community, and pressure from commercial fishermen to increase sea otter hunting and handicraft production has caused additional confusion about what is legal and increased worry about the potential for prosecution. By 2010, the powerful lobbyists of Alaska's fishing industry had determined that increased Alaska Native subsistence harvest of sea otters was the solution to their problem of competition with sea otters, regardless of what Alaska Natives might think of this approach. Fishing groups called upon Representative Don Young (R-Alaska) and Senator Lisa Murkowski (R-Alaska) to introduce federal legislation that would increase the legal harvest of sea otters and lower the standards for meeting the term "significantly altered," if not erase it altogether. They called upon Alaska state legislators to do the same.

Legislative efforts to decrease sea otter abundance in Southeast Alaska began in earnest in March 2011, when Peggy Wilson (R-Wrangell) introduced State of Alaska House Joint Resolution 26, a non-binding state resolution calling on the federal government to implement population control of sea otters by means of the Alaska Native exemption. The resolution urged federal agencies to work with ADF&G, Alaska Natives, and others to develop a plan "for the sustainable management of the reintroduced sea otter population of Southeast Alaska." Purporting to be a call for ecosystem management, the resolution cited the failure of the federal government to develop "an effective management plan for protecting the ecosystems affected by sea otters" and noted that high numbers of "unmanaged" and "reintroduced" sea otters "appear to be contributing to the degradation of the ecological balance in many areas, leading to diminished human harvests of [...] important subsistence and commercial resources." It contained a proposal for federal consideration, which it represented as a winning proposition for Native Alaskans and non-Natives alike. The federal government should "consider broadening the scope of allowable uses for sea otters taken for subsistence purposes to include the use, transfer, and sale of intact sea otter pelts in order to restore to the state's Native people the right to make full use of sea otters harvested for subsistence while expanding and enhancing economic opportunities for residents of Southeast Alaska."

While the resolution represented itself as pro-Alaska Native, it failed to secure Native support. At the 2012 hearing on the resolution, Alaska Federation of Natives board member and Sealaska Heritage Institute President Dr. Rosita Worl testified that it "was received with mixed emotions in the Native community" because Native Alaskans wanted to maintain the protection afforded by their existing exemption and were opposed to allowing the sale of pelts to non-Natives. Worl added that "the language would [...] raise the considerable ire of animal rights groups" (Stigall, 2012). Although the Alaska State House passed the resolution in March 2012, it died in a Senate committee in April of that same year.

Similar bills were introduced at the federal level in July 2011. Senator Murkowski introduced S1453, co-sponsored by Senator Mark Begich (D-Alaska), to amend the MMPA "to allow the transport, purchase, and sale of pelts of, and handicrafts, garments, and art produced from southcentral and southeast Alaska northern sea otters that are taken for subsistence purposes." Representative Young introduced the companion bill in the House. HR 2714, which became known as the "Otter Slaughter Bill," sought to legalize the interstate sale of sea otter pelts and the export of Alaska Native handicrafts. Representative Young explained that he had introduced HR 2714 for two reasons: "to allow the sale of whole sea otter pelts to non-Natives, which is now banned, and as a predator control measure." He echoed the approach of the state's resolution in representing the bill as a "win-win": the legislation would not only protect shellfish fisheries but also expand economic opportunities for Alaska Natives (Burke, 2011). Capitalizing on Alaska Natives' distrust of USFWS, Young's office issued a press release expressing outrage at the federal agency's "blatant abuse of power" and complaining that "[t]oo many of my Alaskan Native hunters and craftsmen fear that their subsistence rights are being infringed upon due to [...] questionable law enforcement techniques" (Young, 2011).

Despite Young's attempts to package the bill as a boon for Alaska Natives, the Alaska Native community remained dissatisfied with its main provision, legalization of the sale of whole sea otter pelts to non-Natives. Although, like Young, they disapproved of what they viewed as USFWS's "overzealous law enforcement and entrapment practices," the Alaska Federation of Natives passed a resolution in October 2011 supporting the bill only if it stated that the MMPA's Native exemption would be maintained, provided for co-management that would allow tribes to determine who could participate in subsistence activities, and disallowed the sale of pelts to non-Natives (Alaska Federation of Natives, 2012). The Alaska Native Tanners' Association was also opposed to allowing non-Natives to legally possess and process raw pelts into products. As one Alaska Native hunter and sewer saw it, the legislation pitted Alaska Native hunters and handicrafters against one another: for hunters, it was an opportunity to kill as many sea otters as they could sell, while for handicrafters, many of whom do not hunt, the bill would expand the market for pelts, increasing the price beyond their reach, and diminish their opportunity to create and sell their products. Confronted with these concerns, Young concluded, "That's just being selfish" (Burke, 2011).

A legislative hearing on HR 2714 was held in October 2011. Testifying before a Subcommittee of the House Natural Resources Committee, USFWS opposed the bill because it could "create an unregulated commercial market" for sea otter hides, affect the recovery of the ESA-listed southwest stock of the northern sea otter, make compliance and enforcement difficult because different provisions would apply to different sea otter stocks, and "result in unsustainable removals from the population" because of the lack of a mechanism under the MMPA to regulate subsistence hunting prior to a finding that the stock is "depleted." The MMC echoed and added to the concerns expressed by USFWS, pointing out that the bill would "open the door to the commercial harvesting of sea otters"; "confound enforcement"; undermine Alaska Native cottage industries because pelts, once sold or resold either nationally or internationally, could be made into products without any restrictions; and "undermine U.S. policy and diplomacy" internationally regarding restrictions on commerce in marine mammal parts. The MMC also questioned the true purpose of the bill, noting that its "impetus [...] may have been an interest to address a fishery management issue," which, if true, "warrants full description and review," including, at the very least, a consideration of "potential ecological interactions between sea otters and fisheries" and "the valuable role that sea otters play in the ecology of nearshore ecosystems" (US Government Printing Office, 2012).

The Indigenous People's Council for Marine Mammals expressed appreciation for Young's efforts and support for the spirit of the legislation but concern "that the language in HR 2714 that allows for the sale of unaltered pelts could only be a short-term fix with a potential for unintended consequences." The Council requested that the sale of unaltered pelts be considered only in accordance with "local harvest management plans as allowed in the MMPA" and only as "consistent with the existing exemptions of the MMPA related to Alaska Natives." It also requested a restriction to ensure that any pelts sold be prohibited from being made into commercial products by people who were not Alaska Natives. The Alaska Sea Otter and Steller Sea Lion Commission, a non-profit consortium representing more than 50 tribes and tribal organizations, submitted written comments that, while similarly appreciative of Young's efforts to expand the market for Alaska Native handicrafts, even more explicitly opposed the sale of raw pelts to non-Natives. The Commission warned that any "benefit gained from increasing markets could be eliminated" because of the "new class of competition" that would result from non-Natives selling products made from raw pelts. It also expressed "deep concern that the potential exists for overharvest to occur," which could harm Native communities if it resulted in a finding of "depleted" status and USFWS regulation of Alaska Native hunting (US Government Printing Office, 2012). HR 2714 died in committee, its only lasting effect a worsening of the already adversarial relationships among fishing groups, Alaska Natives, USFWS, and ADF&G.

Shortly after the hearing on HR 2714, SARDFA sought to renew outrage over the economic losses caused by sea otters with its release of a follow-up to the 2005 McDowell report (McDowell, 2011).[3] Laine Welch, "fish beat" reporter and advocate for Alaska's fishing industry, publicized the new report's finding that sea otter predation "has cost Southeast Alaska's

3. Causality may not be as clear as the 2005 and 2011 McDowell reports suggest. In 2013, 2.6 million pounds of Dungeness crab were landed, well above the anticipated catch for that year and the catch in 2012 (Viechnicki, 2013).

economy more than $28 million in direct and indirect impacts since 1995" and its "grim conclusion" that "commercial dive fishing and large populations of sea otters cannot coexist in the same waters" (Welch, 2011). "Sea otters are cleaning out valuable commercial fisheries in Southeast Alaska," Welch announced, "and they have been at it for decades" (Welch, 2011). However, retired ADF&G biologist Rick Sinnott sought to put this number into perspective. Calling attention to yet another McDowell report, this one commissioned by the Alaska Wilderness League, Sinnott noted that its estimate of total nonresident visitor-related spending in Southeast Alaska for 2010–2011, once multiplier effects were added, was $360 million, of which wildlife viewing (driven by charismatic animals like sea otters) accounted for 42%. This $151 million, Sinnott pointed out, was "89 times the payroll generated by sea urchins, geoducks, and other marine invertebrates harvested by dive fishermen" (Sinnott, 2013).

Southeast Alaskan furor over the economic costs of sea otters, combined with the approaching decision in California on whether to end the sea otter translocation program and its associated "no-otter zone," prompted ecologist Dr. James Estes to write an op-ed that appeared in the *Los Angeles Times* in February 2012. Estes argued that the presumed dichotomy between "ethical and monetary values" whenever the "recovery of a predator is occurring or has been proposed" is faulty, in that it fails to recognize the multiple avenues by which intact ecosystems provide economic value. "Much like the way wolves cull deer and elk populations on land, sea otters eat urchins," Estes noted, and an "ever-growing body of research shows that the ecological and economic influences of predators [. . .], from sea otters to wolves, extend well beyond the things they eat." Citing the benefits of kelp forests—they "support regional fish populations, serving both as a vital habitat and a source of nourishment for many species," and "provide humans with a number of crucial natural services," such as "helping to buffer shorelines from wave exposure" and "reducing rates of coastal erosion and shoreline recession"—Estes stressed that "these ecological services provide very real dollars-and-cents benefits for society that policymakers should take into consideration" (Estes, 2012).

Despite its legislative failures, the Southeast Alaska shellfish industry was unwilling to give up on what it viewed as its most promising avenue for predator control. By early 2012, the Juneau Economic Development Council, a private non-profit corporation, had begun an "Action Initiative" to "Develop a Sea Otter Management Program in Southeast Alaska." Formulated without broad stakeholder involvement under the leadership of the Executive Director of SARDFA, the initiative had a now-familiar goal: "to analyze realistic management approaches to protect important shellfish species and to allow a less restrictive harvest of sea otters by Alaska Natives." Its anticipated product was "an effective Southeast Alaska sea otter management plan for increased subsistence use by Alaska Natives," which "should benefit all users of shellfish resources, protect shellfish resources from depletion, and allow effective subsistence harvest of sea otters by the Alaska Native people." Immediate efforts would be directed at "helping local Native sea otter hunters/artists with the business of making and selling sea otter products" (Juneau Economic Development Council, 2014).

Throughout 2012 and 2013, USFWS engaged with Alaska Native organizations, fishery advocates, academics, environmentalists, and interested citizens to search for a way forward. At a special session on the conflict at a State Board of Fisheries meeting in Petersburg, SARDFA proposed a "scorched earth" management policy: allowing divers to fish out areas completely before sea otters expanded there. This proposal was rejected, and as the session made clear, no resolution could satisfy all stakeholders. In August 2012, USFWS Director Dan Ashe traveled to Juneau and Sitka to discuss the impacts of sea otter predation on subsistence, commercial, and recreational fisheries in Southeast Alaska. Senator Murkowski arranged the meetings with stakeholders for the stated purpose of resolving growing conflicts between USFWS law enforcement officers and Alaska Native handicrafters regarding interpretation and enforcement of the terms, such as "significantly altered," that delimit the Alaska Native exemption. Didrickson was present and determined to make sure that Director Ashe understood the history of the court challenges over the problematic terms. By the end of the meetings, Ashe had agreed that USFWS would clarify the definitions in its MMPA implementing regulations of the terms "significantly altered," "coastal dwelling," and "mass production."

Following another USFWS workshop in October 2012, Alaska Native participants drafted a clarification of the term "significantly altered." After modifying it to conform with statutory language in the MMPA, USFWS released a draft of the clarifying language in March 2013 with a 120-day request for comments from other stakeholders. Few people were pleased. The environmental community reacted with fury, accusing USFWS of crafting the language to increase sea otter hunting by Alaska Natives to placate fishermen. Fishermen were angry that the new language was too restrictive in the types of handicrafts it allowed and would therefore fail to reduce the sea otter population. Some Alaska Natives thought the new language was more restrictive, whereas others thought it was less so. No one agreed on whether the new language was easier to understand. The final language defining the term "significantly altered," with heavy editorial input and approval from Senator Murkowski's office, was released in November 2013 and made available on USFWS's website in January 2014 with little fanfare:

> *A sea otter will be considered "significantly altered" when it is no longer recognizable as a whole sea otter hide, and has been made into a handicraft or article of clothing as is identified below: 1. A tanned, dried, cured, or preserved sea otter hide, devoid of the head, feet, and tail (i.e., blocked) that is*

> *substantially changed by any of the following, but is not limited to: weaving, carving, stitching, sewing, lacing, beading, drawing, painting, other decorative fashions, or made into another material or medium; and cannot be easily converted back to an unaltered hide or piece of hide. 2. Tanned, dried, cured, or preserved sea otter head, tail, or feet, or other parts devoid of the remainder of the hide which includes any of the following, but is not limited to: weaving, carving, stitching, sewing, lacing, beading, drawing, or painting, other decorative fashions, or made into another material or medium. (USFWS, 2014)*

USFWS abandoned its attempt to clarify the remaining two terms, "coastal dwelling" and "mass production," in response to public comments.

The environmental community was now fully engaged with the sea otter–fisheries conflict in Southeast Alaska. Suspicious of USFWS' motives, the Friends of the Sea Otter, Humane Society, Center for Biological Diversity, Oceans Public Trust Initiative, and Alaska Wildlife Alliance jointly filed a Freedom of Information Act request for all USFWS records relating to management of sea otters in Southeast Alaska, including all records that related to Alaska Natives, sea otter harvest, fisheries, and assessment of the sea otter population size. After several months, USFWS handed over thousands of pages of emails, meeting notes, PowerPoint presentations, data, maps, and other files.

In February 2013, Alaska state politicians tried a new approach, attempting to resurrect a wise-use-era method of predator control. SB 60, introduced by Alaska State Senator Bert Stedman (R-Sitka), provided $100 for each sea otter legally killed under the MMPA Alaska Native exemption. Generating international headlines, the "Bounty Bill" was intended to "incentivize the lawful harvest of sea otters by Alaska Natives to, at the very least, reach the potential biological removal" (Stedman, 2013). At the bill's hearing before the Senate Resources Committee, Craig Fleener, then Deputy Commissioner of ADF&G, complained that the "MMPA, blindly and imprudently, provides all-encompassing protections for every marine mammal irrespective of their abundance, impact to other species, or detrimental impacts to humans." He applauded the efforts that the State of Alaska had made to reduce sea otter abundance as reflecting "a holistic and more sensible ecosystem-based approach" (Open States, 2013). Dennis Watson, former commercial dive fisherman and Mayor of Craig, testified that he had "witnessed first-hand the devastation these creatures cause" (Schoenfeld, 2013).

Although the bill had a contingency of vocal supporters, not everyone was in favor of it. Joe Sebastian, a commercial fisherman from Kupreanof, suggested that the bill wrongly blamed sea otters, as overharvesting was likely the cause of shellfish declines, and added, "I find it unprofessional, unscientific, racist and culturally destructive. This particular bill [...] would start the new sea otter gold rush with little or no oversight or scientific direction" (Schoenfeld, 2013). The Alaska Wildlife Alliance observed that the bill had "damaged Alaska's already tarnished image for wildlife management." It argued for an ecosystem view informed by science, noting that by benefiting kelp, sea otters enhance the "productivity of the nearshore, provide habitat for diverse species of invertebrates, fishes—including salmon and herring—and marine birds and mammals," and "reduce greenhouse gases," thereby reducing the threat to shellfish posed by ocean acidification.[4] Alpheus Bullard, attorney for the State of Alaska, sounded the bill's death knell, advising that providing a state bounty for sea otters taken by Alaska Natives likely conflicted with federal law (Open States, 2013). SB 60 was referred to the Judiciary Committee, where it languished until the legislative session closed for the season.

Acknowledging the Bounty Bill's legal issues, Stedman suggested there might be other ways to solve the sea otter problem: "If this mechanism is not palatable with the attorneys [...] we could always move the funds to the tannery [...] to help encourage the same solution" (Miller, 2013). The same month as the hearing, the State of Alaska awarded a three-year grant to the Sealaska Heritage Institute (SHI) to start classes in the art of skin sewing, focusing specifically on sea otter skins. The stated goals of SHI's "Southeast Alaska Sustainable Arts Project" are "to save a nearly lost art form, develop a cottage industry in rural communities, and more fully utilize a sustainable resource." Its website advertises sea otter fur as "the finest in the world" and notes that "SHI is working to open new markets for this luxurious product" (SHI, 2013). As a registered agent with USFWS, SHI may legally buy skins from Native people for skin-sewing programs. These cash-based transactions are intended to help to sustain Native people living in economically depressed areas of Southeast Alaska. Viewing the SHI program as a putatively legal way to pay Alaska Natives to hunt more sea otters, Stedman hoped to revamp the Bounty Bill without the actual bounty: "I need to sit down with [SHI] and have a few more meetings in Juneau to work out what we're actually going to change it to—if it's going to end up trying to be marketing assistance or tanning assistance or something else" (Schoenfeld, 2014).

Even without the assistance of the Bounty Bill, the number of sea otters legally killed and reported to USFWS reached an all-time high in Southeast Alaska in 2013. Approximately 1500 sea otters had been reported as of early March 2014, with more certifications for 2013 arriving (USFWS Marking and Tagging and Reporting Program, unpublished data). This number is far higher than the historical (1989–2012) annual mean of 385 and is fast approaching the Potential Biological Removal (PBR) for the stock. As the sea otter–shellfish fishery conflict in Alaska continues to play out, it remains to be seen whether sea otters in Southeast Alaska will fall victim to

4. Over the last 5 years (2008–2012) the Southeast Alaska sac roe herring fishery took an estimated average of 13,747 tons of fish (Thynes et al., 2013) worth over $10 million annually.

continuing efforts to institute predator control or continue their unprecedented recovery of historical range.

CONCLUSION

The prevailing attitudes toward sea otters in California and Alaska appear to be completely different: Californians, in general, are now fiercely protective of sea otters, whereas a vocal segment of the Southeast Alaskan public and members of the Alaskan legislature whose attention they have captured view sea otters as vermin that should be subjected to predator control. However, as the historical details of the sea otter–shellfish fishery conflicts in each state demonstrate, the pattern has been virtually the same. When sea otters recolonize a new portion of their historical range and enter into competition with a particular fishery, there are calls to "manage" and "control" them for the sake of that fishery. All subsequent decreases in landings tend to be attributed solely to sea otter predation, to the exclusion of other factors such as excessive competition between fishermen, a life history that may predispose certain species to overexploitation and population collapse, poor fishery management, or other biotic or abiotic influences that affect the abundance of prey species. The economic value of the affected fisheries is held up to be of great importance to the state in general, not just to the relatively small proportion of people benefiting from consumptive use of the publicly held marine trust. Other economic benefits provided by sea otters through more complex ecological chains of causation are overlooked and undervalued. The non-economic value of sea otters is particularly undervalued, if acknowledged at all.

The conflict occurring in Southeast Alaska today began, in only slightly different form, in California half a century ago. While not fully resolved, the discord in California has cooled. The coastal regions where sea otters occur have now largely adapted to capitalize on sea otters (sea otters have become a draw for tourism in the former abalone fishery stronghold of Morro Bay, for example), and changes in shellfish fisheries have moderated the intensity of the conflicts associated with sea otter range expansion. The commercial abalone fishery no longer exists, and declines in catch and value have reduced the profitability of the sea urchin industry from its zenith in the early 1990s (CDFG, 2006). Rock crab and lobster fishermen have noticed limited effects from competition with sea otters, dampening their fears about the consequences of range expansion on these fisheries (USFWS 2011; Palomar College Television, 2012). Additionally, sea otter range expansion along the mainland California coastline has slowed, and sea otters have not entered into any new fishing grounds for some time. Finally, many of the fishermen who participated in the lucrative abalone and sea urchin fisheries at their peak and sustained the debate through their personal commitment to the cause have retired or otherwise moved on, and the younger generation seems to have a different view of fisheries as more deeply embedded in ecosystem processes. It remains to be seen if the conflict in Alaska plays out similarly.

There are several important differences in the factors surrounding the conflicts in Alaska and California, the most crucial of which is the MMPA exemption that allows Alaska Natives to hunt sea otters without any limit until the population is deemed "depleted." As the conflict in Alaska has clearly demonstrated, non-Native fishermen and politicians are seeking ways to exploit this exemption to its fullest to further their own interests in predator control. Other differences stem from human population size and the relative importance of fishing and tourism to the state's economy. In Alaska, a state with an estimated population size of approximately 735,000 and a population density of 1.2 persons per square mile, both fishing and tourism constitute a significant proportion of the economy,[5] and the population density remains low enough that Natives and non-Natives alike can engage in fishing as part of a subsistence lifestyle (US Census Bureau, 2014; Alaska Department of Labor, 2014). In California, with an estimated population size of 38,333,000, more than 50 times that of Alaska, and a population density of 239 persons per square mile, the ocean economy is dominated by the tourism and recreation sector, and the shellfish fishing industry constitutes a vanishingly small proportion of the overall state economy (US Census Bureau, 2014; NOEP, 2005; USFWS, 2012). The intense population pressure of California on its natural environment has made many Californians acutely aware of the potential for an oceanic "tragedy of the commons," where a few people take and benefit personally in the short term from the "resources" that belong to everyone. Additionally, the interest of the general public in the marine environment and their familiarity with ecological concepts such as trophic cascades has made them a powerful counterforce to those who would demand some form of sea otter control.

The hope for a positive outcome in such debates about predator control—one that favors biodiversity and intact ecosystems—almost certainly lies in the broadening of the constituency that participates in them and in public education about economic and other benefits resulting from the chain of ecological effects that top predators initiate. Protecting sea otters at the expense of a select number of shellfish fisheries may appear to be privileging sea otters over human beings, but it is actually privileging the long-term good of

5. It should be noted that the number of fishermen affected by competition with sea otters is small compared to the number unaffected (e.g., those fishing for salmon, halibut, pollock, and Pacific cod, which are major economic drivers in Alaska). The number of permits allowed in each affected fishery, with the number actively fished (as of Fall 2011), in parentheses, is as follows: geoduck clams 104 (60), sea cucumbers 436 (180), urchins 95 (10), Dungeness crab 308 (Personal Communication from Phil Doherty, SARDFA). This is a total of only 310 fishermen (in written testimony to the Alaska Senate on 13th March 2013, SARDFA in fact estimated that only about 200 divers were actively fishing Open States (2013), very few compared numerically and economically with other Alaskan fisheries.

the many over the short-term good of the few. As ecological research has shown, top predators provide numerous large-scale benefits through trophic interactions. In the case of the sea otter, these benefits stem from the sea otter's positive effects on the abundance of kelp (see Chapter 2) and seagrass (Hughes et al., 2013), which not only provide habitat for a multitude of other species but also perform other valuable functions, including carbon sequestration (Wilmers et al., 2012). The more that members of the broader public understand their stake in having intact ecosystems and assert their right to have a voice in wildlife policy decisions alongside consumptive users of those ecosystems, the more pro-predator and pro-environment a state's policies will become.

DISCLAIMER

The findings and conclusions in this article are those of the authors and do not necessarily represent the views of the US Fish and Wildlife Service.

ACKNOWLEDGMENTS

We gratefully acknowledge the assistance of Angela Doroff, Carl Benz, Jim Bodkin, and Shawn Larson, whose reviews of the manuscript greatly improved the final product. We would also like to thank Frank Reynolds and Jim Curland, of Friends of the Sea Otter, and Joan Parker, Librarian at Moss Landing Marine Laboratories, for generously providing access to archival materials. Finally, we thank the Bancroft Library, University of California, Berkeley, for permission to quote portions of Margaret Owings' oral history.

REFERENCES

Agler, B.A., Kendall, S.J., Seiser, P.E., Lindell, J.R., 1995. Estimates of Marine Bird and Sea Otter Abundance in Southeast Alaska during Summer 1994. US Fish and Wildlife Service, Anchorage, AK, 102 pp.

Alaska Department of Labor, 2014. Alaska Economic Trends: Employment Forecast for 2014. [Online]. Available from: <http://labor.alaska.gov/trends/> (accessed 13.04.14).

Alaska Federation of Natives, 2012. 2011 Annual Convention Final Resolutions, October 22, 2011. [Online]. Available from: <http://www.nativefederation.org/wp-content/uploads/2012/10/2011-afn-convention-resolutions.pdf> (accessed 13.04.14).

Associated Press, 1967. Wanted: Suggestions for Otter Herding. Reading Eagle, 14 June.

Baur, D.C., Meade, A.M., Rotterman, L.M., 1996. The law governing sea otter conservation. Endangered Species Update 13, 73–78.

Bryant, H.C., 1915. Sea otters near Point Sur. Calif. Dep. Fish Game Bull. 1, 134–135.

Burke, J., 2011. Should Non-Natives Be Allowed to Buy Sea Otter Pelts? Alaska Dispatch. [Online]. 20 October. Available from: <http://www.alaskadispatch.com/article/should-non-natives-be-allowed-buy-sea-otter-pelts> (accessed 13.04.14).

California Department of Fish and Game, 1968. Report on the Sea Otter, Abalone and Kelp Resources in San Luis Obispo and Monterey Counties and Proposals for Reducing the

Conflict between the Commercial Abalone Industry and the Sea Otter. Requested by Senate Concurrent Resolution 74 in the 1967 Legislative Session. 72 pp. + appendices.

California Department of Fish and Game, 1976. A proposal for sea otter protection and research, and request for return of management to the State of California. Unpublished 2 volume report.

California Department of Fish and Game, 2005. Final Abalone Recovery and Management Plan. California Department of Fish and Game, Marine Region.

California Department of Fish and Game, 2006. Review of some California Fisheries for 2005. CalCOFI Reports 47, 9–29.

California Legislature, 1963. Affect [sic] of the Sea Otter on the Abalone Resource. Hearing Transcript of the California State Senate Fact Finding Committee, Subcommittee on Sea Otters, San Luis Obispo, 19 November.

Condit, R., 1970. Sierra club resolution. The Otter Raft 3, 5.

Costa, D.P., Kooyman, G.L., 1984. Contribution of specific dynamic action to heat balance and thermoregulation in the sea otter *Enhydra lutris*. Physiol. Zool. 57, 199–203.

Esslinger, G.G., Bodkin, J.L., 2009. Status and trends of sea otter populations in Southeast Alaska, 1969–2003. US Geological Survey Scientific Investigations Report 2009–5045. 18 pp.

Estes, J.A., 2012. On sea otters, we need to see the big picture. Los Angeles Times, 21 February.

Gibson, J.R., 1976. Imperial Russia in Frontier America: The Changing Geography of Supply of Russian America, 1784–1867. Oxford University Press, New York, NY.

Gross, J.J., Khera, S., 1980. Ethnohistory of the Aleuts. Alaska Historical Commission Studies in History No. 45. Department of Anthropology, University of Alaska, Fairbanks, AK.

Hughes, B.B., Eby, R., Van Dyke, E., Tinker, M.T., Marks, C.I., Johnson, K.S., Wasson, K., 2013. Recovery of a top predator mediates negative eutrophic effects on seagrass. PNAS 110 (38), 15313–15318.

Johnson, A.M., 1982. Status of Alaska Sea Otter Populations and Developing Conflicts with Fisheries. US Fish & Wildlife Publications, Paper 42.

Jones, R.D., 1985. Traditional use of sea otters by Alaskan natives, a literature review. Unpublished report, Division of Wildlife Assistance, US Fish and Wildlife Service, 1011 E. Tudor Road, Anchorage, AK.

Judson, D.L., 1969. Moving day for otters. The Otter Raft 1, 2.

Juneau Economic Development Council, 2014. Southeast Cluster Initiative: Ocean Products. [Online]. Available from: <http://www.jedc.org/southeast-cluster-initiative-ocean-products> (accessed 13.04.14).

Kenyon, K.W., 1969. The Sea Otter in the Eastern Pacific Ocean. Dover Publications, New York, NY.

Klouda, N., 2012. Swimming in Sea Otters. Homer Tribune, 11 July.

Lech, V., Betts, M.W., Maschner, H.D., 2011. An analysis of seal, sea lion, and sea otter consumption patterns on Sanak Island, Alaska: an 1800-year old record on Aleut consumer behavior. In: Braje, T.J., Rick, T.C. (Eds.), Human Impacts on Seals, Sea Lions, and Sea Otters: Integrating Archaeology and Ecology in the Northeast Pacific. University of California Press, Berkeley and Los Angeles, pp. 111–128.

Mader, I., 1993. Sea otter population booms amid Hunting Ban: Alaska: biologists call it an amazing success story. Natives envision a lucrative fur trade. But conservationists fear there will be abuses. Los Angeles Times, 25 July.

Marine Mammal Commission, 1988. Annual Report of the Marine Mammal Commission, Calendar Year 1987, A Report to Congress.

McDowell Group, 2005. Sea Otter Impacts on Commercial Fisheries in Southeast Alaska. McDowell Group, Juneau and Anchorage, Report prepared for the Southeast Alaska Regional Dive Fisheries Association (unpublished).

McDowell Group, 2011. Sea Otter Impacts on Commercial Fisheries in Southeast Alaska. McDowell Group, Juneau and Anchorage, Report prepared for the Southeast Alaska Regional Dive Fisheries Association (unpublished).

Miller, M.D., 2013. Sea otter bounties likely not legal, but popular with some. Juneau Empire. [Online]. 14 March (updated 15 March). Available from: <http://juneauempire.com/state/2013-03-14/sea-otter-bounties-likely-not-legal-popular-some> (accessed 13.04.14).

Morehouse, T.A., McBeath, G.A., Leask, L., 1984. Alaska's Urban and Rural Governments. University Press of America, Lanham, MD.

Nixon, R., 1969. Remarks Following Inspection of Oil Damage at Santa Barbara Beach, March 21, 1969. The American Presidency Project In: Peters, G., Woolley, J.T. (Eds.), [Online]. Available from: <http://www.presidency.ucsb.edu/ws/?pid=1967> (accessed 13.04.14).

NOAA Hazardous Materials Response Assessment Division, 1992. Santa Barbara Oil Spill—January 28, 1969. In "Oil Spill Case Histories, 1967–1991: Summaries of Significant U.S. and International Spills" (unpaginated). NOAA Hazardous Materials Response and Assessment Division, Seattle.

NOEP, 2005. California's Ocean Economy. Report to the Resources Agency, State of California, prepared by the National Ocean Economics Program. [Online]. Available from: <http://resources.ca.gov/press_documents/CA_Ocean_Econ_Report.pdf> (accessed 13.04.14).

Open States, 2013. All documents for SB 60. [Online]. Available from: <http://openstates.org/ak/bills/28/SB60/documents/> (accessed 13.04.14).

Owings, M.W., 1968. Do Sea Otters Have Any Friends? Monterey Peninsula Herald, 11 March.

Owings, M.W., 1970. The outlook. The Otter Raft 4, 1.

Owings, M.W., 1974. Cliffside seat. The Otter Raft 11, 1.

Owings, M.W., 1986. Cliffside seat. The Otter Raft 36, 1.

Owings, M.W., 1991. "Artist, and Wildlife and Environmental Defender," an Oral History Conducted in 1986–1988 by Suzanne Riess and Ann Lage. Regional Oral History Office, The Bancroft Library, University of California, Berkeley, CA.

Palomar College Television, 2012. Threatened: The Controversial Struggle of the Southern Sea Otter. [Online Video]. Available from: <http://vimeo.com/56121277> (accessed 13.04.14).

Pentony, C.C., 1970. The California sea otter petition. The Otter Raft 3, 5.

Pitcher, K.W., 1989. Studies of Southeastern Alaska sea otter populations: distribution, abundance, structure, range expansion and potential conflicts with shellfisheries. Anchorage, Alaska. Alaska Department of Fish and Game, Cooperative Agreement 14-16-0009-954 with US Fish and Wildlife Service, 24 pp.

Robinson, M.J., 2005. Predatory Bureaucracy: The Extermination of Wolves and the Transformation of the West. University Press of Colorado, Boulder, CO.

Roest, A.I., 1973. Subspecies of the sea otter, *Enhydra lutris*. Contributions in Science, Natural History Museum of Los Angeles County 252, 1–17.

Schoenfeld, E., 2013. Otter bounty bill gets good, bad reviews. CoastAlaska News. [Online]. 14 March. Available from: <http://www.kcaw.org/2013/03/14/otter-bounty-bill-gets-good-bad-reviews/> (accessed 13.04.14).

Schoenfeld, E., 2014. Stedman says hydro funds tight, otter bill will change. CoastAlaska News. [Online]. 15 January. Available from: <http://www.kcaw.org/2014/01/15/stedman-says-hydro-funds-tight-otter-bill-will-change/> (accessed 13.04.14).

Sealaska Heritage Institute, 2013. Skin sewing. [Online]. Available from: <http://www.sealaska-heritage.org/programs/Art/SkinSewing.html> (accessed 13.04.14).

Senate Permanent Factfinding Committee on Natural Resources, 1965. Third Progress Report to the Legislature for the 1965 Regular Session, Pursuant to Senate Resolution No. 145, 1963 Regular Session. Senate of the State of California, Sacramento, CA.

Simenstad, C.A., Estes, J.A., Kenyon, K.W., 1978. Aleuts, sea otters, and alternate stable-state communities. Science 200, 403–411.

Sinnott, R., 2013. Waiting for mutiny on proposed Southeast Alaska sea otter bounty. Alaska Dispatch. [Online]. 25 March. Available from: <http://www.alaskadispatch.com/article/20130325/waiting-mutiny-proposed-southeast-alaska-sea-otter-bounty> (accessed 13.04.14).

Stedman, B., 2013. Sea otter bill introduced. Alaska Senator Bert Stedman Blog Archive. [Online]. 4 March. Available from: <http://bertstedman.com/new/?p=2877> (accessed 13.04.14).

Stigall, R., 2012. Opposition to selling otter pelts to non-Natives. Juneau Empire. [Online]. 6 February. Available from: <http://juneauempire.com/state/2012-02-06/opposition-selling-otter-pelts-non-natives> (accessed 13.04.14).

Thynes, T., Gordon, D., Harris, D., Walker, S., 2013. 2013 Southeast Alaska Sac Roe Herring Fishery Management Plan. Alaska Department of Fish and Game, Division of Commercial Fisheries Regional Information Report 1J13-02.

Twiss, J.R., 1980. Letter from Marine Mammal Commission to Mr. Lynn A. Greenwalt, Director, U.S. Fish and Wildlife Service, 2nd December. Final Environmental Impact Statement for Translocation of Southern Sea Otters, Volume III, Comments and Responses (1987). US Fish and Wildlife Service, Sacramento, CA, pp. 15–16.

US Census Bureau, 2014. State and County Quick Facts. Available from: <http://quickfacts.census.gov> (accessed 13.04.14).

US Fish and Wildlife Service, 1982. Southern Sea Otter Recovery Plan. Regional Office, Portland, 70 pp.

US Fish and Wildlife Service, 1984a. Transcript of Scoping Session, Preparation of Environmental Impact Statement, Proposal to Translocate Southern Sea Otters. 23 July, City Council Chambers, Santa Barbara, CA.

US Fish and Wildlife Service, 1984b. Transcript of Scoping Session, Preparation of Environmental Impact Statement, Proposal to Translocate Southern Sea Otters. 25 July, City Council Chambers, Monterey, CA.

US Fish and Wildlife Service, 2003. Final Revised Recovery Plan for the Southern Sea Otter (*Enhydra lutris nereis*). Portland, Oregon. xi + 165 pp.

US Fish and Wildlife Service, 2011. Public hearing: the proposal to end the southern sea otter translocation program. Reporter's Transcript of Proceedings, Taken at Channel Islands National Park Auditorium, Ventura, 27 September.

US Fish and Wildlife Service, 2012. Final Supplemental Environmental Impact Statement on the Translocation of Southern Sea Otters. Ventura Fish and Wildlife Office, Ventura, 348 pp. + front matter and appendices.

US Fish and Wildlife Service, 2013. Stock Assessment Report for the Southeast Alaska Population Stock. Marine Mammals Management, Anchorage, AK.

US Fish and Wildlife Service, 2014. Alaska Sea Otter Hunting and Handicrafting. [Online]. Available from: <http://www.fws.gov/alaska/fisheries/mmm/hunting_seaotter.htm> (accessed 13.04.14).

US Government Printing Office, 2012. Legislative hearing before the Subcommittee on Fisheries, Wildlife, Oceans and Insular Affairs of the Committee on Natural Resources, U.S. House of Representatives, 112th Congress, First Session, October 25, 2011.

Viechnicki, J., 2013. Southeast fall Dungeness crab catch improves. KFSK. [Online]. 24 December. Available from: <http://www.kfsk.org/2013/12/24/southeast-fall-dungeness-crab-catch-improves/> (accessed 13.04.14).

Welch, L., 2011. Study: swelling sea otter population scooping up dollars in Southeast. Alaska Journal of Commerce. [Online]. 15 December. Available from: <http://www.alaskajournal.com/Alaska-Journal-of-Commerce/AJOC-December-18-2011/Study-Swelling-sea-otter-population-scooping-up-dollars-in-Southeast/> (accessed 13.04.14).

Wendell, F.E., Hardy, R.A., Ames, J.A., 1986. An assessment of the accidental take of sea otters, *Enhydra lutris*, in gill and trammel nets. California Department of Fish and Game Marine Resources Technical Report 54.

Wild, P.W., Ames, J.A., 1974. A report on the sea otter, *Enhydra lutris*, in California. California Department of Fish and Game Marine Resources Technical Report 20.

Wilmers, C.C., Estes, J.A., Edwards, M., Laidre, K.L., Konar, B., 2012. Do trophic cascades affect the storage and flux of atmospheric carbon? An analysis of sea otters and kelp forests. Front. Ecol. Environ. 10 (8), 409–415.

Young, D., 2011. Rep. Don Young fights for Alaskan Native hunters and craftsmen. Press release, 25th October. [Online.] Available from: <http://donyoung.house.gov/news/documentsingle.aspx?DocumentID=266127> (accessed 13.04.14).

Chapter 13

Conservation in Practice

Linda M. Nichol
Fisheries and Oceans Canada, Pacific Biological Station, Nanaimo, BC, Canada

INTRODUCTION

Conservation, as a practice, draws upon science to respond to the scope of human impacts to ecosystems, impacts that have changed with growing human population and advancing technology (Vitousek et al., 1997). Conservation practice is influenced by prevailing societal values and advances in our understanding of ecosystems, where both values and knowledge have shifted through history (Meine et al., 2006). Conservation views have changed through time. Perspectives expressed in the United States in the 1800s gave value to plants and animals in accordance with their utility to humans. This gave way to more esthetic views of nature in the late 1800s. In the early 1900s

Sea Otter Conservation. DOI: http://dx.doi.org/10.1016/B978-0-12-801402-8.00013-5

there was a shift back to utilitarian values but with a strong emphasis on the idea of fair distribution of natural resources to humans at the time and in the future. Later, ethical views emerged that recognized the importance of diversity of species and habitats and their relationship to human health (Callicott, 1990).

The sea otter was driven to the brink of extinction in the North Pacific as a result of excessive killing fueled by the demands of commercial markets, as were many other marine mammals including the North Pacific right whale (*Eubalaena japonica*), Pacific gray whale (*Eschrichtius robustus*), and the northern elephant seal (*Mirounga angustirostris*) (Le Boeuf et al., 1974; Rice et al., 1984; Clapham et al., 2004). By 1911, the surviving sea otter population had been reduced to between 200 and 2000 animals, less than 1% of the pre-exploitation global population size (Lensink, 1960; Kenyon, 1969). In Canada, the sea otter was extirpated by 1931. Subsequent reintroductions of sea otters in Canada during the late 1960s and 1970s from Alaska have been successful in re-establishing the population (Watson et al., 1997). As of 2008, the Canadian sea otter population included a minimum of 4700 sea otters along the west coast of Vancouver Island and a small population off the central British Columbia mainland coast (Nichol et al., 2009) (Figure 13.1). The sea otter was listed as Threatened under Canada's Species at Risk Act (SARA), enacted in 2003, and subsequently down-listed to Special Concern in 2009 (SARA, 2009). Oil spills are considered a significant threat to the population given the sea otter's inherent vulnerability to the effects of oiling and because of the proximity of the population to major oil tanker routes (COSEWIC, 2007). Oil spills are regarded as a significant threat to most US sea otter populations as well. Sea otter populations in the United States are protected under the Marine Mammal Protection Act. In addition, two populations (California and southwestern Alaska) are listed as Threatened under the US Endangered Species Act (ESA).

Changing views about conservation of wildlife are mirrored in the fate and recovery of sea otter populations. Species such as the sea otter were historically valued for their fur and its utility to humans and there was a prevailing acceptance of the inevitability of extinction that aligned with a long period of largely unregulated sea otter harvest between 1741 and 1911 (Lensink, 1960; Meine et al., 2006). When early laws were enacted, they were intended to conserve the sea otter for its continued commercial value. Subsequently, increasing rates of species extinction and habitat loss in North America culminated in the 1960s with a shift to a view that recognized the importance of species, ecosystems, and biodiversity and their relationship to human well-being (Caughley, 1994; Meine et al., 2006). What emerged were new laws and approaches to support recovery of depleted species, such as the sea otter in Canada and the United States. These changing attitudes and conservation approaches are reflected in the major events following the near extinction of the sea otter and its subsequent progress toward recovery. As such, a review of the near extinction of the sea otter and subsequent

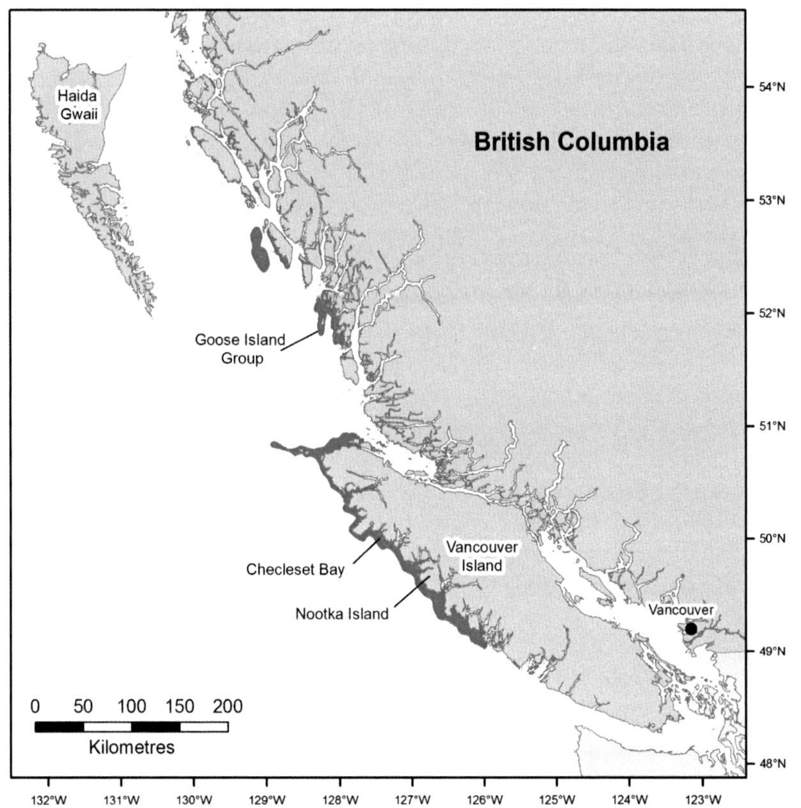

FIGURE 13.1 Range of sea otters in British Columbia, Canada (blue shaded area) as of 2008. *(Adapted from Nichol et al. (2009).)*

conservation efforts provides a unique opportunity to assess the conservation actions that aided or failed to support recovery of the sea otter in the North Pacific and that are relevant to conservation today.

IMPACT OF THE COMMERCIAL FUR TRADE ON SEA OTTER POPULATIONS

Sea otters were hunted intensively for their fur throughout their range in the North Pacific from 1741 to 1911. In 1741 the Bering Expedition returned to Russia with sea otter pelts and in the years following until 1867, Russians hunted sea otters in the Commander and Kurile Islands, the Aleutian Islands, and other areas of western and central Alaska (Lensink, 1960). Meanwhile, along the mainland coast of North America the maritime fur trade in sea otter pelts began following the arrival in 1778 of Captain Cook at Friendly

Cove on Nootka Island off the west coast of Vancouver Island, British Columbia (Dalrymple, 1789). Trading vessels from Britain, the United States, France, Spain, and Portugal came to trade with coastal native people for sea otter pelts. Maritime fur trade activity spread from Vancouver Island northward to include Haida Gwaii (formerly the Queen Charlotte Islands) and the northern mainland of what is now British Columbia and southeastern Alaska and southward from Vancouver Island along the coast of what are now Washington, Oregon, and California.

In the territory of Alaska alone, Russians killed an estimated 800,000 sea otters between 1742 and 1867 (Lensink, 1960). Estimates of pelt landings were less consistently reported from the period of the maritime fur trade along the mainland coast of North America, but it has been suggested that more pelts were obtained in the maritime fur trade from 1778 to 1820 than were taken by the Russians during their entire occupation of Alaska (1741 to 1867) (Lensink, 1960). Incomplete statistics that enumerate pelt landings and the number of trading vessels plying the waters of the west coast of what is now British Columbia and southeastern Alaska convey something of the magnitude of the trade at various places and times. For instance, at Haida Gwaii a single ship trading for less than 2 months in 1787 amassed 1821 pelts and another ship obtained 1400 sea otters in less than 2 months in 1791 (Sloan and Dick, 2012). It is believed that from 1790 to 1800 more than 10,000 pelts a year were obtained by American vessels alone trading with native people along the coast of northern British Columbia and southeastern Alaska (Dmytryshyn and Crownhart-Vaughan, 1976). During the peak of the trade, as many as 20 vessels were trading annually along the northwest coast and, in 1801, 16 ships on the northwest coast amassed about 18,000 sea otter skins (Howay, 1973; Jackman, 1978). By the mid-1800s the maritime fur trade was all but over. The number of trading vessels arriving seeking sea otter pelts diminished as traders moved on to exploit other more plentiful marine mammal resources (Lensink, 1960; Howay, 1973). However, sea otters continued to be hunted opportunistically.

> *Yesterday four valuable otter skins were bought by a furrier from an Indian, who brought them south from Masset, Queen Charlotte Islands. The skins sold for $2,000.*
>
> (*British Colonist,* 1904)

In the same edition of the newspaper, the *British Colonist,* an advertisement listed an eight-room house with barn and stable in Victoria, British Columbia for $2100, underscoring the value of a sea otter pelt at the time.

LAWS AND TREATIES TO PROTECT SEA OTTERS

The first measures taken to conserve sea otters were instituted by the Russians in the territory of Alaska during the period 1820–1867. These measures comprised a system of quotas and restrictions implemented to sustain the sea otter stock for continued harvest. Once the territory of Alaska was ceded to the United States in 1867, the restrictions ended, along with the Russian-American company. A resurgence of largely uncontrolled harvest after 1867 is believed to have led to the near extinction of the sea otter in Alaska by 1911 (Elliott, 1875; Hooper, 1897; Allen, 1942; Lensink, 1960; Kenyon, 1969). Only 13 remnant colonies of approximately 200–2000 animals in aggregate are thought to have remained in the North Pacific by 1911 (Lensink, 1960; Kenyon, 1969). These colonies were in western Alaska, Russia, and California, with one in British Columbia at the north end of Haida Gwaii (Chapter 1). The Haida Gwaii colony disappeared by the 1920s (Kenyon, 1969; Sloan and Dick, 2012).

The International Fur Seal Treaty of 1911

The International Fur Seal Treaty is often referred to as the first protection for sea otters (Kenyon, 1969). The objective of the International Fur Seal Treaty was to restrict hunting of commercially valuable fur-bearing resources, primarily the northern fur seal (*Callorhinus ursinus*) but also the sea otter. At the time, wildlife, including marine mammals, was viewed as a common resource of commercial utilitarian value that was being depleted (Meine et al., 2006). The treaty focused on control of hunting opportunities. Under the terms of the treaty, protection for sea otters was provided both directly and indirectly.

Northern fur seals breed on island rookeries in the Bering Sea in Alaska and Russia. Adult northern fur seals congregate on rookeries for approximately 5 months each year commencing in June, after which they disperse to sea along the continental margins of the North Pacific (Gentry, 1998). The largest breeding rookery is on the Pribilof Islands in Alaska. Intensive hunting had occurred on the rookeries since the late 1700s and an intensive pelagic fur seal hunt commenced in the 1860s (Scheffer et al., 1984). The northern fur seal population is thought to have numbered 2–3 million animals when the Pribilof Islands were discovered by the Russians in 1742. However, by the late 1800s the intensity of the rookery kills and the added impact of pelagic hunting along continental margins had reduced the number of animals on the Pribilof Islands to 300,000 (Lander and Kajimura, 1982). The fur seal population decline led Britain (for Canada), Japan, the United States, and Russia to sign the International Fur Seal Treaty in 1911. The treaty was the culmination of several iterations of treaty attempts which sought to control and restrict hunting of the northern fur seal at rookeries and at sea in the North Pacific (Scheffer et al., 1984; Chapter 14).

The 1911 treaty prohibited pelagic fur sealing and established agreement among the member countries about management and quotas for the commercial hunt of fur seals on the rookeries. The terms of the treaty applied over the North Pacific north of 30°N latitude and included the seas of Bering, Kamchatka, Okhotsk, and Japan, thereby encompassing the range of the fur seal. Direct protection for sea otters is stated in Article V of the treaty which prohibited hunting of sea otters by member countries outside of their respective three-nautical-mile territorial limits. Prior to the treaty, seal hunters took sea otters opportunistically while engaged in pelagic sealing along the continental margins of North America and throughout the Gulf of Alaska. Article V put a direct end to sea otter hunting outside territorial limits and the treaty further reduced opportunistic hunting of sea otters indirectly by ending pelagic sealing and restricting fur seal hunting at the rookeries, all of which served to reduce the presence of potential hunters in the Gulf of Alaska.

Article V addressed sea otter hunting beyond the three-nautical-mile territorial limit, which would have been helpful in protecting sea otters in some areas. For instance, in southwestern Alaska there are areas where sea otters occupied shallow habitat that extends well beyond three nautical miles of shore, but most sea otter populations occupied habitat within three nautical miles of shore. Certainly this is the case along much of the coast of North America from southeastern Alaska to California.

Laws Protecting Sea Otters Within Three-Nautical-Mile Territorial Limits

Laws prohibiting sea otter hunting within three nautical miles of shore were present in California and Alaska by the early 1900s. Even where laws were enacted, it is likely that enforcement would have been minimal over vast and sparsely populated regions and there may have been an acceptance at the time of dwindling wildlife and a belief in the inevitability of extinction (Hooper, 1897; Leopold, 1933).

Japan and Russia

The Japanese hunted sea otters for their pelts beginning in 1873 in the Sea of Okhotsk and the Kuril Islands. Some controls in the form of quotas and restrictions were established after 1912. Commercial hunting of the sea otter by Japanese companies continued until 1945 and sea otter pelts from Japan were traded legally in fur markets in the United States (Anon, 1939; Wada, 1997; Uni, 2001 cited in Hattori et al., 2005). After 1945 sea otters were no longer hunted by the Japanese because the Kuril Islands were considered to be Russian territory. In Russia, sea otters were hunted legally until 1924 (Kirov, 1965).

United States

Beginning in 1868, US federal laws were passed to prohibit sea otter hunting within the limits of the territory of Alaska and its waters (Chapter 14).

However, the law also stated that the federal government could authorize killing by regulation (US Statutes at Large, 1868, 1910). Perhaps this explains the harvest of sea otters that continued in Alaska; 107,121 pelts were shipped from Alaska between 1868 and 1905 (Hooper, 1897; Kenyon, 1969). By 1912 sea otter hunting in Alaska was closed by federal regulation (United States Department of the Interior, 1912). In Washington State, there were no laws against hunting sea otters. On the outer coast of Washington, sea otters were hunted from shore from the mid-1800s to the early 1900s. Hunters built 6-m-high towers called derricks along the shore from which they hunted sea otters using rifles at ranges of 350–550 m (Scheffer, 1940). Carcasses were retrieved when they washed ashore. The last sea otter was reportedly shot in 1910, although there was later a sighting in 1949 (Lantz, 1918; Scheffer, 1940, 1995). There were no laws against hunting sea otters in Oregon, and the last reported sea otter was killed in 1906 (Lantz, 1918; Kenyon, 1969). In contrast, a state law against hunting sea otters was enacted in California in 1913. In 1914 enforcement was strengthened with the addition of a fine of $1000, two-thirds the market value of a pelt at the time (Bryant, 1915).

Canada

In British Columbia, sea otter hunting was not prohibited until 1931 when the British Columbia Game Act was revised to distinguish the sea otter and the land otter as furbearers (Statutes of Province of British Columbia, 1931). The last verified sea otter was shot near Kyuquot on Vancouver Island in 1931 and the incident was reported in local newspapers when the pelt was seized by game wardens. Subsequently it was determined that the animal had been shot before the revision had taken effect.

Reflection on Effectiveness of Early Laws

It is not surprising that the sea otter was extirpated from most of its range in the North Pacific. Sea otters occurred predominantly within three nautical miles of shore throughout most of their range and a pelt was commercially valuable. Where laws prohibiting killing were enacted in sufficient time, sea otters seem to have survived, for example in California and Alaska. Enforcement of the law was likely possible in California because the sea otter colony was small and had a very limited range on the California coast. Conversely, in Alaska law enforcement may not have been as big a factor as the great distance from ports, severity of the climate, and the scarcity of the animals, all of which served to make their pursuit uneconomical and gave the remnant colonies respite from hunting (Lensink, 1960; Baur et al., 1996).

REINTRODUCTION OF SEA OTTERS

After 1911 and the cessation of commercial hunting, most remnant colonies of sea otters in western Alaska, Russia, and California began to recover. By the 1960s the population of sea otters in the Aleutian Islands had increased to about 9700 animals and by the early 1980s to 55,000–73,000 animals (Kenyon, 1969; Calkins and Schneider, 1985). In contrast, the historic sea otter range from central California to the Gulf of Alaska remained unoccupied.

The 1960s saw a period of increasing efforts to conserve and recover species due to greater public awareness of human impacts on the environment and wildlife that accompanied a growing human population and increasing industrialization. It was during this period that sea otters were reintroduced to British Columbia, southeastern Alaska, Washington, and Oregon (Jameson et al., 1982; Caughley, 1994; Meine et al., 2006; Chapter 3). Translocation, the deliberate movement of wild individuals from one part of their range to another in an attempt to re-establish a species in a part of its historical range, was the method used with the goal of speeding recovery of the species (Jameson et al., 1982; Griffith et al., 1989; IUCN, 2012). The period of translocations of sea otters from southwest and south central Alaska to southeastern Alaska, British Columbia, Washington, and Oregon received its early impetus as a result of US government plans for nuclear testing at Amchitka Island. This impending event catalyzed the first efforts to move sea otters from the test zone and made available a variety of logistical supports such as ships and aircraft to assist with transportation of sea otters over great distances. As early as 1965, sea otters were successfully translocated from Amchitka to southeastern Alaska and plans were under way to move sea otters to Washington and Oregon as well (see Chapter 3).

Establishing a satellite population to enhance distribution and thereby reduce the risk of species loss in the event of a catastrophe has also been a goal of sea otter translocation efforts. The southern sea otter was listed as Threatened under the ESA in 1977 and to promote population recovery, a key strategy was to extend the range of the southern sea otter in California in order to reduce the potential impact to the population from an oil spill. Between 1987 and 1990, 139 sea otters were translocated from mainland California to San Nicolas Island, about 220 km south of their established range off the mainland central California coast (VanBlaricom and Jameson, 1982; USFW, 2003; Chapter 3).

Reintroduction of Sea Otters to British Columbia

With the last sea otter reported shot in 1931 in British Columbia there is little evidence of sea otters along Canada's Pacific coast. The opportunity for translocation of sea otters from Alaska to British Columbia emerged in the 1960s.

The effort to return the sea otter to British Columbia was the result of provincial and federal biologists working with Alaska Department of Fish and Game. Ian MacAskie was a marine mammal biologist with Fisheries and Oceans Canada at the time and was one of the people involved and in support of reintroducing sea otters.

> *[M]y whole idea from the beginning was to return to the [British Columbia] coast an animal that we had destroyed*
>
> (Obee, 1984)

Karl Kenyon, biologist with the Alaska Department of Fish and Game, provided advice about suitable habitat in British Columbia for a reintroduction.

> *[T]he outer coast of [Vancouver] island[s] offers ideal locations in which to reintroduce sea otters*
>
> (Kenyon, 1967)

Ultimately Checleset Bay, on the west coast of Vancouver Island, was chosen as the site for the reintroduction of sea otters (Figure 13.1).

> *...[I]t is relatively isolated and offers an abundance of requisite food in 85 square miles of shallow reef-strewn waters. With many islands to provide shelter from heavy seas, the area is considered ideal habitat*
>
> (MacAskie, 1971)

Three reintroduction attempts were made over a period of 4 years. In July 1969, the British Columbia Fish and Wildlife Branch and Alaska Department of Fish and Game captured 30 sea otters at Amchitka Island. One animal died during the air transport, but the remaining 29 were released into Checleset Bay after a long journey with several stops and transfers to different planes (Smith, 1969). Post-release survival is thought to have been poor as a consequence of their long and stressful journey (Bigg and MacAskie, 1978).

In July 1970, 45 animals were captured in Prince William Sound, Alaska, and transported by ship, on a 6-day journey, to Checleset Bay (MacAskie, 1975). During a storm while crossing the Gulf of Alaska, many of the sea otters died from stress and the inability to groom effectively because of the ship's movement in the rough seas. Of the 45 captured, only 14 sea otters survived to be released (MacAskie, 1971).

In 1972 a third and final translocation was organized. This time 47 sea otters from Prince William Sound were transported by aircraft to Vancouver Island. They were held in floating pens in Checleset Bay for several days where they were fed and observed. One female pup died but the remaining 46 animals were released in excellent condition (MacAskie, 1975; Table 13.1).

Although 89 animals were released alive, it is thought that the number of survivors declined to as few as 28 animals by 1973 (Estes et al., 1989).

TABLE 13.1 Summary of Sex, Maturity, and Health of Sea Otters Translocated from Alaska to Checleset Bay, BC

Release Date	Origin	Total	Adult		Immature			Health	Agencies Involved
			♂	♀	♂	♀	?		
July 31, 1969	Amchitka Island	29	9[b]	19[b]			1	Fair-good	BCFW; ADGF
July 27, 1970	Prince William Sound	14[a]	6	8	7[c]	9[c]		Excellent	DFO; BCFW; ADGF
July 15, 1972	Prince William Sound	46	8	22				Excellent	DFO; BCFW; ADGF
Total		**89**	**23**	**49**	**7**	**9**	**1**		

[a] *45 had been captured but only 14 survived.*
[b] *Approximate sex ratio.*
[c] *Included four male and two female pups.*
BCFW: British Columbia Fish and Wildlife Branch; ADGF: Alaska Department of Fish and Game; DFO: Fisheries Research Board of Canada/Fisheries and Oceans Canada.
Source: Adapted from Bigg and MacAskie (1978).

This was likely a result of animals dispersing from the release site and subsequently dying. All of the 1960s and 1970s reintroduction efforts into southeastern Alaska, British Columbia, Washington, and Oregon experienced immediate declines after release. It became apparent from monitoring these translocations that dispersal from release sites was common and that this was probably a result of the sea otter's strong home range fidelity. Selection of high-quality habitat for the site of reintroduction was likely a key contributor to the success of the translocations despite early dispersal (see Chapter 3).

In 1970 a regulation was added under the Canadian Federal Fisheries Act that protected the sea otter from being captured, killed, disturbed, or molested and also prohibited anyone from possessing a sea otter pelt (MacAskie, 1987). Ten years later, sea otters were recognized as a "rare" species by the province of British Columbia and as such protected under the British Columbia Wildlife Act as well (Revised Statutes of British Columbia, 1979).

Along west coast Vancouver Island, the sea otter population increased at a rate of 19.0% per year from 1977 to 1995. Thereafter growth slowed to 8.4% per year (1995–2008) (Figure 13.1) (Nichol et al., 2009). By 2008 the sea otter population was established along most of the west coast of Vancouver Island and the population included a minimum of 4100 animals (Figure 13.2). A high average annual population growth rate until the mid-1990s was likely possible because of the availability of prey and extensive areas of unoccupied habitat within easy dispersal range of the population center. Similar high annual growth rates occurred in the early years following establishment in Washington State and southeastern Alaska. Along west coast Vancouver Island, however, as the occupied range expanded northward and southward from the site of reintroduction, growth slowed. This occurred, at least in part, because the distance from the population center to the edge of the range began to exceed dispersal distance. However, there may have been additional factors at play, such as illegal killing and entanglement in fishing gear, but the extent of these factors and therefore their impact on population growth is not known.

In 1989 a sighting of a small group of sea otters was reported from the Goose Island Group on the central mainland coast of British Columbia (Watson et al., 1997). Along the central mainland coast of British Columbia, the population grew at an annual rate of 11.0% per year from 1990 to 2008 and the population included a minimum of 600 animals by 2008 (Figure 13.3). It is not clear why this population of animals, descendants of the reintroduced animals released in Checleset Bay and/or Washington State, has not shown a pattern of early rapid growth similar to the Vancouver Island population and the translocated populations in Washington and southeastern Alaska (Chapter 3).

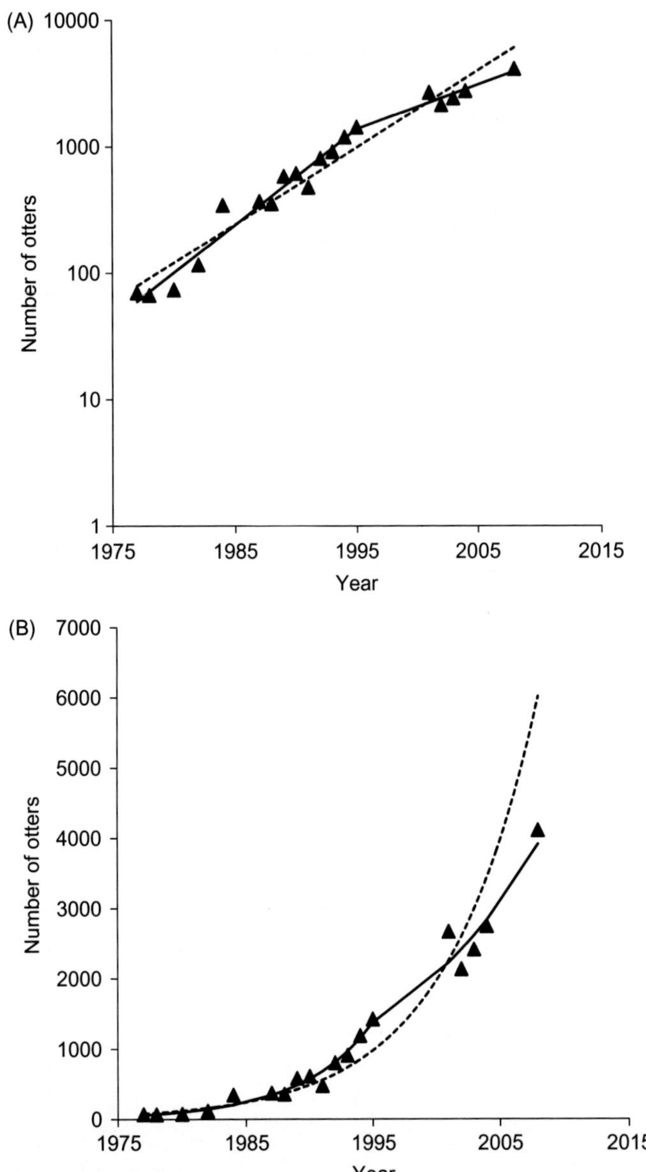

FIGURE 13.2 (A) Trend in sea otter population growth on Vancouver Island. Dashed line represents a simple log-linear regression (growth 15.0% per year $r^2 = 0.950$, $n = 19$). Black line represents a piece-wise regression (growth 19.0% per year[1] from 1977 to 1995 and 8.4% per year 1995 to 2008, $r^2 = 0.978$, $n = 19$). Triangles represent survey counts. (B) Trends and survey counts plotted on an ordinal scale. *(From Nichol et al. (2009).)*

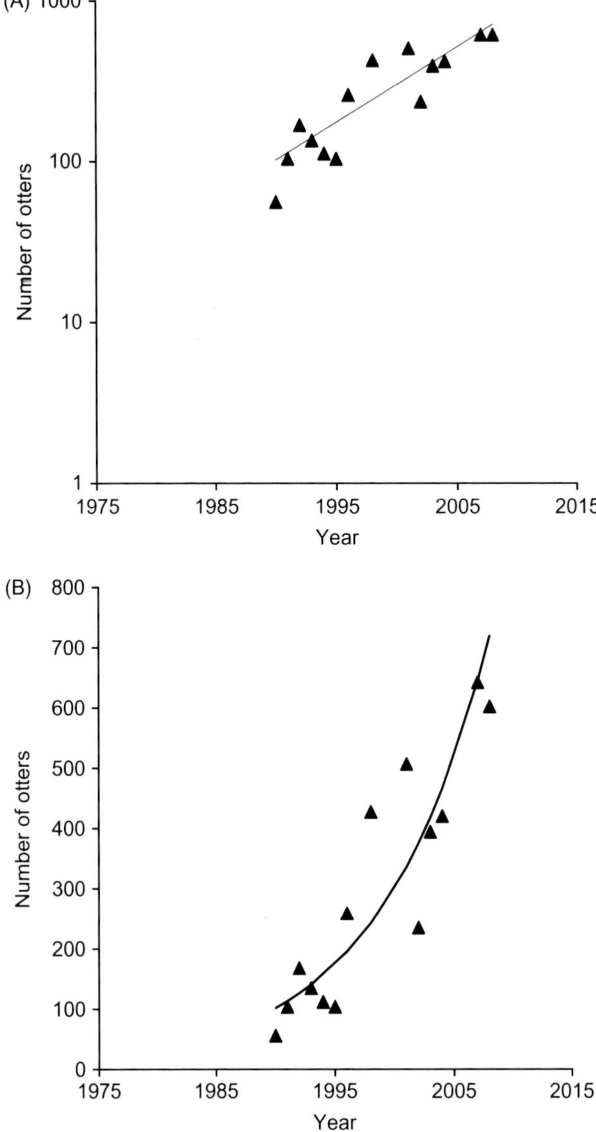

FIGURE 13.3 (A) Trend in sea otter population growth on the central mainland British Columbia coast. Line represents a simple log-linear regression (growth 11.4% per year from 1990 to 2008, $r^2 = 0.801$, $n = 11$). Triangles represent survey counts. (B) Growth trend and survey counts plotted on an ordinal scale. *(From Nichol et al. (2009).)*

PROTECTING HABITAT

Concern about loss of species and loss of habitat emerged and re-emerged at various times in the 1900s. In the United States the "health" of land including soil, water, plants, and animals emerged as an important theme in the 1930s as a result of increasing rates of environmental impacts from human activities (Meine et al., 2006). Ecological concepts such as food webs and interactions among species, and the importance of biodiversity led to a growing awareness of the importance of protecting not only species but the ecosystems upon which they rely and are a part (Meine et al., 2006). Excluding human activities in delineated zones in order to sustain broad ecosystem components has been an important objective of habitat protection (Fanshawe et al., 2003; Meine et al., 2006). The distinction between this approach and early laws that restricted hunting of a particular species, such as the sea otter, is recognition of the need to protect not just a species directly but also entire ecosystem components.

In 1981, 85 km^2 of Checleset Bay on the west coast of Vancouver Island was proposed as an Ecological Reserve under the British Columbia Ecological Reserves Act. The purpose of the reserve was to support recovery of the newly established sea otter colony by excluding human activities. Subsequently, Fisheries and Ocean Canada closed commercial fisheries for invertebrate stocks in the reserve (Jamieson and Lessard, 2000). Other examples of habitat protection that support sea otter population recovery are the Monterey Bay National Marine Sanctuary in California, established in 1992 to protect entire ecosystem components, and the Olympic Coast National Marine Sanctuary in Washington, established in 1994 (Fanshawe et al., 2003; Saunders, 1996). In Alaska, the establishment of the Aleutian Islands Wildlife Refuge in 1913 in western Alaska is a very early example of this approach (Lensink, 1960). In Russia several marine protected areas were established after 1958 including a 30-mile zone around the Commander Islands and several 2- to 12-mile zones around the Kamchatka Peninsula and the Kuril Islands.

Habitat protection is important to support conservation of ecosystem components, but pollution and environmental contamination, which can be transported by air and water and accumulate in the food chain, are examples of threats that are not readily addressed by establishing boundaries. Pollution as a challenge to conservation has led to various scientific advances with regard to understanding impacts but reducing those impacts remains a significant conservation problem. Oil spills also remain a significant threat to sea otter populations (see Chapter 4). Sea otters are particularly vulnerable to oil contamination because they depend entirely upon the integrity of their fur for insulation. Oil destroys the water-repellent nature of the fur, leading to hypothermia (Williams et al., 1988). Further, as nearshore animals that exhibit strong site fidelity and a propensity to rest in large floating aggregations and often in kelp beds, sea otters occupy habitat most likely to collect and retain oil from a marine spill, meaning large numbers of animals could be oiled at the same time. Even lightly oiled animals will ingest oil, which is toxic, while grooming. Sea otters feed on benthic invertebrates, which can accumulate and store toxic hydrocarbons and thereby contribute to long-term exposure in sea otters (Bodkin et al., 2002). Effects on sea otters from the *Exxon Valdez* oil spill in Prince William Sound have persisted for two decades (Bodkin et al., 2012).

The extreme vulnerability of sea otter populations to oil spills was recognized in California by the 1980s because the population was small and occurred along an adjacent oil transportation corridor. The translocation of sea otters from mainland California to San Nicolas Island from 1987 to 1990 was a conservation effort to try to reduce the potential impact to the population from a spill (USFW, 2003). In Canada, oil spills are considered a significant threat to the sea otter population given the species' inherent vulnerability to oil and the proximity of the population to major oil tanker routes (COSEWIC, 2007). Conservation efforts for sea otters will need to continually adapt to address this threat, particularly with projected increases in marine transportation of hydrocarbons (Fisheries and Oceans, 2014).

As sea otter populations continue to recover and as human populations and activities increase, management of human activities to reduce threats by means of regulations and policies will continue to be an important aspect of conservation. For effective conservation there will continue to be the need for long-term data sets and measures with which to detect population responses to threats. In California, a decline in the sea otter population between 1976 and 1984 presents a good example of the importance of long-term data sets for detecting a threat. The cause of the population decline was attributed to the set-net fishery. Subsequent changes to fishery regulation that restricted use of set-nets was a critical conservation measure and was effective in reversing the decline (Estes et al., 2003). Future conservation efforts for sea otters will likely need to involve input to regulations regarding such things as marine transport of hydrocarbons, oil spill response planning, and restricting the use and disposal of various persistent organic pollutants (see Chapter 4).

ENDANGERED SPECIES LEGISLATION

In 1966 the International Union for the Conservation of Nature (IUCN) published its first list of threatened species and in the following years several important environmental laws passed in the United States, one of which was the ESA (Meine et al., 2006). The US ESA was passed in 1973 as a law intended to slow the pace of human-caused species extinction and habitat loss. Three decades later, in 2003, Canada passed the SARA. Both laws express three main objectives: to identify species at risk, to protect them from further harm, and to establish recovery efforts. Both the ESA and the SARA require that science guide all stages of the process (Meine et al., 2006; Waples et al., 2013). Enactment of these laws has been important in supporting the conservation of and recovery efforts for sea otter populations in Canada and the United States.

In Russia, the concept of categorizing and listing species according to their vulnerability to extinction was also adopted in the 1960s, and in 1974 legislation was passed in the USSR that protected endangered species. Currently the sea otter is listed in the Russian Red Book as a category 5 species, which means they are considered to be recovering. As a Red Book species, hunting or capturing of sea otters is prohibited except by special license for scientific or display purposes.

Occasional sightings of sea otters are reported from Hokkaido, northern Japan, south of the Kuril Islands, Russia, where the sea otter population is recovering. Hunting sea otters is prohibited in Japan but there are no other conservation laws to address unintentional capture or harm to sea otters or to support recovery of a population (Hattori et al., 2005). Occasional sightings of sea otters have been reported along the west coast of Baja California Sur, Mexico, as recently as the mid-1990s; however, sea otters have been reported entangled in gill nets and this represents a significant unaddressed threat to any possibility of recovery (Gallo-Reynoso, 1997). Sea otters are legally protected under Mexican law from capture and possession.

Canada's SARA

The SARA provides for the recovery of listed species. Fisheries and Oceans Canada (DFO), the agency responsible for marine mammals such as the sea otter, is required by the SARA to lead recovery planning and recovery actions. The requirement for empirically based science plans that set out recovery and conservation goals for the species in Canada and documentation of approaches to identify and reduce threats to the species have been driven by the SARA. The result has been development of a population assessment program for the species in Fisheries and Oceans Canada since 2001 and an increase in research activities to understand threats to and the ecological role of sea otters in coastal British Columbia.

Prior to the SARA, the plight of endangered wildlife in Canada was taken up by the Committee on the Status of Endangered Wildlife in Canada (COSEWIC). The COSEWIC, an independent national advisory body of scientists, was established in 1977. The COSEWIC assessed and classified species using a process based on the IUCN red-list categories and reported on these species. In so doing, the COSEWIC drew attention to the plight of species and their habitats in Canada, but there were no federal laws that required conservation and protection action based on these assessments (Cook and Muir, 1984; IUCN, 2012). The COSEWIC designated the sea otter as Endangered in Canada in 1978 (Monroe, 1985); subsequent assessment in 1996 led to a change in status to Threatened (Watson et al., 1997).

When the SARA was passed in 2003, species with existing COSEWIC designations were immediately listed. As such the sea otter was legally listed as Threatened. The process of assessing and listing species is a two-part process under the SARA. Part one involves the COSEWIC, which continues to function as the national body of experts that reviews and recommends species listings as part of the SARA process. The COSEWIC prioritizes species assessment efforts, completes assessments, and designates species status, but the decision to legally list under the SARA in accordance with the COSEWIC designation, which is the second part, is made by the governor in council (a subcommittee of cabinet ministers). Once a species is listed, there is a requirement under the SARA for the development of several recovery planning documents and implementation of recovery and conservation actions.

Under the SARA, species listed as Endangered or Threatened are protected from killing, harming, harassment, and capture and there is also a requirement to identify (to the extent possible) and protect the habitat necessary for the survival or recovery of the species, which is defined in the act as Critical Habitat. Species of Special Concern, a lower conservation status designation than Endangered and Threatened, are not protected from killing, harming, harassing, or capture under the SARA, nor is there a requirement to identify and protect Critical Habitat. Species of Special Concern are considered not as vulnerable as Threatened or Endangered species, such that existing protection provisions in other federal or provincial legislation are then relied upon for their management. In 2009 the sea otter in Canada was down-listed to Special Concern. As a species of Special Concern, sea otters are protected from harm by provisions of the Fisheries Act as a marine mammal and by the BC Wildlife Act as a red-listed fur-bearing mammal. Although the SARA does not provide legal protection for species listed as Special Concern, the Act does require the responsible government agencies to develop and implement a forward-looking recovery document called a "management plan." For sea otters, this means Fisheries and Oceans Canada has developed and implemented a management plan for sea otters (Fisheries and Oceans, 2014).

SPECIES RECOVERY VERSUS ECOLOGICAL COMMUNITY RECOVERY

Both the ESA and the SARA are focused on listing and recovery of individual species. However, when the population characteristics of one species are affected by that of another species, recovery expectations and targets need to reflect this. Ecological communities are made up of organisms that associate and interact in complex ways. Species interactions are important forces in shaping ecological communities, particularly when they involve top

FIGURE 13.4 Sea otter eating a northern abalone, *Haliotis kamtschatkana*. *(Photo by Stefan Olcen.)*

predators, yet this is often not accounted for in recovery planning (Soule et al., 2005; Chadès et al., 2012; Figure 13.4).

Recovery planning for sea otters and northern abalone (*Haliotis kamtschatkana*) (Endangered under the SARA) in Canada, two species listed under the SARA and with a predator–prey relationship, presents an interesting example of the difficulties in defining recovery for strongly interacting species. It also draws attention to the possibility of competing objectives of recovery and the need to clarify what is to be recovered. Commercial harvesting of northern abalone began in the 1950s in British Columbia and intensified as a commercial and recreational fishery in the 1970s. Harvest peaked in 1977 but by 1990 the fishery was closed due to a dramatic decline in abundance, as a result of the fishery and poaching (Campbell, 2000). Between 1977 and 2002, mean abalone densities at reference sites declined from 2.4 to 0.27 abalone/m^2 on the central British Columbia coast and from 2.2 to 0.34 abalone/m^2 on Haida Gwaii (Fisheries and Oceans, 2007). The species was designated Threatened in 1999 by COSEWIC, a move that drew attention to the severely depleted state of the species. Under the SARA the species was listed as Threatened in 2003 and subsequently up-listed to Endangered in 2011.

Sea otters prey on abalone. It was the near extirpation of the sea otter throughout its range that allowed abalone and other invertebrate populations to increase rapidly in the absence of predation pressure and subsequently form the basis of lucrative invertebrate fisheries. This phenomenon took place in British Columbia as well as in California and Alaska (Estes and VanBlaricom, 1985; Watson, 2000). Biological assessments of the northern abalone population developed during the decades of the fishery. Recovery planning for the abalone is now taking place in the presence of sea otters, but there are no data on characteristics of the historic abalone population prior to the maritime fur trade with which to inform recovery targets for abalone populations in the presence of their natural predator. Studies at Hopkins Marine Reserve in California, an area that has always been closed to abalone fishing, has a restored sea otter population, and provides insight into the population dynamics of abalone under predation pressure. Red abalone (*Haliotis rufescens*) persists at a stable low population density with a patchy distribution. Abalone are restricted to crevice habitats that provide protection from predation. Recruitment and mortality appear to be balanced and it is thought that crevice-living, which creates aggregations of abalone, may help enhance recruitment success (Lowry and Pearse, 1973; Cooper et al., 1977; Hines and Pearse, 1982; Watson, 2000). Before the maritime fur trade, northern abalone may have been distributed at low densities in cryptic habitat, with availability of cryptic habitat and sea otter predation limiting population growth (Watson, 2000). Abalone sizes and density needed for a fishery are unattainable recovery targets with the sea otter restored (Watson, 2000). For strongly interacting species it has been suggested that recovery targets could be set based on achieving some measure of species interaction that indicates restoration of ecosystem function rather than on achieving a particular population size or range (Soule et al., 2005; Estes et al., 2010).

Although setting recovery targets that reflect a measure of restored species interactions, rather than a particular abundance, would be a useful approach to help focus conservation on achieving ecosystem recovery, it would not address the problem of recovery targets where there are fisheries expectations embedded within it. While commercial fur trades and seal hunts are largely a thing of the past in the North Pacific, harvest of wild stocks of fish and invertebrates for commercial markets remains a significant industry. Fisheries management objectives seek to sustain harvest levels, much like the resource conservation ethic that pervaded wildlife conservation in the mid-1900s (Callicott, 1990). In fact, since the mid-1900s there have been two competing ethics in conservation that have gained favor at various times. One is resource management for sustained human use; the other is conservation of ecological integrity for intrinsic values of nature and for human and environmental health (Callicott, 1990; Meine et al., 2006). These ethics challenge marine conservation in particular where objectives to maintain or restore marine biodiversity conflict with objectives to maintain or increase fisheries (Salomon et al., 2011).

In California, five species of abalone were harvested commercially but serial declines in species abundance, in part due to overfishing, led to closure of commercial and recreational fisheries south of San Francisco in 1997 (Karpov et al., 2000; Micheli et al., 2008). The recovery plan for abalone in California has a goal "to reach sustainable fishery levels in at least part of the range" (ARMP, 2005), but recovery objectives for the southern sea otter population are for a larger population (USFW, 2003).

The abalone–sea otter examples point to the need for new approaches or perspectives when there are competing objectives for restoring ecosystems or restoring fisheries (Campbell, 2000; Fanshawe et al., 2003; Salomon et al., 2011). Conflicts have emerged between sea otter recovery and many invertebrate fisheries including abalone. The issue is often that the invertebrate fishery may be jeopardized and so the question is, what do we want to conserve or recover? As more species are listed and as more top predators recover in abundance, this theme will continue to emerge in marine conservation.

In the case of the sea otter and the abalone, there appear to be two dominant alternative states that could be sought, depending on thc objective. One is to allow for the recovery of abalone fisheries. This would require reducing a sea otter population in an area intended for an abalone fishery to a level at which it no longer contributes to ecosystem structuring. Such a level would likely be to local extirpation, much as it was following the maritime trade. Then it would be necessary to maintain the sea otter population at that low level by removing animals entering the area. The other state is to allow recovery of sea otters and restoration of the full range of ecosystem functions, which would be to the detriment of the fishery.

A third option might be abalone aquaculture. Abalone aquaculture has already developed in other parts of the world in response to market demand in the face of depleted wild stocks (McCormick, 2000). In California, abalone aquaculture efforts have been under way since the 1970s and 14 farm sites were in existence in 1998 (McBride, 1998). Although aquaculture comes with its own suite of challenges, environmental impacts, and conflicts, it does have potential as an alternative means of recovering commercial production of an invertebrate while allowing for recovery of sea otters and ecosystem function.

CONCLUSION

Recovery of sea otter populations has taken place over a period of more than 100 years and has benefited from application of laws and progressive conservation actions as well as from geography and the resilience of the species. The most fundamental action was protection of the species from hunting. Early conservation attitudes saw sea otters as a dwindling resource that might be protected in some regions for potential future profit. Where sea otters continued to be chronically hunted without any controls, they were extirpated.

Later efforts to restore sea otters in their historic range by means of translocation were driven by a desire to restore natural ecosystems for their inherent value. Translocations from Amchitka and Prince William Sound, Alaska, to southeastern Alaska, British Columbia, and Washington in the 1960s and early 1970s were highly successful as a tool to restore sea otter populations. High-quality habitat was likely an important factor that offset the problem of initial dispersal and mortality of translocated individuals. Although sea otter populations are now established in much of their historic range, conservation efforts are still needed, particularly to address emerging threats from human activities. Conservation will need to rely on management of human activities through regulations and policies that reduce threats to the species as new issues emerge in an area, such as entanglement in fishing gear, harm from oil spills from marine transport of hydrocarbons, and mortality from illegal killing and legal hunting.

The recovery of top predators such as the sea otter has drawn attention to the ecosystem structuring role of such animals, the recovery of which can alter the density and distribution of other species upon which it preys. When the prey species is also considered endangered and the subject of recovery efforts, it may be difficult to identify realistic or socially acceptable recovery targets. In such instances, ecosystem function indicators as recovery targets in recovery planning, for example, could help to clarify recovery goals.

The conflict between sea otters and invertebrate fisheries has been a recurring theme throughout the eastern North Pacific as sea otters have recovered. Some fisheries, such as the abalone fisheries in British Columbia, were lost prior to substantial recolonization by sea otters. In essence we have returned to the competing views that have plagued conservation for decades: conservation and recovery of species and ecosystems for inherent and multi-faceted value, or recovery of fisheries to support human use. Marine conservation is at the forefront in this debate.

ACKNOWLEDGMENTS

I wish to thank B. Gisborne for compilation of British Columbia historical materials from archival sources, S. Perks for assistance in interpreting legal materials relating to early wildlife laws in British Columbia, O. Filatova and K. Hattori for assistance with current and historic conservation laws in Russia and Japan, and R. Abernethy for the map figure. I thank J. Bodkin, R. Cameron, J. Ford, S. Larson, and G. VanBlaricom for valuable comments on drafts of this chapter.

REFERENCES

Abalone Recovery and Management Plan (ARMP), 2005. California Fish and Game Commission, <http://www.dfg.ca.gov/marine/armp>.

Allen, G.M., 1942. Extinct and vanishing mammals of the western hemisphere. American Committee for International Wildlife Protection, Special Publication No. 11: 620pp.

Anon, 1939. J. Mammal. 20 (3), 407.

Baur, D.C., Meade, A.M., Rotterman, L.M., 1996. The law governing sea otter conservation. Endangered Species Update Special Issue: Conservation and Management of the Southern Sea Otter 13(12), pp. 73–78.

Bigg, M.A., MacAskie, I.B., 1978. Sea otters re-established in British Columbia. J. Mammal. 59 (4), 874–876.

Bodkin, J.L., Ballachey, B.E., Dean, T.A., Fukuyama, A.K., Jewett, S.C., McDonald, L., et al., 2002. Sea otter population status and the process of recovery from the 1989 "Exon Valdez" oil spill. Mar. Ecol.: Prog. Ser. 241, 237–253.

Bodkin, J.L., Ballachey, B.E., Coletti, H.A., Esslinger, G.G., Kloecker, K.A., Rice, S.D., et al., 2012. Long-term effects of the "Exxon Valdez" oil spill: sea otter foraging in the intertidal as a pathway of exposure to lingering oil. Mar. Ecol.: Prog. Ser 447, 273–287.

Le Boeuf, B.J., Ainley, D.G., Lewis, J.T., 1974. Elephant seals on the Farallones: population structure of an incipient breeding colony. J. Mammal. 55 (2), 370–385.

British Colonist, September 1, 1904. British Columbia Provincial Archives, Victoria, BC.

Bryant, H.C., 1915. Sea otter near Point Sur California. Calif. Fish Game 1, 134–135.

Calkins, D.G., Schneider, K.B., 1985. The sea otter (*Enhydra lutris*). In: Burns, J.J., Frost, K.J., Lowry, L.F. (Eds.), Marine Mammal Species Accounts. Alaska Department of Fish and Game Technical Bulletin 7, pp. 37–45.

Callicott, J.B., 1990. Wither conservation ethics? Conserv. Biol. 4, 15–20.

Campbell, A., 2000. Review of northern abalone, *Haliotis kamtschatkana*, stock status in British Columbia. In: Campbell, A. (Ed.), Workshop on Rebuilding Abalone Stocks in British Columbia. Canadian Special Publication of Fisheries and Aquatic Science 130: pp. 41–50.

Caughley, G., 1994. Directions in conservation biology. J. Anim. Ecol. 63, 215–244.

Chadès, I., Curtis, J.M.R., Martin, T.G., 2012. Managing interacting species at risk: a reinforcement learning decision theoretic approach. Conserv. Biol. 26 (6), 1016–1025.

Clapham, P.J., Good, C., Quinn, S.F., Reeves, R.R., Scarff, J.E., Brownell, R.J., 2004. Distribution of North Pacific right whales (*Eubalaena japonica*) as shown by 19th and 20th century whaling catch and sighting records. J. Cetacean Res. Manage. 6 (1), 1–6.

Cook, F.R., Muir, D., 1984. The Committee on the Status of Endangered Wildlife in Canada (COSEWIC): history and progress. Can. Field-Nat. 98 (1), 63–70.

Cooper, J., Wieland, M., Hines, A., 1977. Subtidal abalone populations in an area inhabited by sea otters. Veliger 20, 163–167.

COSEWIC, 2007. Committee in the Status of Endangered Wildlife in Canada. Response statement—sea otters. <http://www.sararegistry.gc.ca/virtual_sara/files/statements/rs149_211_2007-8_e.pdf> (accessed 12.12.13).

Dalrymple, A., 1789. Plan for promoting the fur-trade and securing it to this country by Uniting the Operations of the East-India and the Hudson's-Bay Company. Published by George Bigg, London, 35pp.

Dmytryshyn, B., Crownhart-Vaughan, E.A.P., 1976. Kyrill T. Khlebnikov's reports, 1817–1832. Colonial Russian America. Translated with Introductions and Notes. By B. Dmytryshyn and E.A.P. Crownhart-Vaughan. Oregon Historical Society, Portland.

Elliott, H.W., 1875. The sea otter and its hunting. In: A Report Upon the Condition of Affairs in the Territory of Alaska. House Executive Document 83, 44th Congress, 1st Session, Washington, pp. 54–62.

Estes, J.A., VanBlaricom, G.R., 1985. Sea otters and shellfisheries. In: Beddington, J.R., Beverton, R.J.H., Lavigne, D.M. (Eds.), Marine Mammals and Fisheries. George Allen and Unwin, London, pp. 187–235.

Estes, J.A., Duggins, D.O., Rathbun, G.B., 1989. The ecology of extinctions in kelp forests. Conserv. Biol. 3, 252–264.

Estes, J.A., Hatfield, B.B., Ralls, K., Ames, J., 2003. Causes of mortality in California sea otters during periods of population growth and decline. Mar. Mammal Sci. 19 (1), 198–216.

Estes, J.A., Tinker, M.T., Bodkin, J.L., 2010. Using ecological function to develop recovery criteria for depleted species: sea otters and kelp forests in the Aleutian Archipelago. Conserv. Biol. 24 (3), 852–860.

Fanshawe, S., VanBlaricom, G.R., Shelly, A.A., 2003. Restored top carnivores as detriments to the performance of marine protected areas intended for fishery sustainability: a case study with red abalones and sea otters. Conserv. Biol. 17, 273–283.

Fisheries and Oceans Canada, 2007. Recovery Strategy for the Northern Abalone (*Haliotis kamtschatkana*) in Canada. *Species at Risk Act* Recovery Strategy Series. Fisheries and Oceans Canada, Vancouver. vi + 31 pp.

Fisheries and Oceans Canada, 2014. Management Plan for the Sea Otter (*Enhydra lutris*) in Canada. *Species at Risk Act* Management Plan Series. Fisheries and Oceans Canada, Ottawa. iv + 50 pp.

Gallo-Reynoso, J.P., 1997. Status of sea otters (*Enhydra lutris*) in Mexico. Mar. Mammal Sci. 13 (2), 332–340.

Gentry, R.L., 1998. Behavior and Ecology of the Northern Fur Seal. Princeton University Press, Princeton, NJ.

Griffith, B., Scott, J.M., Carpenter, J.W., Reed, C., 1989. Translocation as a species conservation tool: status and strategy. Science 345, 447–480.

Hattori, K., Kawabe, I., Mizuno, A.W., Ohtaishi, N., 2005. History and status of sea otters, *Enhydra lutris* along the coast of Hokkaido, Japan. Mammal Study 30, 41–51.

Hines, A.H., Pearse, J.S., 1982. Abalones, shells, and sea otters: dynamics of prey populations in central California. Ecology 63, 1547–1560.

Hooper, C.L., 1897. Sea-otter banks of Alaska. Range and habits of the sea otter—its decrease under American rule, and some of the causes—importance of the sea otter to the natives of Alaska inhabiting the Aleutian Islands—Proposed Legislation for 1898. By C.L. Hooper Captain U.S.R.C.S. Commanding Bering Sea Patrol Fleet 1897. Government Printing Office, Washington.

Howay, F.W., 1973. In: Pierce, R.A. (Ed.), A List of Trading Vessels in the Maritime Fur Trade, 1785–1825. Materials for the Study of Alaskan History. No. 2. Limestone Press, Kingston, ON.

International Union of Conservation of Nature, 2012. Guidelines for Application of IUCN Red-listed Criteria at Regional and National Levels, version 4.0, IUCN, <www.iucnredlist.org/documents/documents/RedListGuidelines.pdf> (accessed 12.12.13).

Jackman, S.W., 1978. The Journal of William Sturgis. The Eighteenth-Century Memoirs of a Sailor Edited with an Introduction and Notes by S.W. Jackman. Sono Nis Press, Victoria, BC.

Jameson, R.J., Kenyon, K.W., Johnson, A.M., Wright, H.M., 1982. History and status of translocated sea otter populations in North America. Wildl. Soc. Bull. 10, 100–107.

Jamieson, G.S., Lessard, J., 2000. Marine protected areas and fishery closures in British Columbia. Can. Special Publ. Fisheries Aquat. Sci. 131, 414.

Karpov, K.A., Haaker, P., Taniguchi, I., Rogers-Bennett, L., 2000. Serial depletion and the collapse of the California abalone (*Haliotis* spp.) fishery. In: Campbell, A. (Ed.), Workshop on Rebuilding Abalone Stocks in British Columbia. Canadian Special Publication of Fisheries and Aquatic Science 130, pp. 11–24.

Kenyon, K.W., 1967. Survey of sea otter (*Enhydra lutris*) habitat in British Columbia, 1–5 June 1967. Unpublished report prepared for the Fish and Wildlife Branch, Department of Recreation and Conservation of British Columbia, 10pp.

Kenyon, K.W., 1969. The sea otter in the eastern North Pacific. North Am. Fauna 68, 1–352.

Kirov, S.V.M., 1965. A natural monument or a commercial species? The future of the Commander Islands sea otters. Priroda 52 (11), 79–83, 1963. Fisheries Research Board of Canada. Translation Series No. 543, 6pp.

Lander, R.H., Kajimura, H., 1982. Status of the northern fur seal. In: Mammals in the Seas, FAO Fisheries Series No. 5. vol. IV. ISBN 92-5-100514-1, FAO, Rome, pp. 319–245.

Lantz, D., 1918. Laws relating to fur-bearing animals, 1918. A summary of laws in the United States, Canada, and Newfoundland, relating to trapping, open seasons, propagation, and bounties. Farmer's Bulletin: 1022, United States Department of Agriculture, 32pp.

Lensink, C.J., 1960. Status and distribution of sea otters in Alaska. J. Mammal. 41 (2), 172–182.

Leopold, A., 1933. Game Management. Reprint 1986 with a new foreword, University of Wisconsin Press. Originally published 1933: New York, Charles Scribner's Sons.

Lowry, L.F., Pearse, J.S., 1973. Abalones and sea urchins in an area inhabited by sea otters. Mar. Biol. 23, 213–219.

MacAskie, I.B., 1971. A sea otter transplant to British Columbia. Fisheries Can. 23 (4), 3–9.

MacAskie, I.B., 1975. Sea otters, a third transplant to British Columbia. Beaver Spring, 9–11.

MacAskie, I.B., 1987. Updated status of the sea otter (*Enhydra lutris*) in Canada. Can. Field-Nat. 101, 279–283.

McBride, S., 1998. Current status of abalone aquaculture in California. J. Shellfish Res. 17, 593–600.

McCormick, T.B., 2000. Abalone (*Haliotis* spp.) aquaculture: present status and a stock enhancement tool. In: Campbell, A. (Ed.), Workshop on Rebuilding Abalone Stocks in British Columbia. Canadian Special Publication of Fisheries and Aquatic Science 130: pp. 55–60.

Meine, C., Soule, M., Noss, R.F., 2006. "A Mission-Driven Discipline": the growth of conservation biology. Conserv. Biol. 20 (3), 631–651.

Micheli, F., Shelton, A.O., Bushinsky, S.M., Chiu, A.L., Haupt, A.J., Heiman, K.W., et al., 2008. Persistence of depleted abalones in marine reserves of central California. Biol. Conserv. 141, 1078–1090.

Monroe, W.T., 1985. Status of the sea otter, *Enhydra lutris* in Canada. Can. Field-Nat. 99, 413–416.

Nichol, L.M., Boogaards, M.D., Abernethy, R., 2009. Recent trends in the abundance and distribution of sea otters (*Enhydra lutris*) in British Columbia. Fisheries and Oceans Canada, Canadian Science Advisor Secretariat Research Document 2009/016. iv + 16 p.

Obee, B., 1984. Sea otters return to BC coast. Can. Geogr. December, 40–43.

Revised Statutes of British Columbia, 1979. Volume 6. Queen's Printer for British Columbia, Victoria.

Rice, D.W., Wolman, A.A., Braham, H., 1984. The gray whale, *Eschrichtius robustus*. Mar. Fish. Rev. 46 (4), 7–14.

Salomon, A.K., Gaichas, S.K., Jensen, O.P., Agostini, V.N., Sloan, N.A., Rice, J., et al., 2011. Bridging the divide between fisheries and marine conservation science. Bull. Mar. Sci. 87 (2), 251–274.

SARA, 2003. Species at Risk Act. Accessed at: <http://www.registrelep-sararegistry.gc.ca/approach/act/default_e.cfm>.

SARA, 2009. Order Amending Schedule 1 to the Species at Risk Act (2009-01-17) <http://www.sararegistry.gc.ca/virtual_sara/files/orders/g1-14303_e.pdf>.

Saunders, R.T., 1996. Does "Sanctuary" mean secure? Endangered Species Update Special Issue: Conservation and Management of the Southern Sea Otter 13(12): pp. 43–46.

Scheffer, V.B., 1940. The sea otter on the Washington coast. Pac. Northwest Q. 3, 370–388.

Scheffer, V.B., 1995. Mammals of the Olympic National Park and vicinity. Northwest Fauna 2.

Scheffer, V.B., Fiscus, C.H., Todd, E.I., 1984. History of scientific study and management of the Alaska fur seal, *Callorhinus ursinus*, 1789–1964. NOAA Technical Report NMFS/SSRF-780, Seattle, WA.

Sloan, N.A., Dick, L., 2012. Sea Otters of Haida Gwaii: Icons in Human-Ocean Relations. Archipelago Management Board and Haida Gwaii Museum, 184pp.

Smith, I., 1969. The sea otter: a fresh start. Western Fish Game 4 (6), 26–28, 48 and 50–52.

Soule, M.E., Estes, J.A., Miller, B., Honnold, D.L., 2005. Strongly interacting species: conservation policy, management, and ethics. BioScience 55, 168–176.

Statutes of the province of British Columbia, 1931. From the 3rd session of the 17th Parliament, Charles F. Banfield, Printer to the King's Most Excellent Majesty 1931, Victoria, British Columbia.

Uni, Y., 2001. The statistics and the materials of the modern hunting of marine mammals near Hokkaido. Res. Rep. Shiretoko Mus. 22, 81–92 (Japanese).

US Statutes at Large, 1868. The Statutes at Large, Treaties and Proclamation of the United States of America, 40th Congress, Session II, Chapter 273: 240–242, 27 July 1868. <http://www.constitution.org/uslaw/sal/sal.htm>. Accessed May 2014.

US Statutes at Large. The Statutes at Large of the United States of America, 62nd Congress, Session II, Chapter 183: 326–328, April 21, 1910. <http://www.constitution.org/uslaw/sal/sal.htm>. Accessed May 2014.

US Department of the Interior, 1912. General Information Regarding the Territory of Alaska. Department of the Interior, Office of the Secretary, Government Printing Office, Washington, DC.

US Fish and Wildlife Service, 2003. Final Revised Recovery Plan for the Southern Sea Otter (*Enhydra lutris nereis*). Portland, Oregon. xi + 165pp.

VanBlaricom, G.R., Jameson, R.J., 1982. Lumber spill in central California waters: implications for oil spills and sea otters. Science 215, 1503–1505.

Vitousek, P.M., Mooney, H.A., Lubchenco, J., Melillo, J.M., 1997. Human domination of Earth's ecosystems. Science 277, 494–499.

Wada, K., 1997. The establishment and succession of the sea otter and northern fur seal sealing industry to resource management (2). Wildl. Conserv. Jpn. 2, 141–163 (in Japanese with English abstract).

Waples, R.S., Nammack, M., Cochrane, J.F., Hutchings, J.A., 2013. A tale of two acts: endangered species listing practices in Canada and the United States. BioScience 63 (9), 723–734.

Watson, J., 2000. The effects of sea otters (*Enhydra lutris*) on abalone (*Haliotis* spp.) populations. In: Campbell, A. (Ed.), Workshop on Rebuilding Abalone Stocks in British Columbia. Canadian Special Publication of Fisheries and Aquatic Science 130: pp. 123–132.

Watson, J.C., Ellis, G.M., Smith, T.G., Ford, J.K.B., 1997. Updated status of the sea otter, *Enhydra lutris*, Canada. Can. Field-Nat. 111 (2), 277–286.

Williams, T.M., Kastelein, R.A., Davis, R.W., Thomas, J.A., 1988. The effects of oil contamination and cleaning on sea otters (*Enhydra lutris*). Thermoregulatory implications based on pelt studies. Can. J. Zool. 66, 2776–2781.

Chapter 14

Synopsis of the History of Sea Otter Conservation in the United States

Glenn R. VanBlaricom
Washington Cooperative Fish and Wildlife Research Unit, USGS, and School of Aquatic and Fishery Sciences, College of the Environment, University of Washington, Seattle, WA, USA

INTRODUCTION

Reviews of the history of sea otter (*Enhydra lutris* [L., 1758]) conservation typically begin with Bering's voyage of 1741–1742 from Kamchatka to areas now known as the Aleutian Islands, the islands of the Bering Sea, and the southern mainland coast of Alaska. The tattered survivors of Bering's crews

Sea Otter Conservation. DOI: http://dx.doi.org/10.1016/B978-0-12-801402-8.00014-7

returned to Kamchatka in 1742, bringing sea otter pelts among other cargo. Fur hunters and traders responded soon after, coming first from Russia and later from Spain, Japan, Great Britain, and the British colonies of Canada and the future United States in search of sea otter pelts. Commercial harvests began in 1741 and evolved into a significant maritime fur trade. The fur trade period effectively ended in ~1910. A modest, legal commercial harvest in Japanese waters was reported in 1939, and government-sponsored harvests in Alaska from 1963 through 1970 incorporated some commercial considerations (Anonymous, 1939; Kenyon, 1969; Abegglen, 1977; Spector, 1998). Legal harvests of sea otters by Alaska Natives, permitted by the Marine Mammal Protection Act of 1972 (MMPA),[1] include a commercial dimension. The annual rate of harvest by Native communities has increased to substantial levels in recent years. However, conventional wisdom holds that sea otters were saved by the International Fur Seal Convention of 1911,[2] credited with imposing the international moratorium on hunting that rescued sea otters from extinction. In the prevailing paradigm the Convention initiated a century and more of sea otter conservation and population growth, and an associated return to more fully functional coastal marine ecosystems.

In this chapter I address two primary goals. First, I trace the sequence of US federal legislation, international treaties, and agreements, and related governmental actions pertaining to sea otter conservation from the purchase of Alaska in 1868,[3] through the passage of MMPA, and on to the final termination of internationally managed harvests of northern fur seals in the Bering Sea in 1985. I consider fur seal management issues here based on my contention that sea otter harvest management during the nineteenth and twentieth centuries was largely done in conjunction with, and in the shadow of, fur seal harvest management. Second, I summarize the historical development of contemporary legislative authorities and administrative frameworks under which conservation activities on behalf of sea otters are now managed.

My historical review emphasizes two categories of legislative, administrative, and legal actions that were beneficial to sea otter conservation regardless of the intended purpose, or at least provided a potential for conservation value. The first is management planning or actions with the intention of explicitly providing conservation benefits to sea otters, as individuals or as populations. The second is designation of protected areas that may benefit individuals or populations of sea otters and other wildlife in residence by virtue of ecosystem-scale protection, either within or adjacent to protected spaces or shorelines.

1. 86 US Statutes at Large 1027–1046, Public Law 92-522, October 21, 1972.
2. 37 US Statutes at Large 1542, Treaty Series 564, December 14, 1911.
3. The purchase followed ratification of the "Treaty concerning the Cession of the Russian Possessions in North America by His Majesty the Emperor of all the Russias to the United States of America" (15 US Statutes at Large, Treaties: 539–544, June 20, 1867).

I assume here that protected spaces in marine benthic and water-column habitats, and protective protocols for marine water quality, have had a significant effect on sea otter conservation for a number of decades. I also assume that shoreline protection generally limits human population densities and activities, with the premise that limitation of human activities in protected areas with marine shorelines reduces, through restriction of access, the probability of actions damaging to conservation values in adjacent marine ecosystems. I suggest that the premise is plausible, but concede that clear documentation is lacking.

Much of my historical narrative concerning the conservation of US sea otter populations, habitats, and ecosystems is based on direct readings of Congressional Acts.[4] I also relied on the United Nations Treaty Series,[5] an archive of international treaties, conventions, and exchanges of notes among national governments.

PROTECTION AND SECURITY OF FUR-BEARING MAMMALS IN ALASKAN TERRITORIAL WATERS IN THE NINETEENTH CENTURY, FOLLOWING THE 1867 TREATY OF CESSION

Dominion over the Territory of Alaska and associated coastal marine waters was transferred from Russia to the US pursuant to the Treaty of Cession proclaimed in 1867[6] and subsequent Congressional action appropriating US$7.2 M in payment to Russia.[7] The US federal government took immediate actions to secure and protect commercially valuable resources in Alaska, including salmon (*Oncorhynchus* spp.) and northern fur seals (*Callorhinus ursinus* [L., 1758]). It is clear from legislative discourse at the time that primary incentives for protection of fur seals, sea otters, and other fur-bearers were economic rather than ecological, and that harvest management concerns relating to northern fur seals were given priority over those of sea otters and other fur-bearers.

US federal statutory involvement with sea otter conservation began on July 27, 1868 with an Act[8] approved on the same day that Congress appropriated funds for the purchase of Alaska from Russia. The Act prohibited killing of "any otter, mink, marten, sable, or fur seal, or other fur-bearing mammal within the limits of said territory, or in the waters thereof." The Act thereby established federal protection, at least in concept, for sea otters within the territorial marine waters of Alaska, defined as waters within 5.6 km (3 nautical miles) of shore. The Act authorized the US Secretary of the Treasury to

4. Principal sources were the 126 existing volumes of the US "Statutes at Large," published by the US federal government from 1848 through 2012 and reflecting congressional actions beginning in 1789.

5. http://treaties.un.org.

6. See note 3.

7. 15 US Statutes at Large 247: 198, July 27, 1868.

8. 15 US Statutes at Large 273: 240–242, July 27, 1868, subsequently incorporated into US Revised Statutes, section 1956, chapter 3, Title XXIII.

establish regulations for a legal harvest of sea otters in Alaska, but the historical record seems to lack indications that harvests were ever formally authorized, or that a harvest management plan was developed.

Congress approved "An Act to provide for the protection of the salmon fisheries of Alaska" in 1889.[9] The Act included a provision extending the spatial scope of protection of fur-bearers from killing, previously afforded by the Act of 1868 described above, "to apply to all the dominion of the United States in the waters of the Behring (*sic*) Sea." The Act mandated annual proclamations by the president warning against entry of unauthorized persons and vessels into US waters within the "Behring (*sic*) Sea" to hunt fur-bearing animals. Seven such proclamations were issued between 1889 and 1896.[10] Each made note of the Act of 1868 and included explicit reference to its prohibitions of killing of fur-bearers in Alaska Territory or the territorial marine waters thereof.

In 1890 Congress approved an Act[11] to fund collection of population status data to obtain "full information as to the impending extinction of the sea-otter industry, and kindred lines of inquiry," along with data on fur seal populations. Thus, Congress explicitly recognized the imperiled status of sea otter populations in Alaska and the tenuous prospects for a sustainable long-term sea otter fur trade.

PROTECTION OF SEA OTTERS IN INTERNATIONAL WATERS

The International Fur Seal Conference held in autumn 1897 was the first attempt at a multilateral convention for protection of fur seals and sea otters. The United States, Russia, and Japan were represented. The delegates approved a Convention[12] containing two articles. The first specified a prohibition of the killing of fur seals and sea otters in international waters of the North Pacific Ocean and adjacent seas by hunters and traders for a period of 12 months. The second article required "adhesion"[13] of Great Britain to the Convention. Great Britain subsequently disapproved the Convention and it was not implemented.

---

9. 25 US Statutes at Large 415: 1009–1010, March 2, 1889.

10. 26 US Statutes at Large, Presidential Proclamations 1: 1543–1544, March 21, 1889; 26 US Statutes at Large, Presidential Proclamations 11: 1558, March 15, 1890; 26 US Statutes at Large, Presidential Proclamations 18: 1565–1566, April 4, 1891; 27 US Statutes at Large, Presidential Proclamations 16: 1008, February 15, 1892. 27 US Statutes at Large, Presidential Proclamations 50: 1070–1071, April 8, 1893; 28 US Statutes at Large, Presidential Proclamations 16: 1258–1259, February 18, 1895; 29 Statutes at Large, Presidential Proclamations 12: 878–880, April 14, 1896.

11. 26 US Statutes at Large 66: 46, April 5, 1890.

12. "Convention for the preservation of the fur seal and sea otter in the North Pacific Ocean and Bering Sea." Unperfected US Treaty Series X-4, November 6, 1897.

13. "Adhesion" is defined as an "agreement to join" by "Webster's ninth new collegiate dictionary" (1988; Merriam-Webster Inc., Springfield, Massachusetts USA. 1563 pages).

The Convention among the United States, Great Britain (also representing Canada), Russia, and Japan for the Preservation and Protection of Fur Seals (hereinafter the Fur Seal Treaty of 1911) was proclaimed in 1911.[14] The principal focus of the Treaty was the prohibition of killing of fur seals in pelagic habitats of the North Pacific Ocean north of 30°N latitude, and in the Bering Sea, the Sea of Japan, the Sea of Okhotsk, and other northern marine waters. Article V of the Treaty[15] provided explicit protection of sea otters from hunting, but only in international waters (≥5.6 km offshore) within the stated geographic boundaries.

In August 1912 Congress passed the Fur Seal Act of 1912,[16] incorporating the Fur Seal Treaty of 1911 as a US statute. The Act affirmed that protection of sea otters was limited to international waters. Independently, the State of California outlawed killing and possession of sea otters within territorial marine waters beginning in 1913 (Wild and Ames, 1974). In autumn 1940 Japan gave notice of abrogation of the Fur Seal Treaty, effecting termination on October 23, 1941. Diplomatic exchanges between the United States and Canada reconstituted the Treaty in December 1942 as the Provisional Fur Seal Agreement (United Nations, 1949). The new agreement prohibited killing of fur seals in the North Pacific but made no mention of sea otters. Congress incorporated the agreement into US statutes with the Fur Seal Act of 1944,[17] restoring language protecting sea otters in international waters. The 1944 Fur Seal Act repealed the Act of 1868 and other extant federal laws relevant to sea otter protection, except for prohibition of sea otter harvest within territorial Alaskan waters by a federally mandated Criminal Code for the Territory of Alaska imposed in 1899.[18]

The 1942 Provisional Fur Seal Agreement was revised through diplomatic exchanges in 1947, 1952, 1957, 1966, and 1983. Japan rejoined the agreement in 1952, the Union of Soviet Socialist Republics in 1957. Sea otters were not mentioned in the agreements of 1947, 1952, or 1957. The Fur Seal Act of 1966[19] apparently was the final Act of Congress explicitly linking sea otter conservation and management with that of fur seals, but the intent of Congress regarding sea otters was ambiguous. The subtitle of the Act included the phrase "to protect sea otters on the high seas," but sea otters were not mentioned in the text of the Act. Internationally managed

---

14. 37 US Statutes at Large 1542, Treaty Series 564, December 14, 1911.

15. Article V reads as follows, in its entirety: "Each of the High Contracting Parties agrees that it will not permit its citizens or subjects or their vessels to kill, capture or pursue beyond three miles from the shorn (*sic*) line of its territories sea otters in any part of the waters mentioned in Article I of this Convention."

16. 37 US Statutes at Large 373: 499–502, August 24, 1912.

17. 58 US Statutes at Large 65: 100–105, February 26, 1944.

18. Code of Criminal Procedures, Territory of Alaska, as imposed by the US Congress in 30 US Statutes at Large 429: 1253–1343, March 3, 1899.

19. 80 US Statutes at Large 1091: 109–121, November 2, 1966.

commercial-scale harvests of northern fur seals in the Bering Sea were terminated after 1985.

PROTECTIVE LAWS AND PROTOCOLS IN THE INDIVIDUAL STATES

Sea otters in US waters were originally distributed among Alaska, Washington, Oregon, and California (the "sea otter states"). Populations in Oregon remain regionally extinct at this writing. The federal Submerged Lands Act of 1953[20] was important to state jurisdiction over sea otter management. The Act specified that submerged lands beneath navigable waters within 5.6 km (3 nautical miles) of the mean high tide line of the mainland shore were within the jurisdiction of the adjoining state, thus allocating to the states the authority to manage territorial waters and associated submerged lands in accordance with applicable state law. The Act affirmed authority of the individual states to manage sea otter populations until the Congressional approval of MMPA in 1972.

Pioneering Conservation Actions in California

Early in the twentieth century, resource agency personnel in California began observing a small surviving group of sea otters off Monterey County (Bryant, 1915). In 1913 the State of California prohibited taking of sea otters and possession of sea otters or their pelts (Wild and Ames, 1974). The state followed with establishment of the California Sea Otter Game Refuge (CSOGR) in 1941. CSOGR included two segments of the Monterey County coast with a combined length of ~38 km (summed great circle lengths[21]). The boundaries of CSOGR were extended in 1959 to a single coastline segment with shoreline length ~165 km (McArdle, 1997).

DEVELOPMENT OF CONTEMPORARY PROTOCOLS OF INDIVIDUAL STATES BENEFITING THE CONSERVATION OF SEA OTTERS

In the early 1970s governments of the four sea otter states developed innovative and far-reaching conservation laws with significant implications for protection of sea otters. The state laws resembled three important new US federal statutes: the National Environmental Policy Act of 1969 (NEPA),[22]

20. 67 US Statutes at Large 65: 29–33, Public Law 31, May 22, 1953.
21. Great circle distances reported herein were calculated with GPS Visualizer (http://www.gpsvisualizer.com).
22. 83 US Statutes at Large 852–856, Public Law 91-190, January 1, 1970.

the Coastal Zone Management Act of 1972 (CZMA),[23] and the Endangered Species Act of 1973 (ESA).[24] The NEPA required published assessments of likely environmental consequences of development actions taken by government agencies. The CZMA mandated regulation of coastal zone development for the benefit of environmental conservation. The ESA provided for protection of plant and animal species based on levels of risk and perceived likelihoods of extinction. For example, the California state legislature passed the California Environmental Quality Act in 1970 (analogous to the federal NEPA), the California Endangered Species Act in 1973 (analogous to ESA), and the California Coastal Act in 1976 (analogous to CZMA). Legislative authorities in Oregon and Washington approved comparable legislation in approximately the same time period. In Alaska a two-tiered listing protocol categorized species at greatest risk as "endangered" and others at lesser risk listed as "species of special concern." The "endangered" list remains in use at present. In 2011 the Alaska Department of Fish and Game (ADF&G) released the Wildlife Action Plan (WAP; first described in ADF&G, 2006). WAP replaced the "species of special concern" protocol with a complex new system for classifying the status of compromised species. In WAP sea otters were categorized as "apparently secure—uncommon but not rare; some cause for long-term concern due to declines or other factors."

All four sea otter states have developed systems of marine protected areas (MPAs). In 1974 the State of California designated Areas of Special Biological Significance (ASBS).[25] Seven ASBS sites encompassed coastal marine waters in the occupied portion of the California sea otter range with a combined marine shoreline length of ~88 km. The California Marine Life Protection Act of 1999 (CMLPA)[26] coordinated and improved MPAs that had developed, largely independently of one another, during preceding decades along the California coast. The CMLPA-mandated system resulted in 124 separate MPAs encompassing 2209 km^2 of surface area, 16.12% of the total surface area in California's territorial waters (California Department of Fish and Game, 2012a,b; California Department of Fish and Wildlife, 2013a,b).

The Oregon Marine Reserves Program was approved by the State Legislature in 2008.[27] Five sites were designated as "marine reserves" with implementation initiated in 2012 and scheduled for completion in 2016. The State of Washington established four different MPA systems beginning in the 1970s. In combination the four programs included 18 areas within historical sea otter range and likely to be used by sea otters given significant population growth and range expansion in the future.

23. 86 US Statutes at Large 1280–1289, Public Law 92-583, October 27, 1972.
24. 87 US Statutes at Large 884–903, Public Law 93-205, December 28, 1973.
25. http://www.waterboards.ca.gov/water.
26. California Fish and Game Code §2850 *et seq.*
27. Oregon Revised Statutes §§196.540–555.

The ADF&G and the Alaska Department of Natural Resources co-managed 32 protected areas in four categories,[28] encompassing a broad range of habitats and ecosystems in terrestrial and coastal marine locations.[29] At least 10 of the locations, representing three of the categories, may have been useful for sea otter conservation. In addition, "Steller sea lion mitigation zones," an array of 38 units associated primarily with rookeries of Steller sea lions (*Eumetopias jubatus* [Schreber, 1776]) were established by federal managers in 2002 (Witherell and Woodby, 2005). The mitigation zones were intended to eliminate disturbances to hauled pinnipeds at the rookeries by transiting vessels. The zones extended 5.6 km offshore at each location and were closed to all vessel activity. Thirty-five of the units likely included resident sea otters, either at implementation in 2002 or prior to the maritime fur trade era.

The *Exxon Valdez* oil spill (EVOS) of 1989 caused significant mortality in sea otter populations of Prince William Sound (PWS) and other locations off the southern and southwestern marine shores of Alaska (Chapter 4). EVOS forced the sea otter states to confront inadequacies in response capabilities on behalf of vulnerable or imperiled marine species and ecosystems, including sea otters and their habitats. An earlier empirically based cautionary note (VanBlaricom and Jameson, 1982) also elaborated oil spill risks to sea otters in California waters. The California State Legislature responded to oil spill events by approving the Lempert-Keene-Seastrand Oil Spill Prevention and Response Act of 1990.[30] The Act created the Office of Spill Prevention and Response (OSPR), managed by the California Department of Fish and Game.[31] The 1990 Act required OSPR to establish and operate rescue and rehabilitation facilities for sea otters, sea birds, and other animals in California's coastal seas known to be vulnerable to oil spills.

In 1994 OSPR initiated the Oiled Wildlife Care Network (OWCN)[32] in California, implemented as a collaborative network of >30 separate entities with interests and capabilities in wildlife conservation. OWCN included public aquaria, universities, scientific organizations, and non-governmental wildlife rehabilitation groups with headquarters at the Wildlife Health Center of the University of California, Davis.

The Prince William Sound Science Center (PWSSC) developed as an informal community association in Cordova, Alaska, in the late 1980s,[33] motivated by interest in greater interaction with scientists working on fishery science and management issues in PWS. PWSSC incorporated formally after the

28. http://www.adfg.alaska.gov/index.cfm?adfg=protectedareas.main.
29. Alaska Statutes 16.20.050-060 and 38.05.
30. Senate Bill 2040, California Statutes 1990, chapter 1248, September 24, 1990.
31. Renamed the California Department of Fish and Wildlife in 2013.
32. http://www.vetmed.ucdavis.edu/owcn.
33. http://pwssc.org

EVOS in March 1989, placing priority on understanding and disseminating research findings regarding oil spill impacts on marine ecosystems, including those occupied by sea otters. The Oil Pollution Liability and Compensation Act of 1989[34] authorized establishment of the Oil Spill Recovery Institute (OSRI) and assigned management responsibility to PWSSC. OSRI was charged with improving the understanding of effects of oil pollution on marine ecosystems including those with sea otters, identifying methods for preventing oil spills, and assessing and enhancing recovery of ecosystems damaged by spills.

Following EVOS, a Natural Resource Damage Assessment procedure was initiated in 1989 to collate information on damage to natural ecosystems for purposes of litigation against responsible parties. As a consequence, a civil settlement was reached in a US federal court in October 1991 pursuant to the Federal Water Pollution Control Act,[35] with the responsible parties agreeing to provide payments totaling US$900 M into a trust fund over a 10-year period to support restoration research and implementation activities. The *Exxon Valdez* Oil Spill Trustee Council (EVOSTC) was created to oversee restoration activities supported by the trust fund on behalf of resources injured by the oil spill, including sea otters.

REGIONAL-SCALE NON-GOVERNMENTAL ORGANIZATIONS WITH FOCI ON THE CONSERVATION OF SEA OTTERS

Friends of the Sea Otter (FSO) was established in 1968.[36] With a politically vocal membership, successful fund-raising strategies, an active professional staff, and the inspiring legacy of co-founder Margaret Owings, FSO has contributed effective support of legislative and management actions on behalf of sea otter conservation. FSO had significant roles in facilitating listing of California sea otters and participating in sea otter recovery planning related to ESA and MMPA (Chapter 12). FSO traditionally focused conservation advocacy on behalf of sea otters in California, expanding its activities and interests in recent years to sea otter populations throughout their geographic range.

In 1977 the Monterey County (California) Chapter, Society for the Prevention of Cruelty to Animals, established a Wildlife Center for rescue and rehabilitation of injured animals. The primary emphasis was terrestrial vertebrates, but the Center staff also provided care for pre-weaning sea otter pups found stranded on the shores of Monterey Bay. The Center developed significant advances in effective treatment for pups in their care. The program was relocated to the Monterey Bay Aquarium (MBA) at the Aquarium's opening in 1984, given MBA's more immediate proximity to Monterey Bay, more

34. 104 US Statutes at Large 484–575, Public Law 101-380, August 18, 1990.
35. 66 US Statutes at Large 758: 1155–1161, June 30, 1948.
36. http://www.seaotters.org/histor/.

elaborate and spacious facilities, and stronger funding base. The program became the Sea Otter Research and Conservation Program (SORAC) at MBA (Chapter 9). The primary purposes of SORAC were rescuing, treating, and releasing injured sea otters, raising and releasing stranded pups (Nicholson et al., 2007; Chapter 9), caring for sea otters that were deemed not releasable back to the wild, and research related to rescue and rehabilitation questions.[37] There are a number of other active and effective non-governmental organizations involved in various aspects of sea otter conservation. A full accounting of all is beyond the scope of this chapter.

THE IUCN RED LIST AND KEY FEDERAL LEGISLATION AS TOOLS FOR SEA OTTER CONSERVATION

In the period from 1963 through 1973, an innovative international ecological accounting system and seven concept-oriented US Congressional Acts revolutionized the relationship of conservation and government. The suite of new laws and protocols reflected an emerging shift in western culture away from the conservation of nature to sustain exploitation, and toward focusing on recovering imperiled species and populations. Subject Congressional actions included the Wilderness Act of 1964,[38] the Endangered Species Preservation Act of 1966,[39] the Endangered Species Conservation Act of 1969,[40] NEPA,[41] MMPA,[42] ESA,[43] and amendments to the Marine Resources and Engineering Development Act of 1966,[44] adding Title III, known as CZMA.[45]

In 1963 the International Union for the Conservation of Nature (IUCN) established the IUCN Red List of Endangered Species™,[46] managed by the Species Survival Commission of the IUCN Global Species Programme. The primary purpose of the List at inception was to promote the conservation of imperiled species and subspecies throughout the world. Red List procedures and protocols were substantially modified beginning in 1994 with the goals of greater objectivity and scientific rigor in the evaluation and listing processes. The revised purpose of the Red List was "to provide information and analyses on the status, trends, and threats to species in order to inform and catalyze action for biodiversity conservation." The most recent categorization

37. http://montereybayaquarium.org/cr/sorac.aspx.
38. 78 US Statutes at Large 890–896, Public Law 88-577, September 3, 1964.
39. 80 US Statutes at Large 926–930, Public Law 89-669, October 15, 1966.
40. Sections 1 through 5, 83 US Statutes at Large 275–283, Public Law 91-135, December 5, 1969.
41. See note 22.
42. See note 1.
43. See note 24.
44. 80 US Statutes at Large 203–208, Public Law 89-454, June 17, 1966.
45. See note 23.
46. http://www.iucn.org/about/red-list-overview.

of sea otters by Red List protocols was "endangered,"[47] based on status reviews in 2000 and 2008. The species-wide trend in numbers of individuals was categorized as "decreasing" in the 2008 status review.

The goal of the US Wilderness Act of 1964 was to create "wilderness areas" that would be valued and preserved on the basis of "their wilderness character." The Act declared prohibitions of developments that might facilitate damaging human incursions, and authorized the Secretaries of Agriculture and the Interior to pursue designations of wilderness areas within other federal protected spaces such as National Wildlife Refuges (NWRs), National Parks and Monuments, and National Forests. Designations of lands with marine shorelines proximate to sea otter habitats as wilderness areas may have contributed to effective conservation of sea otter populations in many locations.

The MMPA and ESA were watershed US Congressional actions with unequivocal benefits to sea otter conservation. MMPA recognized "depleted" populations as those below "optimum sustainable population" size (OSP), while ESA recognized "endangered" and "threatened"[48] species as being in greater peril of extinction than others by virtue of perceptions of likely present and future threats to continued survival. The MMPA and the ESA were functionally interconnected by a protocol that any species or population listed as "endangered" or "threatened" as defined by ESA was generally also classified as "depleted" pursuant to MMPA. However, in current practice a species or population classified as "depleted" pursuant to MMPA is not necessarily listed in one of the ESA categories.

The MMPA was the first US federal conservation law to specify protections that could be applied objectively to sea otters throughout their range in US waters without ambiguity regarding conservation interests of higher priority in other species, and without an arguably questionable focus on minimally utilized marine habitats far from shore. By virtue of MMPA, sea otters decisively emerged from the shadows of conservation interests fixed on northern fur seals as related to federal protected status.

Through MMPA the US Congress established policies stating that marine mammal species and populations should be maintained numerically above OSP, ensuring that species would not be allowed "to diminish beyond the point at which they cease to be a significant functioning element in their ecosystem." Congress further affirmed that marine mammal species and populations should be protected from any "adverse effect of man's actions," and that the "primary objective of their management should be to maintain the

47. http://www.iucnredlist.org.

48. The ESA (see note 21) defines "endangered" as "any species which is in danger of extinction throughout all or a significant portion of its range," excluding certain insect species categorized as pests, and "threatened" as "any species which is likely to become an endangered species within the foreseeable future throughout all or a significant portion of its range."

health and stability of the marine ecosystem." The Act prohibited taking (defined as "to hunt, harass, capture, or kill, or attempt to hunt, harass, capture, or kill"), possessing, or importing marine mammals within the territorial waters of the United States and from international waters, and mandated minimization of bycatch of marine mammals in commercial fisheries. The MMPA established definitions by which species or populations of marine mammals falling below OSP would be categorized as "depleted" and provided guidelines for stringent protection of depleted species. A key initial mandate of MMPA was a moratorium on harvests of marine mammals and prohibitions on importation of marine mammals or associated products from other nations pending establishment of regulations for harvests consistent with MMPA intent and structure.

Four elements of MMPA presented significant implications for sea otter conservation. The first was an exemption allowing taking of marine mammals for subsistence use, or for use in "creating and selling authentic native articles of handicrafts and clothing" by Alaskan Native communities. Such communities have been authorized to legally harvest marine mammals since 1972 under authority of the exemption, and sea otters have been among the most numerically important marine mammals harvested. Harvests of sea otters were initially minimal, but began increasing in the 1980s and increased dramatically in the 1990s, particularly in southeastern Alaska (D.M. Burn, US Fish and Wildlife Service, personal communication, Chapters 4 and 12). Harvest rates allowed by MMPA are now arguably cause for concern in the context of population- and species-level conservation. The second was usurpation by the federal government of management authority over marine mammals in the territorial marine waters of the individual states. The MMPA provided general guidelines for a return of management jurisdiction to the states, conditioned on demonstration that state conservation and management plans for marine mammals are consistent with MMPA. No state has petitioned successfully for permanent return of management authority over marine mammals since the passage of MMPA. The third was affirmation that the scope of species protection should include a focus at the population scale rather than being limited to larger-scale entities such as entire species or sub-species. The fourth was the Marine Mammal Marking, Tagging, and Reporting Program (MTRP), incorporated to MMPA by Congressional amendment in 1981,[49] with effective implementation by the US Fish and Wildlife Service in 1988. The Program requires that hides and skulls from sea otters taken by Alaskan Native hunters be tagged within 30 days of harvest. The purpose of MTRP is to monitor legal subsistence harvests of sea otters (and, in similar fashion, other marine mammals hunted by Alaskan Natives), and to improve the control of illegal take, trade, or transport of specific parts of harvested sea otters.

49. 95 US Statutes at Large 979–987, Public Law 97-58, October 9, 1981.

The protective provisions of ESA were applicable to species regardless of taxonomic affinity. The stated purposes of ESA were conservation of the ecosystems upon which threatened and endangered species are part; facilitation of the conservation of threatened and endangered species; and achievement of the purposes of the Convention on International Trade in Endangered Species of Wild Fauna and Flora (CITES; formally adopted in 1973; United Nations, 1983). Congress mandated that assessments of species or populations pursuant to ESA should be based on a "five-factor analysis": evaluations of threats to current habitat or range, potentially excessive exploitation, effects of disease or predation, potential inadequacy of existing protective regulatory mechanisms, or "other natural or manmade factors." The ESA prohibited transport of "threatened" or "endangered" species or populations across international borders; harvest, possession, selling, or transportation of imperiled species or parts thereof under a number of circumstances; and violations of CITES. The ESA afforded conditional exemptions for harvests of listed species for Native Alaskan communities, similar to those of MMPA.

There are key differences between "depleted" status as defined by MMPA and the categories of "threatened" and "endangered" as defined by ESA. Assignment of "depleted" status to a species or population pursuant to MMPA first requires that numbers of a subject species or population fall below the minimum OSP threshold as defined for the species or population. In contrast, listing under ESA can be anticipatory, focusing on the likelihood that future events may push a species or population toward extinction even if, at the time of listing, the subject species or population is not numerically diminished. Thus, listing as defined by MMPA can be regarded as "reactive," and listing pursuant to ESA as "proactive" (D.M. Burn, US Fish and Wildlife Service, personal communication).

Two groups of sea otters have been listed according to ESA protocols, and consequentially categorized as "depleted" pursuant to MMPA. Sea otters off California were listed as "threatened"[50] in 1977 in response to perceived increasing risks of oil spills, low numbers of animals, and a limited geographic distribution. The translocation of sea otters from coastal mainland populations to San Nicolas Island (SNI),[51] off Southern California, from 1987 through 1990 was a significant response to the 1977 listing (Chapters 12 and 13). Congress resolved legal obstacles to the SNI translocation in 1986 (Baur et al., 1996). Because of listed status pursuant to MMPA and ESA, it was not clear that animals could be taken from groups of sea otters off the California mainland for translocation to SNI without violation of the two statutes. In response, Congress passed an Act "to improve the operation of certain fish and wildlife programs,"[52] exempting the US Fish and Wildlife

50. 42 US Federal Register 2965, January 14, 1977.
51. 33.24°N, 119.46°W.
52. 100 US Statutes at Large 3500–3503, Public Law 99-625, November 7, 1986.

Service from statutory constraints and allowing initiation of the project in 1987. The Southwest Alaska Stock of sea otters was listed as "threatened"[53] in 2005 (Chapter 4). The region included the Aleutian Archipelago, southern Bristol Bay, the south shore of the Alaska Peninsula, the Kodiak Archipelago, and southwestern Cook Inlet and nearby islands. Surveys of sea otters in the area had indicated that numbers declined drastically beginning in the 1980s (Estes et al., 1998; Doroff et al., 2003).

PROTECTION OF FEDERALLY MANAGED HABITAT SPACES AND WATERS UTILIZED BY SEA OTTERS

Section 24 of an 1891 Congressional Act[54] authorized the US President to reserve timbered public lands in any US State or Territory, establishing such lands as "reservations," and became known as the Forest Reserve Act of 1891, a landmark precedent for later actions by the federal government. The Act focused on protection of habitats and spaces rather than particular plant or animal populations for purposes of conservation, first in terrestrial habitats and later in coastal marine ecosystems. The Reclamation Act of 1902[55] expanded the power of the federal government for withdrawals of lands from the public inventory. Congress authorized reservation of lands suitable for establishment of "storage, diversion and development" facilities for water resources supporting reclamation of lands for agriculture. Similarly, the American Antiquities Act of 1906[56] added authority and flexibility to the Executive Branch of the US government by permitting Presidents to withdraw public lands encompassing historic landmarks and objects, and other entities of historic and scientific interest, designating such lands as National Monuments. The three Acts shared a crucially important attribute, each authorizing Presidents to reserve public lands for various purposes without requiring Congressional approval.

The National Wildlife Refuge System

The NWR System developed in several phases, with the creation of individual refuges (originally termed "reservations") predating the emergence of a cohesive collective management system. Two major conservation issues were incentives for the earliest refuge designations. In the late 1880s a combination of unregulated and inappropriate fishery practices and burgeoning cannery operations was leading to severe overexploitation of salmon stocks

53. 70 US Federal Register 46366, August 9, 2005.
54. 26 US Statutes at Large 561: 1095–1103, March 3, 1891.
55. 32 US Statutes at Large 1093: 388–390, June 17, 1902.
56. 34 US Statutes at Large 3060: 225, June 8, 1906.

in Alaska. In 1889 a Congressional Act[57] directed the federal Commissioner of Fish and Fisheries to investigate. The resulting report (subsequently published in McDonald, 1894) included recommendations for establishment of "a national salmon park" to provide protective measures for salmon stocks. President Benjamin Harrison responded by proclaiming the Afognak Forest and Fish Culture Reservation in 1892,[58] using authority provided by the Forest Reserve Act of 1891. The Afognak Reservation was the first federally managed wildlife refuge in the United States and included marine benthic habitats (now part of the Alaska Maritime NWR), apparently the first such habitats in US history to be afforded "refuge" status on behalf of nature by federal authority within US territorial waters.

The second issue involved the depletion of wading birds, especially herons and egrets, in the wetlands of Florida. Hunting for complex feathers known as "nuptial plumes" or "aigrettes," grown by wading birds during courtship and breeding, emerged in the 1870s as a lucrative enterprise satisfying demands of the millinery industry, particularly in affluent societal circles of Boston, New York, and London. Excessive harvest rates and alarming reductions in wading bird numbers soon became apparent. President Theodore Roosevelt responded by designating the first federal bird reservation, at Pelican Island, Florida, in 1903,[59] pursuant to authority granted by the Forest Reserve Act of 1891. The President designated many other wildlife reservations by executive order before leaving office in 1909. He exploited authorities granted to the President by Congress through the Forest Reserve, Reclamation, and American Antiquities Acts to initiate systems of reservations that, together with the precedent-setting Afognak Island Reservation, evolved into the NWR System and the National Forest System (Brasher and Moyle, 2004).

The first five wildlife reservations with marine shorelines adjacent to Alaskan sea otter habitat were established by orders of President Roosevelt in 1909 (Pribilof Islands in the Bering Sea; Chisik, Egg, and Duck Islands in Tuxedni Bay on the western shore of southern Cook Inlet; and Saint Lazaria Island in southeastern Alaska)[60] and by President William Howard Taft in 1912 (Hazy Islands and Forrester Island, both in southeastern Alaska).[61] Because of placement on the geographic fringes of the range of sea otters, extreme exposure to weather and seas, or minimal habitat area, the subject reservations probably provided little value for sea otter conservation. In contrast, President Taft's 1913 Executive Order[62] creating the Aleutian Island Reservation as a "preserve for native birds, animals, and fish" may have

57. 25 US Statutes at Large 415: 1009, March 2, 1889.
58. 27 US Statutes at Large, Presidential Proclamations 39: 1072, December 24, 1892.
59. US Presidential Executive Order, March 14, 1903.
60. US Presidential Executive Orders 1039, 1040, and 1044, all dated February 27, 1909.
61. US Presidential Executive Orders 1459 and 1459, both dated January 11, 1912.
62. US Presidential Executive Order 1733, March 3, 1913.

been among the most significant Acts on behalf of sea otter conservation in US history. President Taft's Order emphasized protection of "seals, sea otter, cetaceans, and other aquatic species" and reiterated Congressional Acts prohibiting the killing of any "otter, mink, marten, sable, or fur seal, or other fur-bearing animals within the limits of Alaska Territory." The Aleutian Islands had been habitat for many of the largest known sea otter populations before and during the maritime fur trade period and represented significant potential for sea otter conservation. The reservation included territorial marine waters of the entire Aleutian Archipelago from False Pass[63] to Cape Wrangell,[64] a great circle distance of 1594 km, along with Bogoslof Island and the Sanak Islands. President Herbert C. Hoover ordered the addition of Amak Island and nearby small islets and rocks to the Aleutian Island Reservation in 1930[65] and authorized establishment of the Semidi Islands Reservation in 1932. The Kodiak NWR was proclaimed in 1941, the Simeonof Island NWR in 1958, and the Izembek National Wildlife Range in 1960. All four additional reservations presented substantial potential for conservation of sea otters. Passage of the Alaska National Interest Lands Conservation Act of 1980 (ANILCA)[66] dramatically modified and expanded the NWRs and other federal conservation-oriented entities in Alaska with a potential for increased levels of protection of sea otter populations, as described below.

The Washington Islands NWR (WINWR) originated as three separate Refuges (NWRs) proclaimed by President Theodore Roosevelt in 1907[67] in historical sea otter habitats off Washington (Flattery Rocks NWR, Quillayute Needles NWR, and Copalis NWR). The WINWR was designated with jurisdiction over land areas (above mean high tide level) of small islands and emergent rocks and reefs in marine waters extending ~12 km offshore. The three original refuges comprising WINWR included a combined total of ~870 emergent rocks, reefs, and small islands with a total land surface area of 2 km^2, with prohibitions against public visitation or close approach by vessel to reduce wildlife disturbance, providing potentially significant habitat for hauling out and resting by sea otters.[68] In 1970[69] most lands in WINWR were designated as "wilderness" pursuant to the Wilderness Act of 1964.[70] Sea otters now occur in or near the waters surrounding the Flattery Rocks and Quillayute Needles units.

There are three NWRs with marine shorelines on the coast of Oregon where sea otters are now regionally extinct. All three have potential for

63. 54.83°N, 163.40°W.
64. 52.92°N, 172.45°E.
65. US Presidential Executive Order 5317, April 7, 1930.
66. 94 US Statutes at Large 2371–2551, Public Law 96-487, December 2, 1980.
67. US Presidential Executive Orders 703, 704, and 705, all issued on October 23, 1907.
68. http://www.fws.gov/refuges/profiles.
69. 84 US Statutes at Large 1104–1106, Public Law 91-504, October 23, 1970.
70. See note 38.

benefiting sea otter conservation should sea otters return to Oregon, either by natural dispersal or from future translocation and restoration projects. The Oregon Islands NWR probably has the greatest potential of the three for contributing to sea otter conservation. The Refuge was established by Executive Order of President Franklin D. Roosevelt in 1935[71] and consists of an estimated 1874 islets, rocks, and reefs extending along the entire 515 km of exposed marine coastline from the Columbia River mouth southward to the California border, excluding only Chiefs Island and the Three Arch Rocks NWR, within a band extending 5.6 km offshore from the mainland. Also included are mainland sites at Coquille Point and Crook Point. Within the Refuge only the Coquille Point site is open to public access. Total land area in the Refuge is 1.5 km^2, and wilderness designation was provided to nearly all refuge lands in 1970,[72] 1978,[73] and 1996.[74]

Nine NWRs are located on the California coast, including the shores of San Francisco Bay, encompassing marine shorelines adjacent to apparently suitable sea otter habitat. Only two, Salinas River (established 1973) and Guadalupe-Nipomo Dunes (2000), have shoreline segments adjoining habitats used currently by sea otters. Ogden (1941) summarized historical records of sea otter distribution in San Francisco Bay, based largely on hunting records. The records confirm that sea otters were abundant and relatively widely distributed in the Bay. The northern limit of the present sea otter population in California is ~70 km south of the entrance to San Francisco Bay (great circle distance). In the likely event that sea otters reoccupy the Bay, at least three of the NWRs within the Bay are located appropriately and contain usable sea otter habitat, and could serve effectively as a source for protection of sea otters.

The names of all wildlife reservations in the United States were changed to "National Wildlife Refuges" (NWRs) in 1940.[75] The National Wildlife Refuge System Act of 1966 was approved as Sections 4 and 5 of the Endangered Species Preservation Act of 1966.[76] The 1966 Act, together with significant amendments in 1997,[77] was the defining legislation for the integrated contemporary NWR System. The Act provided regulatory prohibitions for Refuge lands, forbidding the taking or possessing of "any fish, bird, mammal, or other wild vertebrate or invertebrate animals or part or nest or egg thereof."

---

71. US Presidential Executive Order 7035, May 6, 1935.
72. 84 US Statutes at Large 1104–1106, Public Law 91-504, October 23, 1970.
73. 92 US Statutes at Large 1095–1099, Public Law 95-450, October 11, 1978.
74. 110 US Statutes at Large 4093–4281, Public Law 104-333, November 12, 1996.
75. US Presidential Proclamation 2416, July 25, 1940.
76. See note 39.
77. National Wildlife Refuge System Improvement Act of 1997, 111 US Statutes at Large, Public Law 105-57, October 9, 1997.

The National Park System

Congress passed the "National Park Service Organic Act" in 1916,[78] activating the national parks as a resource management system of administratively interconnected units. The National Park Service (NPS) was charged with managing parks to "conserve the scenery and the natural and historic objects and the wild life therein" among other objectives. The Park Service was also designated to manage national monuments and national seashores, among other categories of federally protected space.

Currently six national parks and four national monuments border occupied sea otter habitats, along with two national parks, a national monument, and a national seashore located adjacent to waters occupied by sea otters prior to the maritime fur trade. Glacier Bay National Park and Preserve (GBNPP) and the California Coastal National Monument (CCNM) are informative examples of entities significant or potentially significant to the protection of sea otters. The GBNPP originated in 1925 as a national monument proclaimed by President Coolidge.[79] The GBNPP's present conservation protocols were defined by the Omnibus Consolidated and Emergency Supplemental Appropriations Act of 1999.[80] The Act imposed regulation of commercial fisheries in marine waters of GBNPP, ranging from fully protected no-take areas to selective prohibitions involving limited participation and sunset clauses to areas open to all harvest activities allowed by the State of Alaska. The Act prohibited development of new commercial fisheries in GBNPP and harvests of marine mammals by Alaskan Native communities and required non-transferable lifetime access permits for fishers working in the Bay. The advantages accruing to sea otters included elimination of directed harvests by Native Alaskans, reduction of fishery bycatch risk, reduction of disturbance rates by commercial vessels, and reduction of competition with certain fisheries, such as for crab, for preferred prey.

The CCNM was proclaimed in 2000[81] by US President William J. Clinton. The proclamation was the culmination of a lengthy administrative history for the protected area, initiated by President Hoover in 1930[82] when the area was withdrawn from the public land inventory pending future designation. The Monument extends along the entire oceanic coastline of California (~1770 km), encompassing the ~20,000 small islands, islets, rocks, and reefs present within 22.2 km (12 nautical miles) of the mainland shore but separated by water from the mainland at mean high tide, with a total land area of 2.46 km^2. All land areas within CCNM are closed to public access for protection of resident wildlife. Waters designated as encompassing

78. 39 Statutes 408: 535–536, August 25, 1916.
79. US Presidential Proclamation 1733, February 26, 1925.
80. 112 US Statutes at Large 2681–3600, Public Law 105-277, October 21, 1998.
81. US Presidential Proclamation 7264, January 11, 2000.
82. US Presidential Executive Oder 5326, April 14, 1930.

CCNM lands effectively include all sea otters dwelling along the California coast. The eight large islands off Southern California are not included in CCNM, but adjacent small islets or rocks are included. The Monument is managed by the Bureau of Land Management.

The National Marine Sanctuary System

Congress explicitly applied the reservation concept to marine habitats with the Marine Protection, Research, and Sanctuaries Act of 1972.[83] Title III of the Act is known as the National Marine Sanctuaries Act of 1972 (NMSA). The NMSA empowered the Secretary of Commerce to designate sanctuaries within territorial waters extending offshore to the outer edge of the continental shelf and in the Great Lakes. Purposes of sanctuary designation included "preserving or restoring such areas for their conservation, recreational, ecological, or esthetic values." The NMSA delegated responsibilities for specification of regulations within sanctuaries to a negotiation process involving the Secretary and the states within whose waters individual sanctuaries were located. The National Marine Sanctuary System (NMSS) has clear conservation benefits to marine species, but has been somewhat controversial by virtue of activities allowed within some Sanctuary boundaries, particularly commercial fisheries.

The NMSS consists of 13 designated sanctuaries and Papahānaumokuākea Marine National Monument, with two units incorporating occupied sea otter habitat. The Olympic Coast National Marine Sanctuary (OCNMS; established 1994) includes virtually the entire current sea otter population(s) off Washington. Much of the mainland shoreline adjacent to OCNMS is within the coastal portion of Olympic National Park,[84] providing an unusual combination of protected upland, shoreline, and offshore habitats and a remarkable level of spatially contiguous protections for sea otters. The Monterey Bay National Marine Sanctuary (MBNMS; established in 1992 and enlarged in 2009) encompasses approximately the northern two-thirds of the sea otter range (2014 data) in California. Three other established National Marine Sanctuaries have significant numbers of sea otters nearby, with immigration of sea otters likely to occur as a consequence of population growth and range expansion in the next several decades. The Gulf of the Farallons National Marine Sanctuary (GFNMS; established 1981) includes a coastal segment ~60 km in length extending northward from the present northern limit of sea otter distribution in California. The Cordell Bank National Marine Sanctuary

83. 86 US Statutes at Large 1052–1063, Public Law 92-532, October 23, 1972.

84. Proclaimed by US President Harry S. Truman, US Presidential Proclamation 3003, January 6, 1953, converting portions of Olympic National Forest (ONF) as an addition to existing lands within Olympic National Park (ONP), which was established in 1938 (52 US Statutes at Large 812: 1241–1242, June 29, 1938).

(CBNMS; established 1989) has no boundaries in contact with shoreline, but is contiguous with the northern offshore part of GFNMS. In combination, GFNMS and CBNMS provide a combined protected area with ~60 km of shoreline, extending offshore ~48 km. The Channel Islands National Marine Sanctuary (CINMS; established in 1980 and modified to incorporate a system of MPAs in collaboration with the State of California in 2007) includes marine waters adjacent to five of the eight large islands of the Southern California continental borderland. The five islands within the Sanctuary are known to have supported resident sea otter populations prior to local extinctions from fur harvests (Fisher, 1930; Ogden, 1941; Kenyon, 1969) and comprise Channel Islands National Park (CINP), which is entirely surrounded by CINMS. CINMS and CINP are relatively close to mainland resident sea otters off Santa Barbara County to the north, and to the restored sea otter population at SNI to the south.

The National Forest System

The National Forest System evolved during the twentieth century along philosophical lines largely consistent with trends in US culture and economic development.[85] The most recently implemented authority for the US Forest Service (USFS) was the National Forest Management Act of 1976.[86] Emphases on ecosystem protection and wilderness values in national forests have provided enhanced protection to interfaces of forested and submerged lands and marine ecosystems, with sea otter conservation a likely beneficiary. The scope of potential benefit to sea otter conservation from the US National Forest System is exemplified by the Tongass and Chugach National Forests (TNF and CNF, respectively) in Alaska.

The TNF was first established as the Alexander Archipelago Forest Reserve (AAFR) in southeastern Alaska by President Theodore Roosevelt in 1902.[87] President Roosevelt proclaimed TNF in 1907,[88] encompassing an area separate from AAFR, with boundaries similar to those of present-day (2014) Misty Fjords National Monument. In 1908 the President ordered the two forests to be combined administratively and named TNF,[89] and in 1909 he proclaimed a major expansion of TNF.[90] Mandates of ANILCA[91] added ~5868 km^2, bringing the total TNF surface area to ~69,000 km^2, encompassing ~80% of the terrestrial lands of southeastern Alaska. TNF includes ~17,700 linear

85. http://www.fs.fed.us/documents/USFS_An_Overview_0106MJS.pdf.
86. 90 US Statutes at Large 2949–2963, Public Law 94-588, October 22, 1976.
87. US Presidential Proclamation 491, August 20, 1902.
88. US Presidential Proclamation 772, September 10, 1907.
89. US Presidential Executive Order 908, July 2, 1908.
90. US Presidential Proclamation 846, February 16, 1909.
91. See note 66.

kilometers of marine shoreline, a significant proportion of which is adjacent to occupied or potential sea otter habitat (Tongass National Forest, 2008).

The TNF includes 18 wilderness areas adjacent to marine waters that are currently occupied or could potentially be occupied by sea otters. Thirteen were established by ANILCA, totaling 21,495 km^2 in surface area, the remaining five by the Tongass Timber Reform Act of 1990[92] with an additional summed surface area of 1138 km^2. Two of the areas, Admiralty Island and Misty Fjords, are also national monuments.[93]

The CNF was established by President Roosevelt in 1907[94] and enlarged in 1909.[95] The enlargement involved incorporation of terrestrial lands on Afognak Island and other islands in the Afognak Forest and Fish Culture Reservation, and added lands to the east of the Copper River as far as Kayak Island and Cape Suckling[96] and to the west as far as the eastern shore of Cook Inlet. In 1980 ANILCA authorized addition of 7690 km^2 of lands to CNF, including large areas of western and northwestern PWS. CNF boundaries extended westward to Resurrection Bay[97] and included virtually the entire marine shoreline of PWS (~5635 km in length) and the Copper River Delta and at least 20 tidewater glaciers (Chugach National Forest, 2002), the latter known to facilitate high rates of biological productivity in adjacent marine ecosystems (Lydersen et al., 2014). Virtually the entire CNF marine shoreline is adjacent to occupied sea otter habitat, and much of the subject shoreline was affected by EVOS. The CNF currently does not include any formally declared wilderness areas, but a "wilderness study area" was designated in 1980 by ANILCA pursuant to the Wilderness Act of 1964,[98] encompassing ~8500 km^2 of terrestrial and marine habitats in CNF. The Study Area in CNF includes much of the western half of PWS and adjacent coastal lands with significant numbers of resident sea otters.

Protected Spaces Managed by the US Department of Defense

Two military installations on the California coast provide potentially significant *de facto* protected areas for sea otters. Vandenberg Air Force Base (VAFB) and SNI (part of Naval Base Ventura County) both have resident sea otters in adjacent marine habitats. The VAFB boundaries include 57 linear kilometers of undeveloped shoreline. The Base originated as a US Army

92. 104 US Statutes at Large 4426–4435, Public Law 101-626, November 28, 1990.
93. Admiralty Island: US Presidential Proclamation 4611; Misty Fjords: US Presidential Proclamation 4623. Both issued December 1, 1978.
94. US Presidential Proclamation 770, July 23, 1907.
95. US Presidential Proclamation 852, February 23, 1909.
96. 60.00°N, 143.91°W.
97. 59.88°N, 149.30°W.
98. See note 38.

Training Center, Camp Cooke, in 1941.[99] Base property and infrastructure were transferred to the US Air Force in 1956 and were renamed VAFB in 1958. The VAFB mission includes launching of rockets carrying satellites into polar orbit and periodical tests of unarmed intercontinental ballistic missiles intended for military applications. Launch operations and other military activities at VAFB require secure perimeters and limitations of public access, as safety precautions and because some activities are classified. Secured perimeters often extend offshore into sea otter habitat, and marine vessel access to the near shore waters may be temporarily denied.

SNI is owned and managed by the US Navy, with 37 linear kilometers of mostly pristine shoreline adjacent to high-quality sea otter habitat. A large sea otter population was present at SNI prior to local extinction from fur hunting in the 1850s (Ogden, 1941). Following translocation activity from 1987 to 1990, SNI now (2014) supports a population of about 70 sea otters (B.B. Hatfield, US Geological Survey, personal communication). In 1933 President Hoover directed[100] the transfer of island jurisdiction to the Navy, rescinding prior Executive Orders reserving the island "for lighthouse purposes."[101] Since 1947 SNI has been used for testing and development of communication and missile-tracking systems as part of the Navy's Pacific Missile Center (Schwartz and Rossbach, 1993). Missile testing at SNI requires a secured perimeter as at VAFB.

A US military airfield was installed at Shemya Island in the Aleutian Archipelago in 1942 during military conflict with Japan in World War II. Similar installations were established at Adak, Amchitka, and Attu Islands. Shemya Island's land space was fully occupied by Shemya Air Force Base following the end of World War II, with the base closing in 1994. Sea otters had been hunted to local extinction at Shemya prior to the War, recolonizing the island without human assistance in the late 1980s and possibly benefiting from associated security measures during the final years of military presence. Space occupied by military installations and consequent potential impact as *de facto* protected areas for sea otters was limited to relatively small portions of Adak, Amchitka, and Attu Islands, and was likely insignificant in the context of sea otter conservation.

The US Departments of Defense and Energy and the US Atomic Energy Commission reoccupied Amchitka Island from 1964 through 1973 for underground testing of strategic nuclear weapons. Three separate tests were conducted at Amchitka from 1965 through 1971, the last and largest known as Project Cannikin. Several hundred sea otters may have been killed by pressure waves generated in ocean waters by Cannikin (Estes and Smith, 1973; Rausch,

---

99. http://www.vandenberg.af.mil/library/factsheets/factsheet.asp?id=4606.

100. US Presidential Executive Order 6009, January 31, 1933.

101. US Presidential Executive Orders, January 26, 1867 (Andrew Johnson) and November 15, 1901 (Theodore Roosevelt).

1973), and a few sea otters on shore apparently died when struck by rockslides caused by the detonation. Estes and Smith (1973) suggested that long-term negative impacts of the sea otter mortality event were likely not significant on a population scale. In addition, aircraft used to transport supplies and equipment used as part of weapons testing at Amchitka were made available, during return trips eastward, to transport sea otters involved in translocation projects to release sites in southeastern Alaska, British Columbia, Washington, and Oregon. Nevertheless, events associated with Cannikin and preceding tests were reminders that *de facto* protected areas controlled by military entities may shift to conservation liabilities based on political circumstances.

The Alaska Native Claims Settlement Act and the Alaska National Interest Lands Conservation Act

Congressional Acts in 1971 and 1980 produced extraordinary potential impacts on conservation of sea otters and other species in Alaskan coastal waters. The Acts resulted from conflict between the State of Alaska and Native Alaskan communities regarding ownership of lands that had been in federal control prior to Alaskan statehood. The Alaska Native Claims Settlement Act of 1971 (ANCSA)[102] specified a process by which Native communities could claim lands in and near existing villages as well as locations deemed culturally significant. ANCSA also authorized the Secretary of the Interior to withdraw up to 323,750 km^2 of unreserved lands in Alaska from the public inventory for possible future designation as national parks or monuments, NWRs, or national forests. In 1978 President Jimmy Carter invoked authorities provided by the 1906 American Antiquities Act[103] to create or expand national monument lands in Alaska, using lands withdrawn pursuant to ANCSA. With approval of ANILCA in 1980,[104] Congress completed establishment of national parks, national preserves, national monuments, NWRs, and wilderness areas initiated by ANCSA (Tables 14.1 through 14.3). Most federally managed protected areas included boundaries adjacent to or encompassing marine shorelines or waters in which sea otters were resident either in 1980 or prior to the maritime fur trade era.

ANILCA resulted in substantial changes to management of NWRs in Alaska, with many existing and new NWRs and additional lands combined into the new Alaska Maritime NWR (AMNWR). AMNWR was organized into five administrative units, four of which included refuge lands in or adjacent to occupied or potential sea otter habitat. ANILCA also mandated return to the NWR system of islands removed from the Aleutian Islands

102. 85 US Statutes at Large 688–716, Public Law 92-203, December 18, 1971.
103. See note 56.
104. See note 66.

TABLE 14.1 NWRs Created, Expanded, or Administratively Reorganized by ANILCA[a] in Alaska with Marine Shoreline or Marine Waters Included, and with Known or Potential Significance for Sea Otter Conservation

Location Name	Original Date Established	Original Status	Authority for Original Designation	Current Status	Total Surface Area at Present (km^2)	Total Area Designated as Wilderness (km^2) with Year of Designation	Wilderness Designated for Marine Shoreline or Marine Waters?	Sea Otters Currently Present?	Sea Otters Historically Present?
Alaska Maritime, all Units	1980	National Wildlife Refuge	ANILCA[a]	As noted to left	17,904	See below by component	Yes	See below by component	See below by component
Aleutian Islands	1913	Aleutian Islands Reservation	Presidential Executive Order 1733, March 3, 1913 (William Howard Taft)	Subunit, Aleutian Islands Unit, Alaska Maritime NWR	12,787	5,369 (1980[a])	Yes	Yes	Yes
Forrester Island	1912	Forrester Island Reservation	Presidential Executive Order 1458, January 11, 1912 (William Howard Taft)	Subunit, Gulf of Alaska Unit, Alaska Maritime NWR	11	11 (1970[b])	Yes	No	Uncertain, possibly intermittent
Hazy Islands	1912	Hazy Islands Reservation	Presidential Executive Order 1459, January 11, 1912 (William Howard Taft)	Subunit, Gulf of Alaska Unit, Alaska Maritime NWR	0.13	0.13 (1970[b])	Yes	No	Uncertain, possibly intermittent

Saint Lazaria Island	1909	Saint Lazaria Reservation	Presidential Executive Order 1040, February 27, 1909 (Theodore Roosevelt)	Subunit, Gulf of Alaska Unit, Alaska Maritime NWR	0.26	0.26 (1970[b])	Yes	Yes	Yes
Semidi Islands	1932	Semidi Islands Wildlife Refuge	Presidential Executive Order 5858, June 17, 1932 (Herbert C. Hoover)	Subunit, Alaska Peninsula Unit, Alaska Maritime NWR	1,039	1,039 (1980[a])	Yes	Yes	Yes
Simeonof Island	1958	Simeonof National Wildlife Refuge	Public Land Order 1749, October 30, 1958 (Secretary of the Interior Fred A. Seaton)	Subunit, Alaska Peninsula Unit, Alaska Maritime NWR	102	102 (1976[c])	Yes	Yes	Yes
Chisik, Duck, and Egg Islands, Tuxedni Bay	1909	Tuxedni Reservation	Presidential Executive Order 1039, February 27, 1909 (Theodore Roosevelt)	Subunit, Gulf of Alaska Unit, Alaska Maritime NWR	23	23 (1970[b])	Yes	Yes	Yes
Walrus and Otter Islands (Pribilof Islands)	1909	Pribilof Reservation	Presidential Executive Order 1044, February 27, 1909 (Theodore Roosevelt)	Subunit, Bering Sea Unit, Alaska Maritime NWR	24.7[d]	None	No	No	Yes

(Continued)

TABLE 14.1 (Continued)

Location Name	Original Date Established	Original Status	Authority for Original Designation	Current Status	Total Surface Area at Present (km^2)	Total Area Designated as Wilderness (km^2) with Year of Designation	Wilderness Designated for Marine Shoreline or Marine Waters?	Sea Otters Currently Present?	Sea Otters Historically Present?
Alaska Peninsula	1980	Alaska Peninsula NWR	ANILCA[a]	As noted to left	14,973	None	No	Yes	Yes
Becharof National Monument	1978	Becharof National Monument	Presidential Proclamation 4613, December 1, 1978 (Jimmy Carter)	Becharof NWR	4,856	1,619 (1980[a])	Yes	Yes	Yes
Izembek Lagoon	1960	Izembek National Wildlife Range	Public Land Order 2216, December 30, 1960 (Secretary of the Interior Fred A. Seaton)	Izembek NWR	1,690	1,246 (1980[a])	Yes	Yes	Yes

Kodiak Island	1941	Kodiak NWR	Presidential Executive Order 8857, August 19, 1941 (Franklin D. Roosevelt)	As noted to left	7,689	None	No	Yes	Yes
Unimak Island	1913	Aleutian Islands Reservation	Presidential Executive Order 1733, March 3, 1913 (William Howard Taft)	Subunit, Aleutian Islands Unit, Alaska Maritime NWR	4,082	3,760	Yes	Yes	Yes

[a] *Alaska National Interest Lands Conservation Act, 94 US Statutes at Large 2371–2551, Public Law 96-487, December 2, 1980.*
[b] *84 US Statutes at Large 1104–1106, Public Law 91-504, October 23, 1970.*
[c] *90 US Statutes at Large 2633–2638, Public Law 94-557, October 19, 1976.*
[d] *Includes 0.69 km^2 on Walrus and Otter Islands as established in 1909, and 24 km^2 on St. Paul and St. George Islands added following ANILCA in 1980.*
Listed refuge entities are managed by the US Fish and Wildlife Service.

TABLE 14.2 National Parks and Monuments Created or Expanded by ANILCA[a] with Marine Shoreline or Marine Waters Included, and with Known or Potential Significance for Sea Otter Conservation

Location Name	Original Date Established	Original Status	Authority for Original Designation	Current Status	Total Surface Area at Present (km^2)	Status of Wilderness Designation	Wilderness Designated for Marine Shoreline or Marine Waters?	Sea Otters Currently Present?	Sea Otters Historically Present?
Aniakchak (ANMP)	1978	National Monument	Presidential Proclamation 4612, December 1, 1978 (Jimmy Carter)	National Monument and Preserve	2,080	None	No	Yes	Yes
Glacier Bay (GBNPP)	1925	National Monument	Presidential Proclamation 1733, February 26, 1925 (Calvin Coolidge)	National Park and Preserve	13,281	10,760 km^2 approved,[a] 186 km^2 eligible[b]	Yes	Yes	Yes
Katmai (KNPP)	1918	National Monument	Presidential Proclamation 1487, September 24, 1918 (Woodrow Wilson)	National Park and Preserve	16,583	13,447 km^2 approved,[a] 2,533 km^2 eligible[b]	Yes	Yes	Yes
Kenai Fjords (KFNP)	1978	National Monument	Presidential Proclamation 4620, December 1, 1978 (Jimmy Carter)	National Park	2,700	None approved, 2,303 km^2 eligible[b]	No	Yes	Yes

Lake Clark (LCNPP)	1978	National Monument	Presidential Proclamation 4622, December 1, 1978 (Jimmy Carter)	National Park and Preserve	16,167	10,408 km^2 approved,[a] 3,857 km^2 eligible[b]	No	Yes	Yes
Wrangell-St. Elias (WSENPP)	1978	National Monument	Presidential Proclamation 4625, December 1, 1978 (Jimmy Carter)	National Park and Preserve	49,844	38,161 km^2 approved,[a] eligibility of other lands unresolved	No	Yes	Yes

[a] *Alaska National Interest Lands Conservation Act, 94 US Statutes at Large 2371–2551, Public Law 96-487, December 2, 1980.*
[b] *Based on information updated to January 2013; Lindholm et al., 2013.*
Indicated park or monument entities are managed by the US National Park Service.

TABLE 14.3 **Wilderness Areas Created by ANILCA[a] on National Forest Lands within the Tongass National Forest[b] in Southeastern Alaska, with Marine Shoreline or Marine Waters Included, and with Known or Potential Significance for Sea Otter Conservation**

Location Name	Current Status	Total Area Designated as Wilderness (km^2)	Sea Otters Currently Present?	Sea Otters Historically Present?
Admiralty Island	Admiralty Island National Monument, Kootznoowoo Wilderness	3,642	No	Yes
Misty Fjords	Misty Fjords National Monument Wilderness	8,644	No	Uncertain
Coronation Island	Coronation Island Wilderness	78	Yes	Yes
Maurille Islands	Maurille Islands Wilderness	20	No	Yes
Petersburg Creek-Duncan Salt Chuck	Petersburg Creek-Duncan Salt Chuck Wilderness	190	No	Uncertain
Russell Fjord	Russell Fjord Wildness	1,411	No	Uncertain
South Baranof Island	South Baranof Wildnerness	1,293	Yes	Yes
South Prince of Wales Island	South Prince of Wales Wilderness	368	Yes	Yes
Stikine-LeConte	Stikine-LeConte Wilderness	1,817	No	Uncertain
Tebenkof Bay	Tebenkof Bay Wilderness	270	No	Uncertain
Tracy Arm-Fords Terror	Tracy Arm-Fords Terror Wilderness	2,643	No	Uncertain
Warren Island	Warren Island Wildnerness	45	Yes	Yes

[a] *Alaska National Interest Lands Conservation Act, 94 Statutes at Large 2371–2551, Public Law 96-487, December 2, 1980.*
[b] *Established by Presidential Proclamation 772, Theodore Roosevelt, September 9, 1907.*
Listed refuge entities are managed by the US Forest Service.

Reservation by President Coolidge in 1928,[105] with the reinstated lands becoming part of the Aleutian Islands Unit of AMNWR. ANILCA added a number of unoccupied islands to AMNWR, and designated that marine benthic habitats that had once been part of the Afognak Forest and Fish Culture Reservation were also to become part of AMNWR. AMNWR also gained jurisdiction over all public lands on named and unnamed islands, islets, rocks, reefs, spires, and designated capes and headlands south of the Alaska Peninsula, from Katmai National Park and Preserve to False Pass, and in the Gulf of Alaska, subject to certain exclusions. ANILCA designated the addition of two new NWRs in or adjacent to Alaskan waters with potential conservation value as sea otter habitat: the Alaska Peninsula NWR and the adjacent Becharof NWR, the latter transferred from National Monument status as declared by President Jimmy Carter in 1978 (Table 14.1). In addition, ANILCA mandated additions of land to the Kodiak NWR and the renamed Izembek NWR. Finally, ANILCA authorized designation of wilderness status for most of the Aleutian Islands and all of the Semidi Islands, all part of AMWNR, along with the Becharof and Izembek NWRs. The Forrester Island, Hazy Islands, St. Lazaria Island, and Tuxedni components of AMNWR had been designated as wilderness in 1970 and Simeonof Island in 1976.

Protection of Water Quality

The discourse concerning water quality issues relating to sea otter conservation has been focused largely on impacts of oil spills. As noted elsewhere (Chapter 7), novel disease has emerged as a potential challenge for the conservation of sea otters. The possible connection of disease transmission to sea otters via exposure to coastal domestic sewage effluents exemplifies the potential importance of water quality issues other than oil spills to sea otter conservation. Congress approved the Water Pollution Control Act[106] in 1948, providing federal assistance to states, municipalities, and industries to identify and characterize locations impaired by water pollution, abate existing crises, and prevent new ones. The mandated actions were deemed necessary to conserve water quality "for public water supplies, propagation of fish and aquatic life, recreational purposes, and agricultural, industrial, and other legitimate uses." Congress updated federal water quality protections with the Federal Water Pollution Control Act Amendments of 1972,[107] commonly known as the Clean Water Act. The Water Pollution Control Act, as

105. Akun, Akutan, Sanak, Sedanka, Tagalda, Umnak, and Unalaska Islands were withdrawn from the Reservation by US Presidential Executive Order 5000 in 1928.

106. See note 35.

107. 86 US Statutes at Large 816–904, Public Law 92-500, October 18, 1972.

amended in 1972, provided significant guidance for governmental responses to the EVOS off 1989 in south-central and southwestern Alaska.

The Oil Pollution Liability and Compensation Act of 1989[108] was the US Congressional response to the *Exxon Valdez* disaster. The most important elements of the 1989 Act involved strengthening of liability provisions for entities responsible for oil spills. The observed decline in oil spill rates in US marine waters in the years after 1989 likely reflected the deterring effect of enhanced potential and scope for liability assessment (Ramseur, 2012). The 1989 Act also imposed new federal requirements for spill prevention and response preparedness.

DISCUSSION

A clear message emerging from material presented here is the exaggerated valuation of the Fur Seal Treaty of 1911 as a tool for sea otter conservation. The "conventional wisdom" (VanBlaricom et al., 2013) that the Treaty ended the commercial harvesting of sea otters is arguably false. Commercially significant, sustained harvests of sea otters did in fact end at about the time of the Fur Seal Treaty of 1911, but the cause was most likely an economically driven response to population depletion rather than any regulatory impact of the Treaty. In addition, harvests of sea otters were reported within Japanese territorial waters of the Kuril Archipelago,[109] certified as compliant with the Treaty by the Japanese and US governments (Anonymous, 1939), with a total of 82 pelts offered for auction at fur exchanges in St. Louis, New York, and London in 1939. Spector (1998) reported that "for the first half of the 1900s, sea otter fur received special attention at Seattle Fur Exchange auctions," but there are no associated data available to evaluate the meaning or consequences of the report in terms of sea otter harvests. The Alaskan Statehood Proclamation of 1959[110] led to transfer of management authority for furbearing animals from the US federal government to the State of Alaska, eliminating the prohibition on harvest of fur-bearers imposed on Alaskan Territory by the US Congress in 1899.[111] Kenyon (1969) reported subsequent harvests by the Alaska Department of Fish and Game (ADF&G) at Amchitka Island in 1962, 1963, and 1967, and at Adak Island in 1967. Abegglen (1977) indicated that ADF&G harvested 2463 sea otters between 1962 and 1970. Although the Alaskan harvest was spread across a time span of 9 years, it is nevertheless worthy of note that the total number harvested was comparable to or larger than

108. See note 34.

109. The Kuril Archipelago was ceded to Japan by Russia in 1875 ("Treaty on the Exchange of Territories," signed April 25, 1875). Military forces of the USSR seized control of the Archipelago in 1945.

110. US Presidential Proclamation 3269. January 3, 1959.

111. See note 18.

some published estimates of sea otter mortalities resulting from the EVOS (Garrott et al., 1993; Garshelis, 1997). Of pelts harvested by ADF&G, 1000 were presented for auction to private-sector furriers at the Seattle Fur Exchange (SFE),[112] the largest commercial fur auction house in the western hemisphere (Lambson and Thurston, 2006) on January 30, 1968. The auction generated moderate interest from the fur industry and resulted in $140,982 in revenues to the General Fund of the State of Alaska (Spector, 1998). The ADF&G harvests were consistent with applicable state and federal laws and were terminated prior to the passage of MMPA in 1972.[113] Harvested material was the basis for significant research on population dynamics of sea otters (von Biela et al., 2009).

There is no basis for claiming the existence of a sustainable commercial harvest of sea otters for fur acquisition after ~1910. The point is simply that the 1911 Fur Seal Treaty did not stop all forms of legal harvest of sea otters, frequent claims to the contrary notwithstanding. It is only by virtue of good fortune and inexorable economic reality, rather than strong policies, protocols, or enforcement, that sea otters escaped significant harvests over most of the twentieth century.

Recognizing that cultural values have changed since the time of the ADF&G harvests, the number of animals killed by design nevertheless warrants contemplation in an ethical context. Such considerations are justified given the intensity of public concern about similar levels of mortality in the aftermath of EVOS, which occurred just 19 years after conclusion of the ADF&G harvests. The ADF&G harvests were motivated by commercial as well as scientific and management interests. Should future sea otter management actions include consideration of large-scale harvests, prior discussion of social and ethical implications must be given priority equivalent to ecological and management considerations, particularly if anticipated culling involves a commercial component.

There is an emerging need for objective scrutiny of Alaskan Native harvest of sea otters, allowed by MMPA as a measure to restore traditional and culturally significant wildlife harvest practices to Native American communities in Alaska. Two issues are of particular concern. First, the cumulative harvest of sea otters taken by Native Alaskan hunters since 1972, pursuant to MMPA, now exceeds 20,000 individuals (Chapter 4), and most animals have been taken within the last two decades. The gap between the current annual harvest rate, now >1000 animals, and those of the last several decades of the maritime fur trade is closing rapidly. Both MMPA and ESA specify that the Secretary of the Interior has authority to impose restrictions on Native harvests when they may threaten the continued survival of the targeted species. The Native harvest includes legal provisions for sale of sea otter parts.

---

112. SFE was purchased by the Mutation Mink Breeders Association in 1973, and through subsequent mergers became the American Legend Cooperative in 1986 (Spector, 1998).

113. See note 1.

Thus, the harvest arguably has a commercial dimension. It seems reasonable to suggest that increased federal scrutiny, possibly including intervention to negotiate reductions in the harvest rate of sea otters, may be appropriate in the immediate future. Second, Native American communities in the other three US "sea otter states" also harvested sea otters historically (Fisher, 1930), and lost the opportunity to do so with "settlement" of the west and associated cultural disruptions in the nineteenth and early twentieth centuries. The legal justification for restoration of sea otter harvesting by Alaskan Natives surely applies as well to the extant Native communities of Washington, Oregon, and California (VanBlaricom, 1996), although MMPA provided no such authority. It seems plausible that increased attention to the burgeoning Native harvest in Alaska could stimulate the political will for expansion of federally authorized Native harvests in sea otter populations south of Alaska. The result, if not properly regulated, could facilitate harvest rates comparable to the unsustainable levels characteristic of the maritime fur trade era.

Neither the Fur Seal Treaty of 1911 nor subsequent international agreements forbade harvests of sea otters in the territorial waters (<5.6 km offshore) of jurisdictions bordering on the pre-harvest geographic range. Riedman and Estes (1990) indicated that sea otters "seldom range more than 1–2 km from shore." Historical records dating from 1799 indicate that the Fur Seal Treaty of 1911 was the endpoint of a series of disputes and rulings primarily intended to ensure participatory and proprietary rights and economic benefits to the four nations most directly involved in fur seal harvests (Great Britain, Japan, Russia, and the United States; Mirovitskaya et al., 1993). One might reasonably argue that the Treaty was *never intended* to provide conservation value in an ecological context.

The inclusion of sea otters as beneficiaries of the Fur Seal Treaty of 1911 reflected information provided by Treasury Agent H.W. Elliott to Secretary of State John M. Hay during preparation of Treaty drafts in 1905.[114] Elliott visited Alaska in the latter nineteenth century and subsequently reported concerns about overexploitation of sea otters in the fur trade (Elliott, 1875). Baur et al. (1996) correctly asserted that "the sea otter survived, not because of the 1911 Treaty or its implementing law,[115] but because the animals were simply too scarce to justify the cost of searching them out and killing them for their skins." In other words, sea otters may have been saved by fundamental principles of economics rather than implementation of any meaningful conservation measures.

Despite the law enacted by the US Congress in 1868[116] to protect sea otters and other fur-bearers in Alaskan territorial waters, intensive sea otter harvests continued until the end of the nineteenth century (Hooper, 1897;

114. http://celebrating200years.noaa.gov/events/fursealtreaty/welcome.html#conserve.
115. See note 16.
116. See note 8.

Cobb, 1906; Jacobi, 1938; Chapter 3). The data clearly indicate overharvest and depletion of Alaskan sea otter populations from the date of the purchase of Alaska from Russia by the United States in 1868 until approximately 1910. The apparent absence of a federal harvest management plan by the US Department of the Treasury, clearly compounded by a lack of effective law enforcement, appears to have been the primary cause for depletion of Alaskan sea otter populations (Elliott, 1875). Elliott's report included recommendations for prohibition of the use of firearms for taking sea otters, and for seasonal restrictions on hunting activity in order to facilitate sustainability of the sea otter harvest, but the suggested measures were not implemented. A more formal proposal for regulations (Hooper, 1897) included restriction of hunting of sea otters to Native vessels (baidarkas and canoes), limitations on types of vessels to be used for transporting Native hunters to and from hunting areas, prohibition of use of nets to capture sea otters, prohibition of foreign vessels from hunting sea otters in US territorial waters, and requirements that federal permits be obtained in advance of sea otter hunting operations. To my knowledge, Hooper's proposed regulations were not incorporated into federal law. It could be reasonably argued that the Hooper proposal was, in any case, too late to be effective in a conservation context for sea otters, given the apparently depleted status of most Alaskan populations in 1897. Historical records suggest that sea otter harvests in the latter nineteenth century exceeded market demand and elicited declining monetary returns per pelt (J. Richardson, as quoted in Coues, 1877), reiterating that, as the harvest rate diminished, the relentless pursuit of sea otter pelts for the maritime fur trade appeared, tragically, to lack any rational direction or purpose.

Several patterns in the historical record may be of particular relevance to future legislative or legal initiatives that seek more comprehensive and effective protection of sea otter populations. For example, entities created for one purpose can, with political creativity and the support of persons in authority, be transformed into protected spaces from which sea otters will benefit. The significant, even redundant array of protective protocols for sea otters off Washington can be traced back to an 1897 proclamation by President Cleveland intended to protect forests from excessive logging, based on authority provided by Congress in the Forest Reserve Act of 1891. Likewise, GPNPP was initially designated as a national monument in 1925,[117] expanded in spatial extent in subsequent years by the Proclamations of three different US presidents,[118] converted to a National Park under ANILCA[119] auspices in 1980, and authorized to implement stringent regulations of fishery practices potentially

117. US Presidential Proclamation 1733, February 26, 1925.

118. US Presidential Proclamations 2330, April 18, 1939; 3089, March 31, 1955; and 4618, December 1, 1978.

119. See note 66.

detrimental to sea otters, based on Congressional Action[120] and the support of President William J. Clinton in 1998.

The period from 1944 to 1972 was a significant gap in sea otter protection in both international and US territorial waters, with minimal US federal involvement. The State of California was effectively alone in sustaining protocols and protected areas explicitly intended for sea otter conservation in the United States. At federal and international levels the absence of progress first reflected a preoccupation with world war and later a focus on economic development and recovery, and associated aspirations for resource exploitation. The first explicit federal recognition of the inherent value of wildlife as a functional component of natural ecosystems was probably the Wilderness Act of 1964.

Application of sea otter translocation as a tool for population restoration was first proposed by Eyerdam (1933). The sea otter translocation projects from the 1950s until the early 1990s, intended for restoration of populations on regional scales, have been the most effective active management programs for the benefit of sea otters in US history (Chapters 3 and 13). At present more than 90% of the sea otters alive along the North American coast east and south of PWS exist in populations restored from regional extinction by translocation of animals from other populations.[121] I suggest that the most effective US federal legislative actions have been the implementations of MMPA, ESA, and ANILCA. A notable strength of both MMPA and ESA, contrasting with many prior laws, treaties, and protocols, is their applicability to US territorial as well as international marine waters. Large-scale government landholdings or sanctuaries that either encompass the waters and benthic habitats used by sea otters, or include extensive coastline segments adjacent to sea otter habitats, are likely to be valuable to sea otter conservation. One might plausibly argue that by virtue of the exemption of Alaskan Natives from prohibitions of marine mammal harvests, MMPA is flawed in a wildlife protection context despite the obvious conservation and cultural restoration benefits of the law.

Given the geographic scale of existing US marine sanctuaries and the types of habitats they protect in the northeastern Pacific, stability and growth in application of the concept seem well suited to improved conservation status for sea otters in the future. Most sanctuaries are large enough to approximate or perhaps exceed the natural spatial scales of sea otter populations, and the protective protocols for resources within sanctuaries extend from the shoreline to the edge of the continental shelf, far beyond the seaward range limits of most sea otters. However, at present the National Marine Sanctuary Program is in crisis. The Program was last reauthorized by Congress in

120. See note 80.

121. Data presented at Sea Otter Conservation Workshop VIII, Seattle Aquarium, Seattle, Washington, USA. March 22–24, 2013.

2000[122] and the reauthorization expired in 2005. Since that time the system has operated with minimal funding and ambiguous prospects for sustained existence. Further, the 2000 Reauthorization Act also placed a moratorium on designation of new sanctuaries.

Morris (2013) provided a detailed review of the history and prospects of the sanctuary system. Morris suggested that the burden of legislative mandates for multiple complex reviews of sanctuary designation proposals has impaired progress of the Program. He noted the congressionally mandated priority role of regional Fishery Management Councils (RFMCs) in determining fishery management policies in new sanctuaries. Creation of the RFMCs resulted from the Fishery Conservation and Management Act of 1976,[123] most recently revised by the Sustainable Fisheries Act of 1996.[124] Morris argued that joint involvement of NOAA and the RFMCs in deliberations over management policies in proposed new sanctuaries was a pathological conflict of mandates, likely prolonging the uncertain status of the sanctuary system. He suggested that US Presidents should exploit the powers granted by Congress via the 1906 American Antiquities Act in pursuing alternative mechanisms for reinvigorating the spirit, if not the letter, of the marine sanctuary concept. Morris's recommendation is an invocation of one of the oldest and most useful legal tools in the history of US conservation. He noted that "a President's exercise of authority under the American Antiquities Act has never been curtailed or invalidated." In 2006 President George W. Bush anticipated Morris's advice in proclaiming the Papahānaumokuākea Marine National Monument in Hawaiian marine waters.[125] A sanctuary proposal for the region had been mired in the political complexities of the approval process mandated by Congress, with no end in sight. The President exploited the authority of the 1906 Antiquities Act to declare the proposed sanctuary a "Marine National Monument" and directed the staff of the National Oceanic and Atmospheric Administration to manage the monument as they would manage a new national marine sanctuary.

The value of MPAs for sea otter conservation remains a compelling but unresolved issue. With intensifying conservation crises such as global climatic trends and ocean acidification that appear intractable given current technologies and divided cultural perceptions, effective direct actions for the benefit of sea otter habitat conservation may become increasingly elusive. MPAs have proliferated rapidly in recent decades within the geographic range of sea otters. Such areas span a range of scales in terms of size, spatial configuration, and dispersion, and represent a broad range of management objectives and protocols for permitted activities. MPA units represented

122. 114 US Statutes at Large 2381–2393, Public Law 106-513, November 13, 2000.
123. 90 US Statutes at Large 331–361, Public Law 94-265, April 13, 1976.
124. 110 US Statutes at Large 3559–3621, Public Law 104-297, October 11, 1996.
125. US Presidential Proclamation 8031, June 15, 2006.

herein as relevant to sea otter conservation number in the hundreds and incorporate large areas of habitat and adjacent shoreline length. However, the effects of MPA system attributes such as number, size, configuration, dispersion, management goals, and protocols on efficacy of sea otter protection remain unknown. It is suggested that implementation of MPAs for the benefit of sea otters holds genuine potential as a successful conservation strategy, but that the effectiveness of such approaches will require rigorous quantitative analyses and modeling of existing and proposed MPA arrays, in addition to empirical studies of existing MPA systems, to characterize their value in sea otter conservation.

ACKNOWLEDGMENTS

Financial support during preparation of this chapter was provided by the Ecosystems Branch of the US Geological Survey, and the School of Aquatic and Fishery Sciences, University of Washington. J.L. Bodkin, D.M. Burn, A.M. Johnson, S.E. Larson, S.G. Lio, and K.K. VanBlaricom provided comments on earlier versions of the chapter manuscript.

REFERENCES

Abegglen, C.E., 1977. Sea mammals: resources and population. In: Merritt, M.L., Fuller, R.G. (Eds.), The Environment of Amchitka Island, Alaska. Technical Information Center, Energy Research and Development Administration, Washington, DC, pp. 493–510.

Alaska Department of Fish and Game, 2006. Our Wealth Maintained: A Strategy for Conserving Alaska's Diverse Wildlife and Fish Resources. A Comprehensive Wildlife Conservation Strategy Emphasizing Alaska's Nongame Species. Alaska Department of Fish and Game, Juneau, AK.

Anonymous, 1939. Comment and news. J. Mammal. 20, 407.

Baur, D.C., Meade, A.M., Rotterman, L.M., 1996. The law governing sea otter conservation. Endangered Species Update 13 (12), 73–78.

Brasher, A., Moyle, P., 2004. Conservation in the USA: legislative milestones. In: Moyle, P., Kelt, D. (Eds.), Essays on Wildlife Conservation—MarineBio.org. MarineBio Conservation Society, Encinitas, CA. <http://marinebio.org/oceans/conservation/moyle/index.asp>.

Bryant, H.C., 1915. Sea otter near Point Sur California. Calif. Fish Game 1, 134–135.

California Department of Fish and Game, 2012a. Guide to the North-central California Marine Protected Areas. Pt. Arena to Pigeon Pt. California Fish and Game Commission, Sacramento, CA.

California Department of Fish and Game, 2012b. Guide to the Southern California Marine Protected Areas. Point Conception to California-Mexico Border. California Fish and Game Commission, Sacramento, CA.

California Department of Fish and Wildlife, 2013a. Guide to the Central California Marine Protected Areas. Pigeon Point to Pt. Conception. California Fish and Game Commission, Sacramento, CA.

California Department of Fish and Wildlife, 2013b. Guide to the Northern California Marine Protected Areas. California-Oregon Border to Pt. Arena. California Fish and Game Commission, Sacramento, CA.

Chugach National Forest, 2002. Revised Land and Resource Management Plan for the Chugach National Forest. US Department of Agriculture, Forest Service, Anchorage, AK.

Cobb, J.N., 1906. The commercial fisheries of Alaska in 1905. In: Bowers, G.M., Commissioner, Report of the Commissioner of Fisheries for the Fiscal Year Ended June 30, 1905 and Special Papers. Bureau of Fisheries Document 603, Government Printing Office, Washington, DC, pp. 124–166.

Coues, E., 1877. Fur-Bearing Animals: A Monograph of North American Mustelidae. US Department of the Interior, Government Printing Office, Washington, DC, Miscellaneous publication No. 8.

Doroff, A.M., Estes, J.A., Tinker, M.T., Burn, D.M., Evans, T.J., 2003. Sea otter population declines in the Aleutian Archipelago. J. Mammal. 84, 55–64.

Elliott, H.W., 1875. A Report Upon the Condition of Affairs in the Territory of Alaska. Government Printing Office, Washington, DC.

Estes, J.A., Smith, N.S., 1973. Amchitka Bioenvironmental Program. Research on the Sea Otter, Amchitka Island, Alaska. Nevada Operations Office, Atomic Energy Commission, Las Vegas, NV, US Atomic Energy Commission Report NVO-520-1.

Estes, J.A., Tinker, M.T., Williams, T.M., Doak, D.F., 1998. Killer whale predation on sea otters linking oceanic and nearshore ecosystems. Science 282, 473–476.

Eyerdam, W.J., 1933. Sea otters in the Aleutian Islands. J. Mammal. 14, 70–71.

Fisher, E.M., 1930. The early fauna of Santa Cruz Island, California. J. Mammal. 11, 75–76.

Garrott, R.A., Eberhardt, L.L., Burn, D.M., 1993. Mortality of sea otters in Prince William Sound following the *Exxon Valdez* oil spill. Mar. Mammal Sci. 9, 343–359.

Garshelis, D.L., 1997. Sea otter mortality estimated from carcasses collected after the *Exxon Valdez* oil spill. Conserv. Biol. 11, 905–916.

Hooper, C.L., 1897. A Report on the Sea Otter Banks of Alaska. US Government Printing Office, Washington, DC, US Department of the Treasury, document number 1977.

Jacobi, A., 1938. Der Seeotter. Monographien der Wildsäugetiere, Band VI. Verlag Dr. Paul Schöps. Leipzig, Germany. 93 pages.

Kenyon, K.W., 1969. The sea otter in the eastern Pacific Ocean. North Am. Fauna 68, 1–352.

Lambson, V.E., Thurston, N.K., 2006. Sequential auctions: theory and evidence from the Seattle Fur Exchange. Rand J. Econ. 37, 70–80.

Lindholm, A., Bruno, B.M., Rego, S., Piercy, J., 2013. Wilderness Stewardship Program—Alaska Region. Annual report 2012. Natural Resource Report NPS/AKSO/NRR-2013/659. Natural Resource Stewardship and Science, National Park Service, US Department of the Interior, Fort Collins, CO.

Lydersen, C., Assmy, P., Falk-Petersen, S., Kohler, J., Kovacs, K.M., Reigstad, M., Steen, H., Strøm, H., Sundfjord, A., Varpe, Ø, Walczowski, W., Weslawski, J.M., Zajaczkowski, M., 2014. The importance of tidewater glaciers for marine mammals and seabirds in Svalbard, Norway. J. Mar. Syst. 129, 452–471.

McArdle, D.A., 1997. California Marine Protected Areas. California Sea Grant College System, University of California, La Jolla, CA, San Diego, CA.

McDonald, M., 1894. Report on the salmon fisheries of Alaska. Bull. US Fish Comm. 12, 1–38.

Mirovitskaya, N.S., Clark, M., Purver, R.G., 1993. North Pacific fur seals: regime formation as a means of resolving conflict. In: Young, O.R., Osherenko, G. (Eds.), Polar Politics: Creating International Environmental Regimes. Cornell University Press, Ithaca, NY, pp. 22–55.

Morris, P.H., 2013. Monumental seascape modification under the Antiquities Act. Environ. Law 43, 173–209.

Nicholson, T.E., Mayer, K.A., Staedler, M.M., Johnson, A.B., 2007. Effects of rearing methods on survival of released free-ranging juvenile southern sea otters. Biol. Conserv. 138, 313–320.

Ogden, A., 1941. The California Sea Otter Trade, 1784–1848. University of California Press, Berkeley, CA.

Ramseur, J.L., 2012. Oil Spills in US Coastal Waters: Background and Governance. Congressional Research Service, Washington, DC, Report for Congress RL33705.

Rausch, R.L., 1973. Post Mortem Findings in Some Marine Mammals and Birds Following the Cannikin Test on Amchitka Island. Nevada Operations Office, Atomic Energy Commission, Las Vegas, NV, US Atomic Energy Commission Report NVO-130.

Riedman, M.L., Estes, J.A., 1990. The sea otter (*Enhydra lutris*): behavior, ecology, and natural history. Biol. Rep. 90 (14), US Department of the Interior, Fish and Wildlife Service, Washington, DC.

Schwartz, S.J., Rossbach, K.A., 1993. A preliminary survey of historical sites on San Nicolas Island. Proc. Soc. Calif. Archaeol. 8, 189–198.

Spector, R., 1998. Seattle Fur Exchange. 100 Years. Documentary Book Publishers, Seattle, DC.

Tongass National Forest, 2008. Tongass National Forest. Land and Resource Management Plan. US Department of Agriculture, Forest Service, Ketchikan, AK.

United Nations, 1949. Exchange of Notes constituting a provisional agreement relating to fur seals. Washington, 8 and 19 December 1942. United Nations Treaty Series 156, 364–377.

United Nations, 1983. Convention on international trade in endangered species of wild fauna and flora (with appendices and Final Act of 2 March 1973). Opened for signature at Washington on 3 March 1973. United Nations Treat Series 993, 243–438.

VanBlaricom, G.R., 1996. Saving the sea otter population in California: contemporary problems and future pitfalls. Endangered Species Update 13 (12), 85–91.

VanBlaricom, G.R., Jameson, R.J., 1982. Lumber spill in central California waters: implications for oil spills and sea otters. Science 215, 1503–1505.

VanBlaricom, G.R., Gerber, L.R., Brownell Jr., R.L., 2013. Extinctions of marine mammals. In: second ed. Levin, S.A. (Ed.), Encyclopedia of Biodiversity, vol. 5. Academic Press, Waltham, MA, pp. 64–93.

von Biela, V.R., Gill, V.A., Bodkin, J.L., Burns, J.M., 2009. Phenotypic plasticity in age at first reproduction of female northern sea otters (*Enhydra lutris kenyoni*). J. Mammal. 90, 1224–1231.

Wild, P.W., Ames, J.A., 1974. A report on the sea otter, *Enhydra lutris* L., in California. Calif. Dep. Fish Game Mar. Resour. Tech. Rep. 20, 1–93.

Witherell, D., Woodby, D., 2005. Application of marine protected areas for sustainable production and marine biodiversity off Alaska. Mar. Fish. Rev. 67, 1–27.

Index

Note: Page numbers followed by "*f*", "*t*" and "*b*" refers to figures, tables and boxes respectively.

B

C

G

H

T

U

V

Printed in the United States
By Bookmasters